Springer Series in
MATERIALS SCIENCE 156

Springer Series in
MATERIALS SCIENCE

Editors: R. Hull   C. Jagadish   R.M. Osgood, Jr.   J. Parisi   Z. Wang

The Springer Series in Materials Science covers the complete spectrum of materials physics, including fundamental principles, physical properties, materials theory and design. Recognizing the increasing importance of materials science in future device technologies, the book titles in this series reflect the state-of-the-art in understanding and controlling the structure and properties of all important classes of materials.

Please view available titles in *Springer Series in Materials Science* on series homepage http://www.springer.com/series/856

Stephen Pearton
Editor

GaN and ZnO-based Materials and Devices

With 328 Figures

 Springer

Editor
Stephen Pearton
University of Florida
Materials Science and Engineering
100 Rhines Hall, 32611 Gainesville, USA
spear@mse.ufl.edu

Series Editors:

Professor Robert Hull
University of Virginia
Dept. of Materials Science and Engineering
Thornton Hall
Charlottesville, VA 22903-2442, USA

Professor Chennupati Jagadish
Australian National University
Research School of Physics and Engineering
J4-22, Carver Building
Canberra ACT 0200, Australia

Professor R. M. Osgood, Jr.
Microelectronics Science Laboratory
Department of Electrical Engineering
Columbia University
Seeley W. Mudd Building
New York, NY 10027, USA

Professor Jürgen Parisi
Universität Oldenburg, Fachbereich Physik
Abt. Energie- und Halbleiterforschung
Carl-von-Ossietzky-Straße 9–11
26129 Oldenburg, Germany

Dr. Zhiming Wang
University of Arkansas
Department of Physics
835 W. Dicknson St.
Fayetteville, AR 72701, USA

Springer Series in Materials Science   ISSN 0933-033X
ISBN 978-3-642-23520-7   e-ISBN 978-3-642-23521-4
DOI 10.1007/978-3-642-23521-4
Springer Heidelberg Dordrecht London New York

Library of Congress Control Number: 2011945030

Springer is part of Springer Science+Business Media (www.springer.com)

Preface

The GaN materials system has applications in visible and UV light emitting devices and in high power, high temperature electronics. On the photonics side, the AlGaInN materials system, consisting of AlGaN/GaN, InAlN/GaN, and InGaN/GaN heterostructures and the GaN, AlN, and InN binaries, is widely used in blue/violet/white/UV light emitting diodes for stoplights and full color displays, blue and green lasers for use in high-density CD-ROM storage and high-resolution printers. The main applications for GaN-based high power microwave transistors are in phased array radar systems and wireless communication systems, while the low noise, radiation hard transistors can be used in high temperature sensors and spaceflight instrumentation. There have been many recent advances in this field, concerning sensors and nanostructure synthesis. Functional nanostructures are attracting much interest for use in sensing, energy harvesting and cell monitoring. In addition, there is a widespread interest in the use of ZnO in UV light emitters and flexible electronics. Flexible displays are attractive for portable devices such as smart phones, laptops, e-books, and wearable devices due to their lightweight, low power consumption, and being bendable. In defense applications, soldiers can use flexible display computers on the battlefield for communication and information access. Thin-film transistors (TFTs) are field effect transistors made by depositing thin films of amorphous Si, organics, or new inorganics based on ZnO for the semiconductor active layer. The channel region of a TFT is deposited onto a substrate such as glass, since the main market is in liquid crystal displays (LCDs). Since these types of substrates do not allow for high temperature processing, the TFT active region must be deposited at low temperature. Recently, indium zinc oxides (IZO) such as 90 wt% In_2O_3–10 wt% ZnO and homogeneous $In_2O_3(ZnO)_k$ ($k = 2, 3, 4, 5, 6, 7, 9, 11, 13$, and 15) compounds have attracted significant attention as new candidates for transparent electrodes due to their good conductivity, high optical transparency, excellent surface smoothness, and low deposition temperature. One of the major challenges in the development of transparent thin-film transistors is to control the carrier concentration with high transparency in the active channel. For active channel materials, oxide semiconductors such as ZnO, $(ZnO)_x(SnO_2)_{1-x}$, 90 wt% In_2O_3–10 wt% ZnO, $In_2O_3(ZnO)_2$, and $InGaZnO_4$ have been reported to be as alternatives to α-Si:H in thin-film transistors. Recent results have shown that InZnO has high electron mobility even for room temperature deposition, allowing for use of cheap substrates such as glass for fabrication of TFTs with superior performance to a-Si(H)TFTs.

This book brings together experts in both the GaN and ZnO areas to provide the most up-to-date information on advances in these fields. The topics begin with growth of advanced III-Nitride structures, specifically non-polar growth and high Al content alloys, leading into devices such as green and UV LEDs, HEMT power devices, and sensors. Advances in nitride nanostructures and understanding of radiation defects in GaN are covered, followed by reviews of developments in synthesis and control of ZnO-based films and nanostructures. Finally, the use of amorphous, transparent conducting oxides for channels in TFTs on novel substrates is reviewed. The purpose of the book is to provide both a summary of the current state-of-the-art and directions for future research.

Gainesville, FL, USA   *S.J. Pearton*
December 2011

Contents

Contributors

Pedro Barquinha CENIMAT-I3N, Departamento de Ciência dos Materiais, and CEMOP/UNINOVA, Faculdade de Ciências e Tecnologia, FCT, Universidade Nova de Lisboa, 2829-516, Caparica, Portugal

Xian-An (Andrew) Cao West Virginia University, Lane Department of Computer Science and Electrical Engineering, Morgantown, WV 26506-6070, USA, xacao@mail.wvu.edu

C.Y. Chang University of Florida, Department of Chemical Engineering, Gainesville, FL 32611, USA

K.H. Chen University of Florida, Department of Chemical Engineering, Gainesville, FL 32611, USA

Ruei-San Chen Graduate Institute of Applied Technology, National Taiwan University of Science and Technology, Taipei 106, Taiwan

Li-Chyong Chen Center for Condensed Matter Sciences, National Taiwan University, Taipei 106, Taiwan, chenkh@pub.iams.sinica.edu.tw

Victor Chen University of Florida, Department of Chemical Engineering, Gainesville, FL 32611, USA

H. Chen Department of Electrical and Computer Engineering, Rutgers University, 94 Brett Road, Piscataway, NJ 0885, USA

Leonid Chernyak Department of Physics, University of Central Florida, Orlando, FL 32816, USA, chernyak@physics.ucf.edu

T. Paul Chow Department of Electrical, Computer and Systems Engineering, Rensselaer Polytechnic Institute, Troy, NY 12180, USA, chowt@rpi.edu

B.H. Chu University of Florida, Department of Chemical Engineering, Gainesville, FL 32611, USA

Zinovy Dashevsky Department of Physics, University of Central Florida, Orlando, FL 32816, USA

Jeff Davis College of Physical and Mathematical Sciences, Australian National University, Canberra, ACT 0200, Australia

Alex Dobrinsky Sensor Electronic Technology, Inc., Columbia, SC 29209, USA, adobrinsky@s-et.com

Elena Flitsyian Department of Physics, University of Central Florida, Orlando, FL 32816, USA

Elvira Fortunato CENIMAT-I3N, Departamento de Ciência dos Materiais, and
CEMOP/UNINOVA, Faculdade de Ciências e Tecnologia, FCT, Universidade Nova de Lisboa, 2829-516, Caparica, Portugal, elvira-fortunato@fct.unl.pt

Abhijit Ganguly Center for Condensed Matter Sciences, National Taiwan University, Taipei 106, Taiwan

Remis Gaska Sensor Electronic Technology, Inc., Columbia, SC 29209, USA

Jung Han Department of Electrical Engineering, Yale University, 15 Prospect Street, BCT 517, New Haven, CT 06511, USA, jung.han@yale.edu

Kuei-Hsien Center for Condensed Matter Sciences, National Taiwan University, Taipei 106, Taiwan, Chen chenlc@ntu.edu.tw
and
Institute of Atomic and Molecular Sciences, Academia Sinica, Taipei 106, Taiwan

C. Jagadish College of Physical and Mathematical Sciences, Australian National University, Canberra, ACT 0200, Australia, Chennupati.Jagadish@anu.edu.au

Wayne Johnson Kopin Corporation, 125 North Drive, Westboro, MA 01581, USA, Wayne_Johnson@kopin.com

Hongxing Jiang Texas Tech University, Electrical and Computer Engineering Department, Lubbock, TX 79409-3102, USA, hx.jiang@ttu.edu

Jihyun Kim Department of Chemical and Biological Engineering, College of Engineering, Korea University, Anam-dong 5-1, Sungbuk-gu, Seoul 136-701, South Korea, hyunhyun7@korea.ac.kr

Z. Li Department of Electrical, Computer and Systems Engineering, Rensselaer Polytechnic Institute, Troy, NY 12180, USA

Jingyu Lin Texas Tech University, Electrical and Computer Engineering Department, Lubbock, TX 79409-3102, USA, jingyu.lin@ttu.edu

Yicheng Lu Department of Electrical and Computer Engineering, Rutgers University, 94 Brett Road, Piscataway, NJ 0885, USA, ylu@ece.rutgers.edu

Rodrigo Martins CENIMAT-I3N, Departamento de Ciência dos Materiais, and
CEMOP/UNINOVA, Faculdade de Ciências e Tecnologia, FCT, Universidade Nova de Lisboa, 2829-516 Caparica, Portugal

Edwin Piner Texas State University, San Marcos, Department of Physics, 601 University Dr., San Marcos, TX 78666, USA, epiner@txstate.edu

B.N. Pantha Texas Tech University, Electrical and Computer Engineering Department, Lubbock, TX 79409-3102, USA

S.J. Pearton Department of Materials Science and Engineering, University of Florida, Gainesville, FL 32611, USA, spear@mse.ufl.edu

Alexander Polyakov Institute of Rare Metals, B. Tolmachevsky 5, 119017 Moscow, Russia, aypolyakov@gmail.com

Fan Ren University of Florida, Department of Chemical Engineering, Gainesville, FL 32611, USA, ren@che.ufl.edu

P.I. Reyes Department of Electrical and Computer Engineering, Rutgers University, 94 Brett Road, Piscataway, NJ 0885, USA

Max Shatalov Sensor Electronic Technology, Inc., Columbia, SC 29209, USA

Michael Shur Department of Electrical, Computer and Systems Engineering, Rensselaer Polytechnic Institute, Troy, NY 12180, USA, shurm@rpi.edu

Michael Snure Nanostructured Materials Research Laboratory, Department of Materials Science and Engineering, University of Utah, Salt Lake City, UT 84112, USA

Paul Slusser Nanostructured Materials Research Laboratory, Department of Materials Science and Engineering, University of Utah, Salt Lake City, UT 84112, USA

Qian Sun Department of Electrical Engineering, Yale University, 15 Prospect Street, BCT 517, New Haven, CT 06511 USA

Ashutosh Tiwari Nanostructured Materials Research Laboratory, Department of Materials Science & Engineering, University of Utah, Salt Lake City, UT 84112, USA, tiwari@eng.utah.edu

David Toledo Nanostructured Materials Research Laboratory, Dept. of Materials Science & Engineering, University of Utah, Salt Lake City, UT 84112 USA

J. Zhong Department of Electrical and Computer Engineering, Rutgers University, 94 Brett Road, Piscataway, NJ 0885, USA

Chapter 1
Heteroepitaxy of Nonpolar and Semipolar GaN

Qian Sun and Jung Han

Abstract There has been increasing research interest in nonpolar and semipolar GaN for high brightness light-emitting diode (LED) and laser diode applications. Due to the very limited supply of GaN bulk substrates, the feasible way of obtaining cost-effective large-area nonpolar and semipolar GaN materials in the foreseeable future is still through heteroepitaxy on foreign substrates. The major challenge in the heteroepitaxy of nonpolar and semipolar GaN is the high density of stacking faults and partial dislocations, which are responsible for the poor performance of heteroepitaxial nonpolar and semipolar LEDs. This chapter presents kinetic Wulff plots (v-plots) of GaN as a novel and powerful methodology to understand and control GaN heteroepitaxy along various crystallographic orientations. Based on the v-plots, a two-step growth scheme is rationally designed and experimentally confirmed in reducing the defect density for nonpolar and semipolar GaN heteroepitaxy on planar substrates. A defect reduction model is proposed based on the correlation between the morphological evolution and the microstructural development. With the orientation of nucleation decoupled from the final film surface orientation, orientation controlled epitaxy has been demonstrated as a very promising approach for device quality nonpolar and semipolar GaN materials. The material research timeline of nonpolar and semipolar GaN is summarized and discussed.

1.1 Introduction

The discovery of a two-step growth procedure for the heteroepitaxy of GaN on c-plane sapphire [1, 2] is widely considered the key breakthrough for GaN-based materials and devices (Fig. 1.1). In the past two decades, the heteroepitaxy of GaN

Q. Sun · J. Han (✉)
Department of Electrical Engineering, Yale University, 15 Prospect Street, BCT 517, New Haven, CT 06511, USA
e-mail: qian.sun@aya.yale.edu; jung.han@yale.edu

S. Pearton (ed.), *GaN and ZnO-based Materials and Devices*,
Springer Series in Materials Science 156, DOI 10.1007/978-3-642-23521-4_1,
© Springer-Verlag Berlin Heidelberg 2012

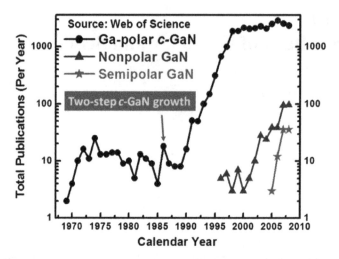

Fig. 1.1 A plot of the total publications per year for Ga-polar c-plane, nonpolar, and semipolar GaN, reflecting the research activities on the individual orientation. The *arrow* indicates the most crucial breakthrough in improving heteroepitaxial c-plane GaN material quality by a two-step growth scheme in the late 1980s [3]

on foreign substrates has been extensively investigated, motivated by the prospects of high-brightness displays and, more recently, energy-efficient illumination using InGaN-based light-emitting diodes (LEDs). The two major challenges in the field of InGaN-based LEDs are the "efficiency droop" under a high injection current density and the "green gap" in the plot of efficiency versus emission wavelength [4]. So far most of InGaN-based LEDs are built along Ga-polar (0001) orientation, which is susceptible to the strong internal electric field induced by the spontaneous and the piezoelectric polarization in Wurtzite III-nitrides. The pronounced tilting of the energy band edges of InGaN quantum wells (QWs) causes a spatial separation between electrons and holes in the QWs and hence reduces the internal quantum efficiency (IQE) of carrier recombination. This effect gets much more pronounced as the emission wavelength is steered from blue to green and yellow ranges due to the increased indium composition in the QWs, giving a great contribution to the well-known "green gap." The tilting of the band edges in the active region also narrows down the effective width of the QWs and hence increases the effective carrier density in the QWs, which favors Auger nonradiative recombination. Meanwhile, the effective barrier height is reduced by the tilting of the band edges, which can facilitate carrier leakage under an applied bias. Both Auger recombination and carrier leakage have been proposed as the likely mechanisms responsible for the universally observed "efficiency droop" phenomenon [5, 6]. To address these challenges, there have been concerted efforts in exploring III-Nitride materials and devices along nonpolar and semipolar crystallographic orientations [7–12]. Very promising reports of LEDs and laser diodes on nonpolar and semipolar GaN bulk substrates, in the longer wavelength of green and yellow, tend to validate the concept of crystallographic engineering [13–18]. The homoepitaxial nonpolar/semipolar

Fig. 1.2 A photo of 2-inch sapphire, 1-inch m-plane SiC, and nonpolar/semipolar GaN substrate with a typical size of 5×10 mm prepared by hydride vapor phase epitaxy (HVPE) [3]

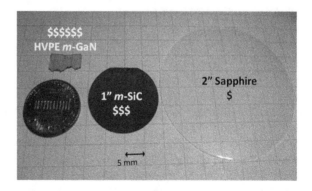

LEDs are mitigating the "green gap" [19] and have also shown significantly reduced "efficiency droop" [20–23].

However, nonpolar and semipolar GaN bulk substrates are presently very small in size and expensive in price (Fig. 1.2), because they are sliced vertically and at an inclined angle, respectively, from a thick c-plane GaN substrate prepared by hydride vapor phase epitaxy (HVPE) [24]. Given the very limited availability of HVPE GaN wafers in the foreseeable future, it is likely that heteroepitaxy on large-area, cost-effective foreign substrates will remain the viable pathway for nonpolar and semipolar devices to reach the mass market. The research activities, including both homoepitaxy and heteroepitaxy, of nonpolar and semipolar GaN are gaining more and more momentum (Fig. 1.1). However, heteroepitaxy of nonpolar (a- and m-plane) and semipolar GaN is still explored empirically and disjointedly, often giving a poor morphology with surface striations and faceted pits, and a defective microstructure, including basal-plane stacking faults (BSFs) bounded by partial dislocations (PDs) [25–33].

In this chapter, we present kinetic Wulff plots (v-plots) as a novel and effective methodology to interpret the major challenges as well as their causes in the heteroepitaxy of nonpolar and semipolar GaN, and to control the heteroepitaxial dynamics on planar substrates with a rational design of multistep growth procedure. Orientation controlled epitaxy (OCE) on nonplanar substrates is introduced and experimentally confirmed as an innovative and promising approach to further boost the material quality of heteroepitaxial nonpolar and semipolar GaN films. The major research progresses in improving nonpolar and semipolar GaN material quality are summarized in a graphic presentation.

1.2 Kinetic Wulff Plot (V-Plot) of GaN

The standard two-step growth procedure of c-plane GaN does not work for nonpolar and semipolar GaN heteroepitaxy. The conventional approach widely taken by the III-Nitride community is to optimize the growth conditions through knob-turning exercises, which normally takes a lot of time and efforts even just for one nonbasal orientation. This time-consuming cycle needs to be repeated for the GaN heteroepitaxy along every different nonbasal orientation. Here we adopted a nonconventional methodology, kinetic Wulff plot, a three-dimensional (3D) polar plot of the growth velocities (v-plots) along the polar, semipolar, and nonpolar orientations of GaN under various MOCVD growth conditions.

The study of shape evolution and growth velocity anisotropy for GaN to date has been largely empirical. Based on the observations of GaN epitaxial lateral overgrowth (ELOG), Hiramatsu et al. [34] compiled the trends of growth velocities for (0001), {10$\underline{1}$1}, and {11$\underline{2}$2} planes. Du et al. [35] proposed a 3D v-plot for GaN using the ratios of growth velocities from the same set of facets formed by selective area growth (SAG). Their study presented a conceptual framework correlating the 3D v-plot with the GaN SAG mesa shape evolution, though the information on the v-plot for nonpolar and the entire N-polar "hemisphere" was largely speculative. In the present work, we take into account the great asymmetry between the Ga-polar and the N-polar "hemispheres" that we have determined in the construction of the 3D v-plot and demonstrated that the full-range v-plot provides crucial insights to unlock the complex heteroepitaxial dynamics, especially at facetted growth fronts of opposite curvature nature. In the case of *convex* growth, such as island formation during the initial phase of heteroepitaxy, the fast-growing facets grow to extinction while the slow-growing facets expand and ultimately define the island shape (Fig. 1.3a); the crystallographic planes at the minima of a v-plot therefore dictate the initial evolution [35, 36]. On the other hand, the presence of *concave* growth fronts is a salient feature in the coalescence stage of heteroepitaxy, as edges of islands come into contact. During *concave* growth, fast-growing facets expand at the expense of slow-growing facets (Fig. 1.3a) and dominate the growth fronts [35, 36]. By employing an annular-ring mask pattern (Fig. 1.3b), we can form convex and concave planes concurrently through SAG. For example, a c-plane GaN SAG mesa (Fig. 1.3d) exhibits six {10$\underline{1}$1} planes on the outside, corresponding to the slow-growing convex planes, and six {11$\underline{2}$2} planes on the inside, corresponding to the fast-growing concave planes, together with the top (0001) [35]. More complicated faceting morphology is exhibited during nonpolar a-plane (11$\underline{2}$0) (Fig. 1.3e) and semipolar (11$\underline{2}$2) GaN SAG (Fig. 1.3f) due to their reduced symmetry (Fig. 1.3c).

Consecutive SAG experiments [36] were employed to monitor the differential advances of both the convex and the concave facets, which gave a direct measure of the relative growth velocities for those orientations. Appropriate trigonometric interpretations were used to bridge the experimental data points for the construct of GaN v-plots. Figure 1.4a shows a 3D v-plot of GaN, viewed along the surface normal of (10$\underline{1}$1), for a growth condition of 1,070°C, 100 mbar, and a V/III ratio of 250. False-colored shading from red to blue is used to depict the measured (or interpolated) growth velocities. We have identified cusp points (local minima) at (0001), {10$\underline{1}$1}, {11$\underline{2}$0}, {10$\underline{1}$0}, and (000$\underline{1}$). These facets are responsible for defining the shape of nucleation islands having a convex curvature on a mesoscopic scale. In addition to plotting the cusps in Fig. 1.4a, we have also included the

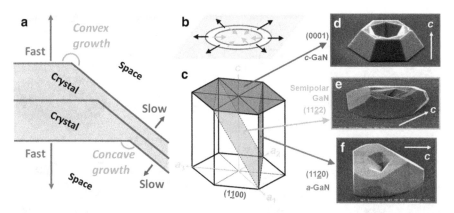

Fig. 1.3 (**a**) Schematic diagram showing the principle of morphological evolution for convex (toward the upper right) and concave (toward the lower left) growth fronts. The convex growth front will be dominated by the slow-growing facets while the concave growth front by the fast-growing facets. (**b**) Schematic of an annular ring opening, which produces both inward concave and outward convex growth fronts. (**c**) Schematic hexagonal lattice showing the c-plane (0001), nonpolar a-plane (11$\underline{2}$0), and semipolar (11$\underline{2}$2). SEM images of GaN SAG mesas grown out of an annular ring opening on (**d**) c-plane (0001), (**e**) semipolar (11$\underline{2}$2), and (**f**) nonpolar a-plane (11$\underline{2}$0) GaN surfaces

measured saddle points (i.e., a local minimum along one direction, but a local maximum along the orthogonal direction) determined by the concave growth fronts in the inner ring.

Given the different atomic bonding configurations on each plane [34], it is expected that the growth velocities and the 3D v-plot depend strongly on the local stoichiometry of reactants. This is confirmed by another set of differential SAG experiments under a condition of much higher V/III ratio $\sim$1,700 and higher reactor pressure 300 mbar. In this case, the relative growth velocities of nearly all the Ga-polar planes (the northern "hemisphere") are suppressed (Fig. 1.4f) while the equatorial portion of the v-plot pertaining to the nonpolar a-planes {11$\underline{2}$0} is increased considerably. These v-plots can be considered "finger prints" of a specific growth condition or procedure (e.g., flow modulation, co-doping, etc.) that can provide insights to a wide range of observed morphology with cross-platform insensitivity.

As discussed earlier, the minima on a 3D v-plot determine the shape evolution during the convex growth (Fig. 1.3a), manifested in the formation of islands bound by the cusp planes (Fig. 1.3d–f). The coalescence stage, in contrast, is governed by the less-understood concave growth when neighboring islands (with convex facets) come into contact. Coalescence produces a growth front that is concave along only *one* of the two spherical axes (azimuth, for instance) while remaining convex along the other orthogonal spherical axis (polar, for instance). The dominant planes during coalescence therefore correspond to saddle points [35, 36] in the v-plots, rather than maxima as one might suppose. The important saddle points are {11$\underline{2}$2} in the

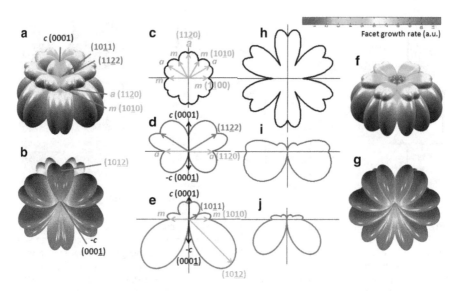

Fig. 1.4 Kinetic Wulff plots (ν-plots) for a low reactor pressure and low V/III ratio condition [(**a**)–(**e**)] and a high reactor pressure and high V/III ratio condition [(**f**)–(**j**)]. For the 3D ν-plots, the [0001] axis is set as $\theta = 0°$, and the m-axis as $(\theta, \varphi) = (90°, 0°)$. The Ga-polar hemisphere [(**a**) and (**f**)] and the N-polar hemisphere [(**b**) and (**g**)] 3D ν-plots are viewed at an angle of $(62°, 0°)$ and $(150°, 0°)$, respectively. (**c**) and (**h**) are the 2D ν-plots for the prism planes mapped onto the basal c-plane; (**d**) and (**i**) the 2D ν-plots for the orientations within the m-plane; (**e**) and (**j**) the 2D ν-plots for the orientations within the a-plane. Note that a sixfold symmetry is assumed for all the ν-plots and the symmetry reduction due to the mask effect during SAG is ignored in this chapter for simplicity

Ga-polar hemisphere (Fig. 1.4a) and $\{10\underline{1}2\}$ in the N-polar hemisphere (Fig. 1.4b). Planes associated with these saddles points are of particular importance in the coalescence stage of nonpolar and semipolar GaN heteroepitaxy to the final surface morphology and the microstructural quality, as will be discussed in detail later.

1.3   Heteroepitaxy of Nonpolar GaN on Planar Substrates

Nonpolar GaN can be epitaxially grown on various foreign substrates, including a-/m-plane SiC [25, 28, 29, 37], (100) LiAlO$_2$ [38–40], r-plane sapphire [26, 27, 30, 41–44], etc. SiC substrates are very expensive and not widely available, especially for nonpolar orientations. (100) LiAlO$_2$ has a small lattice mismatch with nonpolar m-plane GaN, but a poor thermal and chemical stability [45]. In contrast, r-plane sapphire is the most commercially available cost-effective substrates with a good thermal and chemical stability, and hence is widely used for the growth of nonpolar $(11\underline{2}0)$ a-plane GaN (a-GaN).

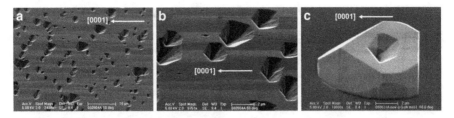

Fig. 1.5 SEM perspective view of nonpolar (11$\bar{2}$0) $a-$ GaN film (**a**) and (**b**), showing a high density of faceted pits and striations, and an SAG $a-$ GaN mesa (grown out of an annular ring opening) with a concave void (**c**). The faceted pits on the $a-$ GaN film surface resemble very much to the concave void of the $a-$ GaN SAG mesa

In the heteroepitaxy of nonpolar a-GaN, there are two major challenges, poor surface morphology and defective microstructure. On the surface of heteroepitaxial a-GaN film, there are often triangular/pentagonal faceted pits and striations along the in-plane c-axis [0001] [27], as shown in Fig. 1.5a, b. The faceted surface pits resemble very much to the concave void of the $a-$GaN SAG mesa (Fig. 1.5c) grown out of an annular ring opening (Fig. 1.3b) [36], indicating that the faceted surface pits are due to incomplete coalescence. To eliminate the surface pits, the vertical out-of-plane a-axis growth needs to be slowed down and the lateral in-plane growth rates be enhanced, which are favored by growth conditions of a low V/III ratio and a low reactor pressure according to the v-plots [36, 42]. The a-GaN surface striations consist of facets vicinal to the surface (11$\bar{2}$0) (Fig. 1.5a, b). The growth rates of the vicinal facets are comparable to that of the vertical a-axis growth under a low V/III and a low reactor pressure, and therefore the striations persist on the surface even after a long growth. A change in growth conditions to enlarge the difference in growth rate between the vicinal facets and the (11$\bar{2}$0) surface has been found effective in relieving the pronounced striations [42].

The other challenge in a-GaN heteroepitaxy on r-plane sapphire lies in their dissimilar structure, including the in-plane stacking mismatch and the lattice mismatch (16 and 1.3% along the m- and c-axes of GaN, respectively), which lead to a high density of microstructural defects, including BSFs bounded by PDs with a typical density of 10^5–10^6 cm^{-1} and 10^{10}–10^{11} cm^{-2}, respectively [46]. These extended defects propagate along the growth axis usually throughout the nonpolar GaN film (Fig. 1.6b), and hence deteriorate material quality and device performance. It has been observed by various groups that BSFs and associated PDs are nucleated at the interface of nonpolar GaN with the underlying template (Fig. 1.6b) [28, 29, 37, 46]. During the initial growth of nonpolar/semipolar GaN on a *heterogeneous* template (foreign substrates, AlN buffer [29, 37], SiO$_2$ [33, 47–49], etc.), rough surface morphology [38, 39], atomic (partial) steps [40], and/or impurities on the surface of the underlying template may trigger nucleation errors at the cubic sites, especially for the N-face lateral growth. BSFs may also be formed during the coalescence of nonpolar GaN islands whose basal-plane stacking sequences are out of registry [33]. I_1 BSFs having a displacement vector with a

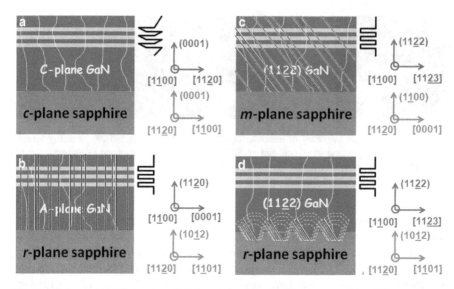

Fig. 1.6 Schematic LED structures build on (**a**) c-plane GaN/c-plane sapphire, (**b**) nonpolar a-plane GaN/r-plane sapphire, (**c**) semipolar $(11\underline{2}2)$ GaN/m-plane sapphire, and (**d**) semipolar $(11\underline{2}2)$ GaN/patterned r-plane sapphire. Perfect threading dislocations are represented by *white curved lines*. BSFs and their associated PDs are indicated by *yellow* and *black straight lines*. An energy diagram drawing is shown on the right side of InGaN MQWs (in *green*) for each structure

c-axis component can contribute to strain relaxation [39, 50, 51]. One effective way to reduce the formation of BSFs in nonpolar GaN due to strain relaxation is to introduce composition-graded AlGaN layers between the AlN buffer and the GaN films [37].

In the study of c-plane GaN (c-GaN) growth, a 3D to 2D transition (a roughening-recovery process) was found crucial in achieving improved crystalline quality [52]. Based on the v-plots (Fig. 1.4), we designed a two-step growth procedure (beyond the initial buffer layer) for a-GaN with different V/III ratios and reactor pressure to effectively reduce the defect density [41–43].

In this study, all the a-GaN samples were grown at $1,050°C$ with an AlN buffer (deposited at $1,150°C$) on nominally on-axis r-plane sapphire in an Aixtron horizontal MOCVD reactor. Trimethylgallium (TMGa), trimethylaluminium (TMAl), and ammonia (NH$_3$) were used as the precursors for Ga, Al, and N, respectively. Hydrogen was used as the carrier gas if not specified separately. For the baseline (sample A), $2 - \mu$m-thick a-GaN was directly grown on the AlN buffer at $1,050°C$, a reactor pressure, P, of 60 mbar and a V/III ratio of 187. For all other samples (B-F) a 3D-growth stage was introduced with various P and V/III (Table 1.1) to control the growth dynamics, before returning to the baseline condition. Sample F, having the longest roughening growth stage, was grown nominally twice as thick as other samples to ensure a fully coalesced pit-free surface.

Table 1.1 Growth parameters and characterization results of nonpolar a-GaN samples A–F

Sample	First-step a-GaN			c-axis LCL (nm)	BSF density $\times 10^5\, cm^{-1}$
	V/III	P(mbar)	t(min)		
A	/	/	0	19	5.3
B	1,440	60	10	24.5	4.1
C	2,162	100	17	30.4	3.3
D	2,162	200	17	36.5	2.7
E	1,682	300	17	45.4	2.2
F (F4)	1,682	300	40	60	1.7

For sample A (one-step growth), the in situ reflectance trace exhibits instant oscillations (Fig. 1.7a) indicating a quasi-2D growth mode. According to the kinetic Wulff plots [36, 42], the vertical growth can be greatly enhanced while the lateral growth (especially along the c-axis) is slowed down under an increased V/III ratio and/or an elevated P, thus facilitating the formation of a-GaN islands with a "tall" aspect ratio. During the first-step growth of samples B–F, as the V/III ratio and/or P increase, surface becomes increasingly rough as revealed by the progressive decay of the reflectance traces, eventually down to nearly background (Fig. 1.7b–f). Once the growth is switched back to the baseline condition (for the second-step growth), island coalescence through lateral growth occurs at a rate that is inversely correlated with the extent of roughening after the first-step growth. The amplitude of reflectance oscillations decreases from sample A to F because of their increased surface striations, which can be relieved by an additional third-step growth [42].

The microstructural extended defects in a-GaN films (Fig. 1.6b) are much more complicated than those in c-GaN due to the reduced symmetry [41, 42]. A-GaN may contain not only $a-$, $c-$, and $a + c$ type dislocations with a Burgers vector (BV) of $1/3 < 11\underline{2}0 >$, $< 0001 >$, and $1/3 < 11\underline{2}3 >$, respectively, but also planar defects including BSFs and prismatic-plane stacking faults (PSFs). I_1, I_2, and E type BSFs are normally bounded by Frank–Shockley, Shockley, and Frank PDs with a BV of $1/6 < 20\underline{2}3 >$, $1/3 < 10\underline{1}0 >$, and $1/2 < 0001 >$, respectively [46]. Given the complexity of TDs/PDs, it is necessary to measure X-ray rocking curves (XRCs) for both on- and *off-axes* planes over a sufficiently large azimuthal and polar angular space (Fig. 1.8a, b) to obtain a comprehensive knowledge of the microstructure of a-GaN [41, 42]. The full widths at half maximum (FWHMs) were obtained through fitting with pseudo-Voigt functions. It is worthy to review the visibility criteria for stacking faults and dislocations under X-ray diffraction. For an XRC with a diffraction vector g to be broadened by a stacking fault with a displacement vector R, the $g \cdot R$ product must be a *noninteger* and the stacking axis has a substantial component along the X-ray rocking direction [42]. A dislocation with a Burgers vector b can broaden an XRC with a diffraction vector g, if the $g \cdot b$ product is *nonzero* (including integers) [42].

Figure 1.8c shows the FWHM of the on-axis $(11\underline{2}0)$ XRCs as a function of the azimuthal angle, φ. The baseline sample A (one-step growth with instant reflectance oscillations) exhibits very anisotropic FWHMs for the on-axis $(11\underline{2}0)$ XRCs, similar

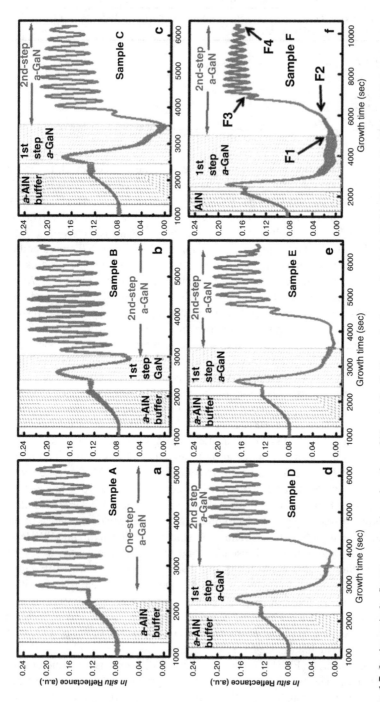

Fig. 1.7 In situ optical reflectance traces (wavelength = 900 nm) of *a*-GaN growths for samples A (**a**), B (**b**), C (**c**), D (**d**), E (**e**), and F (**f**). The reflectance segments shaded in *green* in (**b**)–(**f**) correspond to the first-step 3D *a*-GaN growth for samples B–F. A separate series of *a*-GaN samples F1–F4 were grown by controlled growth interruptions during sample F growth (**f**) [41]

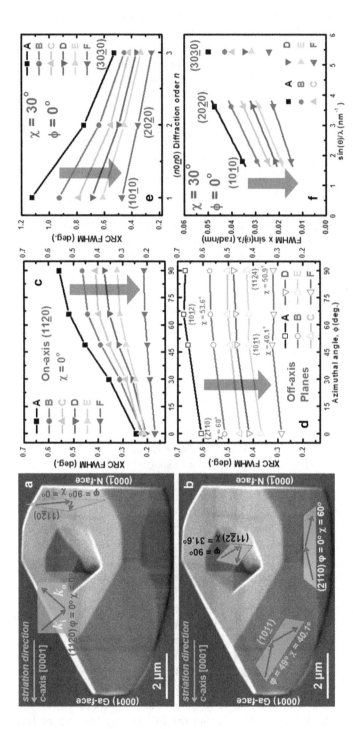

Fig. 1.8 SEM image of an *a*-GaN SAG mesa for the illustration of (**a**) on-axis and (**b**) off-axis XRC configurations. For $\varphi = 0$ and $90°$, the X-ray rocking direction is parallel and perpendicular to the *c*-axis [0001], respectively. The incident and diffracted X-ray beams (*red arrows*) rock within a plane (shaded in *green*) for each diffraction plane with its index and orientation labeled in the vicinity. The (**a**) on-axis ($11\underline{2}0$) ($\chi = 0°$) and (**b**) off-axis planes ($\chi \neq 0°$), such as ($10\underline{1}1$), ($2\underline{1}10$), ($30\underline{3}0$), and ($11\underline{2}2$), were measured in symmetric and skew symmetric geometries, respectively. The FWHM of the (**c**) on-axis and (**d**) off-axis XRCs of *a*-GaN as a function of the azimuthal angle, φ (**e**) The FWHMs of *m*-plane ($n0\underline{n}0$) XRCs ($n = 1, 2,$ and 3; $\varphi = 0°$, $\chi = 30°$); (**f**) the modified WH plots. The *straight lines* in (**f**) are the linear fits to the ($10\underline{1}0$) and ($20\underline{2}0$) data points of each *a*-GaN sample [41]

to many previous reports. However, with the introduction of a 3D growth stage, the FWHM of the on-axis (11$\bar{2}$0) XRC at $\varphi = 90°$ decreases monotonically from sample A to F, and the FWHM-φ plots become less steep, indicative of improved isotropy. Given the insensitivity of (11$\bar{2}$0) diffraction to BSFs, the anisotropic broadening of the on-axis (11$\bar{2}$0) XRCs may be caused by the anisotropy in mosaic tilt and/or domain size, as well as the PSFs within the inclined a-planes with a displacement vector of $1/2 < 10\bar{1}1 >$. Mosaic twist (in-plane misalignment) is another important aspect of the microstructure and normally correlated with dislocations with an edge component. To study twist in the a-GaN films, XRC measurements were implemented in a *skew* symmetric geometry for *off-axis* planes having an inclination angle χ with respect to the on-axis (11$\bar{2}$0) at various azimuths (Fig. 1.8b). As shown in Fig. 1.8d, for all the samples the FWHM-φ plots of the off-axis XRCs are fairly flat with a small anisotropy [unlike the case of on-axis XRCs in Fig. 1.8c), implying a uniform angular distribution of dislocations with an edge component. For sample A, the off-axis XRC FWHMs are very high, ranging from 0.6 to 0.7°. With the insertion of 3D island growths, the off-axis XRC FWHMs decrease *evenly* for all the azimuths from samples B to F, reaching a ~50% reduction for the case with a prolonged 3D growth.

By treating BSFs as boundaries between incoherently scattering domains [53], we can determine the lateral coherence length (LCL) along the c-axis and estimate the BSF density in nonpolar a-GaN film through a modified Williamson-Hall (WH) analysis [41, 42] on m-plane ($n0\bar{n}0$) diffraction ($n = 1, 2,$ and 3; $\varphi = 0°, \chi = 30°$) in a skew symmetric geometry with the X-ray rocking along the c-axis (Fig. 1.8b). The XRC FWHMs for all three diffractions decrease monotonically from sample A to F (Fig. 1.8e), another indicator of the clear trend of improved structural quality. It is noted that for all the samples, the FWHMs of (10$\bar{1}$0) and (20$\bar{2}$0) XRCs are much broader than that of (30$\bar{3}$0) XRC, because the former two XRCs are predominantly broadened by the short c-axis LCLs due to the presence of BSFs, which is more clearly revealed by the WH plots in Fig. 1.8f. Since (30$\bar{3}$0) diffraction is insensitive to the presence of BSFs, the (30$\bar{3}$0) data points in the WH plots (Fig. 1.8f) are significantly lower than the expected values from the linear fits to the corresponding (10$\bar{1}$0) and (20$\bar{2}$0) data points of each sample. The c-axis LCLs can be derived from the y-axis intercepts (y_o) of the linear fits, LCL $= 0.9/(2y_o)$. And the reciprocal of the LCLs gives the density of BSFs. The determined LCL and BSF density of all the samples are summarized in Table 1.1. It is clearly shown that the more the 3D process is involved in the first-step growth, the longer the c-axis LCL and the lower the BSF density (~70% reduction).

To verify the substantial reduction in the densities of both BSFs and dislocations, we examined the microstructure of a-GaN samples A and F by transmission electron microscopy (TEM) images taken from three orthogonal axes, including plan-views (Fig. 1.9a, b, e, f), and two cross-section views along the in-plane c- (Fig. 1.9c, g) and m-axes (Fig. 1.9d, h). Most of the observed BSFs were of I_1 type, normally bounded by Frank–Shockley PDs ($b = 1/6 < 2\bar{2}03 >$). The dislocations observed in the plan-view were all partials according to the one-to-one location correlation between the ends of the BSFs in Fig. 1.9e and the PDs in Fig. 1.9f. No perfect

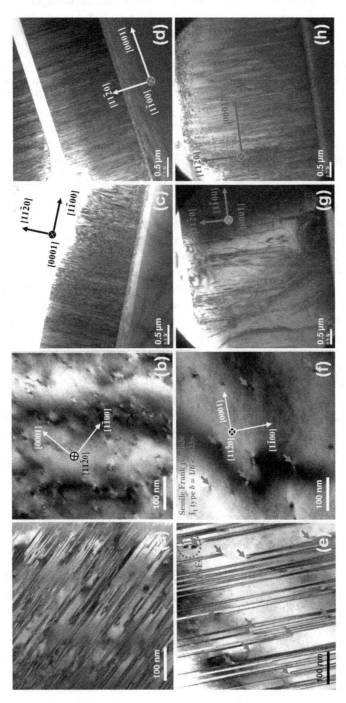

Fig. 1.9 Plan-view TEM images of the one-step grown [(**a**) and (**b**)] and the two-step grown [(**e**) and (**f**)] nonpolar *a*-plane GaN samples. The **g** vectors were $1\bar{1}00$ for (**a**) and (**e**), and 0002 for (**b**) and (**f**), to reveal the SFs and the PDs, respectively. The Frank–Shockley PDs ($b = 1/6 < 20\bar{2}3 >$) are indicated by the *red arrows*. The PSFs observed on the surface of the two-step grown *a*-plane GaN sample are *circled* in (**e**). Cross-section TEM images of the one-step grown [(**c**) and (**d**)] and the two-step grown *a*-GaN samples [(**g**) and (**h**)]. (**c**) and (**g**) were taken near the [0001] axis with **g** = $1\bar{1}00$. (**d**) and (**h**) were taken near the [$1\bar{1}00$] axis with **g** = $11\bar{2}0$ [42]

dislocations were observed in the view although the diffraction configurations adopted in this study did satisfy the visibility criteria for perfect dislocations. As revealed in Fig. 1.9b, f, the PD density of sample F ($\sim 6 \times 10^9\,\text{cm}^{-2}$) was much lower than that of sample A ($\sim 5 \times 10^{10}\,\text{cm}^{-2}$), which was consistent with the 2–3 times difference in the XRC FWHMs between the two samples (Fig. 1.8). By calculating the total length of the observed BSFs divided by the image area, we estimated the BSF density in the a-GaN films to be $\sim 1 \times 10^6$ and $\sim 4 \times 10^5\,\text{cm}^{-1}$ for samples A (Fig. 1.9a) and F (Fig. 1.9e), respectively. The differences between the TEM measurement and the XRC WH analysis can be accounted for by their different nature and principles [42]. But both the methods confirmed the efficacy of the two-step growth process in significantly reducing the BSF density (60–70%).

To unveil the possible mechanisms for the density reduction of BSFs and PDs, controlled growth interruptions [42] were carried out during two-step growth of sample F, giving a series of samples F1–F4 (Fig. 1.7f). Scanning electron microscope (SEM) was utilized to observe the morphological evolution. Sample F1 presented a very rough surface consisting of tall mesas separated by big voids, within which there were many ridge-like islands (roughly aligned along the c-axis) (Fig. 1.10a). After 20 min into the second-step growth (sample F2), the flat top area of the GaN mesas substantially increased, and the previous big voids were largely filled (through concave growths), leaving behind chains of pits along the c-axis (Fig. 1.10b). During the following 17.5-min growth (sample F3), the chains of pits were zipped up by lateral concave growth, producing many striations (Fig. 1.10c). For sample F4, a 90-min growth under the second-step condition fully eliminated the pits (Fig. 1.10d).

A correlation study [42] between the morphological evolution and microstructure development has revealed that Frank–Shockley PDs are strictly confined within the basal plane, because it is energetically unlikely to bend a planar defect BSF out of the basal plane, and the BSFs and the associated PDs projected along the in-plane m-axis are always straight vertical lines (Fig. 1.9d, h). But the PDs can be bent by the inclined $\{10\underline{1}1\}$ and the vertical $\{10\underline{1}0\}$ growth fronts (Figs. 1.10a and 1.11c) toward the m-axis (Figs. 1.9g and 1.11a) [42], which greatly cut down the density of threading PDs while significantly extending the dimension of BSFs on the surface (Figs. 1.9e and 1.11a). For the two-step grown a-GaN sample F, the increased spacing between the BSFs in the plan-view TEM image (Fig. 1.9e) and the reduced density of the vertical lines in the cross-section TEM image along the in-plane m-axis (Fig. 1.9h), together with the knowledge of the v-plots, provided the rational bases of our model for the BSF reduction through the two-step growth process. The first-step growth of a-GaN produces tall convex mesas separated by tiny, incipient ridge-like islands (Figs. 1.10a and 1.11c) containing many BSFs. Coalescence of those tall mesas results in concave growth fronts along $[000\underline{1}]$, with four of the six $\{10\underline{1}2\}$ facets forming an overhanging structure next to the incipient islands (Figs. 1.10a and 1.11c). These *fast-growing* $\{10\underline{1}2\}$ facets advance and sweep laterally along $[000\underline{1}]$ over the incipient islands, burying and blocking the vertical propagation of BSFs as depicted in Fig. 1.11a, b [42]. Our model of BSF and PD reduction has been supported by the one-to-one location correlation

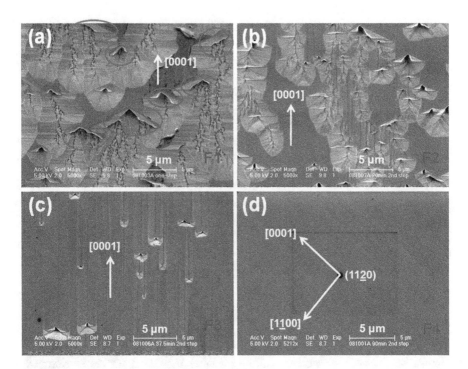

Fig. 1.10 SEM surface images of samples F1–F4 (Fig. 1.7f) showing the evolution of the nonpolar *a*-GaN surface morphology during the two-step growth process. The *red circles* in (**a**) marked out regions showing lateral growth of big mesas along [000$\bar{1}$] over the neighboring small islands [42]

between SEM (Fig. 1.11c) and monochromatic cathodoluminescence (CL) images (Fig. 1.11d) of the same area of *a*-GaN sample F1. Most of the CL emission came from the regions "behind" the over-hanging {10$\bar{1}$2} growth fronts circled by the dash lines, indicating a good material quality with few defects.

1.4 Heteroepitaxy of Semipolar GaN on Planar Substrates

M-plane sapphire (*m*-sapphire) is widely used for semipolar GaN growth, but can produce both nonpolar (10$\bar{1}$0) *m*-GaN and two kinds of semipolar orientations, (11$\bar{2}$2) and (10$\bar{1}$3) [32, 54, 55]. When *m*-sapphire surface is not thoroughly nitridized, the as-grown GaN surface often consists of both (11$\bar{2}$2) orientation and (10$\bar{1}$3) twinned grains (Fig. 1.12). And the grain boundaries prohibited a complete coalescence. *M*-sapphire nitridation is found very effective to suppress the formation of mixed phase (10$\bar{1}$3) grains [32, 54] and consistently obtain (11$\bar{2}$2) GaN [55, 56]. Nitridation breaks the surface symmetry of *m*-sapphire [55] and favors the nucleation of (11$\bar{2}$2) orientation with N-polarity [34]. The

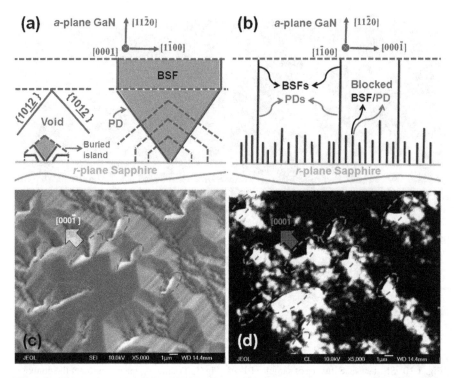

Fig. 1.11 Schematic model of the BSF blocking and the PD bending during the two-step growth of nonpolar a-plane GaN, cross-section viewed along the c-axis [0001] (**a**) and the m-axis [1$\bar{1}$00] (**b**). The *blue dashed lines* represent the a-plane GaN growth fronts, and the *red solid lines* are the PDs. BSFs are shown as the *shaded area* in (**a**) and the *straight black lines* in (**b**). The *vertical black lines* overlap with the *red lines* in (**b**), indicating that most of the BSFs are bounded by PDs. [42] SEM (**c**) and monochromatic (363 nm) CL image (**d**) of the same surface area of sample F1 a-plane GaN islands with the growth intentionally stopped at the end of the first-step roughening stage. The {10$\bar{1}$2} overhanging lateral overgrowth fronts toward the [000$\bar{1}$] circled in (**c**) show a good position correlation with most of the emitting regions (with a low defect density) in (**d**)

in-plane epitaxial relationships between (11$\underline{2}$2) GaN and m-sapphire (Fig. 1.6c) are [1$\underline{1}$00]$_{GaN}$||[11$\underline{2}$0]$_{sapphire}$ and [1$\underline{1}$23]$_{GaN}$||[0001]$_{sapphire}$ with a lattice mismatch of 16.1 and -6.3%, respectively. Semipolar (11$\underline{2}$2) GaN heteroepitaxy shares the same challenge as nonpolar GaN, i.e., the high density of structural defects. Although BSFs and their bounding PDs confined within the basal plane are inclined to the (11$\underline{2}$2) surface by 58.4° (Fig. 1.6c) [33], they can still penetrate the entire semipolar GaN film and affect the material quality and device performance.

The study of semipolar (11$\underline{2}$2) GaN growth on nitridized m-sapphire was not carried out in an empirical manner by turning the knobs for the optimization of growth parameters. Instead, with the knowledge of the v-plots, we designed a two-step growth procedure for (11$\underline{2}$2) GaN to improve the material quality [56], similar to that for nonpolar a-GaN [41, 42]. The two-step growth procedure for (11$\underline{2}$2) GaN

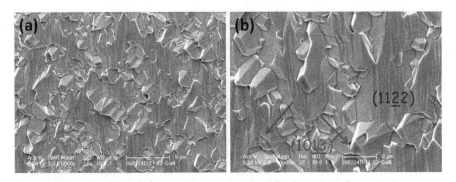

Fig. 1.12 SEM images of a semipolar (11$\underline{2}$2) GaN surface with many grains of mixed phase (10$\underline{1}$3) micro-twins due to insufficient nitridation of the m-plane sapphire substrate. The *arrows* indicate the c-axis direction of the (10$\underline{1}$3) micro-twins

Table 1.2 Growth parameters of semipolar (11–22) GaN samples S1–S4

Sample	The first-step GaN		The second-step GaN	
	P(mbar)	t(min)	P(mbar)	t(min)
S1	/	/	60	27
S2	200	5	60	22
S3	300	5	60	45
S4	300	15	60	60

involves a roughening-recovery process to actively induce dislocation bending and BSF blocking [42]. All the semipolar (11$\underline{2}$2) GaN samples were grown on nominally on-axis m-plane sapphire substrates which were nitridized during the temperature ramping up to $1,050°C$ in a mixture of NH$_3$ and N$_2$. In our previous report, we used a high-temperature AlN buffer for the growth of smooth (11$\underline{2}$2) GaN [56]. Here we show that a low-temperature (LT) GaN buffer (deposited at $600°C$) can also give mirror-like (11$\underline{2}$2) GaN. After a thermal annealing of the LT-GaN buffer, (11$\underline{2}$2) GaN was grown at $1,000°C$ with 2 SLM NH$_3$ and 119 μmol/min TMGa. The difference in the growth parameters for samples S1–S4 is summarized in Table 1.2.

It should be pointed out the annealing behavior of the semipolar LT-GaN buffer is very different from that of the Ga-polar c-plane LT-GaN buffer [52]. As shown in Fig. 1.13, it took much longer time to anneal a semipolar LT-GaN buffer and get the in situ optical reflectance back close to the background level, and no "nose-like" peak was ever observed throughout the annealing process. For sample S1, the HT-GaN layer was grown under the one-step condition with instant oscillations (Fig. 1.13a) reflecting a quasi-two-dimensional (quasi-2D) growth mode. As we increased the reactor pressure from 60 to 200 mbar for the initial HT-GaN growth of sample S2, the reflectance began to decay in both average and amplitude from the very beginning (Fig. 1.13b), an indicator of 3D islanding growth. Once the growth condition was switched back the second-step condition (sample S1 condition), the reflectance was gradually recovered to sustained 2D oscillations. The final

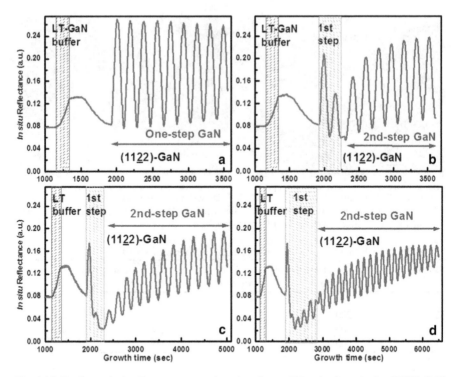

Fig. 1.13 In situ optical reflectance traces (wavelength $= 550\,\mathrm{nm}$) of semipolar (11$\underline{2}$2) GaN samples S1 (**a**), S2 (**b**), S3 (**c**), and S4 (**d**). The reflectance segments shaded in *green* in (**b**)–(**d**) correspond to the first-step 3D GaN growth for samples S2–S4

reflectance average for sample S2 was lower than that of sample S1 because of the striations on the sample S2 surface that were clearly observed under Nomarski optical microscope (not shown). From sample S2 to S3, we further elevated the reactor pressure to 300 mbar for the first-step GaN growth, the reflectance trace decayed in a faster pace to a lower level within the same growth time, and the recovery of the reflectance was accordingly much slower (Fig. 1.13c). For sample S4, the first-step growth time was increased to examine the effect of prolonged island growth (Fig. 1.13d). It is speculated that the difficulty in achieving a rapid recovery of the (11$\underline{2}$2) growth surface when it is very rough is because (11$\underline{2}$2) is a saddle point in the v-plots (Fig. 1.4a), a local minimum along the polar axis but a local maximum within the basal plane [36, 42].

Similar to the case of nonpolar GaN, the configuration of the planar and line defects in (11$\underline{2}$2) GaN is also quite complicated (Fig. 1.6c) [56]. To obtain a comprehensive knowledge of the microstructure, we also measured XRCs for both on-axis and *off-axis* diffraction planes (Fig. 1.14a, b). Figure 1.14c shows the FWHM of the on-axis (11$\underline{2}$2) XRCs as a function of the azimuthal angle, φ. The baseline sample S1 (one-step growth with instant reflectance oscillations) exhibited very anisotropic FWHMs for the on-axis (11$\underline{2}$2) XRCs, similar to previous reports.

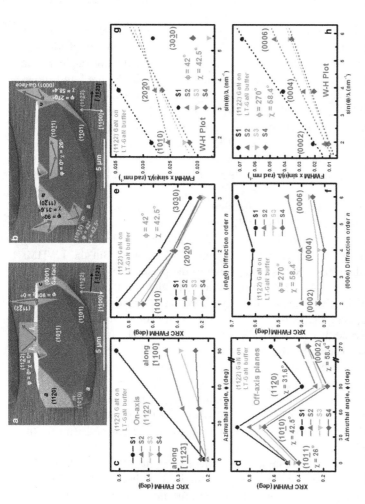

Fig. 1.14 SEM perspective side-view of a semipolar $(11\bar{2}2)$ GaN SAG mesa labeled with orientations for the illustration of (**a**) on-axis and (**b**) off-axis XRC configurations. For $\varphi = 0$ and $90°$, the X-ray rocking direction is parallel to $[\bar{1}\bar{1}23]$ and the in-plane m-axis $[\bar{1}100]$, respectively. The incident and diffracted X-ray beams (*red arrows*) rock within a plane (shaded in *green*) for each diffraction plane with its index and orientation labeled in the vicinity. The (**a**) on-axis $(11\bar{2}2)$ ($\chi = 0°$) and (**b**) off-axis planes ($\chi \neq 0°$), such as $(10\bar{1}1)$, $(10\bar{1}0)$, $(11\bar{2}0)$, and (0002), were measured in symmetric and skew symmetric geometries, respectively. The FWHM of the (**c**) on-axis and (**d**) off-axis XRCs of semipolar $(11\bar{2}2)$ GaN as a function of the azimuthal angle, φ. (**e**) The FWHMs of m-plane $(n0\bar{n}0)$ XRCs ($n = 1$, 2, and 3). (**f**) The FWHMs of c-plane $(000n)$ XRCs ($n = 2$, 4, and 6). (**g**) The modified WH plots of the m-plane $(n0\bar{n}0)$ XRCs. The straight lines in (**g**) are the linear fits to the $(10\bar{1}0)$ and $(20\bar{2}0)$ data points of each $(11\bar{2}2)$ GaN sample. (**h**) The modified WH plots of the c-plane $(000n)$ XRCs

However, with the introduction of a 3D growth stage, the FWHM of the on-axis ($11\bar{2}2$) XRC at $\varphi = 90°$ decreases monotonically from sample S1 to S4 by $\sim$52%, and the FWHM-φ plots become less steep, indicative of improved microstructural quality with a greatly reduced anisotropy. All the BSFs (I_1, I_2, and E) are not visible under ($11\bar{2}2$) diffraction, but the PDs can contribute to the broadening of the on-axis ($11\bar{2}2$) XRCs. In addition, the ($11\bar{2}2$) XRCs at $\varphi = 90°$ can also detect PSFs with a displacement vector of $1/2 < 10\bar{1}1 >$, which may be the dominant broadening factor to the ($11\bar{2}2$) XRCs at $\varphi = 90°$ of sample S1.

In order to evaluate the effect of the two-step growth procedure on the mosaic twist (in-plane misalignment) within the ($11\bar{2}2$) GaN films, we measured XRCs in a *skew* symmetric geometry for *off-axis* planes having an inclination angle χ with respect to the on-axis ($11\bar{2}2$) at various azimuths (Fig. 1.14b). Figure 1.14d shows that from sample S1 to S4, the ($10\bar{1}1$) and ($10\bar{1}0$) XRCs were narrowed down by $\sim$24%, and the ($11\bar{2}0$) and (0002) XRCs by $\sim$50%, which evidenced that the two-step growth technique could significantly improve the microstructural quality. It is noted that for all the samples, the XRC FWHMs of ($10\bar{1}0$) and (0002) XRCs were much larger than that of ($11\bar{2}0$) due to their different broadening factors [56]. It is known that ($11\bar{2}0$) diffraction is not sensitive to the presence of BSFs, in contrast to ($10\bar{1}0$) [41, 42].

As shown in Fig. 1.14e, as the diffraction order n increased from 1 to 3, the ($n0\bar{n}0$) XRC FWHM decreased by $\sim$67% for all the samples. The ($10\bar{1}0$) XRCs were greatly broadened by BSFs [41, 42, 53, 57], although the X-ray rocking direction was not exactly along but at an angle to the c-axis. But the FWHMs ($0.2 \sim 0.3°$) of the ($30\bar{3}0$) XRCs (insensitive to the presence of BSFs according to the visibility criteria [41, 42, 53, 57]) were fairly close to the dynamic range of the other off-axis XRCs linewidths (Fig. 1.14d). In contrast, as the diffraction order n increased, the c-plane ($000n$) XRC FWHMs remained similar (Fig. 1.14f), which was because neither BSFs nor PSFs are visible under the ($000n$) diffraction ($n = 2, 4$, and 6). Therefore, the c-plane XRCs were mainly broadened by PDs and/or perfect dislocations with Burgers vectors having a c-axis component. In the WH plots, the three data points of the ($000n$) diffraction could be linearly fit by a straight line (Fig. 1.14h), but the ($30\bar{3}0$) data points significantly deviated from the linear fits of the corresponding ($10\bar{1}0$) and ($20\bar{2}0$) data points (Fig. 1.14g) [41, 42, 53], which supported our assignment of the broadening mechanisms for the c- and m-plane XRCs.

From samples S1 to S4, the monotonic reduction of the m- and c-plane XRC FWHMs (Fig. 1.14c–f) indicated that a significant decrease in BSF and dislocation densities, respectively. It is believed that during the first-step 3D island growth, some dislocations bent at the inclined growth fronts [42, 58] and annihilated via their interaction with each other. And the reduction of BSF density in the semipolar ($11\bar{2}2$) GaN films also comes from the blocking of BSFs by the lateral growth of tall mesas over small islands. The significant improvement of ($11\bar{2}2$) GaN microstructural quality has led to bright photoluminescence (PL) of blue, green, green-yellow, and yellow ($11\bar{2}2$) InGaN/GaN MQWs grown on GaN template (sample S4), as shown in Fig. 1.15. It should be mentioned that InGaN MQWs

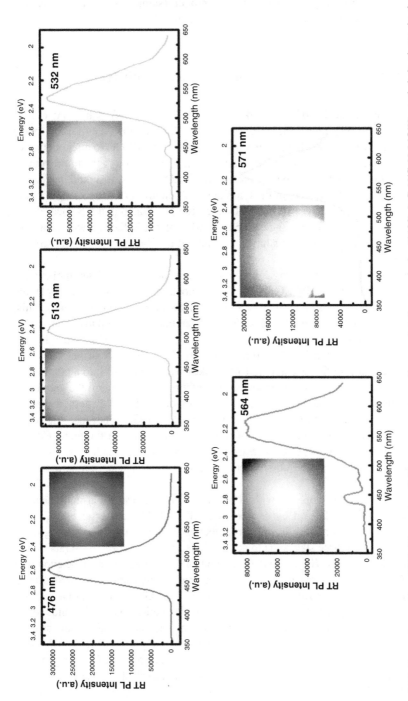

Fig. 1.15 Room temperature (RT) PL spectra of *blue*, *green*, *green–yellow*, and *yellow* semipolar (11$\bar{2}$2) InGaN/GaN MQWs, and their photos in the corresponding inset. The small peak near 445 nm was from the nonlasing line of the He–Cd laser

deposited on one-step grown (11$\underline{2}$2) GaN template (sample S1) did not give any emission because of the very high density of defects.

1.5   OCE of Semipolar GaN on Nonplanar Substrates

Due to the high density of extended defects (BSFs and PDs), semipolar (11$\underline{2}$2) InGaN/GaN MQWs and LEDs are still much dimmer than Ga-polar c-plane structures [3,59,60]. There have been various reports using ELOG to filter out part of the extended defects from the underlying GaN template, although the improvement of material quality is limited to only the Ga-face ELOG wings. It should be pointed out that, however, the defects in the template still penetrate into the mask opening regions, and moreover new BSFs and PDs are often generated in the N-face ELOG wings due to the interaction of the dielectric mask with the N-face growth front [47–49]. Thus, the ELOG GaN films have very limited useable area for device growth and fabrication.

A practical way of obtaining large-area SF-free nonpolar and semipolar GaN film is through OCE (Fig. 1.6d). The key innovation here is to decouple the nucleation orientation and the final coalesced surface orientation. To achieve this, sapphire substrate patterning, which so far has been mainly utilized to enhance light extraction, is required to expose certain facets (often c-plane) of sapphire for the nucleation of c-plane GaN with a good selectivity. Then the GaN stripes grown on the sapphire sidewall facets coalesce (often takes a long growth) into a continuous film with a nonpolar or semipolar surface orientation, depending on the geometric configuration between sapphire substrate and GaN film [61–65]. Patterned r-plane sapphire is an excellent platform for semipolar (11$\underline{2}$2) GaN, because the angle ($\sim 58°$) between (0001) and (10$\underline{1}$2) of sapphire coincides with the angle (58.4°) between (0001) and (11$\underline{2}$2) of GaN (Fig. 1.6d) [64,65]. The OCE idea has also been successfully implemented to obtain nonpolar and semipolar GaN on patterned Si substrates [66–71].

The as-patterned r-sapphire (with no mask on the mesa) was thermally cleaned in H_2 at 1,100°C, followed by a deposition of a LT-AlN buffer at 500°C. The LT-AlN buffer was annealed during the temperature ramping up. The GaN growth was designed with three stages (1) to achieve selective nucleation of c-GaN on the sapphire sidewall facets (Fig. 1.16a), (2) to control the cross-section shape evolution of the GaN stripes to reach an ideal geometric configuration for coalescence (Fig. 1.16b), and (3) to have a rapid coalescence of the GaN stripes into a flat film (Fig. 1.16c). The v-plots of GaN (Fig. 1.4) again provided a very useful guidance in adjusting the growth conditions for the shaping and the coalescence of the stripes. A high reactor pressure and a low V/III ratio were used to promote the growth along the c-axis [0001] to block the BSFs and PDs in the N-face region of the neighboring stripes (Fig. 1.16b). A low reactor pressure and a relatively high V/III ratio were utilized to facilitate the stripe coalescence. The as-grown semipolar (11$\underline{2}$2) GaN film shows very narrow XRCs (not shown here) for all the on- and off-axis diffractions,

Fig. 1.16 OCE growth evolution of semipolar (11$\bar{2}$2) GaN on patterned r-plane sapphire. (**a**) selective nucleation of GaN on the sapphire c-plane sidewalls. (**b**) GaN stripes shaping for effective blocking of the BSFs and PDs in the neighboring stripes. (**c**) full coalescence of GaN stripes into a flat semipolar (11$\bar{2}$2) GaN film

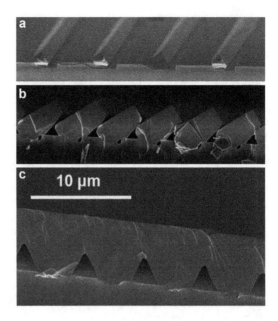

with a linewidth ranging from 0.078 to 0.15°. The linewidths are about two- to fivefold narrower than those of the semipolar (11$\bar{2}$2) GaN grown on planar m-plane sapphire (Fig. 1.14), indicating a remarkable reduction in defect density, which is confirmed by TEM. In fact, the nearly BSF-free semipolar (11$\bar{2}$2) GaN film obtained on patterned r-plane sapphire has a crystalline quality comparable to that of c-plane GaN on c-plane sapphire. High brightness blue LEDs have been achieved on the semipolar (11$\bar{2}$2) GaN grown via OCE with a great spectral purity. The details of the semipolar (11$\bar{2}$2) GaN OCE growth and LED performance will be reported elsewhere.

1.6 Summary and Outlook

In this chapter, we highlighted the major challenges in the heteroepitaxy of nonpolar and semipolar GaN, poor surface morphology and microstructural defects, including BSFs and their associated PDs. Kinetic Wulff plots (v-plots) have been proven as a powerful tool in understanding and controlling GaN heteroepitaxy along various orientations. Based on the v-plots, a two-step growth scheme has been designed and implemented to improve the surface morphology and reduce the defect density through dislocation bending and BSF blocking for both nonpolar a-GaN and semipolar (11$\bar{2}$2) GaN. OCE as an emerging growth technology is producing device-quality heteroepitaxial semipolar GaN film, with a dislocation density comparable to that of c-GaN. Besides the approaches discussed earlier, SiN$_x$ and ScN interlayers [30, 72] have also been reported as an effective method in filtering the BSFs and

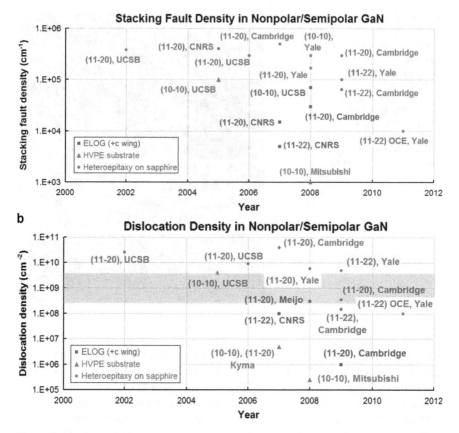

Fig. 1.17 Summary of the research timeline of nonpolar and semipolar GaN material quality: the density of BSFs (**a**) and dislocations (**b**). The *gray bar* in (**b**) indicates the typical range of threading dislocation density in heteroepitaxial *c*-plane GaN film [3]

PDs in nonpolar and semipolar GaN. The research timeline of nonpolar/semipolar GaN material quality by various groups is shown in Fig. 1.17. A combination of OCE and interlayers may be one of the ultimate solutions to further accelerate the research course of nonpolar and semipolar GaN (Fig. 1.1).

The performance gap between homoepitaxial and heteroepitaxial nonpolar/semipolar devices results from mainly the dramatic difference in defect density. The PDs bounding the BSFs are nonradiative recombination centers in nonpolar/semipolar GaN [73]. Therefore, most of the reported heteroepitaxial semipolar InGaN/GaN MQW LEDs were much dimmer than the Ga-polar *c*-plane LED grown under the same condition [3, 60]. The I–V characteristic of heteroepitaxial semipolar LED often exhibited a strong leakage behavior [3], which is probably related to the presence of the structural defects. In the long wavelength nonpolar/semipolar InGaN/GaN MQWs, the high strain can trigger the formation of new BSFs and PDs for both homoepitaxial and heteroepitaxial GaN substrates

[50, 51]. Moreover, in the semipolar (Al,In)GaN heterostructures, the (0001) slip plane is inclined with respect to the surface normal, and hence there is a component of shear stress on the slip plane which can cause misfit dislocation formation at the heterointerfaces [74–76]. The misfit dislocations together with the BSFs and PDs can further compromise the semipolar LED performance. Heteroepitaxial semipolar green LED wafer surface often appeared darkish when the growth temperature of p-GaN was high, indicating that the semipolar InGaN/GaN MQWs were not as robust as the c-plane MQWs against baking probably also due to the high density of extended defects. In addition, excess flow of Cp_2Mg is needed to achieve p-type (11$\underline{2}$2) GaN because of the lower incorporation efficiency of Mg [77].

Acknowledgements This work was supported by the US DOE under Contract DE-FC26-07NT43227 and the US DOE office of Basic Energy Sciences under Contract DE-SC0001134. Christopher D. Yerino helped compiling the v-plots of GaN through consecutive SAG experiments under various growth conditions. Benjamin Leung carried out the OCE experiments for the growth of semipolar (11$\underline{2}$2) GaN on patterned r-plane sapphire. The authors are thankful to Prof. Hyung Koun Cho and Mr. Bo Hyun Kong at Sungkyunkwan University (Korea) for their great help with the TEM measurements, and Dr. T.-S. Ko at National Chiao Tung University (Taiwan) for his help with the CL measurement.

References

1. H. Amano, N. Sawaki, I. Akasaki, Y. Toyoda, Appl. Phys. Lett. **48**, 353 (1986)
2. S. Nakamura, Jpn. J. Appl. Phys. Part 2 – Lett. **30**, L1705 (1991)
3. Q. Sun, J. Han, Proc. SPIE **7617**, 761717 (2010)
4. M.R. Krames, O.B. Shchekin, R. Mueller-Mach, G.O. Mueller, L. Zhou, G. Harbers, M.G. Craford, J. Disp. Technol. **3**, 160 (2007)
5. Y.C. Shen, G.O. Mueller, S. Watanabe, N.F. Gardner, A. Munkholm, M.R. Krames, Appl. Phys. Lett. **91**, 141101 (2007)
6. M.H. Kim, M.F. Schubert, Q. Dai, J.K. Kim, E.F. Schubert, J. Piprek, Y. Park, Appl. Phys. Lett. **91**, 183507 (2007)
7. P. Waltereit, O. Brandt, A. Trampert, H.T. Grahn, J. Menniger, M. Ramsteiner, M. Reiche, K.H. Ploog, Nature **406**, 865 (2000)
8. T. Takeuchi, H. Amano, I. Akasaki, Jpn. J. Appl. Phys. Part 1 – Regul. Pap. Short Notes Rev. Pap. **39**, 413 (2000)
9. S.-H. Park, J. Appl. Phys. **91**, 9904 (2002)
10. A.E. Romanov, T.J. Baker, S. Nakamura, J.S. Speck, J. Appl. Phys. **100**, 023522 (2006)
11. S.-H. Park, D. Ahn, Appl. Phys. Lett. **90**, 013505 (2007)
12. H. Shen, M. Wraback, H. Zhong, A. Tyagi, S.P. DenBaars, S. Nakamura, J.S. Speck, Appl. Phys. Lett. **95**, 033503 (2009)
13. M.C. Schmidt, K.C. Kim, H. Sato, N. Fellows, H. Masui, S. Nakamura, S.P. DenBaars, J.S. Speck, Jpn. J. Appl. Phys. Part 2 – Lett. Express Lett. **46**, L126 (2007)
14. H. Sato, R.B. Chung, H. Hirasawa, N. Fellows, H. Masui, F. Wu, M. Saito, K. Fujito, J.S. Speck, S.P. DenBaars, S. Nakamura, Appl. Phys. Lett. **92**, 221110 (2008)
15. N. Fellows, H. Sato, H. Masui, S.P. DenBaars, S. Nakamura, Jap. J. Appl. Phys. **47**, 7854 (2008)
16. K. Okamoto, J. Kashiwagi, T. Tanaka, M. Kubota, Appl. Phys. Lett. **94**, 071105 (2009)

17. Y. Enya, Y. Yoshizumi, T. Kyono, K. Akita, M. Ueno, M. Adachi, T. Sumitomo, S. Tokuyama, T. Ikegami, K. Katayama, T. Nakamura, Appl. Phys. Express **2**, 082101 (2009)
18. Y. Yoshizumi, M. Adachi, Y. Enya, T. Kyono, S. Tokuyama, T. Sumitomo, K. Akita, T. Ikegami, M. Ueno, K. Katayama, T. Nakamura, Appl. Phys. Express **2**, 092101 (2009)
19. J.S. Speck, S.F. Chichibu, MRS Bull. **34**, 304 (2009)
20. X. Li, X. Ni, J. Lee, M. Wu, U. Ozgur, H. Morkoc, T. Paskova, G. Mulholland, K.R. Evans, Appl. Phys. Lett. **95**, 121107 (2009)
21. J. Lee, X. Li, X. Ni, U. Ozgur, H. Morkoc, T. Paskova, G. Mulholland, K.R. Evans, Appl. Phys. Lett. **95**, 201113 (2009)
22. S.-P. Chang, T.-C. Lu, L.-F. Zhuo, C.-Y. Jang, D.-W. Lin, H.-C. Yang, H.-C. Kuo, S.-C. Wang, J. Electrochem. Soc. **157**, H501–H503 (2010)
23. S.-C. Ling, T.-C. Lu, S.-P. Chang, J.-R. Chen, H.-C. Kuo, S.-C. Wang, Appl. Phys. Lett. **96**, 231101 (2010)
24. K. Fujito, S. Kubo, I. Fujimura, MRS Bull. **34**, 313 (2009)
25. M.D. Craven, F. Wu, A. Chakraborty, B. Imer, U.K. Mishra, S.P. DenBaars, J.S. Speck, Appl. Phys. Lett. **84**, 1281 (2004)
26. M.D. Craven, S.H. Lim, F. Wu, J.S. Speck, S.P. DenBaars, Appl. Phys. Lett. **81**, 469 (2002)
27. X. Ni, Y. Fu, Y.T. Moon, N. Biyikli, H. Morko, J. Cryst. Growth **290**, 166 (2006)
28. Q. Sun, S.Y. Kwon, Z.Y. Ren, J. Han, T. Onuma, S.F. Chichibu, S. Wang, Appl. Phys. Lett. **92**, 051112 (2008)
29. Q. Sun, C.D. Yerino, Y. Zhang, Y.S. Cho, S.-Y. Kwon, B.H. Kong, H.K. Cho, I.-H. Lee, J. Han, J. Cryst. Growth **311**, 3824 (2009)
30. C.F. Johnston, M.A. Moram, M.J. Kappers, C.J. Humphreys, Appl. Phys. Lett. **94**, 161109 (2009)
31. T.J. Baker, B.A. Haskell, F. Wu, P.T. Fini, J.S. Speck, S. Nakamura, Jpn. J. Appl. Phys. Part 2 – Lett. Express Lett. **44**, L920 (2005)
32. T.J. Baker, B.A. Haskell, F. Wu, J.S. Speck, S. Nakamura, Jpn. J. Appl. Phys. Part 2 – Lett. Express Lett. **45**, L154 (2006)
33. P. Vennegues, Z. Bougrioua, T. Guehne, Jpn. J. Appl. Phys. Part 1 – Regul. Pap. Brief Commun. Rev. Pap. **46**, 4089 (2007)
34. K. Hiramatsu, K. Nishiyama, A. Motogaito, H. Miyake, Y. Iyechika, T. Maeda, Phys. Stat. Sol. A **176**, 535 (1999)
35. D.X. Du, D.J. Srolovitz, M.E. Coltrin, C.C. Mitchell, Phys. Rev. Lett. **95**, 155503 (2005)
36. Q. Sun, C.D. Yerino, T.S. Ko, Y.S. Cho, I.H. Lee, J. Han, M.E. Coltrin, J. Appl. Phys. **104**, 093523 (2008)
37. Y.S. Cho, Q. Sun, I.H. Lee, T.S. Ko, C.D. Yerino, J. Han, B.H. Kong, H.K. Cho, S. Wang, Appl. Phys. Lett. **93**, 111904 (2008)
38. Y.J. Sun, O. Brandt, U. Jahn, T.Y. Liu, A. Trampert, S. Cronenberg, S. Dhar, K.H. Ploog, J. Appl. Phys. **92**, 5714 (2002)
39. T.Y. Liu, A. Trampert, Y.J. Sun, O. Brandt, K.H. Ploog, Philos. Mag. Lett. **84**, 435 (2004)
40. A. Trampert, T.Y. Liu, O. Brandt, K.H. Ploog, J. Phys. IV, **132**, 221 (2006)
41. Q. Sun, T.S. Ko, C.D. Yerino, Y. Zhang, I.H. Lee, J. Han, T.-C. Lu, H.-C. Kuo, S.-C. Wang, Jap. J. Appl. Phys. **48**, 071002 (2009)
42. Q. Sun, B.H. Kong, C.D. Yerino, T.-S. Ko, B. Leung, H.K. Cho, J. Han, J. Appl. Phys. **106**, 123519 (2009)
43. J.L. Hollander, M.J. Kappers, C. McAleese, C.J. Humphreys, Appl. Phys. Lett. **92**, 101104 (2008)
44. T.S. Ko, T.C. Wang, R.C. Gao, H.G. Chen, G.S. Huang, T.C. Lu, H.C. Kuo, S.C. Wang, J. Cryst. Growth **300**, 308 (2007)
45. Y. Dikme, P.V. Gemmern, B. Chai, D. Hill, A. Szymakowski, H. Kalisch, M. Heuken, R.H. Jansen, Phys. Stat. Sol. (c) **2**, 2161 (2005)
46. D.N. Zakharov, Z. Liliental-Weber, B. Wagner, Z.J. Reitmeier, E.A. Preble, R.F. Davis, Phys. Rev. B **71**, 235334 (2005)

47. B.A. Haskell, F. Wu, M.D. Craven, S. Matsuda, P.T. Fini, T. Fujii, K. Fujito, S.P. DenBaars, J.S. Speck, S. Nakamura, Appl. Phys. Lett. **83**, 644 (2003)
48. B.A. Haskell, T.J. Baker, M.B. McLaurin, F. Wu, P.T. Fini, S.P. DenBaars, J.S. Speck, S. Nakamura, Appl. Phys. Lett. **86**, 111917 (2005)
49. T. Guhne, Z. Bougrioua, P. Vennegues, M. Leroux, M. Albrecht, J. Appl. Phys. **101**, 113101 (2007)
50. Z.H. Wu, T. Tanikawa, T. Murase, Y.Y. Fang, C.Q. Chen, Y. Honda, M. Yamaguchi, H. Amano, N. Sawaki, Appl. Phys. Lett. **98**, 051902 (2011)
51. F. Wu, Y.-D. Lin, A. Chakraborty, H. Ohta, S.P. DenBaars, S. Nakamura, J.S. Speck, Appl. Phys. Lett. **96**, 231912 (2010)
52. Q. Sun, Y.S. Cho, I.H. Lee, J. Han, B.H. Kong, H.K. Cho, Appl. Phys. Lett. **93**, 131912 (2008)
53. M.B. McLaurin, A. Hirai, E. Young, F. Wu, J.S. Speck, Jpn. J. Appl. Phys. **47**, 5429 (2008)
54. T. Wernicke, C. Netzel, M. Weyers, M. Kneissl, Phys. Stat. Sol. (c) **5**, 1815 (2008)
55. P. Vennegues, T. Zhu, D. Martin, N. Grandjean, J. Appl. Phys. **108**, 113521 (2010)
56. Q. Sun, B. Leung, C.D. Yerino, Y. Zhang, J. Han, Appl. Phys. Lett. **95**, 231904 (2009)
57. M.A. Moram, C.F. Johnston, J.L. Hollander, M.J. Kappers, C.J. Humphreys, J. Appl. Phys. **105**, 113501 (2009)
58. X. Ni, U. Ozgur, A.A. Baski, H. Morkoc, L. Zhou, D.J. Smith, C.A. Tran, Appl. Phys. Lett. **90**, 182109 (2007)
59. M.J. Kappers, J.L. Hollander, C. McAleese, C.F. Johnston, R.F. Broom, J.S. Barnard, M.E. Vickers, C.J. Humphreys, J. Cryst. Growth **300**, 155 (2007)
60. P.D. Mierry, T. Guehne, M. Nemoz, S. Chenot, E. Beraudo, G. nataf, Jpn. J. Appl. Phys. **48**, 031002 (2009)
61. N. Okada, Y. Kawashima, K. Tadatomo, Appl. Phys. Express **1**, 111101 (2008)
62. K. Okuno, Y. Saito, S. Boyama, N. Nakada, S. Nitta, R.G. Tohmon, Y. Ushida, N. Shibata, Appl. Phys. Express **2**, 031002 (2009)
63. Y. Saito, K. Okuno, S. Boyama, N. Nakada, S. Nitta, Y. Ushida, N. Shibata, Appl. Phys. Express **2**, 041001 (2009)
64. N. Okada, A. Kurisu, K. Murakami, K. Tadatomo, Appl. Phys. Express **2**, 091001 (2009)
65. P. de Mierry, N. Kriouche, M. Nemoz, S. Chenot, G. Nataf, Appl. Phys. Lett. **96**, 231918 (2010)
66. Y. Honda, N. Kameshiro, M. Yamaguchi, N. Sawaki, J. Cryst. Growth **242**, 82 (2002)
67. T. Hikosaka, T. Tanikawa, Y. Honda, M. Yamaguchi, N. Sawaki, Phys. Stat. Sol. (c) **5**, 2234 (2008)
68. N. Sawaki, T. Hikosaka, N. Koide, S. Tanaka, Y. Honda, M. Yamaguchi, J. Cryst. Growth **311**, 2867 (2009)
69. M. Yang, H.S. Ahn, T. Tanikawa, Y. Honda, M. Yamaguchi, N. Sawaki, J. Cryst. Growth **311**, 2914 (2009)
70. T. Tanikawa, D. Rudolph, T. Hikosaka, Y. Honda, M. Yamaguchi, N. Sawaki, J. Cryst. Growth **310**, 4999 (2008)
71. N. Sawaki, Proc. SPIE **7279**, 727902 (2009)
72. C.F. Johnston, M.J. Kappers, M.A. Moram, J.L. Hollander, C.J. Humphreys, J. Cryst. Growth **311**, 3295 (2009)
73. A.Y. Polyakov, N.B. Smirnov, A.V. Govorkov, H. Amano, S.J. Pearton, I.H. Lee, Q. Sun, J. Han, S.Y. Karpov, Appl. Phys. Lett. **98**, 072104 (2011)
74. A. Tyagi, F. Wu, E.C. Young, A. Chakraborty, H. Ohta, R. Bhat, K. Fujito, S.P. DenBaars, S. Nakamura, J.S. Speck, Appl. Phys. Lett. **95**, 251905 (2009)
75. F. Wu, A. Tyagi, E.C. Young, A.E. Romanov, K. Fujito, S.P. DenBaars, S. Nakamura, J.S. Speck, J. Appl. Phys. **109**, 033505 (2011)
76. E.C. Young, F. Wu, A.E. Romanov, A. Tyagi, C.S. Gallinat, S.P. DenBaars, S. Nakamura, J.S. Speck, Appl. Phys. Express **3**, 011004 (2010)
77. S.C. Cruz, S. Keller, T.E. Mates, U.K. Mishra, S.P. DenBaars, J. Cryst. Growth **311**, 3817 (2009)

Chapter 2
High-Quality Al-Rich AlGaN Alloys

B.N. Pantha, J.Y. Lin, and H.X. Jiang

Abstract A review is given of the synthesis and characterization of high Al-content AlGaN thin films, including optical properties, bandgap bowing, exciton localization, and n- and p-type doping. A summary of energy levels for various acceptors in AlN is also given.

2.1 Introduction

AlGaN alloys have the capability of tuning the direct bandgap over a large energy range, from around 3.4–6.1 eV, making them very useful for ultraviolet (UV) and deep ultraviolet (DUV) optoelectronic device applications with operating wavelength down to 200 nm [1–7]. The bandgap of AlGaN alloys can be adjusted by varying Al content so the visible/solar-blind UV detectors with variable cutoff wavelengths can be fabricated without the use of extra filters. Other potential applications of Al-rich AlGaN-based devices include high-density optical data storage, fluorescence detection of chemical and biological agents, water and air purification, medical research, and healthcare. GaN-based devices have already been commercialized and are important materials in the fields of compound semiconductors and optoelectronic devices. However, many fundamental properties of Al-rich AlGaN alloys are still not fully understood.

The growth of Al-rich AlGaN alloys is challenging due to the large bond strength between the Al and N atoms compared to that between the Ga and N atoms. This strong bonding results in reduced surface mobility of the Al species, making it challenging to achieve step flow growth [8, 9]. Because of this, AlN- and Al-rich AlGaN alloys are generally synthesized at higher growth temperatures than GaN

B.N. Pantha · J.Y. Lin (✉) · H.X. Jiang (✉)
Department of Electrical and Computer Engineering, Texas Tech University, Lubbock, TX 79409, USA
e-mail: hx.jiang@ttu.edu; jingyu.lin@ttu.edu

S. Pearton (ed.), *GaN and ZnO-based Materials and Devices*,
Springer Series in Materials Science 156, DOI 10.1007/978-3-642-23521-4_2,
© Springer-Verlag Berlin Heidelberg 2012

using metal organic chemical vapor deposition (MOCVD). Low growth pressure and V/III ratio are normally chosen to avoid undesired parasitic reactions at high temperatures. Indium as a surfactant has also been investigated to increase the surface mobility of Al atoms, reduce dislocation density, and enhance the n-type conductivity in Al-rich AlGaN alloys [10–15].

III-nitride materials and devices are usually grown on c-plane sapphire and SiC substrates. Due to the large piezoelectric field, quantum well (QW) structures grown on c-plane sapphire or SiC substrates with certain well thickness may suffer from the quantum confined stark effect (QCSE) [16–21]. This in turn causes spatial separation of electron and hole wave functions in QWs and reduces radiative recombination efficiency. Furthermore, devices consisting of Al-rich AlGaN and grown on polar substrates significantly reduce the surface emission intensity as Al content increases since the dominant emission in Al-rich AlGaN has a polarization of $\mathbf{E}\|\mathbf{c}$[22, 23]. This difference in band structures between GaN and AlN causes the polarization dependence emission in AlGaN alloys. Therefore, growth of DUV devices on nonpolar AlN templates such as a-plane AlN/r-plane sapphire, m-plane AlN/m-plane SiC, and m-plane AlN/m-plane AlN substrates have attracted large attention because of the absence of polarization fields and enhanced surface emission [16, 24–37].

AlN and GaN are completely miscible. Thus, AlGaN alloys in the entire compositional range have been successfully synthesized and investigated [38–43]. Compositional and temperature dependences of the energy bandgap bowing parameter and optical transitions have been extensively studied [44–49]. Due to the advances in growth and characterization methods, these parameters are now known more accurately. For example, recent studies have found the value of the bandgap bowing parameter was in a narrower variation range compared to those from previous reports ($b \sim$ 0.6–1.3 eV vs. –0.8 to 2.6 eV) [41, 50]. This accuracy is attributed to the availability of wide compositional range AlGaN alloys with high-quality epitaxial layers. The knowledge of compositional dependence of AlGaN alloys is essential to the design of AlGaN/GaN QW devices.

Other important properties which affect the optical and electrical properties of AlGaN alloys are carrier and exciton localizations. Exciton localization energy and PL emission linewidth yield information about compositional and potential fluctuations occurring in semiconductor alloys. Fundamental optical properties such as optical transitions, carrier dynamic processes, and carrier–phonon interactions predominantly determine the device performance. Therefore, understanding fundamental optical properties is crucial for the development of suitable material qualities and device structures [22, 50–62].

Control of electrical conductivity in Al-rich $Al_xGa_{1-x}N$ alloys ($x > 0.7$) is still challenging. Silicon has been widely studied for n-type doping in Al-rich AlGaN alloys, which are intrinsically insulating. The deepening of Si donor levels is one of the main reasons for low electrical conductivity of high Al content AlGaN [63–69]. Another major difficulty in obtaining highly conductive n-type Al-rich AlGaN alloys is compensation of electrons by cation vacancies $(V_{III})^{3-}$ and their complexes. It has been recognized that suppressing these intrinsic defects could significantly improve the conductivity and material quality of Al-rich AlGaN alloys [63–69]. Resistivity as low as 0.0075Ω cm has been obtained in $Al_{0.7}Ga_{0.3}N$: Si by suppressing these defects [65, 66].

Mg is the most widely investigated dopant for p-type doping in III-nitride semiconductors. However, the deep nature of Mg acceptors, (160 meV for GaN and 510–600 meV for AlN) especially in Al-rich AlGaN alloys, is the main reason of not achieving room temperature p-type conductivity [70–76, 78]. Searches for alternative doping elements such as Be, Zn, and C have also been attempted with limited success [62, 63]. In this chapter, we cover the recent progresses in growth, optical and electrical properties of high-quality Al-rich AlGaN alloys.

2.2 Growth of AlGaN

2.2.1 Typical Growth Condition of AlGaN

High-quality AlGaN alloys with high Al contents are generally grown by MOCVD on c-plane sapphire (Al_2O_3) substrates. SiC, bulk AlN, and Si substrates have also attracted research interest in the recent years. Trimethyl aluminum (TMAl), trimethyl gallium (TMGa), and NH_3 are used as Al, Ga, and N sources, respectively. General MOCVD growth conditions for InN, GaN, and AlN are illustrated in Fig. 2.1. Compared to InN or GaN, AlN is grown at a higher growth temperature ($\sim1,300\,^{\circ}C$) due to short migration of Al species at low growth temperature [8]. Low growth pressure and V/III ratio are normally chosen to avoid unwanted parasitic reaction between metal organic precursors and NH_3, which is prominent at high temperatures.

Okada et al. [9] studied the thermodynamic aspects of growing high Al content AlGaN by MOCVD. They investigated the effect of V/III ratio and growth temperature on incorporation of Al content in AlGaN. The results of their investigation are shown in Fig. 2.2. Lines denote the calculation results while scattered points are

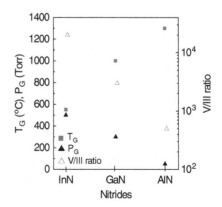

Fig. 2.1 Typical growth conditions for InN, GaN, and AlN. Growth conditions such as high growth temperature, low growth pressure, and low V/III ratio are the most conventionally used conditions to grow Al-rich AlGaN by MOCVD

Fig. 2.2 Effect of V/III ratio
on Al incorporation in GaN
(after [9])

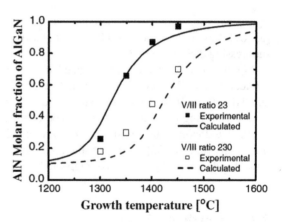

experimental data. They found that incorporation of Al content is strongly affected
by V/III ratio. With fixed growth temperature, lower V/III ratio resulted AlGaN in
higher Al content. In order to get similar Al content with higher V/III ratios, the
growth temperature needs to be raised. The experimental results matched well with
the calculated results. The reason for low incorporation efficiency of Al with higher
V/III ratios is that the equilibrium partial pressure of Ga increases with an increase
of V/III ratio compared to that of Al. On the other hand, equilibrium constants
(see [9] for details) increase with an increase of growth temperature so that the
partial pressure of Ga deceases. At high growth temperatures, the partial pressure
of Ga is much lower than that of Al and hence, Al content increases with increasing
of temperature. The AlGaN alloys grown at high temperatures were found to be of
high quality.

2.2.2 Effect of In as Surfactant in Al-Rich AlGaN Alloys

2.2.2.1 Electrical Conductivity

Highly conductive n-type Al-rich AlGaN alloys are very difficult to achieve due
to the higher activation energy of Si impurity (donor) and the generation of cation
vacancy $(V_{III})^{3-}$ and cation vacancy complexes $(V_{III}\text{-complex})^{2-}$ during the growth
[51,59,60,64–68,70–73,79]. Dislocations may also introduce acceptor-like centers
through dangling bonds along the dislocation line [80]. Indium–silicon codoping
of Al-rich AlGaN layers at relatively low temperatures $\sim$ (920–950°C) have been
reported to result in high electron concentrations and low resistivity [10, 12].
It is believed that indium incorporation can reduce threading dislocation (TD)
density [13], resulting in suppression of dislocation-induced compensation sites and
thus increase electrical conductivity. Indium may also counteract the incorporation
of defects responsible for self-compensation in Al-rich AlGaN layers, such as DX
centers, cation vacancies, and their complexes [14].

Fig. 2.3 Free electron concentration as a function of $Si/(Al + Ga)$ for different Al contents. Sample series: $Al_{0.24}Ga_{0.76}N : Si$ (*filled square*), $Al_{0.49}Ga_{0.51}N : Si$ (*filled triangle*), $Al_{0.62}Ga_{0.38}N$ (*inverted filled triangle*) samples grown at $1,150°C$, and In–Si codoped $Al_{0.65}Ga_{0.35}N$ (o) samples grown at $920°C$ (after [10])

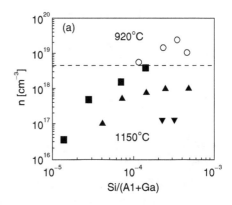

Cantu et al. [10] have studied the Si incorporation efficiency in $Al_xGa_{1-x}N$ alloys. Figure 2.3 shows free electron concentration n measured at $300\,K$ as a function of $Si/(Al+Ga)$ ratio in gas phase for each different $x(x = 0.24, 0.49,$ and $0.62)$. It was observed that Si doping efficiency strongly depends on x in $Al_xGa_{1-x}N$ grown at $1,150°C$. The free electron concentration, n, was found to decrease as x increased, as expected. For the highest Al content studied ($x = 0.62$), carrier concentration saturated at $n \sim 1.3 \times 10^{17}\,cm^{-3}$. In contrast, a remarkable increase of n was observed in In–Si codoped $Al_{0.65}Ga_{0.35}N$ grown at $920°C$. This layer exhibited n as high as $2.5 \times 10^{19}\,cm^{-3}$, with an electron mobility $\mu = 22\,cm^2/Vs$, corresponding to a resistivity of 0.014Ω cm. The free electron concentration was two orders of magnitude higher in In–Si codoped samples compared to samples without In codoping. The authors argued that In may occupy group III sites in In–Si codoped $Al_{0.65}Ga_{0.35}N$ epilayers, and thus suppress deep acceptor center formation, thus reducing self-compensation.

2.2.2.2 Surface Morphology and Material Quality

As mentioned previously, the larger bond strength between the Al and N atoms compared to that between the Ga and N atoms results in reduced surface mobility of the Al species, making it challenging to achieve step flow growth [8] in Al-rich AlGaN alloys. Using indium as a surfactant allows the group III atom to diffuse to an energetically favorable incorporation site by saturating the nitrogen bonds with indium. It has been reported that a slight incorporation of In plays a beneficial role on structural or electrical properties of nitride materials grown by MOCVD [10, 11, 81–83]. AlGaN epilayers and AlN/GaN multi-QWs (MQWs) both exhibited improved surface morphology using indium as a surfactant.

Figure 2.4 compares the atomic force microscopy (AFM) images of the surfaces of $Al_{0.75}Ga_{0.25}N$ epilayers grown on AlN/SiC templates by MOCVD [11]. Trimethyl indium (TMIn) flow rates for images (a), (b), (c), and (d) were 0, 18, 24, and $30\,ml/min$, respectively. The measured RMS roughness over the $10 \times 10\,\mu m^2$ scans range from 1.2 to 1.5 nm. The surface of the epilayer without TMIn flow in (a)

(a) TMIn = 0 ml/min   (b) TMIn = 18 ml/min

(c) TMIn = 24 ml/min   (d) TMIn = 30 ml/min

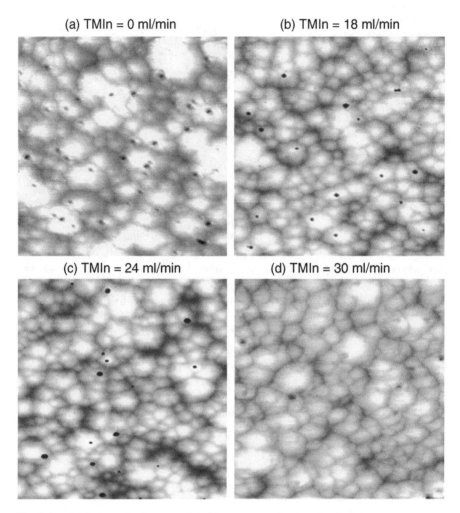

Fig. 2.4 AFM images of $0.8 - \mu$m-thick Si-doped $Al_{0.75}Ga_{0.25}N$ showing change in density of surface pits with TMIn flow rate (**a**) 0, (**b**) 18, (**c**) 24, and (**d**) 30 ml/min (after [11])

is characterized by a high density of nanometer-scale surface pits (dark spots). These surface pits are associated with the surface termination of TDs. As seen in the AFM images from (a) to (d), increasing TMIn flow rate reduces the density of surface pits. This could be a result of the enhanced adatom mobility of group III atoms. The effects of In as surfactant on optical, electrical, and structural properties of Si-doped $Al_{0.75}Ga_{0.25}N$ have also been studied. The results are summarized in Fig. 2.5. Improvement in overall material quality is evident. It has been observed that the full width at half maximum (FWHM) of X-ray diffraction (XRD) (0 0 2) rocking curves and PL intensity of deep level impurity transition at 3.2 eV related to cation vacancy complex $(V_{III}$-complex$)^{2-}$ [11] decrease and free electron density n increases with increasing TMIn flow rate.

Fig. 2.5 Variations of (a) the electron concentration n, (b) intensity of the deep level impurity transition at 3.2 eV I_{imp}, (c) electron mobility μ, (d) resistivity ρ, and (e) XRD FWHM of (0 0 2) rocking curve of Si-doped $Al_{0.75}Ga_{0.25}N$ alloys with the TMIn flow rate (after [11])

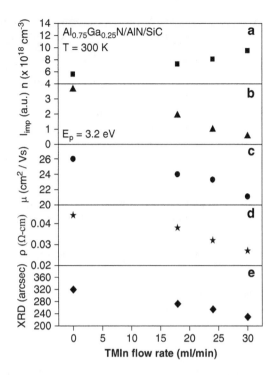

2.2.3 AlN/Al$_x$Ga$_{1-x}$N Quantum Well Structures Grown on Substrates with Different Orientations

2.2.3.1 Polarization and Piezoelectric Fields

The total polarization in a crystal in the absence of external fields is the sum of spontaneous and piezoelectric polarizations. The deviation of unit cells from the ideal hexagonal wurtzite geometry and the strong ionic character of the III-N bond are the origins of spontaneous polarization in III-nitride materials while the piezoelectric polarization is caused by strain-induced deformation of lattice parameters a, c and u, where a is in-plane lattice constant, c is lattice constant in c-axis, and u is an internal parameter defined as the anion–cation bond length along the (0 0 0 1) axis in units of c. In the linear regime, the piezoelectric polarization is quantified by the following equation:

$$\delta p_i = \sum_j e_{ij} \in_j,$$

(2.1)

where e_{ij} are piezoelectric tensors and $\in_j$ are strains.

Table 2.1 Calculated spontaneous polarization and piezoelectric constants (in units of C/m^2) for wurtzite III-nitrides, II-VI oxides, and zinc-blende III-V compound semiconductors

	P^{sp}	e_{33}	e_{31}		e_{33}	e_{31}
AlN	−0.081	1.46	−0.60	GaP	−0.07	0.03
GaN	−0.029	0.73	−0.49	AlAs	−0.01	0.01
InN	−0.032	0.97	−0.57	InAs	−0.03	0.01
ZnO	−0.057	0.89	−0.51	AlSb	−0.04	0.02
BeO	−0.045	0.02	−0.02	GaSb	−0.12	0.06
InP	−	0.04	−0.02	InSb	−0.06	0.03
AlP	−	0.04	−0.02	GaAs	−0.01	0.01

In wurtzite lattice, the piezoelectric polarization p_{pe} along the c-axis depends on two independent piezoelectric coefficients e_{31} and e_{33} as,

$$\delta p_3 = e_{33} \in_3 + e_{31}(\in_1 + \in_2) = e_{33}\frac{c - c_o}{c_o} + 2e_{31}\frac{a - a_o}{a_o}, \qquad (2.2)$$

where a_o and c_o are the lattice constants in unstained conditions, $\in_3$ is strain along the c-axis, and $\in_1 = \in_2$ is in-plane strain and assumed to be isotropic. The polarization due to shear strain is not considered here as it is not present in epitaxial structures.

The spontaneous polarization p_{sp} and piezoelectric constants for wurtzite III-nitrides, II–VI wurtzite oxides, and conventional III–V semiconductors are listed in Table 2.1 [20, 21].

The spontaneous polarization is very large in III-nitrides. The absolute values of piezoelectric constants are up to about ten times larger than those in other III–V compounds. In fact, the values of constants e_{33} and e_{21} of AlN are the largest among the tetrahedral bonded semiconductors, even larger than those of ZnO [20, 21].

The build-in electric potential due to the polarization related field in Al$_x$Ga$_{1-x}$N alloys is given by the following formula:

$$E(x) = -\frac{p}{\varepsilon_{r(x)}\varepsilon_o}, \qquad (2.3)$$

where $\varepsilon_r(x)$ and ε_o are relative and absolute dielectric constants, respectively.

2.2.3.2   Effect of Polarization in AlN/Al$_x$Ga$_{1-x}$N QW Devices

Nitride-based device structures are usually grown along the c-crystallographic direction. QWs have been used for such device structures as light emitting diodes (LEDs), laser diodes (LDs), and other devices. Although polarization-induced electric fields have been utilized to design new structures and devices, such as dopant-free high electron mobility transistors (HEMTs), the huge built-in electric field created within QWs of optoelectronic devices leads to a spatial separation of electron and hole wavefunctions, and results in a quantum-confined Stark effect (QCSE). The consequences of this effect are decreased recombination efficiency and a redshift of emission wavelength which limits the performance of the optoelectronic devices. Thus, the polarization of wurtzite nitrides has a deleterious effect on the performance of optoelectronic devices.

2.2.3.3 Growth of a-Plane AlN/AlGaN QWs

There has been increasing interest in growth of a-plane GaN/AlGaN QW structures on r-plane sapphire substrates [16, 24–37]. Optical characterization of these structures has shown that the effects due to the polarization-induced electric fields are greatly reduced in nonpolar QWs. Recently, nonpolar QWs structures are being extensively employed in DUV emitters to avoid the detrimental effect due to the polarization field that occurs in c-plane QW structures [16,24–37]. AlN/Al$_x$Ga$_{1-x}$N ($x \sim 0.65$) MQWs grown on r- and c-plane sapphire substrates by MOCVD have been investigated in the author's laboratory. Growth conditions and layer structures are described elsewhere [18]. The optical characteristics of MQWs have been performed by DUV PL spectroscopy, which consists of a frequency quadrupled 100 fs Ti:sapphire laser with an average power of 3 mW and a repetition rate of 76 MHz at 196 nm, a 1.3 m monochromator with a detection capability ranging from 185 to 800 nm, and a streak camera detector with 2 ps time resolution.

The (10 K) PL spectra of the c- and a-plane AlN/Al$_{0.65}$Ga$_{0.35}$N QWs are shown in Fig. 2.6a, b, respectively. Independent of the crystal orientation, the QW PL emission peak shifts to lower energies with increasing well width L_w, due to the weakening of the quantum confinement effect and increased piezoelectric field. In particular, the emission peak energies of the a-plane QWs steadily approach, but do not redshift beyond the band-edge transition of the Al$_{0.65}$Ga$_{0.35}$N epilayer with increasing L_w. Conversely, the c-plane QW emission peak energy redshifts with L_w and becomes even lower than the band-edge transition peak of the Al$_{0.65}$Ga$_{0.35}$N epilayer at $L_w > 2$ nm. This strong dependence of the PL emission energy on L_w in c-plane QWs is due to the strong spontaneous and strain-induced piezoelectric fields of $\sim$4 MV/cm [27], in addition to the quantum confinement effect. These polarization fields are much weaker in a-plane QWs [84].

The normalized PL emission intensity of the a- and c-plane QWs as a function of L_w is plotted in Fig. 2.7 (each orientation's intensity was normalized separately). It was observed that high emission intensity can be obtained in a-plane QWs with $L_w > 2$ nm and in c-plane QWs with $L_w = 2$ nm. In c-plane QWs, the balance between the reduced radiative recombination efficiency in thick wells due to the polarization fields and in thin wells due to enhanced carrier leakage to the barrier layer [85] results in an optimal choice of L_w for obtaining high emission efficiency. In contrast, since the polarization fields are much weaker in nonpolar QWs, high emission efficiency is obtained in QWs with $L_w > 2$ nm. This suggests that the a-plane AlN/AlGaN QW potentially provides much greater flexibility for the device structural design than its c-plane counterparts.

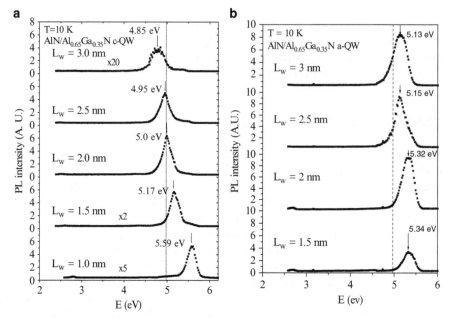

Fig. 2.6 Low temperature (10 K) PL spectra of (**a**) c-plane AlN/Al$_{0.65}$Ga$_{0.35}$N QWs with well width L_w, varying from 1 to 3 nm and (**b**) a-plane AlN/Al$_{0.65}$Ga$_{0.35}$N QWs with well width L_w, varying from 1.5 to 3 nm. All samples have a fixed barrier width of 10 nm. The *vertical dashed lines* represent the emission peak position of Al$_{0.65}$Ga$_{0.35}$N epilayers (after [18])

Fig. 2.7 Normalized low temperature PL intensity plotted as a function of well width L_w for both a- and c-plane AlN/Al$_{0.65}$Ga$_{0.35}$N QWs (after [18])

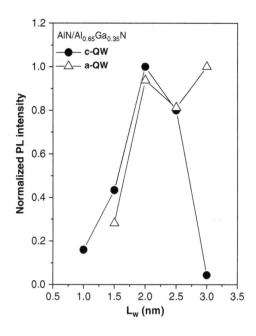

Fig. 2.8 Low temperature
(10 K) PL decay transients of
two representative
AlN/Al$_{0.65}$Ga$_{0.35}$N
(**a**) a-plane QWs and
(**b**) c-plane QWs with
$L_w = 1.5$ and 3 nm
(after [18])

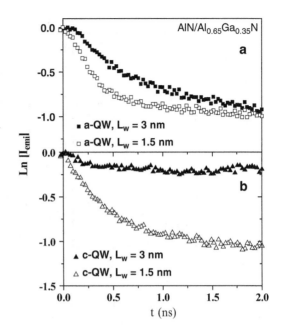

To further investigate the differences between polar and nonpolar QWs, the PL decay characteristics of a- and c-plane QWs for two representative well widths ($L_w = 1.5$ and 3 nm) were measured. The results are shown in Fig. 2.8. The PL decay transients show nonexponential decay with a slower component at longer decay times. For c-plane QWs, the PL decay time strongly depends on L_w. This can be explained by the presence of the strong polarization fields (~ 4 MV/cm) in the polar QWs. These fields spatially separate the electron and hole wave functions, thereby reducing the oscillator strength for their radiative recombination. At low temperatures, the measured decay time corresponds mainly to the radiative lifetime, which is inversely proportional to the oscillator strength [86]. Conversely, for a-plane QWs the PL decay time exhibits only a weak dependence on L_w.

2.2.3.4 Growth of AlN Epilayers on Substrates with Different Orientations

The dominant emission in Al-rich AlGaN alloys grown on c-plane substrates has a polarization of $\mathbf{E} \| \mathbf{c}$. This unique optical property significantly reduces the surface emission intensity of DUV emitters utilizing c-plane AlGaN alloys as active layers. Although growth of nonpolar nitride epilayers and devices on GaN templates has been reported by several groups [87–93], the growth of DUV devices on nonpolar AlN templates is rarely investigated [18, 29, 94, 95]. For optoelectronic devices operating at short wavelengths ($\lambda < 300$ nm), nonpolar AlN would provide reduced

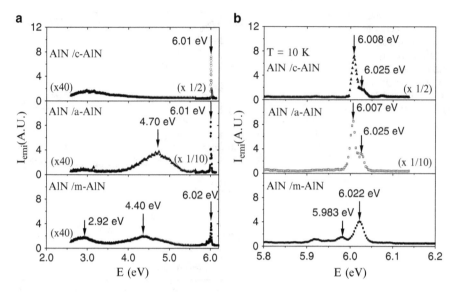

Fig. 2.9 Low temperature (10 K) PL spectra of c-plane, a-plane, and m-plane AlN homoepilayers measured in a (**a**) wide spectral range from 2 to 6.2 eV and (**b**) narrow spectral range from 5.8 to 6.2 eV (after [19])

effects of polarization fields as well as enhanced surface emission. It has also been reported that the Mg energy level in a-plane GaN is smaller than that in c-plane GaN [96]. If this trend holds true for $Al_xGa_{1-x}N$ alloys and AlN, it will help to improve p-type conductivity in Al-rich AlGaN alloys.

a-Plane (11–20) AlN can be grown on r-plane (1–102) sapphires or a-plane SiC or a-plane bulk AlN. Similarly, growth of m-plane (1–100) AlN is reported on m-plane SiC, m-plane bulk AlN, and m-plane ZnO [19, 70–73]. The overall quality, however, of nonpolar AlN epilayers is still inferior to that of c-plane AlN epilayers due to much less matured growth technology.

A systematic study of photoluminescence (PL) properties of AlN homoepilayers grown on different orientations of AlN bulk substrates (a-, c-, and m-plane) has been carried out in the authors' laboratory [19]. Figure 2.9 compares the low temperature (10 K) PL spectra of AlN homoepilayers with three different orientations (a-, c-, and m-plane) (a) in a broad spectral range from 2.0 to 6.2 eV and (b) in a narrower spectral range from 5.8 to 6.2 eV. The weak and deep impurity related emission bands (1% of the corresponding band-edge emissions) centered around 2.9 and 4.4 eV from the m-plane sample are related to O impurities [97] while the emission at 4.70 eV in a-plane sample is related to Al vacancy complex with one-negative charge state (V_{Al}-complex)$^{1-}$ [98]. Negligible emission from impurities ensures the high quality of these homoepilayers. The transition types, peak emission intensity (I_p), peak position (E_p), and binding energy of free exciton (E_x) are summarized in Table 2.2. The binding energy of free exciton (E_x) was obtained by the Arrhenius plot of the PL intensity I by using the following equation:

Table 2.2 Comparison of PL emission properties of $a-$, $c-$, and m-plane AlN homoepilayers, $c - \text{AlN}/c - \text{Al}_2\text{O}_3$, and $a - \text{AlN}/r - \text{Al}_2\text{O}_3$ heteroepilayers including emission peak position E_p, peak intensity I_p, full width at half maximum FWHM, and binding energy E_b for donor bound exciton (I_2), and free exciton (FX) (after [19])

Plane		Transition type	10 K				300 K
			E_p(eV)	I_p (A.U.)	FWHM (meV)	E_b (meV)	E_p (eV)
Homoepilayers	c-plane	FX	6.025	24	16.0	74	5.961
		I_2	6.008	136	13.1	17	
	a-plane	FX	6.025	320	13.0	74	5.961
		I_2	6.007	820	11.6	18	
	m-plane	FX	6.022	35	21.0		5.961
Heteroepilayers (Al_2O_3)	c-plane	FX	6.063	89	22	78	5.984
		I_2	6.045	63	19.2	18	
	r-plane	FX	6.030	38	33		5.956
		I_2	6.013	23	20	17	
AlN-bulk	c-plane	FX					5.961

$$I(T) = \frac{I_o}{1 + C\, e^{\frac{E_x}{K_B T}}}. \qquad (2.4)$$

The band-edge emission peak intensity ratio for a-plane, c-plane, and m-plane AlN homoepilayers is approximately 32:5:1. Due to the unique band structure of AlN near the Γ point, the PL intensity in a-plane AlN is enhanced by the fact that AlN exhibits a maximum emission in the directions perpendicular to the c direction (or in the $\mathbf{E}\|\mathbf{c}$ measurement geometry) [23]. The c direction of a-plane AlN lies in the substrate plane so that the maximum emitted light propagation direction is normal to the surface. Furthermore, homoepilayers are nearly stress free [99] and possess an identical bandgap of about 6.099 (6.035) eV at 10 (300 K).

2.2.3.5 Nonpolar AlN Epilayers on SiC

The growth of polar or nonpolar AlN on SiC substrates offers several advantages over sapphire such as better wetting of AlN and a much smaller lattice mismatch [31, 35]. The first growth of nonpolar AlN on nonpolar SiC was performed by molecular beam epitaxy (MBE) by Stemmer et al. [36]. Horita et al. [37] reported the comparable crystalline quality of 2H-AlN (11–20) grown by MBE on 6H-SiC (11–20) to that grown on GaN/sapphire (1–102) templates. The FWHM of XRD of ω-scan was between 240 and 3,000 arcsec. Similarly, AlN grown on 4H-SiC (11–20) exhibited much better crystalline qualities, as evident by narrow FWHM of XRD rocking curve which was only about 90 arcsec [31].

In a recent study [31], Horita et al. reported the growth of superior crystalline quality 4H-AlN on 4H-SiC (11–20) substrate by preparing an atomically flat and damage-free surface and by optimizing V/III ratio [31]. A very smooth AlN

Fig. 2.10 Cross-sectional
high-resolution transmission
electron microscopy (HR
TEM) images of AlN grown
on $4H$-SiC (11–20) in (a) the
present study and
(b) previous study (after [31])

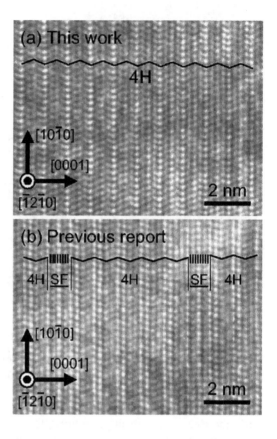

epilayer with an rms roughness of 0.15 nm scanned in an area of $2 \times 2 \, \mu m^2$ has been obtained. FWHMs of rocking curve of planes (11–20), (03–20), and (11–24) have been measured as small as 39.8, 45.5, and 76.1 arcsec, respectively. Cross sectional high-resolution transmission electron microscopy (HRTEM) images taken in (−12−10) zone axis shown in Fig. 2.10 revealed that 4H-AlN exhibits a highly ordered crystalline structure with a lattice constant of 1.03 nm. The stacking fault (SF) density was measured at $\sim 2 \times 10^{-5} \, cm^{-1}$. TD density was measured by plane view transmission electron microscopy (TEM). Bright field images of a $4H$-AlN epilayer with $g = 1$–100 and $g = 0004$, respectively, are shown in Fig. 2.11. Dark lines running along [1–100] correspond to SFs in the TEM image under the $g = 1$–100 diffraction condition. In Fig. 2.11b, TDs are observed in the TEM image under the $g = 0004$ diffraction condition. The density of TDs having the [0 0 0 1] component is about $1 \times 10^7 \, cm^{-2}$. This number is extremely small for any heteroepitaxial AlN epilayers. For example, the densities of SF and TD of $4 \times 10^5 \, cm^{-1}$ and $3 \times 10^{10} \, cm^{-2}$, respectively, were reported for MOCVD-grown GaN on r-plane sapphire (1–102) [100]. The SF and TD densities of $4H$-AlN were reduced by two orders of magnitude, compared to that of the $4H$-AlN previously reported [37].

Fig. 2.11 Bright field
plan-view TEM images of
AlN grown on 4H-SiC
(11–20) under (**a**) $g = 11–00$
and (**b**) $g = 0004$ diffraction
conditions (after [31])

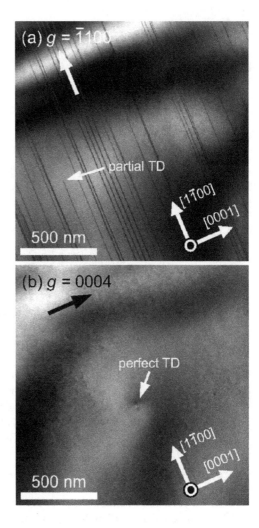

2.3   Fundamental Properties of $Al_x Ga_{1-x} N$ Alloys

2.3.1   Band Structures of $Al_x Ga_{1-x} N$ Alloys

Although the band structure of III-nitrides in the wurtzite (WZ) structure is quite
similar to that of direct bandgap zinc blend semiconductors such as GaAs, nitrides
have no inversion symmetry along and perpendicular to the direction of the c-axis.
The valence band degeneracy is lifted by the crystal–field interaction, and the strain
has a less dramatic effect on the valence band dispersion than that in GaAs. Also, the
valence band is highly nonparabolic. This anisotropy results in various phenomena
like the presence of a spontaneous electric polarization, the splitting of the valence

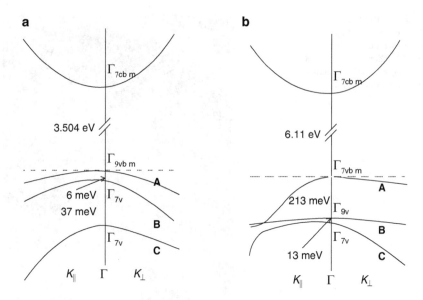

Fig. 2.12 Band structure of WZ (**a**) GaN and (**b**) AlN near the Γ point (after [5,42])

bands, and polarized spontaneous emission. Among the III-nitride family, GaN has been well studied theoretically and experimentally. Its key parameters which describe the band structure near the Γ point, including the value of the valence band splitting and the effective hole masses are known. But there is a difference in the valence band dispersion between GaN and AlN [5, 101]. The band structure of GaN near the Γ point is shown in Fig. 2.12a [42]. The order of the A-, B-, and C-valence bands are Γ_9, Γ_7, and Γ_7, respectively. For the splitting between the A and C valence band maxima (E_{AC}), values of 28 meV [102], 24 meV [103], and 43 meV [42] have been published, while the value of E_{AB} is about 6 meV. The crystal–field splitting in WZ GaN is positive (+38 meV) and the dominant optical emission is $\mathbf{E} \perp \mathbf{c}$, while in WZ AlN it is negative (−219 meV) [5]. Due to the negative crystal–field splitting, the order of the valence band in AlN is Γ_7, Γ_9, and Γ_7, respectively, and the dominant optical emission is $\mathbf{E} \parallel \mathbf{c}$. Band structure of WZ AlN near the Γ point is shown in Fig. 2.12b [5].

2.3.2 *Bandgap Bowing in* Al$_x$Ga$_{1-x}$ *N Alloys*

The bandgap energy of Al$_x$Ga$_{1-x}$N alloys exhibits a nonlinear variation with Al composition x as described by following equation:

$$E_g(x) = (1 - x)E_g(\text{GaN}) + xE_g(\text{AlN}) - bx(1 - x), \qquad (2.5)$$

in which b is the bowing parameter. Knowledge of bandgap energy in relation to x is crucial for the design of $Al_xGa_{1-x}N/GaN$ QW structures and many DUV photonic devices. Bandgap discontinuity and band-offset ratio are very important in predicting the behavior of two-dimensional electron gas, which is also affected by piezoelectric effects, which in turn is related with Al content x. With knowing the energy bandgap of GaN and AlN, the bandgap bowing parameter b determines the bandgap energy of the $Al_xGa_{1-x}N$ alloys. Values of b reported by various researchers are scattered from –0.8 eV (upward bowing) to +2.6 eV (downward bowing) ([43], and reference therein). The reason for such a wide dispersion in b is most likely emanated from $Al_xGa_{1-x}N$ alloys prepared by different techniques, different material quality, and different range of alloy compositions investigated by different groups.

Koide et al. [44] have reported value of $b = 1 \pm 0.3$ eV from $Al_xGa_{1-x}N(0 \leq x \leq 0.4)$ alloys grown on c-plane sapphire by MOCVD. Bandgap was extrapolated from absorption spectra. Angerer et al. [45] estimated the value of $b = 1.3$ eV from AlGaN alloys prepared by plasma-induced MBE in the entire composition range. Takeuchi et al. [46] have determined bandgap of strained $Al_xGa_{1-x}N(0 \leq x \leq 0.25)$ layers using PL measurements. The value of b was estimated to be 0.25. Jiang et al. [47] used reflectance method to determine the bandgap of AlGaN from AlGaN/GaN heterostructure and XRD to determine Al content. The value of b was estimated to be 0.53 eV from $Al_xGa_{1-x}N(0 \leq x \leq 0.4)$ alloys. Lee et al. [48] reported a bowing parameter of 0.69 eV from $Al_xGa_{1-x}N(0 \leq x \leq 0.45)$ epilayers. The bandgap and Al compositions were measured by PL and XRD, respectively. Yun et al. [43] measured bandgaps of $Al_xGa_{1-x}N(0 \leq x \leq 1)$ alloys grown by MBE by optical reflectance. The Al composition was measured by secondary ion mass spectrometry (SIMS) and Rutherford backscattering spectrometry (RBS). Bandgap energy as a function of x is plotted in Fig. 2.13 and fitted by (2.5) which results bowing parameter b of 1 eV. The dashed line shows the case of zero bowing. They

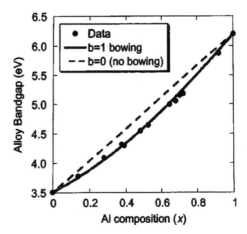

Fig. 2.13 Bandgap energy of $Al_xGa_{1-x}N$ ($0 \leq x \leq 1$) as a function of Al composition (after [43])

took the bandgaps of GaN as $E = 3.505\,\text{eV}$ (at 4 K) and AlN as $E = 6.20\,\text{eV}$ (at 2 K).

2.3.3 Unique Optical Properties of AlGaN Alloys

It is well documented that the surface emission efficiency in $Al_xGa_{1-x}N$ alloys decreases with increasing x [40, 51] and hence the efficiency of UV emitters using AlGaN alloys as active layers is lower than that of blue emitters using InGaN alloys as active layers [50, 52–155]. Recently, the properties of the fundamental optical transitions in AlN have been reported [5, 50, 55]. The band structure of wurtzite AlN near the Γ point was found to be very different from GaN as described in Sect.(3.1). The recombination between the conduction band electrons and the holes in the top most valence band is polarized along the direction of $E\|c$ ($E\perp c$) in AlN (GaN) [5, 50, 57]. The unusual valence band structure of AlN gives rise to unique optical properties of AlGaN alloys and DUV emitters.

Figure 2.14 shows the low temperature (10 K) PL spectra for $Al_xGa_{1-x}N$ alloys ($0 \le x \le 1$). Dominant emission lines are attributed to the localized exciton recombination [40]. The dotted lines indicate the emission spectra collected with the polarization of $E\|c$. The solid lines indicate the emission spectra collected with the polarization of $E\perp c$. Several features are evident: (a) the emission energy position increases with increasing x for both polarization components and (b) the PL emission intensity, I_{PL}, decreases with increasing x for $E\perp c$ polarization

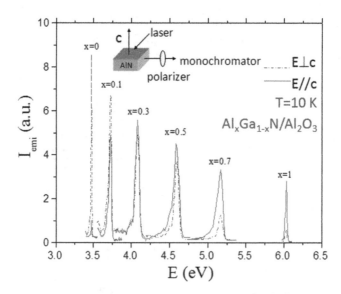

Fig. 2.14 Low temperature (10 K) PL spectra of $Al_xGa_{1-x}N$ alloys of varying x, from $x = 0$ to 1. The experimental geometry is depicted in the inset, where the electrical field of PL emission (**E**) can be selected either parallel ($\|$) or perpendicular ($\perp$) to the c axis (after [23])

component. The PL emission component evolves for $\mathbf{E} \perp \mathbf{c}$ being dominant for GaN to $\mathbf{E} \| \mathbf{c}$ for AlN.

The bowing parameter for $Al_x Ga_{1-x} N$ alloys was obtained by fitting data with (2.5) and was determined $b = 0.86 \, eV$, which is in agreement with previously reported values [43–57, 79]. PL spectral peak positions measured for both $\mathbf{E} \| \mathbf{c}$ and $\mathbf{E} \perp \mathbf{c}$ components are exactly lined up for each x. This is primarily due to the fact that the exact selection rule only applies to the free-hole transition at $\Gamma = 0$, without considering the excitonic effect. However, the transition is not totally forbidden in the presence of the excitonic effect. Instead, the forbidden transition could appear at the same energy position as the allowed transition with weaker emission intensity [107]. Moreover, strain effect and substrate misorientation can also relax the selection rule.

The degree of polarization P is deduced by:

$$P = \frac{(I_\perp - I_\|)}{[[(I)]]_\perp + I_\|]},$$ (2.6)

where $I_\perp$ and $I_\|$ are the integrated PL intensities for the polarization components of $\mathbf{E} \perp \mathbf{c}$ and $\mathbf{E} \| \mathbf{c}$, respectively. Figure 2.15 plots P as a function of x. P decreases almost linearly with increasing x, and $P = 0$ at $x = 0.25$. When Al content is increased from 0 to 1 in $Al_x Ga_{1-x} N$ alloys, crystal–field splitting decreases from $+38$ to $-238 \, meV$ as the split-off band moves up to be the top most valence band. Therefore, in Al-rich AlGaN alloys grown on c-plane substrates, light emission is polarized in $\mathbf{E} \| \mathbf{c}$.

However, a recent theoretical study [58] predicted that $\mathbf{E} \perp \mathbf{c}$ emission in c-plane $Al_x Ga_{1-x} N/AlN$ QW can be remarkably enhanced by valence band engineering

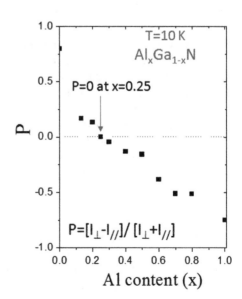

Fig. 2.15 The degree of polarization P of band-edge emission vs. x in $Al_x Ga_{1-x} N$ alloys (after [23])

Fig. 2.16 Polarization
resolve PL spectra of 1.5-nm
well width $Al_xGa_{1-x}N/AlN$
MQWs with different x
($x = 0.91$, 0.82, and 0.69).
The inset shows the
polarization PL resolve
spectra of AlN (after [22])

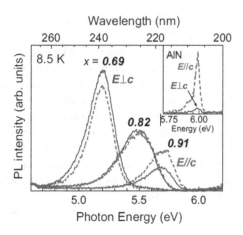

Fig. 2.17 Well
width-dependent polarization
resolved PL spectra of 1.5-nm
well width
$Al_{0.82}Ga_{0.18}N/AlN$ MQWs
(after [22])

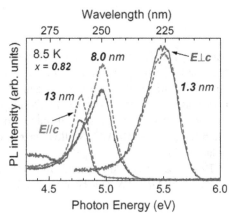

through QW width and/or in-plane compressive strain. Later, Banal et al. [22] reported experimentally that predominant polarization direction in QWs with a well width of ~1.5 nm switched from $E\perp c$ to $E\|c$ at Al content as high as $x \sim 0.83$. This Al content is considerably higher than previously reported values in $Al_xGa_{1-x}N$ epilayers ($x \sim 0.25$) [23].

Optical polarization anisotropy in $Al_xGa_{1-x}N/AlN$ ($x > 0.69$) QWs grown on c-plane sapphire substrate has been systematically investigated by Banal et al. [22]. Figure 2.16 shows the PL spectra of 1.5 nm well width $Al_xGa_{1-x}N/AlN$ QWs with various x values. It clearly shows that as x increases from 0.69 to 0.91, PL emission intensity of the $E\perp c$ polarization component decreases. The predominant polarization switched from $E\perp c$ to $E\|c$ at $x = 0.82$. Similarly, well width-dependent polarization studies in $Al_{0.82}Ga_{0.18}N/AlN$ QW reveal that well width is a key parameter for enhancing PL emission along $E\perp c$ polarization in Al-rich AlGaN/AlN QWs. Figure 2.17 shows that $E\|c$ polarization is dominant in large well width QWs. This implies that quantum confinement affects the valence band order and hence favors $E\perp c$ polarization in Al-rich AlGaN/AlN QWs. Additionally it has been proposed that compressive strain also plays an important role in enhancing $E \perp c$ polarization emission in Al-rich AlGaN alloys grown on c-plane sapphire. In an unstrained AlGaN, the Al content that switches $E \perp c$ to $E \| c$ polarization was estimated to be 0.044 from the condition that crystal–field splitting is zero [22]. Thus, previously measured value of $x = 0.25$ in $1 - \mu$m-thick $Al_x Ga_{1-x} N$ epilayer grown on sapphire is also due to the compressive strain.

2.3.4 Exciton Localization in AlGaN Alloys

2.3.4.1 Compositional and Temperature Dependence of AlGaN Energy Bandgaps

Determination of the bandgap and its variation with temperature and composition are fundamentally important to the design of practical devices based on semiconductor alloys. The temperature dependent variation of bandgap energy in semiconductors is usually described by two empirical models. The first one is the Varshni equation:

$$E_g(T) = E_g(0) - \frac{\alpha T^2}{\beta + T},\tag{2.7}$$

where α and β represent the Varshni coefficients and $E_g(0)$ is the bandgap energy at 0 K. The second model is based on Bose–Einstein statistics and expressed as:

$$E_g(T) = E_g(0) - \frac{2\alpha_B}{\exp\left(\frac{\Theta_B}{T}\right) - 1},\tag{2.8}$$

where a_B is a constant and Θ_B is the Einstein's characteristics temperature. The composition dependence of bandgap energy is described in (2.5). Studies show that values of Varshni coefficients depend on the alloy compositions as expected. Thus, the composition dependence of bandgap energy of semiconductor alloys can be determined accurately only if Varshni coefficients at that composition are known.

Compositional dependence of the Varshni coefficients α and β in ZnCdSe alloys between the temperature range 10 and 250 K has been investigated by Kuo et al. [109]. Shih et al. [105] studied the temperature dependence of the bandgap in $Cd_x Zn_{1-x} Te$ alloys ($x < 0.58$). They found a trend of compositional dependence of α and β. Generally, Varshni coefficients are obtained by fitting the measured bandgap energy at different temperatures by the Varshni equation. Because of the large localization energies of excitons in AlGaN alloys, experimental values of the band-edge emission peaks deviate from the value predicted by the Varshni equation in the low temperature region (< 200 K) [40] and measurements at higher temperatures are necessary for an accurate determination of α and β. Equation (2.7) has been employed to fit PL data of band-edge transition of AlGaN alloys obtained

Fig. 2.18 Temperature dependence of the band-edge emission peak between 250 and 800 K for $Al_xGa_{1-x}N$ alloys with different x ($0 \leq x \leq 1$). The solid curves are the least squares fit of data with (2.7). The fitted vales of $\acute{a}$ and $\hat{a}$ are also indicated (after [41])

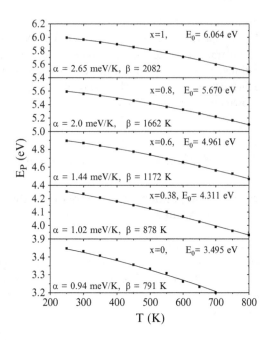

in the temperature range 200–800 K. The variation of bandgap energy as a function of temperature for different Al content (covering the whole composition range) of AlGaN alloys has been plotted in Fig. 2.18. The solid lines are the least squares fit of data by Varshni equation (2.7). From Fig. 2.18, the fitted values of $E_g(0)$ are 3.495 eV for $x = 0$, 4.311 eV for $x = 0.38$, 4.961 eV for $x = 0.6$, 5.649 eV for $x = 0.8$, and 6.064 eV for $x = 1$, whereas values of $\alpha(\beta)$ increase from 0.94 meV/K (791 K) for $x = 0$ to 2.63 meV/K (2,082 K) for $x = 1$.

Variation of Varshni coefficients as a function of x for $Al_xGa_{1-x}N$ is shown in Fig. 2.19. A parabolic relationship has been observed, and empirical formulae for composition-dependent Varshni equations have been proposed as follows:

$$\alpha(x) = (1 - x)\alpha(\text{GaN}) + x\alpha(\text{AlN}) - cx(1 - x), \qquad (2.9)$$

$$\beta(x) = (1 - x)\beta(\text{GaN}) + x\beta(\text{AlN}) - dx(1 - x), \qquad (2.10)$$

here, αs and βs are Varshni coefficients of GaN and AlN, c and d, are related to quadratic terms of $\alpha(x)$ and $\beta(x)$, respectively. Fitted values of c for $\alpha(x)$ is 2.15 meV/K and d for $\beta(x)$ is 1,561 K. Thus, compositional and the temperature dependence energy gap, $E_g(x, T)$ of $Al_xGa_{1-x}N$ alloys for the entire alloy range $0 \leq x \leq 1$ can be described by the following expression:

$$E_g(T, x) = E_g(0, x) - \frac{\alpha(x)T^2}{\beta(x) + T}. \qquad (2.11)$$

Fig. 2.19 Al composition
dependence of the Varshni
coefficients (**a**) $\acute{a}(x)$ and
(**b**) $\hat{a}(x)$ in $Al_xGa_{1-x}N$ for
$0 \leq x \leq 1$. *Solid curves* are
the least squares fit of data
with (2.9) and (2.10) for $\acute{a}(x)$
and $\hat{a}(x)$, respectively
(after [41])

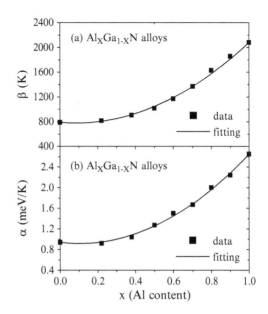

2.3.4.2 Large Exciton Localization Energy in AlGaN Alloys

The exciton localization energy is normally estimated from temperature-dependent
PL emission energy and the modified Varshni equation (2.11). Temperature-
dependent bandgap energies of an alloy measured at high temperature ($T > 200$ K)
can be well fitted by this equation; however, bandgap energy predicted by the
Varshni equation at low temperatures deviates from the measured values by an
amount of exciton localization energy E_{Loc} [40]. Thus, E_{Loc} is defined as an energy
difference between the predicted energy from Varshni equation and the PL peak
energy at low temperatures. Recently, excitonic properties of AlGaN have been
studied by other means such as time resolve PL, cathodoluminescence (CL), and
optical reflectance (OR), etc. [50, 106–110, 112, 113].

Localization of carriers and excitons in semiconductor alloys, in particular,
is caused by the composition fluctuation induced potential fluctuation. Carrier
and exciton localization could prevent the carriers moving to the nonradiative
recombination centers. It is thus believed that the high efficiency of nitrides' emitters
even with high densities of dislocation is due partly to the effect of localized exciton
recombination. Localized states act as traps for radiative recombination and prevent
the carriers from reaching the dislocation centers.

Coli et al. [114] found that the emission linewidth of the excitonic transition in
$Al_xGa_{1-x}N$ alloys increases with Al content and reaches a maximum at $x \sim 0.7$.
Increase in potential fluctuation in $Al_xGa_{1-x}N$ alloys with x has been reported for
$x \leq 0.25$ [115]. Increase in exciton localization energy with x in $Al_xGa_{1-x}N$ alloys
grown on GaN templates for $x \leq 0.76$ has been reported by fitting data with the
Bose–Einstein expression (2.8) in the temperature range between 200 and 300 K

Fig. 2.20 PL spectra of
Al$_{0.5}$Ga$_{0.5}$N alloy measured
from 10 to 800 K (after [49])

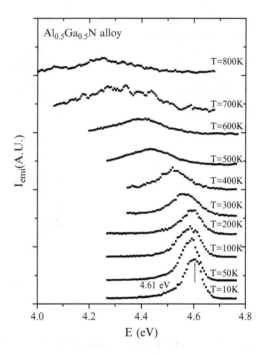

Fig. 2.20 PL spectra of
Al$_{0.5}$Ga$_{0.5}$N alloy measured
from 10 to 800 K (after [49])

[39]. An increase in exciton localization energy (E_{Loc}) from 7 to 34 meV as x increased from 0.05 to 0.35 was also measured by fitting the exciton emission peak energy with the Varshni equation in the temperature range of $T \leq 300°C$ [40]. The direct measurement of E_{Loc} from the Varshni equation and its relation with localized exciton activation energy and low temperature emission linewidth in the entire compositional range of Al$_x$Ga$_{1-x}$N$(0 \leq x \leq 1)$ alloys has been investigated in the authors' laboratory [49].

The temperature evolutions of the PL spectra of Al$_x$Ga$_{1-x}$N alloys have been measured between 10 and 800 K for the entire composition range. Figure 2.20 shows the temperature evolution of the PL spectra for one representative Al$_x$Ga$_{1-x}$N sample with $x = 0.5$. The dominant emission line at 10 K (4.61 eV) is attributed to the localized exciton transition [39, 40, 114–118]. The PL emission intensity I_{emi} decreases with increasing temperature due to thermal activation of the localized excitons. The spectral peak position is redshifted from 4.61 eV at 10 K to 4.25 eV at 800 K as described by (2.9–2.11).

Figure 2.21 shows temperature dependence of the localized exciton peak emission energy measured between 10 and 600 K for different $x(0 \leq x \leq 1)$. The *solid lines* are the least squares fit of the experimental results with the modified Varshni equation (2.11) for $T > 200$ K, for Al$_x$Ga$_{1-x}$N alloys. At low temperatures, the exciton localization dominates and the PL peak emission energy is lower than the energy value predicted by (2.11) by an amount of the localization energy. At higher temperatures, the PL emission peak follows the temperature dependence of bandgap

Fig. 2.21 Temperature dependence of the excitonic emission peak energy E_p between 10 and 600 K for different x ($0 \leq x \leq 1$). The *solid lines* are the least squares fit of data with (2.11). Deviations from (2.11) at the lowest measurement temperatures are the values of the exciton localization energies, E_{Loc} (after [49])

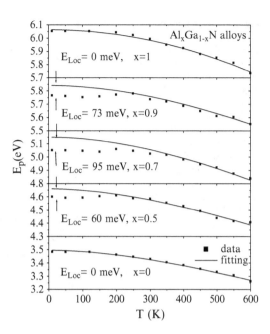

energy as described by (2.11). Thus, the deviation at the lowest measurement temperature provides a direct measure of the exciton localization energy.

Figure 2.22a shows E_{Loc} as a function x for $Al_xGa_{1-x}N$ alloys, which increases with x and reaches a maximum at $x = 0.7$. The measured value of E_{Loc} in $Al_{0.7}Ga_{0.3}N$ alloys is about 95 meV, which is the largest exciton localization energy ever reported for semiconductor alloys. In semiconductor alloys, the excitons are localized at low temperatures and free at higher temperatures. The onset temperature for the localized excitons to become free depends on the degree of localization. Due to the large values of the exciton localization energies, this onset temperature is much higher in $Al_xGa_{1-x}N$ than in other semiconductor alloys and increases with x (for $x < 0.7$). Thus, a wide measurement temperature range is required to measure E_{Loc} accurately.

2.3.4.3 Thermal Activation of Localized Exciton

Thermal activation energy E_{act} of the localized exciton is the reason of reduced emission efficiency at high temperatures. E_{act} can be obtained using (2.4) by replacing E_x with E_{act}.

E_{act} obtained by fitting I as a function of the reciprocal of T (T between 150 and 400 K) for samples with different x ($0 \leq x \leq 1$) have been plotted in Fig. 2.22b, similar to the behavior of E_{Loc} shown in Fig. 2.22a the activation energy increases with x for $x \leq 0.7$ and decreases with x for $x \geq 0.8$.

Fig. 2.22 Variations of the (a) exciton localization energy (E_{Loc}), (b) thermal activation energy (E_{act}), and (c) full width at half maximum (σ) of the PL emission line with Al content (x) in $Al_xGa_{1-x}N$ alloys. E_{Loc}, E_{act}, and σ, all increase with x for $x \leq 0.7$, and decrease with x for $x \geq 0.8$, and have maxima at $x \sim 0.7$ [The solid curve in (c) is a calculation result of (2.13)] (after [49])

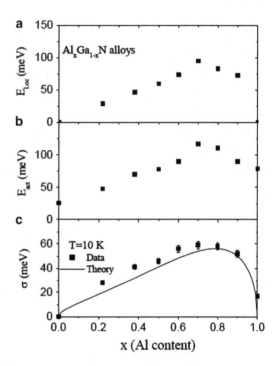

2.3.4.4 Linewidth of Localized Exciton

It is well known that exciton linewidth of semiconductor alloys is inhomogeneously broadened. The spectral broadening $\sigma(x)$ of an alloy of type $A_xB_{1-x}N$ is given by [116, 118]:

$$\sigma(x) = 0.41 \frac{dE_g(x)}{dx} \sqrt{\frac{8\ln(2)x(1-x)(V_o(x))}{\frac{4\beta a_{ex}^3(x)}{3}}}, \qquad (2.12)$$

here, $dE_g(x)/dx$ describes the variation of the direct bandgap energy with alloy composition, $V_c(x) = a_o^3(x)/(\sqrt{2})$ is the volume of primitive cell, and $a_{ex} = (\in \hbar^2)/(\mu e^2)$ is the exciton Bohr radius. $E_g(x)$ is given by (2.5). The PL exciton linewidth σ measured at 10 K of $Al_xGa_{1-x}N(0 \leq x \leq 1)$ alloys as a function of x is shown in Fig. 2.22c. To calculate the value of $\sigma(x)$ in $Al_xGa_{1-x}N$ alloys, the following relations and physical quantities were used. Lattice constant, $\alpha_o(x) = 3.162(1-x)+3.112x$ (Å); dielectric constant $\in (x) = 9.6(1-x)+6.3x$; effective masses of electrons and holes, $(m_e)/(m_o) = 0.22(1-x)+0.33x$; and $m_h = 1.5m_o$, where m_o is the electron mass in the free space. The variation of the bandgap energy is given by (2.5). Low temperature ($T = 10$ K) bandgaps E_g(GaN)= 3.5 eV, E_g(AlN)= 6.1 eV, and $b = 1$ eV have been used. The error bars are indicated for each x value. In the experimental data, σ of GaN has been subtracted at 10 K to consider only the compositional disorder. Experimental results appear to be in

Fig. 2.23 Full width at half maximum (σ) of the localized exciton emission line as a function of Al content x in $Al_xGa_{1-x}N$ alloys (after [49])

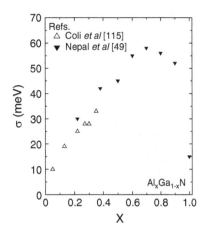

agreement with the calculations. The variation of σ with x follows the same trend as the localization energy and thermal activation energy shown in Figs. 2.22a and 2.14b and σ increases with x for $x \leq 0.7$ and decreases with x for $x \geq 0.8$. The alloy fluctuation (or σ) reaches a maximum at $x \sim 0.7$ instead of $x \sim 0.5$. This is because $\sigma(x)$ depends on two terms: $\sigma(x) \propto \sqrt{x(1-x)}^{dE_g(x)}/dx$. $\sqrt{x(1-x)}$ is symmetric and maximizes around $x = 0.5$, while $dE_g(x)/dx$ increases linearly with x. This combination makes $\sigma(x)$ nonsymmetric at about $x \sim 0.5$ and has a maximum at $x \sim 0.7$ instead of $x \sim 0.5$. Several groups have found very similar results on the variation of the excitonic linewidth of PL transitions with alloy composition x in $Al_xGa_{1-x}N$ alloys, which are summarized together in Fig. 2.23.

2.3.4.5 Correlation Between Exciton Localization Energy, Linewidth, and Activation Energy

A clear correlation between the localization parameters E_{Loc} σ, and E_{act} has been observed. A plot of E_{Loc} and σ as functions of E_{act} is shown in Fig. 2.24a. The solid line is a linear fit of the experimental data with a slope of 1. This establishes the identical characteristic of E_{Loc} and E_{act} in $Al_xGa_{1-x}N$ alloys. Similarly, from Fig. 2.24b, the solid line is a linear fit of the experimental values with a slope of 0.4 indicating a linear correlation between the E_{Loc} and σ in $Al_xGa_{1-x}N$ alloys. Lankes et al. [119] found a linear relation ($\sigma = 0.26E_{Loc}$) between the localization energy and linewidth in $ZnSe_xTe_{1-x}$ alloys.

2.3.4.6 Linewidth of Localized Exciton in Si-Doped AlGaN Alloys

The PL properties of $Al_xGa_{1-x}N$ alloys with fixed doping levels ($N_D \approx 2 \times 10^{18} \, cm^{-3}$), were studied by James et al. [56]. The FWHM of 12 K PL emission

Fig. 2.24 (**a**) Exciton localization energy (E_{Loc}) as a function of the thermal activation energy (E_{act}) of the localized exciton in AlGaN alloys. *Solid line* is a linear fit with slope of 1. (**b**) Full width at half maximum (σ) of the localized exciton emission line as a function of E_{act} in AlGaN alloys. *Solid line* is a linear fit with slope of 0.4 (after [49])

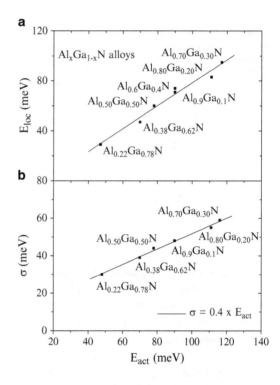

Fig. 2.25 PL FWHMs as functions of x for undoped and Si-doped $Al_xGa_{1-x}N$ (after [56])

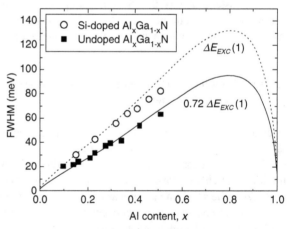

peak of $Al_xGa_{1-x}N$ alloys as a function of x is plotted in Fig. 2.25. PL FWHMs of Si-doped AlGaN alloys were found to be greater than that of undoped AlGaN as expected. This is attributed to the band filling effects, as reported for GaN [120, 121]. Similarly, exciton localization energy has also been observed to increase in Si-doped AlGaN alloys for $x > 0.3$ as shown in Fig. 2.26. It has been proposed that a sudden increase in the exciton localization for $x > 0.3$ could be related to the

Fig. 2.26 Exciton localization energy as a function of x for undoped and Si-doped $Al_xGa_{1-x}N$ (after [56])

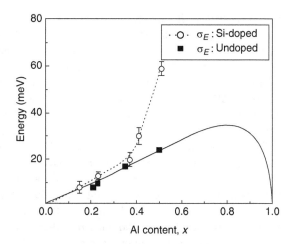

$\Gamma_9 \rightarrow \Gamma_7$ valence band crossover, which occurs at roughly the same composition $(0.2 < x < 0.3)$.

2.4   Optical Properties of $Al_xGa_{1-x}N$

2.4.1   Impurity Transitions in $Al_xGa_{1-x}N$ Alloys

AlGaN alloys are recognized as promising materials for applications in optoelectronic devices in the DUV spectral range. Highly conductive (p- and n-type) AlGaN alloys are essential for device applications. Although n-type AlN has been obtained by Si doping with Si energy level ranging from 86 to 250 meV [64–67], further improvements in the material quality and conductivity are still needed for many device applications. One of the major difficulties in obtaining highly conductive n-type Al-rich AlGaN alloys is the effect of compensation of electrons by cation vcancies $(V_{III})^{3-}$ and their complexes. It has been recognized that suppressing such intrinsic defects could significantly improve the conductivity and other material qualities of Al-rich AlGaN alloys [60,64,65,67–73]. Theoretical and experimental studies have been carried out to explore different intrinsic defects in III-nitrides. Calculations have indicated that the cation vacancy and its complex in undoped and Si-doped AlGaN alloys (particularly in Al-rich alloys) have small formation energies and are easily formed during the crystal growth [67,70–73]. Capacitance deep-level transient spectroscopy (DLTS) studies have revealed that there are at least three dominant deep levels with the same origin in low Al content n-type AlGaN alloys [122].

Two groups of deep impurity-related transitions in $Al_xGa_{1-x}N$ alloys have been reported, which were assigned to the recombination between shallow donors and two different deep acceptors [60], one of which was identified as an isolated cation vacancy with three-negative charges $(V_{cation})^{3-}$ and the other V_{cation} complex with two-negative charges $(V_{cation}\text{-complex})^{2-}$. The energy levels of these deep acceptors in AlGaN alloys are pinned to common energy levels in a vacuum [73]. By minimizing the densities of these defects, room temperature n-type resistivity of about $40\,\Omega$ cm with a free electron concentration of about $1 \times 10^{17}\,\text{cm}^{-3}$ in Si-doped AlN epilayers grown on sapphire substrates has been achieved [65, 66]. More recently, a room temperature resistivity as low as 1Ω cm with an electron concentration up to $7.4 \times 10^{17}\,\text{cm}^{-3}$ has also been reported for Si-doped AlN epilayers grown on SiC substrates by MBE [123]. However, crystalline and optical qualities of these epilayers were not reported.

Room temperature (300 K) PL spectra for a set of $Al_xGa_{1-x}N$ alloys covering the entire alloy range are shown in Fig. 2.27 [59, 60]. The band edge transition peak energy blueshifts from 3.42 eV for $x = 0$ to 5.96 eV for $x = 1$ in $Al_xGa_{1-x}N$ alloys. In addition to the band edge transition, each spectrum also exhibits deep impurity transitions. The impurity emission peak at around 2.15 eV (yellow line or YL) in GaN has been extensively studied. It was previously attributed to the transition from a shallow donor to a deep acceptor [70, 71, 124–127]. It was suggested that the deep acceptor involved in YL may be linked to $V_{Ga}-O_N$ and $V_{Ga}-2O_N$ based on local-density-functional calculation [127]. Since O_N sits next to V_{Ga} to reduce the coulomb energy, V_{Ga} forms defect complexes with it [70, 71]. More recent studies have suggested that the deep acceptor involved in the YL is related to V_{Ga} complexes with two-negative charges $(V_{Ga}\text{-complex})^{2-}$ such as $(V_{Ga}-O_N)^{2-}$ or $(V_{Ga}-Si_{Ga})^{2-}$ [60, 72, 73]. The PL emission peak at 3.4 eV (violet line or VL) in AlN was first observed by Youngman and Harris [129]. Nepal et al. [59] have suggested that the VL in AlN is due to the transition from shallow donors to isolated Al vacancies with three-negative charges $(V_{Al})^{3-}$. Al-rich AlGaN alloys with reduced emission intensities associated with deep impurity transitions generally show improved conductivity. By examining the PL spectra shown in Fig. 2.27, besides $(V_{III})^{3-}$ and $(V_{III} - \text{complex})^{2-}$, a group of impurity transitions (indicated with *dark arrows*) with emission energies above those of $(V_{III})^{3-}$ and $(V_{III}-\text{complex})^{2-}$ related transitions have been observed in $Al_xGa_{1-x}N$ alloys. The spectral peak positions of this group of impurity transitions blueshift from 2.86 eV in GaN to 4.71 eV in AlN. The blue band (2.8 eV) has been observed in Si and undoped GaN in PL, CL, and photoconductivity measurements [130, 131] and was attributed to a $V_{Ga}-O_N$ complex [130].

The room temperature PL spectral peak position (E_{imp}) of impurity transition corresponds to 4.71 eV in AlN is plotted in Fig. 2.28 for $Al_xGa_{1-x}N$ alloys as a function of x (*closed squares*). The energy position of this impurity transition continuously increases with x and follows the same trend as those of $(V_{III})^{3-}$ and $(V_{III}-\text{complex})^{2-}$ related impurity transitions, indicating that they are of the similar nature of donor–acceptor pair (DAP) type transitions involving shallow donors and deep acceptors.

The energy level, E_A, of this deep acceptor with the conduction and valence band edges (E_c and E_v) as function of Al content x is plotted in Fig. 2.29, which

Fig. 2.27 Room temperature PL spectra of $Al_xGa_{1-x}N$ alloys with x between 0 and 1, showing the band edge and impurity transitions. A group of impurity transitions has been identified and their peak positions are marked with *bold arrows* (after [59])

Fig. 2.28 PL peak position of impurity transition (E_{imp}) obtained from Fig. 2.27 (*closed squares*) and those of $(V_{III})^{3-}$ and $(V_{III} - complex)^{2-}$ related transitions from [60] (*open circle* and *squares*) in $Al_xGa_{1-x}N$ as functions of Al content (x). *Dotted lines* are guide to the eyes (after [59])

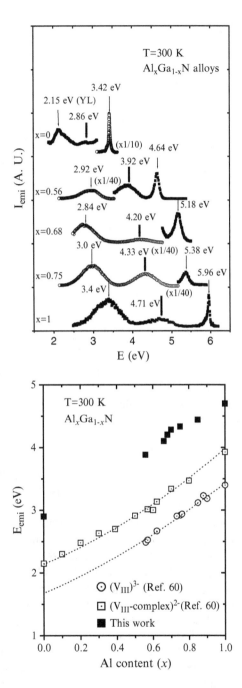

further identifies the origin of this level. The conduction and valence band offset parameters have been assumed to be $\Delta E_c = 70\% \Delta E_g$ and $\Delta E_v = 30\% \Delta E_g$ for $Al_xGa_{1-x}N(0 \leq x \leq 1)$ alloys, respectively. The valence band maximum of GaN

Fig. 2.29 The deep acceptor energy levels (E_A) associated with (V_{III} − complex)$^{1-}$ (*closed squares*), (V_{III} − complex)$^{2-}$ (*open squares*) and (V_{III})$^{3-}$ (*open circles*) obtained from Figs. 2.27 and 2.28 and [60] plotted together with E_c, E_v, and E_D^0 (shallow donor level) as functions of Al content x (after [59, 60])

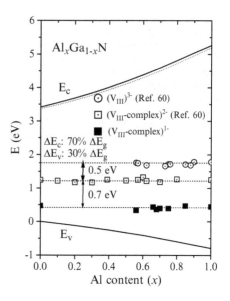

is chosen as $E = 0$. The chemical origin of the shallow donors involved here is believed to be either Si or O in $Al_xGa_{1-x}N$ alloys. Assuming the ionization energies of the shallow donors (E_D^0) increase linearly from 25 to 86 meV with varying x from 0 to 1 in $Al_xGa_{1-x}N$ alloys [26,27], the acceptor level (E_A) deduced from Figs. 2.27 and 2.28 by neglecting the Coulomb interaction between the ionized donors and acceptors is then given by:

$$E_A = E_g(x) - E_{imp} - E_D + E_v, \qquad (2.13)$$

where $E_g(x)$ is the bandgap of $Al_xGa_{1-x}N$ alloys, $E_v = -0.3\Delta E_g(x)$, $E_c = E_g(GaN) + 0.7\Delta E_g(x)$, and E_D is the shallow donor energy level. E_A is plotted in Fig. 2.29 (*closed squares*) with E_c and E_v as functions of x. The result clearly shows that the observed deep acceptor has an energy level of 1.25 eV (0.55 eV) in AlN (GaN). The striking feature is that, similar to (V_{III})$^{3-}$ and (V_{III}-complex)$^{2-}$ deep acceptor levels in $Al_xGa_{1-x}N$ alloys, the deduced energy level of this deep acceptor as a function of x is also a horizontal line in the whole range of x. This clearly indicates that the energy levels of these deep acceptors in $Al_xGa_{1-x}N$ alloys are also pinned to a common energy level in a vacuum with energy separations of 0.7 eV and 1.2 eV from those of (V_{III}-complex)$^{2-}$ and (V_{III})$^{3-}$, respectively.

The formation energies (E_f) of cation vacancies and vacancy complexes have been calculated in GaN and AlN, showing that E_f of (V_{III})$^{3-}$ and (V_{III} − complex)$^{2-}$ are lower than that of (V_{III}-complex)$^{1-}$ and anion vacancies cannot be formed in Si-doped GaN and AlN [71–73]. Due to the energy position relative to the those of (V_{III})$^{3-}$ and (V_{III} − complex)$^{2-}$, it is believed that the deep level acceptor involved in this impurity transition is a cation vacancy complex with one-negative charge (V_{III} − complex)$^{1-}$ such as (V_{Ga}-2O$_N$)$^{1-}$. The possibility of isolated (V_{III})$^{1-}$ as the origin of this deep acceptor in $Al_xGa_{1-x}N$ alloys can be excluded due to its much higher formation energy than $(V_{III}-complex)^{1-}$. The energy levels of $(V_{Ga}-2O_N)^{1-}$ deep acceptors increase almost linearly from about 0.55 to 1.25 eV with varying x from 0 to 1 in $Al_xGa_{1-x}N$ alloys. The calculated energy level of $(V_{Al}-2O_N)^{1-}$ ranges from 1.0 to 1.12 eV in AlN [71–73]. The calculated differences in binding energies between $(V_{III})^{3-}$ and $(V_{Al}-O_N)^{2-}$ and between $(V_{Al}-O_N)^{2-}$ and $(V_{Al}-2O_N)^{1-}$ in AlN, respectively, are about 0.5 and 0.7 eV [71–73], which agree perfectly with the measured values shown in Fig. 2.29 (0.5 and 0.7 eV).

Since $(V_{III}-2O_N)^{1-}$ captures only one electron, its presence will not detriment the conductivity as severely as the presence of $(V_{III})^{2-}$ and $(V_{III}-O_N)^{2-}$ impurities, which is consistent with the experimental results presented in Fig. 2.30. Figure 2.30 illustrates the evolution of room temperature (300 K) PL spectra and resistivity of $Al_xGa_{1-x}N$ alloys ($x \sim 0.66$) grown under various conditions. The arrows indicate the peak positions of the spectra. All samples exhibit a band-edge transition at 5.10 eV and the band-edge emission intensity increases with decreasing resistivity.

As shown in Fig. 2.30a, for an early grown Si-doped $Al_{0.66}Ga_{0.34}N$ alloy with high resistivity, the dominant impurity peak at 2.84 eV is related to the isolated cation vacancy $(V_{III})^{3-}$ [60]. The triply negatively charged cation vacancies capture three electrons and thus their presence significantly increases the resistivity. As illustrated in Fig. 2.30b, suppression of $(V_{III})^{3-}$ related transition results in a reduction in resistivity to 1.8 Ω cm and simultaneously an emission peak at 3.31 eV associated with the presence of $(V_{III}-complex)^{2-}$ emerges [60]. Since $(V_{III})^{3-}$ and $(V_{III}-complex)^{2-}$ defects in AlGaN alloys capture three and two electrons,

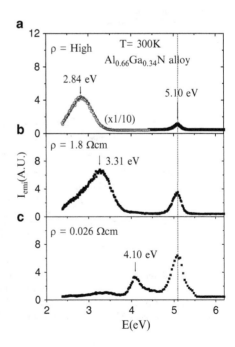

Fig. 2.30 Room temperature PL spectra of $Al_{0.66}Ga_{0.34}N$ alloys with different resistivities (after [59])

respectively, these defects act as compensation centers for n-type doping, and their densities have to be minimized in order to further enhance the conductivity. As illustrated in Fig. 2.30c, once $(V_{III})^{3-}$ and $(V_{III} - \text{complex})^{2-}$ related transitions were suppressed by varying the growth conditions, we have significantly reduced the $Al_{0.66}Ga_{0.34}N$'s resistivity to $0.026\ \Omega$ cm. However, the impurity transition involving $(V_{III} - \text{complex})^{1-}$ at 4.10 eV is evident. It is believed that the conductivities of Al-rich AlGaN alloys and pure AlN will continue to improve by further reducing the densities of the intrinsic defects – cation vacancies (and complexes) with triple-, double-, and single-negative charges.

2.4.2 Impurity Transition in Mg-Doped AlGaN Alloys

PL studies of impurity transitions in Mg-doped $Al_xGa_{1-x}N$ alloys grown on c-plane (0 0 0 1) sapphire substrates by MOCVD for x ranging from 0 to 1 have been studied in the authors' laboratory [61]. Figure 2.31a shows the low temperature (10 K) PL

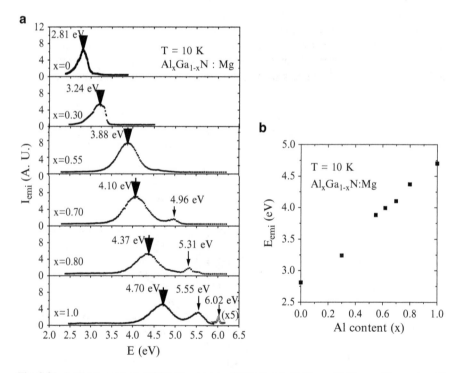

Fig. 2.31 Low temperature (10 K) PL spectra of Mg-doped $Al_xGa_{1-x}N$ alloys with varying Al contents (x) from 0 to 1. A group of impurity transitions, which are highlighted by the bold arrows, is dominant over that of the band edge. (**b**) PL emission peak position (E_{imp}) of the group of deep impurity transitions in $Al_xGa_{1-x}N$ alloys (after [61])

spectra of Mg-doped $Al_xGa_{1-x}N$ alloys of varying x values ($x = 0, 0.3, 0.55, 0.7$, 0.8, and 1.0). In all samples, a group of impurity transitions, which are highlighted by the *bold arrows*, is dominant over that of the band edge. The spectral peak positions (E_{emi}) of this group of impurity transitions are blueshifted from 2.81 eV in GaN to 4.7 eV in AlN and are plotted in Fig. 2.31b. Linear variation in E_{emi} indicates that these impurity transitions are of the same physical origin. Although there is a debate concerning the origin of the 2.8 eV emission peak in Mg-doped GaN, various investigations support the model of a DAP transition involving a deep donor and Mg acceptor [133–137].

The formation energies (E_f) of nitrogen vacancies with triple positive charges $(V_N)^{3+}$ were predicated to be relatively low in Mg-doped GaN and AlN [72, 138]. Thus, the theoretical calculations suggested that the generation of V_N^{3+} is favorable during the growth of Mg-doped AlGaN alloys. The energy level of V_N^{1+} is shallower than V_N^{3+}, and the observed emission lines are not consistent with the presumed transitions involving V_N^{1+} and Mg acceptors. Furthermore, the observed impurity transitions are completely different from the impurity transitions in n-type AlGaN alloys [59, 60]. Thus, these impurity emissions have been related to the transitions between electrons bound to V_N^{3+} and Mg acceptors.

The emission peaks at 4.96, 5.31, and 5.55 eV for $x = 0.7$, 0.8, and 1, respectively, are attributed to the transition of electrons from the conduction band (or bound to shallow donors) to neutral Mg acceptors [138]. This assignment provides an energy level of Mg acceptors in AlN of about 0.5 eV, which is consistent with the previously reported result [139].

To provide a more comprehensive picture of the nitrogen vacancy levels in AlGaN alloys, alloy composition dependence of the conduction band edge (Ec) and valence band edge (E_v), Mg level, and V_N^{3+} level in $Al_xGa_{1-x}N$ alloys have been plotted, as shown in Fig. 2.32. In obtaining Fig. 2.32, bandgap bowing parameter b and bandgaps of GaN and AlN were taken as $b = 1$ and 3.50 and 6.11 eV, respectively, at 10 K and the band offset of the AlGaN alloy system was taken as 70% (30%) for the conduction (valence) band. The valence band maximum of GaN was chosen as $E = 0$. The Mg acceptor levels in $Al_xGa_{1-x}N$ alloys for various xs are taken from the published reports and summarized in Fig. 2.33. Based on these results, the variation of the Mg energy level with x has been traced. From the Mg acceptor energy levels $E_A(x)$ and the observed impurity emission peaks $E_{emi}(x)$, the energy levels of V_N^{3+} in $Al_xGa_{1-x}N$ alloys as a function of x can be written as

$$E_{V_N^{3+}} = E_g(x) - E_{emi}(x) - E_A(x) + E_v. \qquad (2.14)$$

The Coulomb interaction term between ionized donors and acceptors has been neglected in the calculation. Here, $E_v = -0.3\Delta E_g(x)$. The three arrows drawn from the V_N^{3+} level to the neutral Mg acceptor level represent the impurity emission energy peak positions in GaN, AlGaN, and AlN. From Fig. 2.32, we obtain energy levels of V_N^{3+} of about 0.53 eV in GaN and 0.88 eV in AlN. Calculated energy level

Fig. 2.32 The nitrogen vacancy (V_N^{3+}) and Mg acceptor energy levels in $Al_xGa_{1-x}N$ alloys as functions of x. The conduction and valence band offsets of the AlGaN alloys are taken to be 70% and 30%, respectively. The valence band maximum of GaN is taken as $E = 0$ (after [61])

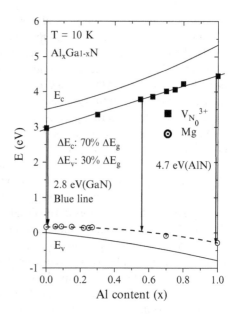

of V_N^{3+} is in the range of 0.4–0.59 eV in GaN 0.9–1.1 eV in AlN [139, 140]. Hence, the energy levels of V_N^{3+} obtained here experimentally for GaN and AlN agree reasonably well with the previous calculations.

2.4.3 Energy Level of Various Acceptors in AlN

In Fig. 2.33, continuous-wave PL spectra of undoped, Mg-, Zn-, and Be-doped AlN epilayers measured at 10 K have been plotted together [62, 63]. Two band-edge peaks were well resolved in undoped AlN with virtually no impurity-related emission lines. The dominant emission peak at 6.063 eV is due to the recombination of free excitons (FX) while the peak at 6.045 eV is from a donor-bound exciton transition (the I_2 line). In Mg-doped AlN epilayers, FX transition disappears and a new emission line with a peak position at 6.022 eV is observed which is attributed to the recombination of excitons bound to neutral Mg acceptors (A_0, X) or the I_1 line in Mg-doped AlN [55]. Two additional impurity emission lines at 4.70 and 5.54 eV in Mg-doped AlN are also observable and believed to be DAP transitions involving two different donors (deep and shallow level donors) and Mg acceptors [139]. The deep level donors participating in the 4.70 eV transition in Mg-doped AlN were identified as nitrogen vacancies with three positive charges (V_N^{3+}) that act as compensating centers for p-type doping [75].

Compared to the PL spectrum of Mg-doped AlN, the PL spectrum of Zn-doped AlN has a very similar line shape with impurity transitions at 4.50 and 5.40 eV. The emission line at 4.50 eV in Zn-doped AlN has been assigned as a DAP transition

Fig. 2.33 Low temperature
(10 K) PL spectra of undoped,
Mg-, Zn-, and Be-doped AlN
epilayers (after [62,63])

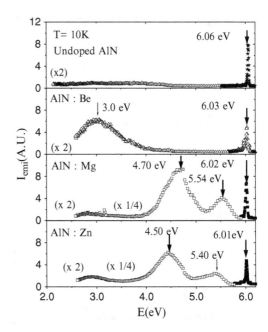

of electrons bounded to nitrogen vacancies with three positive charges (V_N^{3+}) to neutral Zn acceptors, The band edge emission line at 6.01 eV in Zn-doped AlN can be indisputably assigned to the recombination of excitons bound to neutral Zn acceptors (I_1 transition). By comparing PL spectra of Zn- and Mg-doped AlN epilayers with undoped epilayers, it was deduced that Zn energy level is about 0.74 eV, which is about 0.23 eV deeper than Mg energy level (0.51 eV) in AlN. It is thus concluded that Zn would not be a better candidate than Mg as an acceptor dopant in AlN [63]

In Be-doped AlN, weak emission lines observed at 6.08, 5.97, and 5.86 eV are 2LO, 3LO, and 4LO phonon replicas of excitation lasers (6.30 eV) with LO phonon energy of 110 meV in AlN. The emission line at 6.030 eV in Be-doped AlN is due to the recombination of excitons bound to neutral Be impurities (the I_1 line). The energy difference between the FX peak position in undoped AlN and the I_1 line in Be-doped AlN gives the BE of the excitons bound to Be impurities to be 33 meV (=6.063–6.030 eV). According to Haynes' rule, the BE of the exciton-neutral impurity complex is about 10% of the impurity BE, neglecting the central cell correction [141]. It is thus implied that Be acceptor BE is about 0.33 eV, which agrees quite well with the calculation value of ~0.34 eV for the substitution Be at the Al site [142].

The energy levels of different acceptors and compensating donors for GaN and AlN obtained from optical measurements are depicted in Fig. 2.34. It has been found that Be has the lowest activation energy among the three acceptors studied to date.

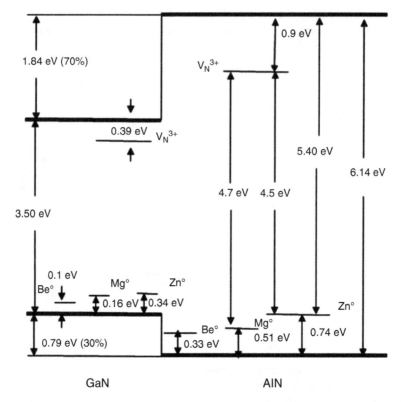

Fig. 2.34 Energy diagram showing the energy levels of Mg, Zn, and Be acceptors in AlN and GaN ([62,63] and reference therein)

2.5 Electrical Properties of AlGaN

2.5.1 n-Type $Al_x Ga_{1-x}N$ Alloys

In the past, AlN- and Al-rich AlGaN alloys grown by several methods have been found to be highly resistive [143–145]. Theoretical studies predicted that it is essentially impossible to overcome the insulating properties of AlN [45, 146–148]. According to these studies, triply charged Al vacancies (V_{Al}^{3-}), which form during the Si doping due to the low formation energy, tend to compensate the Si donors. Recent studies revealed, however, that by suppressing these vacancies one can make conductive AlN by Si doping [10, 51, 52, 65–67, 123, 149–151]. This argument has been widely accepted now and backed by results from many groups including ours.

A number of factors compensating for the Si donors in $Al_x Ga_{1-x}N$ alloys have been extensively explored as DUV optical devices and wide bandgap electronic devices have witnessed an increasing demand in recent years. Increase of Si ionization energy with increasing Al content x (especially $x > 0.7$) [156], formation of a DX center by Si at $x > 0.60$ [152], acceptor-like defects such as impurities, dislocations, and native defects [153], formation of oxygen DX center ($x > 0.27$) [69], carbon C_N [154], and cation vacancies (V_{III}) and (V_{III} − complex) [153], etc., are the most frequently referred compensating factors. Recently, however, there have been several reports on highly conductive n-type Si-doped Al-rich AlGaN alloys [10, 51, 52, 65–68, 123, 149, 150]. The main problem has been identified and methods to suppress these compensating centers have been developed. Resistivity as low as 0.075 and 1 Ω cm were reported for AlGaN and AlN, respectively [65, 66, 123].

2.5.1.1 Si Level in AlGaN

Silicon is a commonly used donor impurity in the AlGaN system. Its energy level is relatively shallow in $Al_xGa_{1-x}N$ for x up to 0.7 [51, 150], however, its level unexpectedly increases for Al content higher than 0.7. Figure 2.35 shows the experimental results for Si activation energy as a function of Al content obtained from temperature-dependent Hall carrier density. Nakarmi et al. [65, 66] found the activation energy E_D remains as low as 23 meV until $x = 0.7$, while Borisov et al. [150] found that E_D is fairly constant up to $x = 0.85$. The reason for sudden jump in E_D in $Al_xGa_{1-x}N$ with $x > 0.7$ or 0.85 is not fully understood. Borisove et al. found that E_D in $Al_xGa_{1-x}N(x > 0.85)$ is correlated with increased incorporation of oxygen and carbon. Oxygen is believed to be involved in formation of the DX center and carbon may act as an acceptor in AlN [69, 154]. The formation of defect complexes with these impurities could be the reason of deepening of Si donor impurity level in Al-rich AlGaN [150].

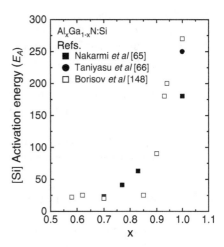

Fig. 2.35 Activation energy of Si donors in $Al_xGa_{1-x}N$ alloys

Fig. 2.36 The temperature
dependence of the free
electron concentration (n) in
an unintentionally doped
$Al_{0.67}Ga_{0.33}N$ epilayer. The
solid line is a linear fit of the
experimental data with the
relation
$n(T) \propto \exp(-E_D/kT)$
(after [68])

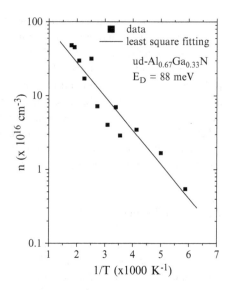

2.5.1.2 Unintentionally Doped n-Type AlGaN

As-grown undoped GaN is n-type. However, conductivity of undoped AlGaN alloys
decreases with increasing Al content. It was proposed that one of the reasons of
decreasing conductivity with increased Al content in undoped $Al_xGa_{1-x}N$ is due
to the transition of oxygen from shallow donors into DX center [69]. However,
Nakarmi et al. [68] have achieved room temperature n-type conductivity of an
undoped $Al_{0.67}Ga_{0.33}N$ alloy by optimizing MOCVD growth conditions. A shallow
donor level responsible for n-type conductivity has been identified by both Hall and
PL measurement which was about ~90 meV below the conduction band.

Figure 2.36 shows the Arrhenius plot of free electron concentration n of an
undoped $Al_{0.67}Ga_{0.33}N$ epilayer. The free electron concentration increases exponentially with increasing temperatures. At room temperature, n-type resistivity of
~85Ω cm was measured. The solid line is the least square fitting with the relation,
$n(T) \sim \exp(-E_D/kT)$, where $n(T)$ is the free electron concentration at temperature
T and E_D is the donor activation energy. The fitted E_D is found to be 88 meV. This
implies that there must be an unintentional shallow donor in the $Al_{0.67}Ga_{0.33}N$ alloy
contributing to n-type conductivity. Temperature-dependent PL measurement also
confirmed the existence of shallow donors with comparable E_D as obtained by Hall-
effect measurement. Isolated oxygen impurities are believed to be a strong candidate
of this shallow donor which remains shallow until $x = 0.7$ [68].

2.5.1.3 Highly Conductive Al_xGa_{1-x} N Alloys ($x > 0.7$)

Recently, there have been several reports for highly conductive n-type $Al_xGa_{1-x}N$
alloys for x up to 0.7 [10, 51, 65–67, 123, 149, 150]. Nam et al. [51] reported a

Fig. 2.37 Room temperature
Hall measurement results:
(**a**) resistivity, (**b**) electron
concentration, and
(**c**) electron mobility of n-type
$Al_xGa_{1-x}N(x \geq 0.7)$ vs. Al
content x (after [65,66])

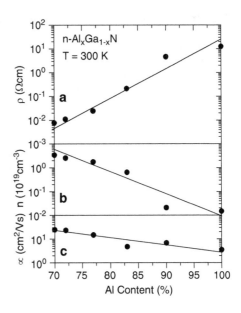

resistivity value of $0.15\,\Omega$ cm (with a free electron concentration of $2.1 \times 10^{18}\,cm^{-3}$
and mobility of $20\,cm^2/V\,s$) for Si-doped $Al_{0.65}Ga_{0.35}N$. Cantu et al. [10] have
obtained n-type $Al_{0.65}Ga_{0.35}N$ with an electron concentration of $2.5 \times 10^{19}\,cm^{-3}$
and mobility of $22\,cm^2/V\,s$, corresponding to a resistivity of $0.011\,\Omega$ cm using n–Si
codoping. There are also several studies on Si-doped AlN [65–67,123]. An electron
concentration of $7.4 \times 10^{16}\,cm^{-3}$ and resistivity as low as $1\,\Omega$ cm in Si-doped AlN
have been reported [123].

Figure 2.37 presents the room temperature Hall measurement results of Si-doped
$n-Al_xGa_{1-x}N(x > 0.7)$ alloys grown on sapphire substrates by MOCVD [65,66],
showing the Al content (x) dependent resistivity (a), electron concentration (b),
and electron mobility (c), respectively. The targeted Si-doping concentration (N_{Si})
was around $4 \times 10^{19}\,cm^{-3}$ in all samples. Resistivities of $0.0075\,\Omega$ cm with electron
concentrations of $3.3 \times 10^{19}\,cm^{-3}$ and mobility of $25\,cm^2/V\,s$ have been obtained
for Si-doped $Al_{0.7}Ga_{0.3}N$. The measured resistivity increases with increasing x very
rapidly and the dependence can be described by the following empirical equation:

$$\rho(Al_xGa_{1-x}N) = \rho(AlN)10^{-\frac{(1-x)}{0.8}}, \qquad (2.15)$$

from this, one can deduce that the resistivity of $n-Al_xGa_{1-x}N$ ($x > 0.7$) increases
by about one order of magnitude when x is increased by about 8%. This rapid
increase in resistivity is predominantly due to the increase in donor ionization energy
with x.

The temperature-dependent ρ for $n-Al_xGa_{1-x}N(x > 0.7)$ alloys in the temperature range of 70 to 650 K is shown in Fig. 2.38. Strong temperature dependence
is observed, especially for samples with high Al content ($x > 0.8$). The value of
E_D obtained by fitting the temperature-dependent data has already been plotted in

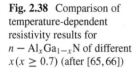

Fig. 2.38 Comparison of temperature-dependent resistivity results for $n - Al_xGa_{1-x}N$ of different $x (x \geq 0.7)$ (after [65, 66])

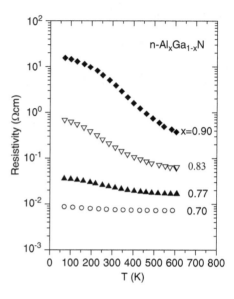

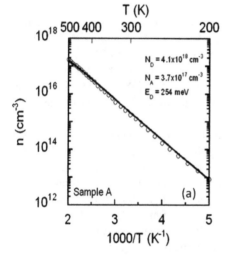

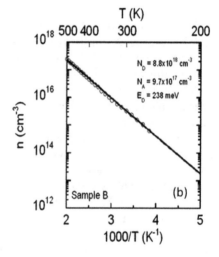

Fig. 2.39 The temperature dependence of (**a**) electron concentration and (**b**) electron mobility. The solid line shows the calculated electron mobility (after [7, 67])

Fig. 2.35 from available references for comparison. Deepening of Si donor impurity at $x > 0.7$ is the predominant reason for higher resistivity of n-type $Al_xGa_{1-x}N$ alloys ($x > 0.7$).

Taniyasu et al. [67] have obtained room temperature mobility as high as $125 \, cm^2/Vs$ in Si-doped AlN grown on 4H SiC (0 0 1) substrates by low pressure MOCVD. Free electron concentration n as a function of the inverse of the temperature for samples A and B is plotted in Fig. 2.39a, b. Sample A has lower

Table 2.3 Room temperature electrical properties for Si-doped AlN layers with two different Si doping concentrations (samples A and B). E_D is donor energy level. N_D and N_A are donor and acceptor concentrations, respectively (after [67])

Samples	n_{300K} (10^{15} cm^{-3})	μ_{300K} (cm^2/V s)	E_D (meV)	N_D (10^{18} cm^{-3})	N_A (10^{17} cm^{-3})
A	1.75	−125	254	4.1	3.7
B	3.23	−56	238	8.8	9.7

doping concentration than that of sample B and shows that n was lower in the entire temperature range. At 300 K, n was 1.75×10^{15} and 3.23×10^{15} cm^{-3} for samples A and B, respectively. Higher n were obtained at higher temperature. The *solid lines* were obtained from least squares fit of data by the charge neutrality equation:

$$\frac{n(n + N_A)}{N_D - N_A - n} = \frac{N_c}{g} \exp\left(-\frac{E_D}{kT}\right), \qquad (2.16)$$

in which N_D, N_A, E_D, and N_c, g, are donor concentration, acceptor concentration, donor ionization energy effective density of state in the conduction band, and degeneracy factor, respectively. The fitted values are summarized in Table 2.3. The mobility limiting scattering factor was identified as neutral impurity scattering rather than ionized impurity scattering or lattice scattering because of the large donor ionization energy ($\sim$250 meV). More recently, the same authors have reported even a higher value of μ (426 cm^2/V s) in Si-doped AlN with an electron concentration 7.3×10^{14} cm^{-3} at room temperature [7].

2.5.2 p-Type Al$_x$Ga$_{1-x}$N

p-Type GaN was thought impossible until Amano et al. in 1989 realized it could be achieved via Mg doping and postgrowth low-energy electron-beam irradiation (LEEBI) treatment [155]. Later, Nakamura et al. [157] discovered that postgrowth thermal annealing in hydrogen free ambient is the most effective and reliable method of achieving p-type GaN. However, free hole concentration in p-type GaN is relatively low compared to other semiconductor materials because of the high activation energy (E_A) of Mg acceptors ($\sim$160 meV). The typical (or benchmark) p-type resistivity of best Mg-doped GaN is around 1 Ω cm. Free hole concentration is even lower as x increases in Mg-doped Al$_x$Ga$_{1-x}$N since E_A of Mg acceptors increases with increasing x values. Therefore, it is very difficult to achieve p-type AlGaN with high Al content.

E_A as a function of Al content x in Al$_x$Ga$_{1-x}$N in the entire range of composition, as reported by various authors, is shown in Fig. 2.40 [7, 74–78]. It is clearly shown that E_A increases from 160 meV to > 500 meV as x increases from 0 to 1. This implies that only a small fraction ($\sim$10^{-8}) of Mg impurities can be activated at room temperature in AlN. Assuming 10^{20} cm^{-3} is the upper limit

Fig. 2.40 Activation energy
of Mg acceptors in
$Al_xGa_{1-x}N$ for the whole
composition range

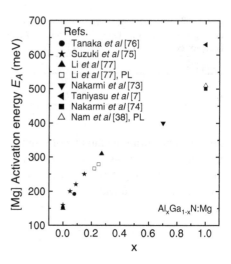

of Mg doping concentration N_A in AlN, the corresponding hole p concentration in Mg-doped AlN would not exceed 10^{12} cm^{-3} according to $p = N_A \exp(-E_A/kT)$. This number is too low for any practical device applications. Nevertheless, Mg is the most promising acceptor in AlN. Beryllium has also been investigated for the possible acceptor dopant in AlN. A previous calculation using effective mass theory predicted that Be occupying Al site (Be$_{Al}$) acts as an acceptor with activation energy of 0.26–0.47 eV [158]. More recent calculation also suggested a lower activation energy ($\sim$0.34 eV) for Be than that of Mg [142]. Zn doping has also been studied experimentally and conformed that Zn is a deeper acceptor than Mg in AlN [63].

2.5.2.1 Electrical Properties of Al-Rich Mg-Doped p-Type Al$_{0.7}$Ga$_{0.3}$N

Nakarmi et al. [74, 159, 160] studied the electrical properties of Mg-doped p-type AlGaN epilayers grown on AlN/sapphire templates by MOCVD. As-grown AlGaN epilayers with high Al contents were highly resistive, and postgrowth rapid thermal annealing was done to activate Mg acceptors. Ni/Au was used as the p-contacts for the standard Van der Pauw configuration for Hall-effect measurements. The room temperature resistivity of a typical Mg-doped Al$_{0.7}$Ga$_{0.3}$N epilayer was about $10^5\,\Omega$ cm (semi-insulating). Figure 2.41a shows the variation of resistivity with temperature ranging between 450 and 800 K of a typical Mg-doped Al$_{0.7}$Ga$_{0.3}$N epilayer. The resistivity decreases exponentially with increasing temperature. At high temperatures (> 700 K), it consistently shows a signature of p-type conduction. At 800 K, the measured p-type resistivity is about 40 Ω cm. The activation energy of the Mg acceptor is estimated by using the following equation:

$$\rho(T) = \rho_0 e^{\frac{E_A}{kT}}, \tag{2.17}$$

where $\rho(T)$ is the resistivity at temperature T, and E_A is the activation energy of the Mg acceptor. Figure 2.41b shows the semilog plot of the resistivity vs. $1/T$ ($\ln \rho$ vs. $1/T$). The fitted value of E_A is about 320 meV. To obtain E_A more accurately, E_A is also estimated by including the temperature dependence of the mobility at high temperatures ($\mu \propto T^{-3/2}$). The fitted E_A value increased to 396 meV, which is close to the previously estimated value, by extracting Mg levels between GaN and AlN [159]. With this result, we can now estimate the theoretical limit of the resistivity of Mg-doped Al-rich $Al_xGa_{1-x}N$ alloys.

$$p = N_A e^{\frac{-E_A}{kT}}, \quad \sigma = ep\mu_p, \quad \rho = \frac{1}{\sigma}, \tag{2.18}$$

$$\sigma(Al_xGa_{1-x}N : Mg) = ep(Al_xGa_{1-x}N : Mg) \ \mu_p(Al_xGa_{1-x}N : Mg), \tag{2.19}$$

where N_A is the Mg-doping concentration. With experimental results, hole concentration can be estimated as (assuming the same doping concentration for GaN):

$$
\begin{aligned}
p(Al_xGa_{1-x}N : Mg) &= N_A \exp\left\{-\frac{E_A(Al_xGa_{1-x}N : Mg)}{kT}\right\} \\
&= N_A \exp\left\{-\frac{E_A(GaN : Mg)}{kT}\right\} \cdot \exp\left\{-\frac{\Delta E_A}{kT}\right\}, \tag{2.20} \\
&= p(GaN : Mg) \cdot \exp\left\{-\frac{\Delta E_A}{kT}\right\}
\end{aligned}
$$

where $\Delta E_A = E_A (Al_xGa_{1-x}N : Mg) - E_A (GaN:Mg)$ is the Mg energy level difference between $Al_xGa_{1-x}N$ and GaN.

The hole mobility μ_p depends on its effective mass and alloy scattering. As alloy scattering increases with Al content, μ_p in $Al_{0.7}Ga_{0.3}N:Mg$ is smaller than that in GaN:Mg. The mobility of $Al_{0.7}Ga_{0.3}N : Mg$ is estimated (from low Al content AlGaN:Mg) http://apl.aip.org/applab/v86/i9/p092108_s1?view=fulltext - c4 in the range of $(1–3) \, cm^2/V \, s$,

$$\mu_p (Al_{0.7}Ga_{0.3}N : Mg) \approx (1 - 3) \, cm^{-2}/V \, s \approx (0.1 - 0.3) \, \mu_p(GaN : Mg), \tag{2.21}$$

Inserting (2.20) and (2.21) into (2.19), the conductivity of Mg-doped $Al_{0.7}Ga_{0.3}N$ alloys is

$$\sigma(Al_{0.7}Ga_{0.3}N : Mg) \approx (0.1 - 0.3) \, \sigma(GaN : Mg) \times \exp\left\{-\frac{\Delta E_A}{kT}\right\}, \tag{2.22}$$

The resistivity of $Al_{0.7}Ga_{0.3}N : Mg$ is thus:

$$\rho(Al_{0.7}Ga_{0.3}N : Mg) \approx (3 - 10) \, \rho (GaN : Mg) \times \exp\left\{\frac{\Delta E_A}{kT}\right\}, \tag{2.23}$$

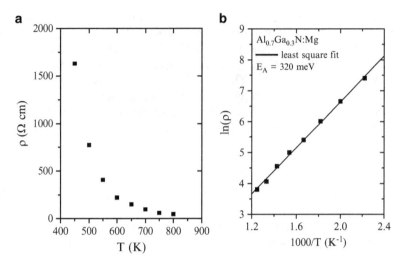

Fig. 2.41 (a) Temperature variation of resistivity of an Mg-doped $Al_{0.7}Ga_{0.3}N$ epilayer in the temperature range of 450–800 K, (b) is the semilog plot of ρ vs. $1/T$. The *solid line* is the least squares fit of data with (2.17) (after [74])

Using typical values for p-GaN, $E_A = 160\,$meV and $\rho = 1\,\Omega\,$cm (at room temperature), we have the room temperature resistivity

$$\rho\,(Al_{0.7}Ga_{0.3}N : Mg) \approx (3 - 10)\ e^{236/kT} = (3 - 10)\,e^{236/25\ meV} \qquad (2.24)$$

$$= (4 - 13) \times 10^4\ \Omega\ cm.$$

Thus, ρ is in the range of $(4–13) \times 10^4\,\Omega\,$cm (semi-insulating) at room temperature for Mg-doped $Al_{0.7}Ga_{0.3}N$ alloys. The resistivity we obtained for $Al_{0.7}Ga_{0.3}N : Mg$ is in the range of the value estimated by (2.24). Taking into account of the hole mobility variation with temperature ($\mu_p \propto T^{-3/2}$), the resistivity of $Al_{0.7}Ga_{0.3}N :$ Mg at 800 K is estimated by (2.25) to be in the range of $(9–30)\,\Omega\,$cm as a lower limit, which is consistent with measured value at 800 K ($40\,\Omega\,$cm). Similar ρ ($\sim10\,\Omega\,$cm at 670 K) of Mg-doped p-type $Al_{0.69}Ga_{0.31}N$ grown SiC has been reported by Chakraborty et al. [161] as well. A ρ as high as $10^8\,\Omega\,$cm and hole concentration as low as $10^{10}\,cm^{-3}$ has been reported in AlN [7]. The p-type conductivity in Mg-doped AlN has been measured by Taniyasu et al. [7] and also in authors' laboratory in 2006 [75]. With the improved p-type conduction in Mg-doped AlN, Taniyasu et al. have demonstrated a 210-nm aluminum nitride LED for the first time [7].

Temperature dependence of p and μ_p for Mg-doped AlN with an Mg-doping concentration of $2 \times 10^{19}\,cm^{-3}$ is shown in Fig. 2.42 [7]. The scattering of measured values is because of the high resistivity (low mobility). As the measurement temperature increased, the hole concentration increased exponentially. They have measured E_A of $\sim630\,$meV by assuming that the hole concentration follows

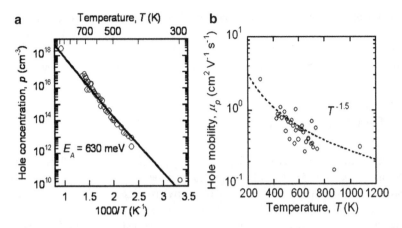

Fig. 2.42 Temperature dependence of hole concentration in Mg-doped AlN. Solid lines are obtained by fitting the data with expression; $\exp(-E_A/k_B T)$, where E_A is the acceptor ionization energy level, k_B the Boltzmann constant, and T the sample temperature. (b) Temperature dependence of hole mobility. The broken line is least squares fit of data with $\mu \alpha T^{-1.5}$. The solid line shows the calculated mobility (after [7])

$\exp(-E_A/kT)$, which is in a reasonable agreement with the values obtained by optical and electrical measurements of about 510 meV [38, 75] in authors' lab. The room temperature, p, was only in the order of 10^{10} cm^{-3}. The hole mobility shown decreased as temperature increased and fitted well to $\mu \alpha T^{-1.5}$ (broken line). This is typical of phonon scattering, which has the strongest influence on mobility in semiconductors at high temperature. Additionally, μ_p has also been verified by calculation using the effective mass theory.

2.5.2.2 Codoping O with Mg

Oxygen codoping with Mg has been proposed and applied to improve p-type conductivities in some wide bandgap semiconductors [156, 162, 163]. There is some experimental evidence that codoping enhanced the hole concentration (p) in GaN and ZnO. The conductivity of Mg-doped GaN could be significantly enhanced by annealing in an O environment. The hole concentration in ZnO has shown improvement from 10^{13} cm^{-3} in N-monodoped ZnO to 10^{17} cm^{-3}, or even higher in (Al, Ga, or In)-N codoped ZnO [164–166]. Therefore, this approach could be an alternative to increase p in Mg-doped AlN.

Recently, Wu et al. have studied the effect of oxygen codoping in AlN in detail by using density functional theory [167]. It has been argued that if the concentration of oxygen [O] is the same as the concentration of magnesium [Mg], Mg might be completely compensated by the formation of Mg–O complex. However if [Mg] is much higher than [O], complexes like Mg$_2$–O, Mg$_3$–O, and Mg$_4$–O would be likely to form and the energy level of these complexes might be shallower than that of a single Mg atom. They have found that Mg_n–O ($n = 1, 2, 3$, and 4) complexes are energetically favored and have an activation energy at least 0.23 eV lower than that of single Mg. The shallower level of Mg–O–Mg complex has also been confirmed by another independent study using hybrid density functional calculation [168].

2.6   Concluding Remarks

The aim of this chapter is to review growth, optical, and electrical properties of high-quality Al-rich AlGaN alloys, and to provide an overview on contributions of such studies to the understanding of many aspects of this material system. Recent extensive studies on Al-rich AlGaN alloys are driven primarily by technological importance and needs of these materials for a variety of DUV devices. It has been witnessed that many issues regarding the fundamental properties, origin/formation/nature of impurities and defects, and control of conductivity in Al-rich AlGaN have been largely addressed. Bandgaps and/or band structures, classification of impurities (vacancies and their complexes), and methods to grow high-quality AlGaN alloys in the entire composition range are now well understood.

Growth of nonpolar AlGaN/AlN-based devices has recently attracted much attention due to the absence of polarization fields and enhanced surface emission. Recent developments in the growth of nonpolar AlN in different orientation substrates have been reviewed. However, the growth of AlGaN on nonpolar substrates is still in its embryonic state.

Fundamental optical properties such as optical transitions, carrier dynamic processes, and carrier–phonon interactions predominantly determine the device performance. Therefore, understanding of fundamental optical properties is crucial for the development of suitable material qualities for device applications. Band structure, bandgap bowing parameter, bandgap, Varshni coefficients, exciton localization energy, and PL emission line width as a function of Al content in AlGaN have been reviewed. Due to the advances in growth and characterization methods, these parameters are now known more accurately. Measured exciton localization energy as high as 95 meV has been reported in $Al_{0.7}Ga_{0.3}N$, which is the largest exciton localization energy ever reported for semiconductor alloys.

In the past, the insulating nature of $Al_xGa_{1-x}N$ ($x > 0.5$) alloys was the major problem in the development DUV devices based on these materials. However, now causes of such insulating nature and methods of obtaining highly conductive $Al_xGa_{1-x}N$ ($x > 0.5$) alloys have been identified. The deepening of Si donor levels and compensation of electrons by cation vacancies $(V_{III})^{3-}$ and their complexes $(V_{III} - \text{complex})^{2-/1-}$ have been identified as major reasons behind the insulating nature of Si-doped Al-rich AlGaN alloys. DUV PL studies are found to be a highly effective method to identify and monitor such intrinsic defects. Suppressing such defects by optimizing the growth conditions or by using In as a surfactant has proved effective methods of achieving highly conductive Al-rich AlGaN alloys. Resistivity as low as 0.0075 Ω cm has been obtained in $Al_{0.7}Ga_{0.3}N$: Si by suppressing these defects. However, p-type conductivity of Mg-doped Al-rich AlGaN is still too low and remains as a key challenge for achieving efficient practical DUV devices.

Acknowledgements We would like to thank Dr. J. Li, Dr. K.B, Nam, Dr. K.H. Kim, Dr. M.L. Nakarmi, Dr. N. Nepal, Dr. T.A. Tahtamouni, Dr. R. Dahal, and Dr. Z.Y. Fan and A. Sedhain for their contributions to this work. The advancements made in authors' laboratory would not have been possible without the continuous support of DOE, NSF, ARO, and DARPA. H.X. Jiang and J.Y. Lin are also grateful to the AT&T foundation for their support of Edward Whitacre and Linda Whitacre Endowed Chairs.

References

1. I. Akasaki, H. Amano, Jpn. J. Appl. Phys. **36**, 5393 (1997)
2. J. Han, M.H. Crawford, R.J. Shul, J.J. Figiel, M. Banas, L. Zhang, Y.K. Song, H. Zhou, A.V. Nurmikko, Appl. Phys. Lett. **73**, 1688 (1998)
3. T. Nishida, N. Kobayashi, Phys. Stat. Sol. A **176**, 45 (1999)
4. V. Adivarahan, W.H. Sun, A. Chitnis, M. Shatalov, S. Wu, H.P. Maruska, M.A. Khan, Appl. Phys. Lett. **85**, 2175 (2004)
5. J. Li, K.B. Nam, M.L. Nakarmi, J.Y. Lin, H.X. Jiang, P. Carrier, S.H. Wei, Appl. Phys. Lett. **83**, 5163 (2003)
6. J. Li, Z.Y. Fan, R. Dahal, M.L Nakarmi, J.Y. Lin, H.X. Jiang, Appl. Phys. Lett. **89**, 213510 (2006)
7. Y. Taniyasu, M. Kasu, T. Makimoto, Nature **441**, 325 (2006)
8. M.D. Bremser, W.G. Perry, T. Zheleva, N.V. Edwards, O.H. Nam, N. Parikh, D.E. Aspnes, R.F. Davis, MRS Internet. J. Nitride Semicond. Res. **1**, 8 (1996)
9. N. Okada, N. Fujimoto, T. Kitano, G. Narita, M. Imura, K. Balakrishnan, M. Iwaya, S. Kamiyama, H. Amano, I. Akasaki, K. Shimono, T. Noro, T. Takagi, A. Bandoh, Jpn. J. Appl. Phys. **45** 2502 (2006)
10. P. Cantu, S. Keller, U. Mishra, S. DenBaars, Appl. Phys. Lett. **82**, 3683 (2003)
11. T.M. Al tahtamouni, A. Sedhain, J.Y. Lin, H.X. Jiang, Appl. Phys. Lett. **92**, 092105 (2008)
12. V. Adivarahan, G. Simin, G. Tamulaitis, R. Srinivasan, J. Yang, A. Khan, M. Shur, R. Gaska, Appl. Phys. Lett. **79**, 1903 (2001)
13. H. Kang, S. Kandoor, S. Gupta, I. Ferguson, S.P. Guo, M. Pophristic, Phys. Stat. Sol. (c) **2**, 2145 (2005)
14. C. Stampfl, C. Van de Walle, Appl. Phys. Lett. **72**, 459 (1998)
15. M. Kanechika, T. Kachi, Appl. Phys. Lett. **87**, 132106 (2005)
16. F.A. Ponce, D.P. Bour, Nature (London) **386**, 351 (1997)
17. F. Bernardini, V. Fiorentini, D. Vanderbilt, Phys. Rev. B **56**, R10024 (1997)
18. T.M. Al Tahtamouni, A. Sedhain, J.Y. Lin, H.X. Jiang, Appl. Phys. Lett. **90**, 221105 (2007)
19. A. Sedhain, N. Nepal, M.L. Nakarmi, T.M. Al tahtamouni, J.Y. Lin, H.X. Jiang, Z. Gu, J.H. Edgar, Appl. Phys. Lett. **93**, 041905 (2008)
20. F. Bernardini, V. Fiorentini, D. Vanderbilt, Phys. Rev. B **56**, R10024 (1997)
21. A.D. Corso, R. Resta, S. Baroni, Phys. Rev. B **47**, 16252 (1993)
22. R.G. Banal, M. Funato, Y. Kawakami, Phys. Rev. B **79**, 121308 (R) (2009)
23. K.B. Nam, J. Li, M.L. Nakarmi, J.Y. Lin, H.X. Jiang, Appl. Phys. Lett. **84**, 5264 (2004)
24. A.A. Yamaguchi, Phys. Stat. Sol (b), **247**, 1717 (2010)
25. H. Hirayama, N. Noguchi, N. Kamata, Appl. Phys. Express **3**, 032102 (2010)
26. A.A. Yamaguchi, Appl. Phys. Lett. **96**, 151911 (2010)
27. T.M. Al Tahtamouni, N. Nepal, J.Y. Lin, H.X. Jiang, W.W. Chow, Appl. Phys. Lett. **89**, 131922 (2006)

28. J. Shakya, K. Knabe, K.H. Kim, J. Li, J.Y. Lin, H.X. Jiang, Appl. Phys. Lett. **86**, 091107 (2005)
29. T. J. Badcock, P. Dawson, M.J. Kappers, C. McAleese, J.L. Hollander, C.F. Johnston, D.V. Sridhara Rao, A.M. Sanchez, C.J. Humphreys, J. Appl. Phys. **105**, 123112 (2009)
30. K. Ueno, A. Kobayashi, J. Ohta, H. Fujioka, Appl. Phys. Lett. **91**, 081915 (2007)
31. M. Horita, J. Suda, T. Kimoto, Appl. Phys. Lett. **89**, 112117 (2006)
32. R. Armitage, J. Suda, T. Kimoto, Appl. Phys. Lett. **88**, 011908 (2006)
33. N. Onojima, J. Suda, H. Matsunami, Jpn. J. Appl. Phys. **41**(Part 2), L1348 (2002)
34. D.M. Schaadt, O. Brandt, A. Trampert, H.P. Schönherr, K.H. Ploog, J. Cryst. Growth **300**, 127 (2007)
35. N. Onojima, J. Suda, H. Matsunami, Jpn. J. Appl. Phys. **42**(Part 2), L445 (2003)
36. S. Stemmer, P. Pirouz, Y. Ikuhara, R.F. Davis, Phys. Rev. Lett. **77**, 1797 (1996)
37. N. Onojima, J. Suda, T. Kimoto, H. Matsunami, Appl. Phys. Lett. **83**, 5208 (2003)
38. K.B. Nam, J. Li, J.Y. Lin, H.X. Jiang, Appl. Phys. Lett. **85**, 3489 (2004)
39. B.K. Meyer, G. Steude, A. Goldner, A. Hofmann, H. Amano, I. Akasaki, Phys. Stat. Sol. (b) **216**, 187 (1999)
40. H.S. Kim, R.A. Mair, J. Li, J.Y. Lin, H.X. Jiang, Appl. Phys. Lett. **76**, 1252 (2000)
41. N. Nepal, J. Li, M.L. Nakarmi, J.Y. Lin, H.X. Jiang, Appl. Phys. Lett. **87**, 242104 (2005)
42. G.D. Chen, M. Smith, J.Y. Lin, H.X. Jiang, S.H. Wei, M.A. Khan, C.J. Sun, Appl. Phys. Lett. **68**, 2784 (1996)
43. F. Yun, M.A. Reshchikov, L. He, T. King, H. Morkoç, S.W. Novak, L. Wei, J. Appl. Phys. **92**, 4837 (2002)
44. Y. Koide, H. Itoh, M.R.H. Khan, K. Hiramatu, N. Sawaki, I. Akasaki, J. Appl. Phys. **61**, 4540 (1987)
45. H. Angerer, D. Brunner, F. Freudenberg, O. Ambacher, M. Stutzmann, R. Hopler, T. Metzger, E. Born, G. Dollinger, A. Bergmaier, S. Karsch, H.J. Korner, Appl. Phys. Lett. **71**, 1504 (1997)
46. T. Takeuchi, H. Takeuchi, S. Sota, H. Sakai, H. Amano, I. Akasaki, Jpn. J. Appl. Phys. **36**(Part 2), L177 (1997)
47. H. Jiang, G.Y. Zhao, H. Ishikawa, T. Egawa, T. Jimbo, M. Umeno, J. Appl. Phys. **89**, 1046 (2001)
48. S.R. Lee, A.F. Wright, M.H. Crawford, G.A. Petersen, J. Han, R.M. Biefeld, Appl. Phys. Lett. **74**, 3344 (1999)
49. N. Nepal, J. Li, M.L. Nakarmi, J.Y. Lin, H.X. Jiang, Appl. Phys. Lett. **88**, 062103 (2006)
50. K.B. Nam, J. Li, M.L. Nakarmi, J.Y. Lin, H.X. Jiang, Appl. Phys. Lett. **82**, 1694 (2003)
51. K.B. Nam, J. Li, M.L. Nakarmi, J.Y. Lin, H.X. Jiang, Appl. Phys. Lett. **81**, 1038 (2002)
52. V. Adivarahan, J. Zhang, A. Chitnis, W. Shuai, J. Sun, R. Pachipulusu, M. Shatalov, M.A. Khan, Jpn. J. Appl. Phys. **41**(Part 2), L435 (2002)
53. A. Chitnis, J. Zhang, V. Adivarahan, W. Shuai, J. Sun, R. Pachipulusu, M. Shatalov, J.W. Yang, G. Simin, M.A. Khan, Jpn. J. Appl. Phys. **41**(Part 2), L450 (2002)
54. A. Yasan, R. McClintock, K. Mayes, D. Shiell, L. Gautero, S. R. Darvish, P. Kung, M. Razeghi, Appl. Phys. Lett. **83**, 4701 (2003)
55. J. Li, K.B. Nam, M.L. Nakarmi, J.Y. Lin, H.X. Jiang, Appl. Phys. Lett. **81**, 3365 (2002)
56. G.R. James, A.W.R. Leitch, F. Omnes, M. Leroux, Semicond. Sci. Technol. **21**, 744 (2006)
57. D.C. Reynolds, D.C. Look, B. Jogai, A.W. Saxler, S.S Park, J.Y. Hahn, Appl. Phys. Lett. **77**, 2879 (2000)
58. A.A. Yamaguchi, Phys. Stat. Sol. C **5**, 2364 (2008)
59. N. Nepal, M.L. Nakarmi, J.Y. Lin, H.X. Jiang, Appl. Phys. Lett. **89**, 092107 (2006)
60. K.B. Nam, M.L. Nakarmi, J.Y. Lin, H.X. Jiang, Appl. Phys. Lett. **86**, 222108 (2005)
61. M.L. Nakarmi, N. Nepal, J.Y. Lin, H.X. Jiang, Appl. Phys. Lett. **94**, 091903 (2009)
62. A. Sedhain, T.M. Al Tahtamouni, J. Li, J.Y. Lin, H.X. Jiang, Appl. Phys. Lett. **93**, 141104 (2008)
63. N. Nepal, M.L. Nakarmi, H.U. Jang, J.Y. Lin, H.X. Jiang, Appl. Phys. Lett. **89**, 192111 (2006)
64. Y. Taniyasu, M. Kasu, N. Kobayashi, Appl. Phys. Lett. **81**, 1255 (2002)

65. M.L. Nakarmi, K.H. Kim, K. Zhu, J.Y. Lin, H.X. Jiang, Appl. Phys. Lett. **85**, 3769 (2004)
66. K. Zhu, M.L. Nakarmi, K.H. Kim, J.Y. Lin, H.X. Jiang, Appl. Phys. Lett. **85**, 4669 (2004)
67. Y. Taniyasu, M. Kasu, T. Makimoto, Appl. Phys. Lett. **85**, 4672 (2004)
68. M.L. Nakarmi, N. Nepal, J.Y. Lin, H.X. Jiang, Appl. Phys. Lett. **86**, 261902 (2005)
69. M.D. McCluskey, N.M. Johnson, C.G. Van de Walle, D.P. Bour, M. Kneissl, W. Walukiewicz, Phys. Rev. Lett. **80**, 4008 (1998)
70. J. Neugebauer, C.G. Van de Walle, Appl. Phys. Lett. **69**, 503 (1996)
71. T. Mattila, R.M. Nieminen, Phys. Rev. B **55**, 9571 (1997)
72. C. Stampfl, C.G. Van de Walle, Phys. Rev. B **65**, 155212 (2002)
73. I. Gorczyca, N.E. Christensen, A. Svane, Phys. Rev. B **66**, 075210 (2002)
74. M.L. Nakarmi, K.H. Kim, M. Khizar, Z.Y. Fan, J.Y. Lin, H.X. Jiang, Appl. Phys. Lett. **86**, 092108 (2005)
75. M.L. Nakarmi, N. Nepal, C. Ugolini, T.M. Al Tahtamouni, J.Y. Lin, H.X. Jiang, Appl. Phys. Lett. **89**, 152120 (2006)
76. M. Suzuki, J. Nishio, M. Onomura, C. Hongo, J. Cryst. Growth **189/190**, 511 (1998)
77. T. Tanaka, A. Watanabe, H. Amano, Y. Kobayashi, I. Akasaki, S. Yamazaki, M. Koike, Appl. Phys. Lett. **65**, 593 (1994)
78. J. Li, T.N. Oder, M.L. Nakarmi, J.Y. Lin, H.X. Jiang, Appl. Phys. Lett. **80**, 1210 (2002)
79. S.T. Bradley, S.H. Goss, L.J. Brillson, J. Hwang, W.J. Schaff, J. Vac. Sci. Technol. B **21**, 2558 (2003)
80. B. Pödör, Phys. Stat. Sol. **16**, K167 (1966)
81. S. Keller, S. Heikman, I. Ben-Yaacov, L. Shen, S.P. DenBaars, U.K. Mishra, Appl. Phys. Lett. **79**, 3449 (2001)
82. S. Yamaguchi, M. Kosaki, Y. Watanabe, S. Mochizuki, T. Nakamura, Y. Yukawa, S. Nitta, H. Amano, I. Akasaki, Phys. Stat. Sol. A **188**, 895 (2001)
83. S. Nicolay, E. Feltin, J.F. Carlin, M. Mosca, L. Nevou, M. Tchernycheva, F.H. Julien, M. Ilegems, N. Grandjean, Appl. Phys. Lett. **88**, 151902 (2006)
84. N. Akopian, G. Bahir, D. Gershoni, M.D. Carven, J.S. Speck, S.P. DenBaars, Appl. Phys. Lett. **86**, 202104 (2005)
85. K.C. Zeng, J. Li, J.Y. Lin, H.X. Jiang, Appl. Phys. Lett. **76**, 3040 (2000)
86. C.J. Collins, A.V. Sampath, G.A. Garrett, W.L. Saeney, H. Shen, M. Wraback, A.Yu. Nikiforov, G.S. Cargill, V. Dierolf, Appl. Phys. Lett. **86**, 031916 (2005)
87. H. Murotani, T. Kuronaka, Y. Yamada, T. Taguchi, N. Okada, H. Amano, J. Appl. Phys. **105**, 083533 (2009)
88. H.M. Ng, Appl. Phys. Lett. **80**, 4369 (2002)
89. T.J. Baker, B.A. Haskell, F. Wu, P.T. Fini, J.S. Speck, S. Nakamura, Jpn. J. Appl. Phys. **44**, L920 (2005) (Part 2)
90. M.C. Schmidt, K.-C. Kim, R.M. Farrel, D.F. Feezell, D.A. Cohen, M. Saito, K. Fujito, J.S. Speck, S.P. DenBaars, S. Nakamura, Jpn. J. Appl. Phys. **46**, L190 (2007)
91. K. Okamoto, H. Ohta, S.F. Chichibu, J. Ichihara, H. Takasu, Jpn. J. Appl. Phys. **46**, L187 (2007)
92. M. Kubota, K. Okamoto, T. Tanaka, H. Ohta, Appl. Phys. Express **1**, 011102 (2008)
93. K. Okamoto, T. Tanaka, M. Kubota, Appl. Phys. Express **1**, 072201 (2008)
94. M.R. Laskar, T. Ganguli, A.A. Rahman, A.P. Shah, M.R. Gokhale, A. Bhattacharya, Phys. Stat. Sol. **4**, 163 (2010)
95. A. Saxler, P. Kung, C.J. Sun, E. Bigan, M. Razeghi, Appl. Phys. Lett. **64**, 339 (1993)
96. Y. Tsuchiya, Y. Okadome, A. Honshio, Y. Miyake, T. Kawashima, M. Iwaya, S. Kamiyama, H. Amano, I. Akasaki, Jpn. J. Appl. Phys. **44**, 1516 (2005)
97. M. Bickermann, B.M. Epelbaum, A. Winnacker, Phys. Stat. Sol. C **0**(7), 1993–1996 (2003)
98. P. Lu, R. Collazo, R. Dalmau, X. Li, J. Tweedie, Z. Sitar, MRS Symposia No. 1040E (Materials Research Society, Pittsburgh, 2007)
99. B.N. Pantha, N. Nepal, T.M. Al Tahtamouni, M.L. Nakarmi, J.Y. Lin, H.X. Jiang, Appl. Phys. Lett. **91**, 121117 (2007)

100. H. Wang, C. Chen, Z. Gong, J. Zhang, M. Gaevski, M. Su, J. Yang, M.A. Khan, Appl. Phys. Lett. **84**, 499 (2004)
101. S.L. Chuang, C.S. Chang, Appl. Phys. Lett. **68**, 1657 (1996)
102. R. Dingle, D.D. Sell, S.E. Stokowski, M. Ilegems, Phys. Rev. B **4**, 1211 (1971)
103. M. Monemar, Phys. Rev. B **10**, 676 (1974)
104. M.C. Kuo, K.C. Chiu, T.H. Shin, Y.J. Lai, C.S. Yang, W.K. Chen, D.S. Chuu, M.C. Lee, W.C. Chou, S.Y. Jeng, Y.T. Shih, W.H. Lan, Jpn. J. Appl. Phys. **43**, 5145 (2004)
105. Y.T. Shih, W.C. Fan, C.S. Yang, M.C. Kuo, W.C. Chow, J. Appl. Phys. **94**, 3791 (2003)
106. T. Onuma, S.F. Chichibu, T. Sota, K. Asai, S. Sumiya, T. Shibata, M. Tanaka, Appl. Phys. Lett. **81**, 652 (2002)
107. J.A. Freitas Jr., G.C.B. Braga, E. Silveira, J.G. Tischler, M. Fatemi, Appl. Phys. Lett. **83**, 2584 (2003)
108. E. Silveira, J.A. Freitas Jr., M. Kneissl, D.W. Treat, N.M. Johnson, G.A. Slack, L.J. Schowalter, Appl. Phys. Lett. **84**, 3501 (2004)
109. E. Silveira, J.A. Freitas Jr., O.J. Glembocki, G.A. Slack, L.J. Schowalter, Phys. Rev. B **71**, 041201(R) (2005)
110. T. Koyama, M. Sugawara, T. Hoshi, A. Uedono, J.F. Kaeding, R. Sharma, S. Nakamura, S.F. Chichibu, Appl. Phys. Lett. **90**, 241914 (2007)
111. E. Silveira, J.A. Freitas, S.B. Schujman, L.J. Schowalter, J. Cryst. Growth **310**, 4007 (2008)
112. T. Onuma, T. Shibata, K. Kosaka, K. Asai, S. Sumiya, M. Tanaka, T. Sota, A. Uedono, S.F. Chichibu, J. Appl. Phys. **105**, 023529 (2009)
113. L. Chen, B.J. Skromme, R.F. Dalmau, R. Schlesser, Z. Sitar, C. Chen, W. Sun, J. Yang, M.A. Khan, M.L. Nakarmi, J.Y. Lin, H.X. Jiang, Appl. Phys. Lett.**85**, 4334 (2004)
114. G. Coli, K.K. Bajaj, J. Li, J.Y. Lin, H.X. Jiang, Appl. Phys. Lett. **80**, 2907 (2002)
115. A. Bell, S. Srinivasan, C. Plumlee, H. Omiya, F.A. Ponce, J. Christen, S. Tanaka, A. Fujioka, Y. Nakagawa, J. Appl. Phys. **95**, 4670 (2004)
116. G. Coli, K.K. Bajaj, J. Li, J.Y. Lin, H.X. Jiang, Appl. Phys. Lett. **78**, 1829 (2001)
117. H.X. Jiang, L.Q. Zu, J.Y. Lin, Phys. Rev. B **42**, 7284 (1990)
118. M.S. Lee, K.K. Bajaj, J. Appl. Phys. **73**, 1788 (1993)
119. S. Lankes, H. Stanzl, K. Wolf, M. Giegler, W. Gebhardt, J. Phys. Condens. Matter **7**, 1287 (1995)
120. M.C. Wagener, G.R. James, F. Omnès, Appl. Phys. Lett. **83**, 4193 (2003)
121. M. Yoshikawa, M. Kunzer, J. Wagner, H. Obloh, R. Schmidt, N. Herres, U. Kaufmann, J. Appl. Phys. **86**, 4400 (1999)
122. J. Osaka, Y. Ohno, S. Kishimoto, K. Maezawa, T. Mizutani, Appl. Phys. Lett. **87**, 222112 (2005)
123. T. Ive, O. Brandt, H. Kostial, K.J. Friedland, L. Daweritz, K.H. Ploog, Appl. Phys. Lett. **86**, 024106 (2005)
124. T. Ogino, M. Aoki, Jpn. J. Appl. Phys. **19**, 2395 (1980)
125. S.J. Rhee, S. Kim, E.E. Reuter, S.G. Bishop, Appl. Phys. Lett. **73**, 2636 (1998)
126. P. Perlin, T. Suski, H. Teisseyre, M. Leszczynski, I. Grzegory, J. Jun, S. Porowski, P. Bogustawski, J. Bernholc, J.C. Chervin, A. Polian, T.D. Moustakas, Phys. Rev. Lett. **75**, 296 (1995)
127. B.J. Skrome, J. Jayapalan, R.P. Vaudo, V.M. Phanse, Appl. Phys. Lett. **74**, 2358 (1999)
128. J. Elsner, R. Jones, M.I. Heggie, P.K. Sitch, M. Haugk, T. Frauenheim, S. Oberg, P.R. Briddon, Phys. Rev. B **58**, 12571 (1998)
129. R.A. Youngman, J.H. Harris, J. Am. Ceram. Soc. **73**, 3238 (1990)
130. H.C. Yang, T.Y. Lin, Y.F. Chen, Phys. Rev. B **62**, 12595 (2000)
131. A.V. Fomin, A.E. Nikolaev, I.P. Nikitina, A.S. Zubrilov, M.G. Mynbaeva, N.I. Kuznetsov, A.P. Kovarsky, B. Ja. Ber, D.V. Tsvetkov, Phys. Stat. Sol. (a) **188**, 433 (2001)
132. I. Shai, C.E.M. de Oliviera, Y. Shapira, J. Salzman, Phys. Rev. B **64**, 205313 (2001)
133. F. Shahedipour, B.W. Wessels, Appl. Phys. Lett. **76**, 3011 (2000)
134. U. Kaufmann, M. Kunzer, M. Maier, H. Obloh, A. Ramakrishnan, B. Santic, P. Schlotter, Appl. Phys. Lett. **72**, 1326 (1998)

135. U. Kaufmann, M. Kunzer, H. Obloh, M. Maier, Ch. Manz, A. Ramakrishnan, B. Santic, Phys. Rev. B **59**, 5561 (1999)
136. Y.H. Kwon, S.K. Shee, G.H. Gainer, G.H. Park, S.J. Hwang, J.J. Song, Appl. Phys. Lett. **76**, 840 (2000)
137. M.A. Reshchikov, G.C. Yi, B.W. Wessels, Phys. Rev. B **59**, 13176 (1999)
138. C.G. Van de Walle, J. Neugebauer, J. Appl. Phys. **95**, 3851 (2004)
139. K.B. Nam, M.L. Nakarmi, J.Y. Lin, H.X. Jiang, Appl. Phys. Lett. **83**, 878 (2003)
140. C.G. Van de Walle, J. Neugebauer, J. Appl. Phys. **95**, 3851 (2004)
141. J.R. Haynes, Phys. Rev. Lett. **4**, 361 (1960)
142. R.Q. Wu, L. Shen, M. Yang, Z.D. Sha, Y.Q. Cai, Y.P. Feng, Z.G. Huang, Q.Y. Wu, Appl. Phys. Lett. **91**, 152110 (2007)
143. K.B. Nam, M.L. Nakarmi, J. Li, J.Y. Lin, H.X. Jiang, Appl. Phys. Lett. **83**, 2787 (2003)
144. J. Li, K.B. Nam, J.Y. Lin, H.X. Jiang, Appl. Phys. Lett. **20**, 3245 (2001)
145. R. Zeisel, M.W. Bayerl, S.T.B. Goennenwein, R. Dimitrov, O. Ambacher, M.S. Brandt, M. Stutzmann, Phys. Rev. B **61**, Rl6283 (2000)
146. M. Kasu, N. Kobayashi, J. Cryst. Growth **221**, 739 (2000)
147. A. Fara, F. Bernardini, V. Fiorentini, J. Appl. Phys. **85**, 2001 (1999)
148. S.B. Zhang, S.H. Wei, A. Zunger, Phys. Rev. Lett. **84**, 1232 (2000)
149. Y. Taniyasu, M. Kasu, N. Kobayashi, Appl. Phys. Lett. **81**, 2155 (2002)
150. B. Borisov, V. Kuryatkov, Yu. Kudryavtsev, R. Asomoza, S. Nikishin, D.Y. Song, M. Holtz, H. Temkin, Appl. Phys. Lett. **87**, 132106 (2005)
151. A.Y. Polyakov, N.B. Smirnov, A.V. Govorkov, M.G. Milidivskii, J.M. Redwing, M. Shin, M. Skowronski, D.W. Greve, R.G. Wilson, Solid-State Electron. **42**, 627 (1998)
152. P. Boguslawski, J. Bernholc, Phys. Rev. B **56**, 9496 (1997)
153. C.G. Van de Walle, C. Stampfl, J. Neugebauer, M.D. McCluskey, N.M. Johnson, MRS Internet. J. Nitride Semicond. Res. 4S1, **G10** 4 (1999)
154. S. Fisher, C. Wetzel, E.E. Haller, B.K. Meyer, Appl. Phys. Lett. **67**, 1298 (1995)
155. H. Amano, M. Kitoh, K. Hiramatsu, I. Akasaki, Jpn. J. Appl. Phys. **28**(Part 1),L2112 (1989)
156. T. Yamamoto, H. Katayama-Yoshida, Jpn. J. Appl. Phys. **36**(Part 2), L180 (1997)
157. S. Nakamura, T. Mukai, M. Senoh, N. Iwasa, Jpn. J. Appl. Phys. **31**(Part 1), L139 (1992)
158. F. Mireles, S.E. Ulloa, Phys. Rev. B **58**, 3879 (1998)
159. Z.Y. Fan, J.Y. Lin, H.X. Jiang Proc. SPIE **6479**, I4791 (2007)
160. M.L. Nakarmi, K.H. Kim, J. Li, J.Y. Lin, H.X. Jiang. Appl. Phys. Lett. **82**, 3041 (2003)
161. A. Chakraborty, C.G. Moe, Y. Wu, T. Mates, S. Keller, J.S. Speck, S.P. DenBaars, U.K. Mishra, J. Appl. Phys. **101**, 053717 (2007)
162. H. Katayama-Yoshida, T. Yamamoto, T. Nishimatsu, Phys. Stat. Sol. B **210**, 429 (1998).
163. T. Yamamoto, H. Katayama-Yoshida, Phys. B **274**, 113 (1999)
164. J.G. Lu, Z.Z. Ye, F. Zhuge, B.H. Zhao, L.P. Zhu, Appl. Phys. Lett. **85**, 3134 (2004)
165. J.M. Bian, X.M. Li, X.D. Gao, W.D. Yu, L.D. Chen, Appl. Phys. Lett. **84**, 541 (2004)
166. M. Joesph, H. Tabata, H. Saeki, K. Ueda, T. Kawai, Phys. B **302**, 140 (2001)
167. R.Q. Wu, L. Shen, M. Yang, Z.D. Sha, Y.Q. Cai, Y.P. Feng, Z.G. Huang, Q.Y. Wu, Phys. Rev. B **77**, 073203 (2008)
168. Á. Szabó, N.T. Son, E. Janzén, A. Gali, Appl. Phys. Lett. **96**, 192110 (2010)

Chapter 3
Deep Ultraviolet Light-Emitting Diodes

Michael Shur, Max Shatalov, Alex Dobrinsky, and Remis Gaska

Abstract A review is given of the current state-of-the-art in nitride-based UV light-emitting diodes. We begin with a summary of applications for these compact light sources and then detail recent improvements in the quality of epitaxial wafers and optimization of the structure design that have enabled increases in the DUV LED device area while retaining reasonable quantum efficiency at low current densities.

3.1 Introduction

Compact, solid-state deep ultraviolet (DUV) light-emitting diodes (LEDs) [1] have found applications in biomedical and analytical instrumentation, defense, biotechnology, medicine, air, water and surface sterilization and decontamination, bio-agent detection and identification, radiation hard UV sources and UV curing, (see Figs. 3.1–3.4).

AlN/GaN/InN compounds used for DUV LEDs allow covering the UV band from UVA (320–400 nm) and UVB (280–320 nm) all the way into UVC range (100–280 nm) with the shortest achievable wavelength on the order of 210 nm [2]. As seen from Fig. 3.5, this technology is very effective in reducing *Escherichia coli* bacteria concentration in water to the levels approved for safe drinking water.

Deep UV might also stimulate a new green revolution by increasing crop yields and nutritional value of plants and revolutionizing postharvest storage and treatment – see Fig. 3.6 [4] and Figs. 3.7–3.9.

M. Shur (✉)
Department of Electrical, Computer and Systems Engineering, Rensselaer Polytechnic Institute, Troy, NY 12180, USA
e-mail: shurm@rpi.edu

Max Shatalov · Alex Dobrinsky · Remis Gaska
Sensor Electronic Technology, Inc., Columbia SC, 29209, USA
e-mail: adobrinsky@s-et.com

S. Pearton (ed.), *GaN and ZnO-based Materials and Devices*,
Springer Series in Materials Science 156, DOI 10.1007/978-3-642-23521-4_3,
© Springer-Verlag Berlin Heidelberg 2012

Fig. 3.1 Disinfection pen
UVCLEAN®

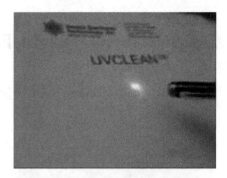

Fig. 3.2 All-LED
Spectrometer operating at
> 250 nm ([1], IEEE© 2010)

Fig. 3.3 Ultraviolet all-LED
water disinfection unit

Fig. 3.4 UV LED lamps in
water disinfection chamber

Fig. 3.5 *E. coli* reduction in
UV LED disinfection
chamber as a function of flow
rate [3]

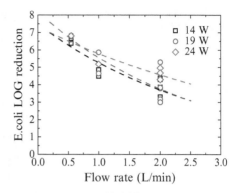

Fig. 3.6 Red leaf lettuce after illumination with deep ultraviolet light after harvesting. Plants A, D
were not irradiated as controls. Plants B, C were irradiated for approximately 12 h per day for each
of the 7 days. Plant B was irradiated with blue/UV-A radiation ($\lambda > 350$ nm), while plant C was
irradiated with blue/UV-A/UV-B radiation ($\lambda > 290$ nm). As illustrated, plant C has substantially
more red color, which results from an accumulation of anthocyanin cyandin-3-malonylglucoside,
a type of flavonoid [4]

Fig. 3.7 UV LED based
greenhouse mobile light track
system for supplemental UV
irradiation

Fig. 3.8 Light track based on
three UV LED lamps

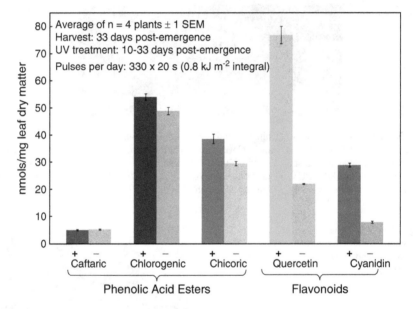

Fig. 3.9 Phenolic compound levels in Red Leaf "Galactic" Lettuce harvested with (marked with *plus*) and without (marked with *minus*) supplemental UV LED irradiation

Band-aid, tooth brush, vacuum cleaner, sterilizing blankets [5, 6], and skin treatment using DUV LEDs have been proposed and demonstrated (see Figs. 3.10–3.13).

With improved performance, the DUV LED technology could help eliminating the practice of using antibiotics for preventing infections and enable new testing procedures based on UV fluorescence. These applications are enabled by the DNA damage caused by deep UV radiation (see Fig. 3.14).

Medical applications of UV LED sources include treatments for psoriasis, vitiligo, topic dermatitis, and other skin diseases (see Fig. 3.15). Detection of forged documents (see Fig. 3.16) is another important application.

Fig. 3.10 UV LED
self-sanitizing band aid [7]

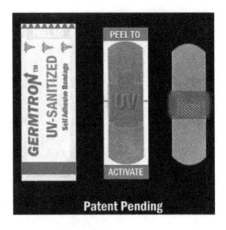

Fig. 3.11 Phillips
UV-sanitized toothbrush
using a small mercury lamp
that can be replaced by DUV
LED

Fig. 3.12 Vacuum cleaner
could benefit from installing
DUV LEDs for sterilization
of air

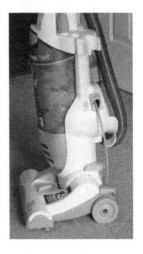

Fig. 3.13 Schematic representation of sterilizing blanket with DUV LEDs (from [6])

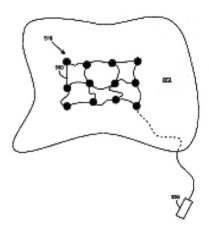

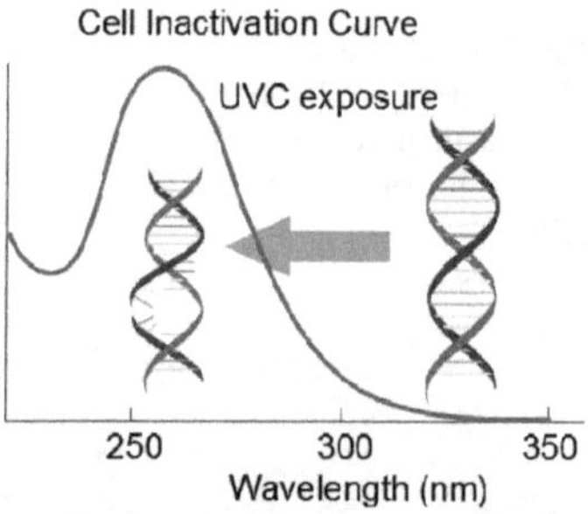

Fig. 3.14 Cell inactivation curve

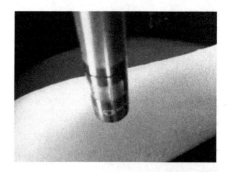

Fig. 3.15 Skin light therapy with DUV LEDs

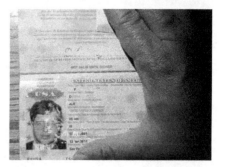

Fig. 3.16 UV LED used in portable forgery detector

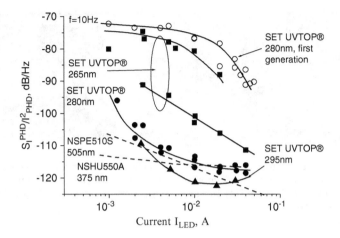

Fig. 3.17 Dependence of relative noise spectra on LED current, I_{LED}, for different LEDs. Frequency of analysis $f = 10\,Hz$. *Dashed lines* show the noise level for NICHIA NSHU550 A (375 nm) and NICHIA NSPE510 S (505 nm). Reprinted with permission from [8], Copyright© 2006, American Institute of Physics

Fig. 3.18 Radiation hardness of DUV LEDs compared to more conventional devices (after [9])

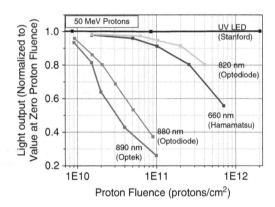

DUV LEDs have low noise [8] (see Fig. 3.17), high modulation frequencies, exceptional radiation hardness (see Fig. 3.18), and small form factor. In contrast to mercury vapor or other types of lamps that require long warm-up time, the LED can offer nearly instant on/off switching, which opens new range of applications related to sensing and disinfection. They could be combined into high power lamps, which demonstrated UV output powers as high as 100 mW at 275 nm [1].

As seen from Fig. 3.17, the 295-nm SET DUV LEDs has low noise comparable to visible LEDs.

The relative spectral noise density of the light intensity fluctuations decreased with an increase of the LED current. At high currents, the difference in the noise level for LEDs with different wavelengths is small and is of the same order of magnitude or even smaller than for visible LEDs.

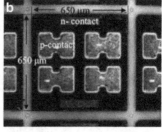

Fig. 3.19 DUV LED chips with junction area of 0.5 mm² (**a**) and 1 mm² (**b**) operating at 273 nm and 247 nm. The inset shows a CCD image of operating chip. (**a**) [10] and CCD image of 2 × 2 monolithic 245 nm LED array [11] Reprinted with permission from [11], Copyright© 2010, American Institute of Physics

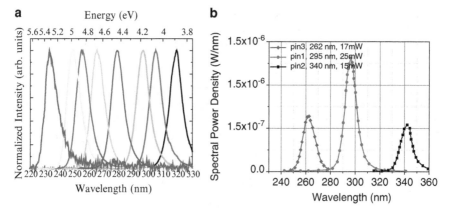

Fig. 3.20 (**a**) normalized room-temperature electroluminescence (EL) spectra of DUV LEDs with peak emissions from 235 nm to 320 nm ([1], IEEE© 2010) and (**b**) multi-wavelength UV LED assemblies coupled to optical fiber

Recent improvements in the quality of epitaxial wafers and optimization of the structure design enabled increase in the DUV LED device area retaining reasonable quantum efficiency at low current densities, and high power DUV LEDs based on a single large area chip and DUV LED arrays have been demonstrated [10, 11] (see Fig. 3.19).

DUV LEDs allow for a variety of spectral and space power distributions (see Fig. 3.20a). In contrast to gas lamps for which emission wavelength is predetermined by fundamental transitions in gas or vapor, LED sources offer flexibility to combine different emission wavelengths in single device. Multiwavelength DUV LEDs have been directly coupled to an optical fiber, enabling spectroscopic and fluorimeter applications (see Fig. 3.20b). DUV LEDs have potential for achieving high internal quantum efficiency (see Fig. 3.21) and high wall plug efficiency. At the present time, the highest achieved wall plug efficiency is around 2–3% [1, 14, 15].

Fig. 3.21 Internal quantum efficiency as a function of dislocation density extracted from photoluminescence data (Plotted after [12]). The data are for weak excitation of 10^{18} cm^{-3}. The theoretical line is based on the model of [13]

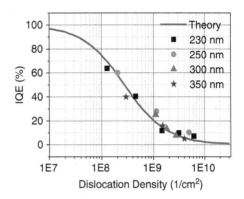

Recent ambitious DARPA project [16] aims to increase DUV LED efficiency up to 20%. Achieving this goal will accelerate the DUV LED penetration into consumer markets (now estimated to grow at 30% a year) [17].

DUV LEDs have many advantages compared to more conventional DUV sources – mercury lamps. Compact and flexible DUV LED form factor contrasts with bulky mercury lamps, requiring high voltage supplies. They have a much smaller power requirements, could be turned on and off nearly instantaneously (in contrast to slow mercury lamps), and are environmentally benign (do not contain mercury).

3.2 Materials Properties

DUV LEDs designs use ternary or quaternary AlN/GaN/InN alloys. Narrow quantum wells (QWs) separated by barriers form light generating active layers. The quantum shift of the electron ground state in these narrow QWs leads to higher transition energy and a shorter emission wavelength, see Fig. 3.22, showing the wavelengths corresponding to different compositions of Al$_x$Ga$_{1-x}$N.

The calculated wavelength are based on the relationships of the band gaps of AlN/InN/GaN alloys on composition given by [18–21]

$$Ga_x In_{1-x} N : E_g(eV) = 0.77 + 2.65x - 2.4x(1-x),$$

$$Al_x Ga_{1-x} N : E_g(eV) = 3.42 + 2.86x - x(1-x),$$

$$Al_x In_{1-x} N : E_g(eV) = 0.77 + 5.51x - 3.0x(1-x),$$

and on the electron and hole effective masses given in Appendix 1.

As seen, AlN/GaN/InN compounds allow covering the UV band from UVA (320–400 nm) and UVB (280–320 nm) all the way into UVC range (100–280 nm) with the shortest achievable wavelength on the order of 210 nm [2, 22].

Appendices A.1–A.3 list material parameters used in DUV LED simulation. These parameters have been compiled using available nitride semiconductor data [21, 23, 24].

3.3 Materials Growth

As seen from Fig. 3.22, DUV LEDs require AlGaN epilayers with high aluminum content, which is a challenge for AlGaN epitaxy. These epilayers are typically grown on sapphire substrates (transparent to DUV) that have a large lattice mismatch with AlGaN and GaN films. Blue LEDs have been also grown on SiC and even Si substrates (see Fig. 3.23), which absorb UV radiation. However, a recent report on a transparent p-type contact might enable a DUV LED design emitting via the top layer using these substrates [25].

Other proposed substrates are lithium aluminate [26] GaN, AlGaN, AlON, and LiGaO$_2$.

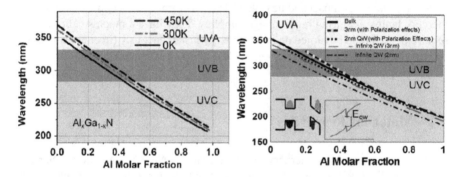

Fig. 3.22 Wavelengths for Al$_x$Ga$_{1-x}$N quantum wells versus Al molar fraction. The *inset* shows the blue shift due to quantization and the red shift and diminished overlap of the wave functions due to polarization field (figures are partially based on [1]),. Note that E_{qw} may be smaller than the corresponding band gap in bulk material due to polarization

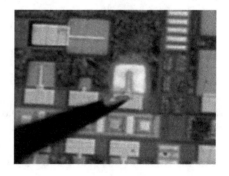

Fig. 3.23 Blue LED on Si substrate

As a consequence of heteroepitaxial growth on foreign substrates with a large lattice mismatch, threading dislocation densities (TDD) in DUV LED structures are very high, and AlN and AlGaN films grown on such substrates beyond critical thicknesses suffer from cracking due to the strong tensile strain. Native $Al_xGa_{1-x}N$ substrates (or AlN substrates for wavelengths of 260 nm or shorter) might alleviate this issue [27–32]. However, relatively small area and high cost hinder commercial use of AlN substrates. Presently commercially available AlN substrates are produced by very few companies around the world [33].

Metal Organic Chemical Vapor Deposition (MOCVD) [34] is the standard technique of growing nitride semiconductor films and heterostructures. Other growth techniques used include Molecular Beam Epitaxy (MBE) [35] and Reactive Molecular-Beam Epitaxy (RMBE) [36]. MBE allows for growth at lower substrate temperatures but this technique is difficult to adapt for film growth with high Al content. MOCVD requires a higher growth temperature. Thus, this technique is less adequate when growth of nitride heterostructures containing both indium and aluminum is required. Pulsed Atomic Layer Epitaxy (PALE) [37] and Migration Enhanced Metalorganic Chemical Vapor Deposition (MEMOCVD®) [26] are improved versions of MOCVD. PALE uses sequential metal organic precursor pulses of Al–, In–, Ga–, and NH_3, with the NH_3 pulse always following each metalorganic pulse of the fixed duration. In MEMOCVD®, the durations and waveforms of precursor pulses are optimized and the pulses might overlap, providing a continuum of growth techniques ranging from PALE to conventional MOCVD. In a MOCVD mode (at higher growth temperature), MEMOCVD® has a high growth rate. This growth mode is used for buffer layers. Active layers are grown at lower temperatures (by ~150°C) and have a much better quality due to a better mobility of precursor species on the surface, better atomic incorporation, and improved surface coverage. MEMOCVD® grown layers exhibit about an order of magnitude reduction in the dislocation densities, and, as a consequence, have longer lifetimes and narrower photoluminescence (PL) lines (see Fig. 3.24).

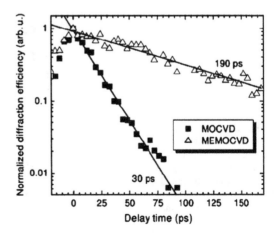

Fig. 3.24 Light Induced Transient Grating (LITG) decay in MOCVD and MEMOCVD8Λ™-grown AlGaN epilayers for the grating period of 7.7 μm. Carrier lifetimes were estimated by fitting the decay transients with single exponents (*lines*) ([1], IEEE© 2010 after [38])

Further improvements in materials quality are achieved by combining MEMOCVD® with the Epitaxial Lateral Over Growth (ELOG) resulting in Migration Enhanced Lateral Over (MELEO) growth [39]. In this process, trimethylaluminum (TMAl), trimethylgallium (TMGa), and ammonia (NH_3) are used as the precursors for the layer growth performed under low pressure with H_2 as the carrier gas; then AlGaN layers of 2–5 μm thickness are deposited on c-plane sapphire substrates. Such AlGaN layers grown by conventional MOCVD have TDD on the order of 10^{10}–10^{11} cm^{-2}. The MELEO layers used linear patterns with 2 − μm-wide mesas and 5 − μm-wide trenches (fabricated using conventional photolithography and reactive ion etching). The AlN and high Al-content AlGaN grown over these templates at temperatures 1, 100–1, 200°C with the growth rate of 1–2 μm/h had a dislocation density reduced by orders of magnitude (see Fig. 3.25).

Superlattice buffers can also reduce strain as was first predicted by Bykhovski et al. [40, 41] (see Fig. 3.26).

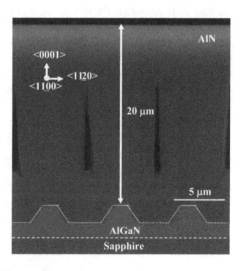

Fig. 3.25 SEM micrographs of a fully coalesced 20 − μm-thick AlN sample grown by MELEO technique. Reprinted with permission from [39], Copyright© 2008, American Institute of Physics

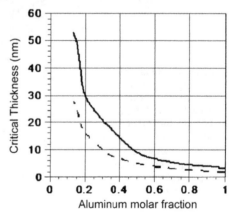

Fig. 3.26 Critical thickness versus Al mole fraction for AlGaN/GaN: superlattice (*solid line*) and SIS structure (*dashed line*). Reprinted with permission from [40], Copyright© 1997, American Institute of Physics

Fig. 3.27 Schematics of an AlGaN structure showing domains, their boundaries, and preferential dislocation sites. Reprinted with permission from [42], Copyright© 2011, American Institute of Physics

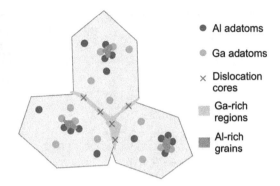

Nonuniformity of AlGaN composition in semiconductor layers has a complex effect on performance of LED devices [42]. On one hand, carrier localization at potential fluctuations has been found to enhance the room temperature carrier lifetime by inhibiting carrier diffusion into nonradiative recombination sites [43,44]. On the other hand, domains in AlGaN QWs with a lower Al content have been found to cause current crowding and decrease the LED lifetime [45].

During growth process of AlGaN semiconductor layers, small islands with high aluminum content are formed as shown in Fig. 3.27 [42] Grains with high aluminum content are separated by domain boundaries that contain extended defects which are formed in order to accommodate the relative difference in crystal orientation among the islands.

These defects have high gallium content. Thus, emission spectra in these regions are red shifted. Further the emission intensity is reduced at dislocation cores due to enhanced nonradiative recombination. PL measured using scanning near-field optical microscopy (SNOM) reveals nonuniform emission pattern from AlGaN layer as shown in Fig. 3.28.

Analysis of PL spectra allows evaluating average potential fluctuations due to inhomogeneous growth of AlGaN layers. The findings [42] suggest that there are two spatial scales of potential fluctuation – large scale of order of 1 μm and small scale that is less than 100 nm. Potential fluctuations reach amplitudes of few tens of meV at each scale.

Findings of Pinos et al. [42] agree with other PL measurements [38,46] that the potential profile in AlGaN epilayers has two nearly equal fluctuation scales.

These scales were also obtained from the temperature dynamics of the PL linewidth using Monte Carlo simulations of localized exciton hopping [38,46].

Exciton hopping occurs within the potential fluctuations (dispersed on the first scale) in isolated regions with the average energy of the localized states dispersed on the second scale. Both scales are almost independent of the growth techniques used. Time-resolved PL and light-induced transient grating technique (LITG) measurements established the correlation of this potential profile with the measured lifetimes [38].

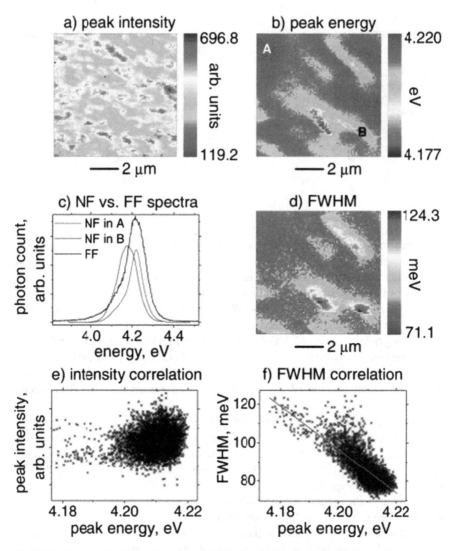

Fig. 3.28 (Color online) Near-field PL maps of the peak intensity (**a**), peak energy (**b**), and FWHM (**d**) for the $Al_{0.30}Ga_{0.70}N$ layer. (**c**) The far-field spectrum (*black*) and the near-field spectra at the spots marked A and B in the energy map. The correlations graphs between the peak intensity and peak energy and between the FWHM and energy are shown in (**e**) and (**f**). Reprinted with permission from [42], Copyright© 2011, American Institute of Physics

3.4 Design of Deep UV LEDs

Figure 3.29 shows a typical DUV LED design.

Figure 3.30 compares typical values of different efficiencies contributing to the total external efficiencies of visible (blue) and UV LEDs (defined in Appendix A.5)

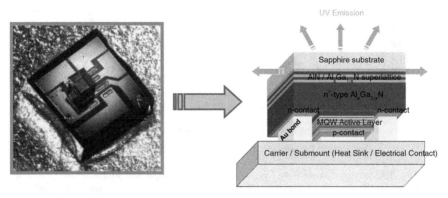

Fig. 3.29 Photograph and typical deep UV LED design

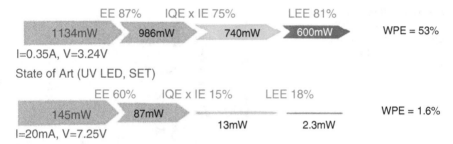

Fig. 3.30 Typical values of different efficiencies contributing to the total external efficiencies of visible (*blue*) and UV LEDs. (The values for visible LEDs are after [47]). Here EE is electric efficiency, IQE is internal quantum efficiency, LEE is light extraction efficiency, WPE is wall plug efficiency, see Appendix A.5 for definitions

Figure 3.30 clearly illustrates the challenges of DUV LED technology. The biggest improvements might be achieved by improving materials quality (first of all reducing the dislocation density), using transparent superlattice structures for the p-contact and by increasing light extraction efficiency.

Figures 3.31 and 3.32 present the simulation results showing the effect of the TDD on IQE.

Figure 3.32 shows different layers and interfaces in the DUV LED design playing a role in limiting LEE.

Table 3.1 shows the fraction of light transmitted through interface.

A large polarization charge at the AlGaN heterointerfaces due to the differences in spontaneous polarization [48] and strain-induced polarization [41] is the key design factor. Due to a large polarization induced field, band bending in the barriers and in the QWs leads to the separation of the electron and hole wave functions (see

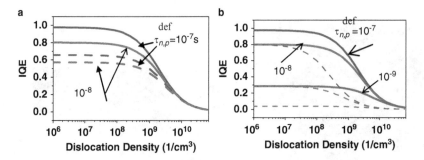

Fig. 3.31 IQE versus < *inline* − *fig*/ > TDD for different Auger and point <inline-fig/>defect recombination lifetimes. (a) *Solid lines* – no Auger recombination, Auger constant < *inline* − *fig*/ > C = 0; *dashed lines*: C = 5 × 10³¹ cm⁶/s. *Arrows* indicate corresponding lifetimes $\tau_{n,p}^{def}$. (**b**) effect of lifetime $\tau_{n,p}^{def}$ on IQE for carrier concentration $n = p = 10^{19}$ cm⁻³ (*solid curve*) and $10^{18} \times 1/$cm⁻³ (*dashed curve*)

Fig. 3.32 Different layers and interfaces in the DUV LED design, which play a role in limiting LEE. *SL* superlattice, *QW* quantum well region

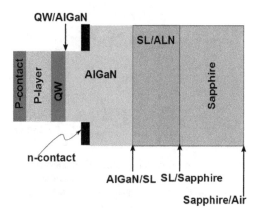

Table 3.1 Fraction of light transmitted through interface

Interface	Fraction of transmitted light
QW/AlGaN	0.78
AlGaN/superlattive	0.70
Superlattice/sapphire	0.52
Sapphire/air	0.52
Total	0.78 × 0.7 × 0.52 × 0.52 = 0.148

Fig. 3.22). This effect reduces the probability of the radiative recombination. The piezoelectric polarization caused by strain is [41]

$$P_{\text{pz},z} = 2 \left(e_{31} - \frac{c_{13}}{c_{33}} e_{33} \right) u_{xx}. \tag{3.1}$$

Assuming that the cladding layer is relaxed, the strains $u_{xx,\text{b}}$ and $u_{xx,\text{qw}}$ for the barrier and the QWs are

$$u_{xx,\text{b}} = \frac{a_{\text{b}}}{a_{\text{c}}} - 1; \; u_{xx,\text{qw}} = \frac{a_{\text{w}}}{a_{\text{c}}} - 1. \tag{3.2}$$

Here a_{b}, a_{w}, and a_{c} are the lattice constants for the barrier, QW, and cladding layer, respectively, $e_{\text{ij}}-$ components of piezoelectric tensor, and $c_{\text{ij}}-$ components of elasticity tensor.

Spontaneous polarization in $\text{Al}_x\text{Ga}_{1-x}\text{N}$ (in C/m^2) is

$$P_{\text{sp},\text{Al}_x\text{Ga}_{1-x}\text{N}} = -0.09x - 0.034(1 - x) + 0.019x(1 - x). \tag{3.3}$$

Ambacher et al. [48]. In (3.3), the spontaneous polarization in AlN $P_{\text{sp, AlN}} = -0.090\,C/m^2$ but a much smaller value $P_{\text{sp, AlN}} = -0.040\,C/m^2$ was reported by Park and Chuang [49].

The electric field, F_{QW}, in the QW is the sum of the polarization induced electric field, F_{QWP}, and the build-in field of the $p - i - n$ structure, F_{PIN},

$$F_{\text{QW}} = F_{\text{QWP}} + F_{\text{PIN}}. \tag{3.4}$$

F_{QWP} is determined by the spontaneous and piezoelectric polarization charges at the heterointerfaces. Under zero bias, $F_{\text{PIN}} \approx E_{\text{g}}/L$, where E_{g} is the energy gap of the contact layers and L is the distance between the contacts. Under large forward bias, $F_{\text{PIN}} \approx 0$ (see Fig. 3.33). The PL peak is affected by the electric field in the QW (see the inset in Fig. 3.22). Under high illumination intensity, the concentrations

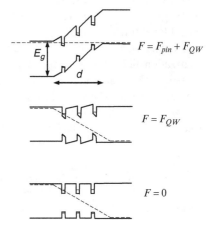

Fig. 3.33 Band diagrams of a $p - i - n$ QW structure (**a**) without external bias and illumination, (**b**) with the $p - i - n$ field compensated by external bias, and (**c**) with total electric field compensated by external bias and intense optical excitation. Reprinted with permission from Marcinkevicìus et al. [50], Copyright© 2007, American Institute of Physics

Fig. 3.34 Spatial fluctuations of the Al molar fraction and/or well width that might relieve strain (from [1], IEEE© (2010)

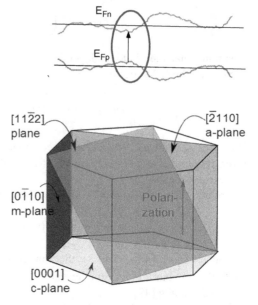

Fig. 3.35 Polar plane [0001], non-polar (a-plane, m-plane), and semi-polar ([11$\bar{2}$2]) in wurtzite crystal structure

of electrons and holes in the QWs are high enough to screen the electric field. Therefore, the dependences of the PL spectra on the applied bias and illumination intensity allow extracting the electric field in the QWs (see Fig. 3.33 from) [50].

According to [50], the polarization fields are weaker than expected from (3.1) to (3.4). The studies of screening dynamics of intrinsic electric field in AlGaN QWs using transient PL [51] also confirm this conclusion.

It is possible that the spontaneous polarization is smaller than calculated by Bernardini and Fiorentini [52]. Also, PL studies point to Al clustering, see Fig. 3.34 [53, 54], that might partially relieve strain.

Dislocations might also partially relieve strain. Indeed, thicker layers of AlGaN can be grown on top of GaN buffer layer than are expected from estimates of the critical layer thickness [55].

Growing GaN in nonpolar orientations (see Fig. 3.35) [56–58] can reduce the polarization field. For example, a-plane GaN is grown on r-plane sapphire and m-plane GaN is grown on $\gamma - \text{LiAlO}_2$ or m-plane SiC. Semipolar [11$\bar{2}$2] plane orientation has also been used [56].

Gaska et al. [59], Taniyasu et al. [2], and Gaska et al. [60] reported on reduced band bending in QWs using very narrow QWs, thereby pushing the electron ground state above or close to the top energy of the band–band region in the QW (see Fig. 3.36).

Figures 3.36 and 3.37 show another innovative feature of the SET, Inc. DUV LED design [59] facilitating the emission of polar optical phonons by electrons in the light emitting structure. Since the optical polar phonon energy ($\sim$90 meV) in

Fig. 3.36 Phonon-assisted electron capture in QWs [59]

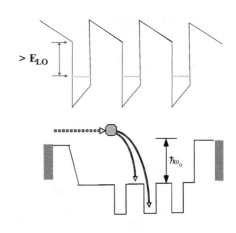

Fig. 3.37 Schematic band diagram of DUV LED for capturing electrons in the light emitting region (from [1], IEEE© 2010)

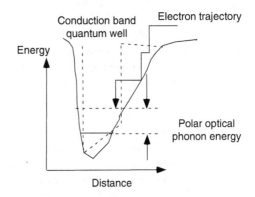

Fig. 3.38 Effect of compositional grading for allowing electrons to emit polar optical phonons to be confined in the quantum wellQW for more effective radiative recombination [59]

AlGaN is much larger than the LED operating temperature, electrons get trapped in the light emitting structure after the polar optical phonon emission and are available to form excitons and recombine with much slower holes. A relatively wide energy well with the depth larger than energy of a polar optical phonon containing QWs retains the injected electrons in the light generating structure longer and enhances radiative recombination. Further improvements of nitride-based DUV emitting heterostructures [59] resulted from using graded composition, such that electrons lose energy entering a QW. This design (see Fig. 3.38) alleviates the need for the barrier layer for the electrons.

Using superlattice buffers for strain relief allows for increasing the thickness of the n-type AlGaN contact layer and achieving high doping levels. This results in a higher conductivity and a lower contact resistance. Typical parameters for the AlGaN contact layer are $\sim 6\,\mu$m layer thickness with the electron mobility close to $110\,\text{cm}^2/\text{V}\,\text{s}$ and electron concentration $1.5 \times 10^{18}\,\text{cm}^{-3}$.

3.4.1 DUV LED Fabrication

For fabrication of DUV LEDs, plasma reactive ion etching (RIE) is used to access the bottom n-AlGaN layer. LED devices are then fabricated on the etched mesa. Depending on the wavelength and the application, sizes might range from $150 \times 150 \,\mu m^2$ devices to mm size chips [61], with $300 \times 300 \,\mu m^2$ being a typical size. The Ti/Al/Ti/Au n-type ohmic contact metallization is annealed at 900°C. Transfer length measurements (TLM) determine the contact resistance. The specific ohmic contact resistance is larger for layers with a higher Al molar fraction because of difficulties of doping AlGaN with a high Al molar fraction. For example, the sheet resistance of the n-AlGaN layer for 247 nm DUV LEDs was $950 \,\Omega^{-2}$ [62], compared to $250 \,\Omega^{-2}$ for the 265 nm DUV LEDs [63]. The specific contact resistance to n-type AlGaN DUV LEDs was $3.3 \times 10^{-3}\Omega$ cm^2 [39] for 247 nm LEDs and $1.06 \times 10^{-5}\Omega$ cm^2 for 265 LEDs [63].

For the same 247 nm devices, the p-type specific contact resistance and sheet resistance values (also measured by TLM) were $7.27 \times 10^{-5}\Omega$ cm^2 and $58 \,k\Omega^{-2}$, respectively, similar to those measured for the 265 nm LEDs [63]. Figure 3.39 shows representative current–voltage characteristics for the 275 nm LEDs.

As seen, the 247 nm devices have a much higher leakage (at -5 V reverse bias the device leakage current is approximately 20 nA) and a higher forward operating voltage ($\sim$7.6 V at 20 mA with a series differential resistance of 28Ω). The device forward characteristics indicated behavior typical of two (or more) diodes connected in series, with different ideality factors at various regimes of the current–voltage curve. These parasitic diodes might arise from the p-contact or p-GaN/p-AlGaN interface and from a possible Schottky formation at the n-contact [39].

DUV LED mounting and packaging are crucial fabrication steps as well. Electric surges and/or electrostatic discharges (ESD) can cause LED failure. Often, a protective circuit element, such as a Zener diode, a $p - i - n$ diode, or a Schottky diode is added to protect a more sensitive electronic device. An alternative solution for DUV LEDs is to use the submount structure itself for the ESD protection [64].

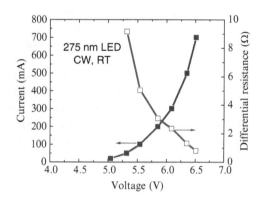

Fig. 3.39 Current–voltage characteristics of 275 nm DUV LEDs

Fig. 3.40 Qualitative equivalent circuit (**a**) and current–voltage characteristics (**b**) of DUV LED and submount providing ESD protection ([1], IEEE© 2010 after [64])

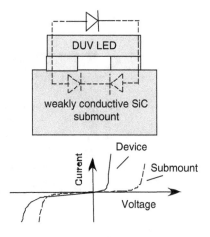

A weakly conductive semiconductor material of the mounting structure (typically, SiC) forms highly nonideal Schottky contacts with DUV LED contact pads; therefore limiting the voltage across the DUV LED terminals (see Fig. 3.40).

Both flat and semispherical lens packages have been developed for DUV LEDs (see Fig. 3.41).

The pad layout is very important for controlling heat sinking. The optimized "H-shape" layout proposed in Bilenko et al. [67] has led to a considerable increase in the output power (see Fig. 3.42).

3.4.2 DUV LED Performance

Traditionally, LED devices with small active area of $300 \times 300 \, \mu m^2$ were manufactured to maintain high current density at low operating currents in order to enhance radiative recombination and thus to achieve a higher quantum efficiency. These devices are typically driven at 20 mA continuous wave (CW) with output power of about 1.5 mW. Using an array of LED chips, high power UV lamps were developed that could reach powers of 100 mW at wavelength of 275 nm. However, assembly of multiple chips into a single package did not provide a uniform emission pattern making these devices less suitable for imaging optics [10]. To address this problem, an alternative approach based on a single large area chip (with chip area reaching $1 \times 1 \, mm^2$) was developed. Larger size DUV LEDs ($1 \times 1 \, mm^2$) provide more uniform emission pattern then array of smaller LED chips [10], making large DUV LEDs more suitable for imaging optics. A large chip size also enhances heat removal from the junction to the heat sink thereby improving CW operation [10].

In this section, we first present results of LED performance for traditional geometry ($300 \times 300 \, \mu m^2$ area). We discuss radiated power versus wavelength for

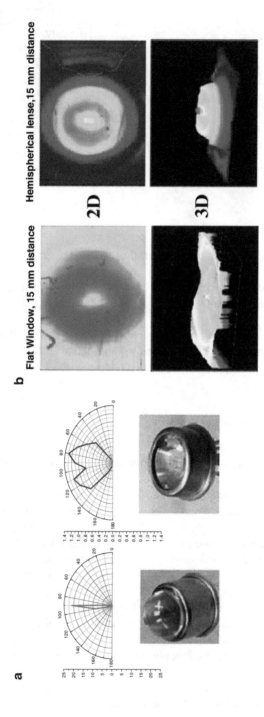

Fig. 3.41 DUV LEDs with ball and flat lenses (**a**) [65] and light distribution of single chip DUV LEDs with different packaging options (**b**) [66]

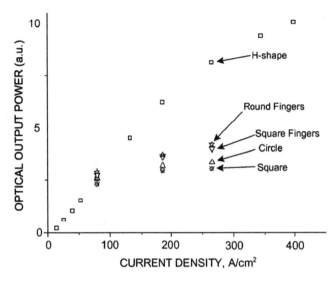

Fig. 3.42 Output characteristics of 340 nm LEDs with differently shaped contacts (*inset* shows H-shape contact.) [68]

different operation conditions and present spectral characteristics of the emitted UV light. Second, we present state-of-the-art performance results for large area LED device. Finally, we consider how LED efficiency can be improved by decreasing absorption of p-cladding and increasing contact reflectance.

For typical DUV LED size ($300 \times 300 \, \mu m^2$), Fig. 3.43 shows the output powers for the 235–310 nm LEDs at 20 mA DC. We have also fabricated and packaged DUV LED devices with emission at 227 nm. Under short pulse current pumping ($\sim$50 ns) pulsed power as high as 220 μW was measured at $\sim$4 A drive current [61].

Figures 3.44–3.46 present electroluminescence spectral power distributions for 270, 249, and 227 nm DUV LEDs of conventional size ($300 \times 300 \, \mu m^2$). A typical line width is on the order of 10–30 nm, with peak to the background ratio of several orders of magnitude.

One way to increase power is to use device arrays that have a larger light emitting area and can operate at lower current density (see Fig. 3.47). Combining many (up to a hundred) DUV LEDs in one package can provide very large output powers (up to 100 mW).

Another way to produce large output powers is to use large chip geometry [10]. Figure 3.48 shows I–V characteristics of 100 mW 275 nm single chip LED based on a large chip geometry.

Figure 3.49 shows the output power for large area 275 nm LED with CW output power larger than 100 mW at 2.8 A.

Similar efficiency is achieved for smaller output powers for the same large chip geometry. Figure 3.50 shows the output power for large area 1.1×1.1 mm

Fig. 3.43 Output optical power of single-chip DUV LEDs and DUV LED lamps for CW and pulsed modes ([1], IEEE© 2010)

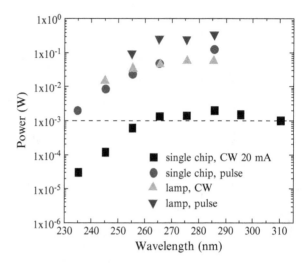

Fig. 3.44 Spectral power distribution for 270 nm DUV LEDs at two driving currents (as shown)

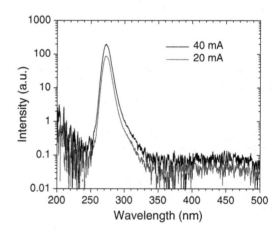

Fig. 3.45 Spectral power distribution for 249 nm DUV LED ([1], IEEE© 2010)

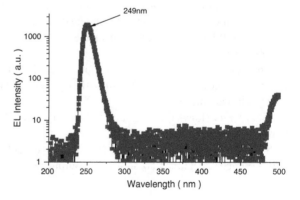

Fig. 3.46 Spectral power distribution for 227 nm DUV LEDs ([1], IEEE© 2010)

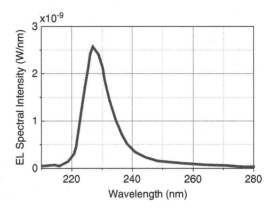

Fig. 3.47 DUV LED power versus current for a single device and for 3 × 3 device array. From [1], IEEE© 2010

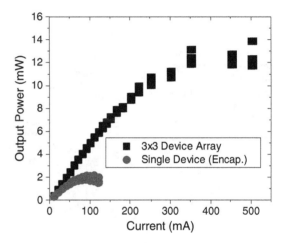

248 nm LED with CW output power of 0.2 mW at 20 mA and ~6 mW at 300 mA. Figure 3.51 shows their spectral power distribution.

A major improvement in the DUV LED design was achieved by using new reflecting p-contact design using patent pending superlattice structure and novel reflecting metal composition (see Figs. 3.52 and 3.53). This design increased the output power, EQE, and WPE by 2–2.5 times (see Figs. 3.54 and 3.55) at the expense of a slight increase in forward voltage (see Fig. 3.56). Results in Figs. 3.54–3.56 are for a typical LED geometry of $300 \times 300\,\mu m^2$.

PL studies and simulations show that the achieved internal quantum efficiency of DUV LEDs might be quite high (between 15 and 70%). However, the external quantum efficiency and the wall plug efficiency of DUV LEDs is below 2% (the highest for 280 nm LEDs and smaller for LEDs with a shorter wavelength) [61]. The

Fig. 3.48 Differential resistance as a function of applied current for large area single chip 100 mW, 275 nm LED. Insert shows IV characteristics of the device

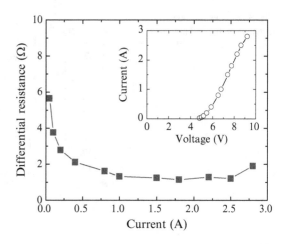

Fig. 3.49 Output power and emission peak position as a function of driving current for large area 275 nm LED. *Inset* shows spectral power density as a function of wavelength

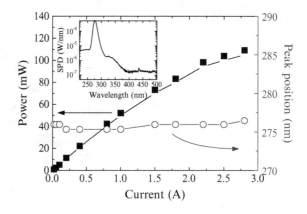

Fig. 3.50 Driving current dependence of output power for large area 248 nm LED

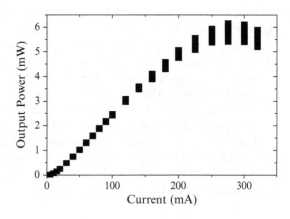

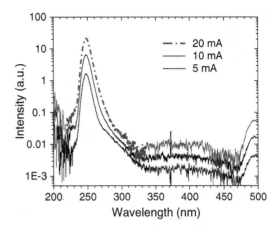

Fig. 3.51 Spectral power distribution of 248 nm LED at different driving currents (indicated)

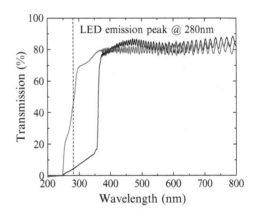

Fig. 3.52 Comparison of optical transmittance spectra of transparent and conventional p-contact structures

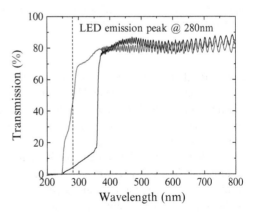

Fig. 3.53 Reflectivity spectra of conventional and reflective contacts

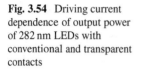

Fig. 3.54 Driving current dependence of output power of 282 nm LEDs with conventional and transparent contacts

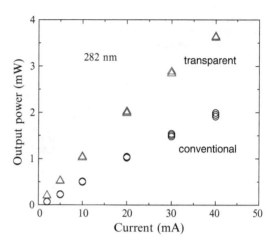

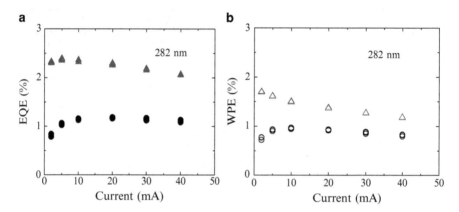

Fig. 3.55 External quantum efficiency (**a**) and wall plug efficiency (**b**) of 282 nm LEDs with conventional and reflective contacts

reason for this is a very low light extraction efficiency limited by internal reflection from the sapphire substrate and by strong absorption in the p-GaN.

However, technological approaches to remedy these problems have been developed for visible and near UV LEDs and this problem has a good chance to be fully or partially solved.

Another aspect of DUV LED performance is related to LED life time and LED heating. DUV LED performance as well as lifetime is affected by heating and formation of conduction channels [69]. In the next section, LED aging process is illustrated by analyzing the radiated power as a function of time. Further we illustrate the effect of heating by analyzing emission power from pulse and CW modes of operation.

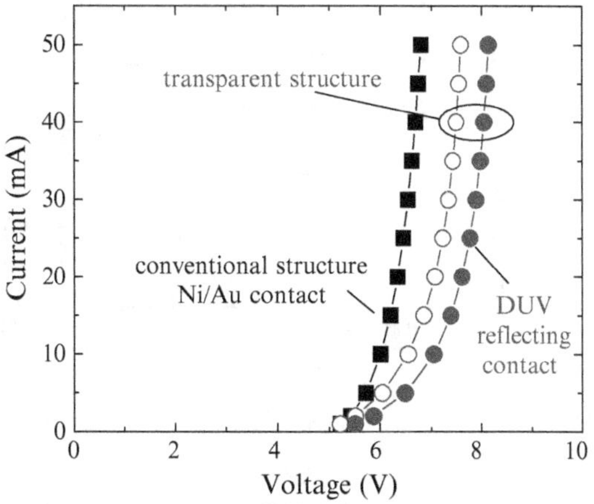

Fig. 3.56 *I–V* characteristics of LEDs with conventional and reflective contacts

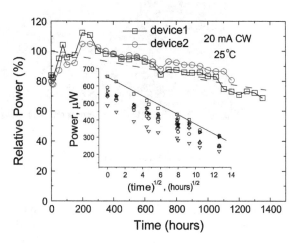

Fig. 3.57 Relative optical power as a function of time for 310 nm DUV LED under 20 mA CW [39]. The *inset* shows 280 nm DUV LED aging data for older generation of 280 nm LEDs [70] Reprinted with permission from [39], Copyright© 2008, American Institute of Physics

As mentioned earlier, recent improvement in materials growth, using the MELEO growth technology [39] allowed us to greatly improve the DUV LED life time (see Fig. 3.57, which compares the power degradation data for the recent and older generations of DUV LEDs).

The DUV LED performance is negatively affected by self-heating, especially at high current densities. Driving LEDs with short pulses leads to a tremendous increase of the output power (see Fig. 3.58).

Pinos et al. [69] studied high current stress degradation of DUV LEDs with wavelengths of 285 and 310 nm by measuring electroluminescence, time-resolved PL, and current–voltage characteristics. In the shorter wavelength device, the presence of the emission band at about 40 nm below the main LED peak and a large leakage current in a virgin device indicated a high initial concentration of nitrogen vacancies and open core screw dislocations. The main role in the device degradation was ascribed to the formation of highly conductive channels, probably, via activation of the closed core screw dislocations involving the nitrogen vacancies.

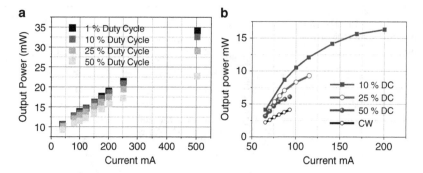

Fig. 3.58 Output power versus current for CW and pulsed modes of operation for DUV LEDs. From [1], IEEE© 2010

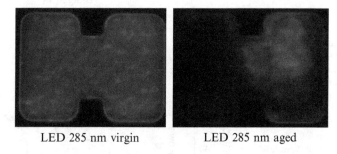

LED 285 nm virgin   LED 285 nm aged

Fig. 3.59 DUV LEDs before and after degradation. Reprinted with permission from Pinos et al. 2011b, Copyright© 2011, American Institute of Physics

Carrier lifetimes in the QWs and *p*-cladding layer were found to be unaffected by the aging process, suggesting that the nonradiative recombination has a lesser influence on the degradation process (see Fig. 3.59).

The operating lifetime improves with wavelength up to ~300 nm and then is nearly constant (see Fig. 3.60) with little change in the output power (see Fig. 3.61).

3.4.3 Conclusion

DUV LEDs have yet to reach their full potential. Great improvements have been made in materials growth, design, fabrication, characterization, and understanding of their device physics but many problems related to insufficient light extraction, high dislocation density, and high contact resistance still need to be solved [73–77]. Future research directions include the development of novel epitaxial materials with reduced defect density (below 10^7 cm^{-2}), improved light extraction (for 2–3 times increase in LEE), improved doping efficiency of high Al-content p-type and n-type AlGaN, improved packaging, and better thermal management. Still, even the

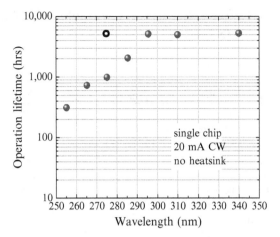

Fig. 3.60 Operating lifetime versus wavelength for commercially available devices (*red*) and engineering samples (*blue*)

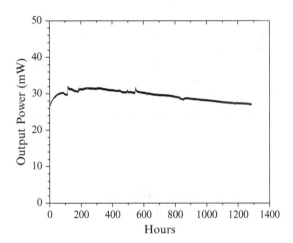

Fig. 3.61 Output power versus time. 10% duty cycle, 400 μs pulse width, 400 mA operating current

current generation of commercial DUV LEDs devices has impressive performance, long lifetimes, and output powers sufficient for many applications.

Acknowledgements We are grateful to talented and dedicated scientists and engineers of Sensor Electronic Technology, Inc.: Drs. Jinwei Yang, Wenhong Sun, Yuri Bilenko, Olga Bilenko, Alex Lunev, Xuhong Hu, and Jianyu Deng, Ajay Sattu, Mr. Rakesh Jain, Mr. Robert Kennedy, and Mr. Igor Shturm; to our colleagues from Vilnius University, Professors Arturas Zukauskas, Gintautas Tamulaitis, Edmundas Kuokstis, from Rensselaer Polytechnic Institute: Drs. Sergey Rumyantsev, Kai Liu, Dmitry Veksler, Bill Stillman, and Professor Shayla Sawyer, to Professor Saulius Marcinkevicius from Royal Institute of Technology in Stockholm, Sweden, and to Professor Grigory Simin from the University of South Carolina for their hard work and inspiration. Our sponsors from the NSF, ARL, ARO, NASA, DARPA, DHS, USDA, and ONR have made this work possible. The work at RPI has been partially supported by the National Science Foundation (NSF) Smart Lighting Engineering Research Center (#EEC-0812056).

A.1   Band Structure Parameters for AlN, InN, and GaN (STR 2011)

Table A.1 Band structure parameters for AlN, InN, and GaN used for LED simulations

Parameter	Units	Parameter value		
Semiconductors	–	AlN	InN	GaN
Energy band gap	eV	6.25	0.69	3.51
Varshni parameter a	meV/K	1.799	0.245	0.909
Varshni parameter b	K	1462	624	830
Crystal field splitting	meV	−93.2	37.3	22.3
Spin-orbital splitting	meV	11.1	11.1	11.1
Electron affinity	eV		3.85	1.96
Dielectric constant	–	8.5	15.3	8.9
Electron effective mass along axis a	m0	0.26	0.1	0.2
Electron effective mass along axis c	m0	0.25	0.1	0.2
Heavy Hole (HH) effective mass along axis a	m0	2.58	1.45	1.65
HH effective mass along axis c	m0	1.95	1.35	1.1
Light Hole (LH) effective mass along axis a	m0	0.27	0.1	0.15
LH effective mass along axis c	m0	1.95	1.35	1.1
Split-off hole effective mass along axis a	m0	1.95	1.54	1.1
Split-off hole effective mass along axis c	m0	0.27	0.1	0.15

A.2   Mechanical and Polarization Properties of AlN, InN, and GaN (STR 2011)

Table A.2 Mechanical and polarization properties of AlN, InN, and GaN used for LED simulations

Parameter	Units	Parameter value		
Semiconductors	–	AlN	InN	GaN
Lattice constant c	nm	0.4982	0.5705	0.5186
Lattice constant a	nm	0.3112	0.354	0.3188
Stiffness Constant (SC) C11	Gpa	395	225	375
SC C12	Gpa	140	110	140
SC C13	Gpa	115	95	105
SC C33	Gpa	385	200	395
SC C44	Gpa	120	45	100
Piezoelectric constant (PC) e15	C/m^2	−0.48	−0.18	−0.27
PC e31	C/m^2	−0.58	−0.22	−0.33
PC e33	C/m^2	1.55	0.43	0.65
Spontaneous polarization	C/m^2	−0.045	−0.032	−0.018

A.3   Ionization Energies and Nonradiative Recombination Constants for AlN, InN, and GaN (STR 2011)

Table A.3 Ionization energies and nonradiative recombination constants for AlN, InN, and GaN used for LED simulation

Parameter	Units	Parameter value		
Semiconductors	–	AlN	InN	GaN
Electron g-factor	–	2	2	2
Acceptor ionization energy	meV	470	170	170
Donor ionization energy	meV	15	15	15
Hole g-factor	–	4	4	4
Radiative recombination constant	10^{-11} cm^3/s	2.00	2.00	2.00
Auger coefficient Cn	10^{-31} cm^6/s	5.00	5.00	5.00
Auger coefficient Cp	10^{-31} cm^6/s	5.00	5.00	5.00

A.4   Optical Constants for Materials Used for LED Design

Table A.4 Optical constants for AlGaN semiconductor alloy

Material	Refractive index (n)	Absorption/extinction coefficient
AlN	2.3[2]	100[1] 1/cm
n-AlGaN (Superlattice)	2.3[2]	100 1/cm
n-AlGaN $x = 0.6$	2.5[1]	100 1/cm
n-metal	1.15[3]	1.31 ext. coef. k
n-AlGaN $x = 0.6$	2.5[1]	100 1/cm
P-layer/MQW (P-region)	2.56[2]	1000 1/cm
p-metal	0.79[3]	2.67 ext. coef. k
Sapphire	1.824[3]	1 1/cm

A.5   Definition of Efficiencies of Ultra-Violet Light Emitting Diode (UV LED)

Radiated power can be expressed in terms of radiative recombination rate of electrons and holes, and *light extraction efficiency* (LEE) as:

$$P_{\text{rad}} = \text{LEE} \cdot \int \hbar\omega R_{\text{rad}}(x)\mathrm{d}x^3.$$

Here, R_{rad} is *radiative recombination rate* giving the number of radiatively recombining carrier pairs per unit time, $\hbar\omega^-$ radiated photon energy and LEE is a fraction of radiated energy that is escaped from the device.

The fraction of radiative recombination rate to overall recombination rate is called *internal quantum efficiency* (IQE)

$$IQE = \int \frac{R_{rad}(x)dx^3}{\int R(x)dx^3}.$$

Injection efficiency (IE) is defined as

$$IE = \frac{I_{inj}}{I}.$$

Here I is a total current through the device and $I_{inj}-$ is the current due to electron–hole recombination in the active region.

Efficiency of LEDs is best characterized by *wall plug efficiency*(WPE) given by the ratio of radiated power P_{rad} to electrical power P_{elec} drawn by the device

$$WPE = \frac{P_{rad}}{P_{elec.}}$$

Here $P_{rad} = LEE \cdot \hbar\omega \cdot IQE \cdot \frac{1}{q} \cdot IE \cdot I$. Electric power drawn by the device is $P_{elsc} = IV$, therefore the WPE is given by

$$WPE = LEE \cdot IQE \cdot IE \cdot \left(\frac{\hbar\omega}{qV}\right)$$

The last term $\hbar\omega/qV$ is defined as *feeding efficiency* (FE)*or electrical efficiency* (EE) (see [71]). FE is the ratio of the mean energy of the photons emitted and the total energy that an electron–hole pair acquires from the power source. Above definitions follow standard LED convention [72].

References

1. M.S. Shur, R. Gaska, Deep-ultraviolet light-emitting diodes. IEEE Trans. Electron Dev. **57**, 12–25 (2010)
2. Y. Taniyasu, M. Kasu, T. Makimoto, An aluminium nitride light-emitting diode with a wavelength of 210 nanometres. Nature **441**, 325–328 (2006) doi: 10.1038/nature04760
3. I. Gaska, O. Bilenko, I. Shturm, Y. Bilenko, M. Shatalov, R. Gaska, Efficiency of point-of-use water disinfection using deep UV light emitting diode technology. Cleantech, June, 13–16, 2011, Boston, MA
4. S.J. Britz, Deep Ultraviolet Light-Emitting Diodes – A Transformative Technology to Enhance the Nutritional Value of Plants. Workshop on Frontiers in Electronics (2009)
5. R. Gaska, Y. Bilenko, M.S. Shur, Organism growth suppression using ultraviolet radiation. United States Patent 7,553,456 (2009a)
6. R. Gaska, M.S. Shur, Y. Bilenko, Ultraviolet radiation sterilization, United States Patent US7634996 (2009b)

7. Germtron UV LED Bandage (2011). http://www.germtron.com/UV%20LED%20Bandage%20Page.html
8. S. Sawyer, S. Rumyantsev, M.S. Shur, N. Pala, Y. Bilenko, J. Zhang, X. Hu, A. Lunev, J. Deng, R. Gaska, Current and optical noise of GaN/AlGaN light emitting diodes. J. Appl. Phys. **100**, 4504 (2006)
9. K.X. Sun, N. Leindecker, S. Higuchi, J. Goebel, S. Buchman, R.L. Byer, UV LED operation lifetime and radiation hardness qualification for space flights. J. Phys.: Conf. Ser. **154**, 012028 (2009)
10. M. Shatalov, W. Sun, Y. Bilenko, A. Sattu, X. Hu, J. Deng, J. Yang, M.S. Shur, C. Moe, M. Wraback, R. Gaska, Large chip high power deep ultraviolet light-emitting diodes. Appl. Phys. Exp. **3**, 2101 (2010)
11. W. Sun, M. Shatalov, J. Deng, X. Hu, J. Yang, A. Lunev, Y. Bilenko, M.S. Shur, R. Gaska, Efficiency droop in 245–247 nm AlGaN light-emitting diodes with continuous wave 2 mW output power. Appl. Phys. Lett. **96**, 1102 (2010)
12. K. Ban, J. Yamamoto, K. Takeda, K. Ide, M. Iwaya, T. Takeuchi, S. Kamiyama, I. Akasaki, H. Amano, Internal quantum efficiency of whole-composition-range AlGaN multiquantum wells. Appl. Phys. Exp. **4**, 2101 (2011)
13. S.Y. Karpov, N. Makarov, Dislocation effect on light emission efficiency in gallium nitride. Appl. Phys. Lett. **81**, 4721 (2002)
14. A. Fujioka, T. Misaki, T. Murayama, Y. Narukawa, T. Mukai, Improvement in output power of 280-nm deep ultraviolet light-emitting diode by using AlGaN multi quantum wells. Appl. Phys. Exp. **3**(4), 041001 (2010)
15. C. Pernot, M. Kim, S. Fukahori, T. Inazu, T. Fujita, Y. Nagasawa, A. Hirano, M. Ippommatsu, M. Iwaya, S. Kamiyama, I. Akasaki, and H. Amano, Improved efficiency of 255–280 nm AlGaN-based light-emitting diodes, Appl. Phys. Exp. **3**, 061004 (2010)
16. DARPA Compact Mid-Ultraviolet Technology, DARPA-BAA-10–45, (2010)
17. D. Yole, (2011) UV LED Market and Technology Report http://www.i-micronews.com/upload/Rapports/Yole_UV_LED_&_AlN_Substrates_market_report_March_2011.pdf.
18. N. Nepal, J. Li, M.L. Nakarmi, J.Y. Lin, H.X. Jiang, Temperature and compositional dependence of the energy band gap of AlGaN alloys. Appl. Phys. Lett. **87**, 2104 (2005)
19. E.F. Schubert, Light-emitting diodes, 2nd edn. (Cambridge University Press, Cambridge, 2006)
20. I. Vurgaftman, J.R. Meyer, Band parameters for nitrogen-containing semiconductors. J. Appl. Phys. **94**, 3675–3696 (2003)
21. R.S. Rumyantsev, M.S. Shur, M. Levinshtein, Nitride materials properties, in GaN-based materials and devices growth, fabrication, characterization and performance. World Scientific, Singapore; River Edge, NJ (2004)
22. M. Shatalov, M. Gaevski, V. Adivarahan, A. Khan, Room-temperature stimulated emission from AlN at 214 nm. Jpn. Appl. Phys. **45**(49), L1286–L1288 (2006)
23. M. Levinshtein, S. Rumyantsev, M.S. Shur, Properties of advanced semiconductor materials: GaN, AlN, InN, BN, SiC, SiGe. (Wiley, New York, 2001)
24. S.K. O'Leary, B.E. Foutz, M.S. Shur, L.F. Eastman, Steady-state and transient electron transport within the III–V nitride semiconductors, GaN, AlN, and InN: a review. J. Mater. Sci: Mater. Electron **17**, 87–126 (2006) doi: 10.1007/s10854–006–5624–2
25. M. Shatalov, J. Yang, Y. Bilenko, M.S. Shur, R. Gaska, High power III-nitride UV emitters, in CLEO:2011 – laser applications to photonic applications, OSA Technical Digest (CD) (Optical Society of America, 2011), paper Joint Symposium on Semiconductor Ultraviolet LEDs and Lasers: Semiconductor Mid-UV LEDs and Lasers, (2011)
26. Q. Fareed, R. Gaska, M.S. Shur, Methods of growing nitride-based film using varying pulses. US Patent 7192849, (2007)
27. R. Gaska, Q. Fareed, G. Tamulaitis, I. Yilmaz, M.S. Shur, C. Chen, J. Yang, E. Kuokstis, A. Khan, J.C. Rojo, L.J. Schowalter, (2003) Stimulated emission at 258 nm in AlN/AlGaN quantum wells grown on bulk AlN substrates. Mat. Res. Symp. Proc. **743**, L6.30 (2003)

28. M. Kneissl, Z. Yang, M. Teepe, C. Knollenberg, O. Schmidt, P. Kiesel, N.M. Johnson, S. Schujman, L.J. Schowalter, Ultraviolet semiconductor laser diodes on bulk AlN. J. Appl. Phys. **101**, 3103 (2007)

29. F. Mymrin, K.A. Bulashevich, N.I. Podolskaya, S.Y. Karpov, Bandgap engineering of electronic and optoelectronic devices on native AlN and GaN substrates: a modelling insight. J. Cryst. Growth **281**, 115–124 (2005)

30. T. Nishida, T. Makimoto, H. Saito, T. Ban, AlGaN-based ultraviolet light-emitting diodes grown on bulk AlN substrates. Appl. Phys. Lett. **84**, 1002 (2004)

31. Z. Ren, Q. Sun, S.Y. Kwon, J. Han, K. Davitt, Y.K. Song, A.V. Nurmikko, H.K. Cho, W. Liu, J.A. Smart, L.J. Schowalter, Heteroepitaxy of AlGaN on bulk AlN substrates for deep ultraviolet light emitting diodes. Appl. Phys. Lett. **91**, 1116 (2007)

32. G. Tamulaitis, I. Yilmaz, M.S. Shur, Q. Fareed, R. Gaska, M.A. Khan, Photoluminescence of AlGaN grown on bulk AlN substrates. Appl. Phys. Lett. **85**, 206 (2004) doi: 10.1063/1.1771804

33. Compound Semiconductors (2011) AlN: can it become a universal substrate for III-nitrides? http://compoundsemiconductor.net/csc/features-details.php?id=19732954

34. H. Amano, I. Akasaki, Novel aspects of the growth of nitrides by MOVPE. J. Phys.: Condens. Mater. **13**, 6935–6944 (2001)

35. M.E. Aumer, S.F. Leboeuf, F.G. McIntosh, S.M. Bedair, High optical quality AlInGaN by metalorganic chemical vapor deposition. Appl. Phys. Lett. **75**, 3315 (1999)

36. H. Gotoh, T. Suga, H. Suzuki, M. Kimata, Low temperature growth of gallium nitride. Jpn. J. Appl. Phys. **20**, L545–L548 (1981)

37. J. Zhang, E. Kuokstis, Q. Fareed, H. Wang, J. Yang, G. Simin, M.A. Khan, R. Gaska, M.S. Shur, Pulsed atomic layer epitaxy of quaternary AlInGaN layers. Appl. Phys. Lett. **79**, 925 (2001)

38. K. Kazlauskas, A. Zkauskas, G. Tamulaitis, J. Mickevicius, M.S. Shur, Q. Fareed, J. Zhang, R. Gaska: Exciton hopping and nonradiative decay in AlGaN epilayers. Appl. Phys. Lett. **87**, 172102 (2005) doi: 10.1063/1.2112169

39. R. Jain, W. Sun, J. Yang, M. Shatalov, X. Hu, A. Sattu, A. Lunev, J. Deng, I. Shturm, Y. Bilenko, R. Gaska, M.S. Shur, Migration enhanced lateral epitaxial overgrowth of AlN and AlGaN for high reliability deep ultraviolet light emitting diodes. Appl. Phys. Lett. **93**, 1113 (2008)

40. A.D. Bykhovski, B.L. Gelmont, M.S. Shur, Elastic strain relaxation and piezoeffect in GaN-AlN, GaN-AlGaN and GaN-InGaN superlattices. J. Appl. Phys. **81**, 6332–6338 (1997)

41. A.D. Bykhovski, B.L. Gelmont, M.S. Shur, The influence of the strain-induced electric field on the charge distribution in GaN-AlN-GaN structure. J. Appl. Phys. **74**, 6734–6739 (1993)

42. A. Pinos, V. Liuolia, S. Marcinkevicius, J. Yang, R. Gaska, M.S. Shur, Localization potentials in AlGaN epitaxial films studied by scanning near-field optical spectroscopy. J. Appl. Phys. **109**(11), 113516 (2011a)

43. C.J. Collins, A.V. Sampath, G.A. Garrett, W.L. Sarney, H. Shen, M. Wraback, A.Y. Nikiforov, G.S. Cargill, V. Dierolf, Enhanced room-temperature luminescence efficiency through carrier localization in $Al_xGa_{1-x}N$ alloys. Appl. Phys Lett. **86**(3), 031916 (2005)

44. Y.B. Lee, T. Wang, Y.H. Liu, J.P. Ao, H.D. Li, H. Sato, K. Nishino, Y. Naoi, S. Sakai, Fabrication of high-output-power AlGaN/GaN-Based UV-light-emitting diode using a Ga droplet layer. Jpn. J. Appl. Phys. **41**(Part 2, No. 10A), L1037–L1039 (2002)

45. A. Pinos, S. Marcinkevicius, J. Yang, Y. Bilenko, M. Shatalov, R. Gaska, M.S. Shur, Aging of AlGaN quantum well light emitting diode studied by scanning near-field optical spectroscopy. Appl. Phys. Lett. **95**(18), 181914 (2009)

46. G. Tamulaitis, K. Kazlauskas, A. Žukauskas, J. Mickeviius, M.S. Shur, Q. Fareed, J. Zhang, R. Gaska, Study of exciton hopping in AlGaN epilayers by photoluminescence spectroscopy and Monte Carlo simulation. Phys. Stat. Sol. (c). **3**, 2099–2102 (2006) doi: 10.1002/pssc .200565334

47. A. Laubsch, M. Sabathil, J. Baur, M. Peter, B. Hahn, High-power and high-efficiency InGaN-based light emitters. IEEE Trans. Electron Dev. **57**, 79–87 (2010). doi: 10.1109/TED.2009 .2035538

48. O. Ambacher, J. Smart, J.R. Shealy, N.G. Weimann, K. Chu, M. Murphy, W.J. Schaff, L.F. Eastman, R. Dimitrov, L. Wittmer, M. Stutzmann, W. Rieger, J. Hilsenbeck, Two-dimensional electron gases induced by spontaneous and piezoelectric polarization charges in N- and Ga-face AlGaN/GaN heterostructures. J. Appl. Phys. **85**, 3222–3233 (1999)
49. S.H. Park, S.L. Chuang, Spontaneous polarization effects in wurtzite GaN/AlGaN quantum wells and comparison with experiment. Appl. Phys. Lett. **76**, 1981 (2000)
50. S. Marcinkevicìus, A. Pinos, K. Liu, D. Veksler, M.S. Shur, J. Zhang, R. Gaska, Intrinsic electric fields in AlGaN quantum wells. Appl. Phys. Lett. **90**, 081914 (2007) doi: 10.1063/1.2679864
51. A. Pinos, S. Marcinkevicìus, K. Liu, M.S. Shur, E. Kuokštis, G. Tamulaitis, R. Gaska, J. Yang, W. Sun, Screening dynamics of intrinsic electric field in AlGaN quantum wells. Appl. Phys. Lett. **92**, 061907 (2008) doi: 10.1063/1.2857467
52. F. Bernardini, V. Fiorentini, Macroscopic polarization and band offsets at nitride heterojunctions. Phys. Rev. B (Condensed Matter and Materials Physics). **57**, 9427 (1998)
53. J. Mickevicius, E. Kuokstis, V. Liuolia, G. Tamulaitis, M.S. Shur, J. Yang, R. Gaska, Photoluminescence dynamics of AlGaN quantum wells with built-in electric fields and localized states. Phys. Status Solidi (a). **207**, 423–427 (2010) doi: 10.1002/pssa.200925227
54. J. Mickevicìus, G. Tamulaitis, E. Kuokštis, K. Liu, M.S. Shur, J.P. Zhang, R. Gaska. Well-width-dependent carrier lifetime in AlGaN/AlGaN quantum wells. Appl. Phys. Lett. **90**, 131907 (2007) doi: 10.1063/1.2717145
55. R. Gaska, A.D. Bykhovski, M.S. Shur, Piezoelectric doping and elastic strain relaxation in AlGaN–GaN heterostructure field effect transistors. Appl. Phys. Lett. **73**, 3577 (1998)
56. M.D. Craven, S.H. Lim, F. Wu, J.S. Speck, S.P. DenBaars, Structural characterization of nonpolar (1120) a-plane GaN thin films grown on (1102) r-plane sapphire. Appl. Phys. Lett. **81**, 469–471 (2002)
57. E. Kuokstis, J. Zhang, Q. Fareed, J.W. Yang, G. Simin, M.A. Khan, R. Gaska, M.S. Shur, C. Rojo, L.J. Schowalter, Near-band-edge photoluminescence of wurtzite-type AlN. Appl. Phys. Lett. **81**, 2755 (2002)
58. P. Waltereit, O. Brandt, A. Trampert, H.T. Grahn, J. Menniger, M. Ramsteiner, M. Reiche, K.H. Ploog, Nitride semiconductors free of electrostatic fields for efficient white light-emitting diodes. Nature **406**, 865–868 (2000)
59. R. Gaska, J. Zhang, M.S. Shur, Heterostructure including light generating structure contained in potential well. US patent number 7619238 (2009d)
60. R. Gaska, J. Zhang, M.S. Shur, Light emitting heterostructure. US patent number 7,537,950 (2009c)
61. M. Shatalov, J. Yang, W. Sun, R. Kennedy, R. Gaska, K. Liu, M.S. Shur, G. Tamulaitis, Efficiency of light emission in high aluminum content AlGaN quantum wells. J Appl. Phys. **105**, 073103–073106 (2009)
62. J. Deng, Y. Bilenko, A. Lunev, X. Hu, T.M. Katona, J. Zhang, M.S. Shur, R. Gaska, 247 nm Ultra-Violet Light Emitting Diodes. Jpn. J. Appl. Phys. **46**, L263–L264 (2007)
63. Y. Bilenko, A. Lunev, X. Hu, J. Deng, T.M. Katona, J. Zhang, R. Gaska, M.S. Shur, W. Sun, V. Adivarahan, M. Shatalov, Khan A. 10 Milliwatt Pulse Operation of 265 nm AlGaN Light Emitting Diodes. Jpn. J. Appl. Phys. **44**, L98–L100 (2005)
64. Y. Bilenko, R. Gaska, M.S. Shur, G. Simin, Mounting structure providing electrical surge protection. Pub. No US 2007/0151755 (2007)
65. M.S. Shur, R. Gaska, III-nitride based deep ultraviolet light sources. Gallium Nitride materials and devices III. Ed by Morkoc 6894, 689419 (2008)
66. R. Gaska, J. Zhang, Y. Bilenko, A. Lunev, X. Hu, J. Deng, M.S. Shur, AlGaN-based deep ultraviolet light source. In: Proceedings of WOCSDICE, Venice, Italy (2007a)
67. Y. Bilenko, R. Gaska, M.S. Shur, Nitride-based light emitting heterostructure, Pub. No US 2008/0081390 (2008)
68. Y. Bilenko, R. Gaska, M.S. Shur, Shaped contact layer for light emitting heterostructure, U.S. Patent 7928451, (2011)

69. A. Pinos, S. Marcinkevicìus, M.S. Shur, High current-induced degradation of AlGaN ultraviolet light emitting diodes. J. Appl. Phys. **109**, 103108–10 (2011b)
70. S. Sawyer, S. Rumyantsev, M.S. Shur, Degradation of AlGaN-based ultraviolet light emitting diodes. Solid State Electron **52**, 968–972 (2008)
71. A.I. Zhmakin, Enhancement of light extraction from light emitting diodes. Phys. Rep. **498**, 189–241 (2010) doi: 10.1016/j.physrep.2010.11.001
72. LED Magazine (2004) LED efficiencies http://www.ledsmagazine.com/features/1/1/10
73. X. Hu, J. Deng, J. Zhang, A. Lunev, Y. Bilenko, T. Katona, M.S. Shur, R. Gaska, M. Shatalov, A. Khan, Deep ultraviolet light-emitting diodes. Physica Status Solidi (A). **203**, 1815–1818 (2006)
74. A.H. Johnston, T.F. Miyahira, Characterization of proton damage in light-emitting diodes. IEEE Trans. Nucl. Sci. **47**, 2500–2507 (2000)
75. M.A. Khan, R. Gaska, M.S. Shur, J. Yang, Method of producing nitride-based heterostructure devices. US patent 6764888 (2004)
76. Philips (2011) Philips UV Sanitizer. http://www.p4c.philips.com/cgi-bin/dcbint/cpindex.pl?ctn=HX7990/02&slg=en&scy=US.
77. SiLENSE Manual: STR, SiLENSE Manual (2011)

Chapter 4
Green Nitride LEDs

Xian-An Cao

Abstract The "green gap" is a major unsolved obstacle in current efforts to create high-efficiency solid-state solutions. The gap gets deeper at high injection currents, as green LEDs exhibit more pronounced efficiency droop than other visible LEDs. This chapter reviews recent advances in the development of high-brightness InGaN-based green LEDs on c-plane substrates as well as nonpolar and semipolar substrates. The influences of piezoelectric polarization and carrier localization effects on the LED performance are discussed. Recent work on efficiency droop in green LEDs is reviewed and a summary of the understanding of this problem is given.

4.1 The "Green Gap"

The development of solid-state lighting (SSL) based on light-emitting diodes (LEDs) offers an opportunity to significantly lower worldwide energy consumption [1–3]. The total annual energy consumption for lighting in the United States is approximately 800 terawatt-hours, accounting for $\sim 22\%$ of all the electricity generated. If conventional light sources such as incandescent and fluorescent lamps are replaced by highly efficient SSL sources with luminous efficacy $> 150\,\text{lm/W}$, more than 50% of the electricity consumption currently used for lighting can be saved [2]. The potential energy saving relies on effective concepts of generating solid-state white light. The most energy-efficient way to produce white light is by color mixing, which mixes light from InGaN-based blue and green LEDs with light emission of AlInGaP-based red LEDs [3]. The combination of red, green, and blue LEDs also completes the primary color spectrum and enables the fabrication of large-scale full-color displays.

X.-A. Cao (✉)
West Virginia University, Department of Computer Science and Electrical Engineering, Morgantown, WV 26506-6109, USA
e-mail: xacao@mail.wvu.edu

S. Pearton (ed.), *GaN and ZnO-based Materials and Devices*,
Springer Series in Materials Science 156, DOI 10.1007/978-3-642-23521-4_4,
© Springer-Verlag Berlin Heidelberg 2012

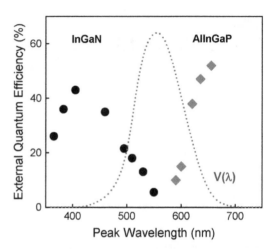

Fig. 4.1 External quantum efficiencies of the state-of-the-art high-brightness LEDs as a function of emission wavelength

Since the early 1990s, InGaN alloys have been the primary focus in pursuing high-brightness blue and green LEDs, which have undergone a phenomenal development effort [4]. However, further improvement is required in order to implement LEDs in general illumination and display applications [1, 2]. As seen in Fig. 4.1, the external quantum efficiencies (EQEs) of InGaN-based LEDs are still lower than those of AlInGaP-based red LEDs. In addition, the performance of InGaN-based LEDs is strongly wavelength dependent. The EQE decreases with increasing wavelength from the near-ultraviolet (near-UV) spectrum, and drops by over 50% at green wavelengths. On the other hand, the EQE of AlInGaP-based LEDs decreases with decreasing wavelength from the visible red spectrum. These trends result in very low EQEs in the green spectral range, often referred to as the "green gap" [4, 5]. The gap must be closed by improving green light emission so as to develop environmentally benign lighting solutions that offer high-quality spectral quality and promise substantial energy saving.

Cost-effective SSL applications require high-efficiency as well as high-power LEDs. However, the state-of-the-art InGaN-based LEDs are not suitable for high-current operation as they suffer from a significant drop in efficiency at high injection levels. This problem is commonly referred to as "efficiency droop" [6–8]. The magnitude of efficiency droop is also wavelength dependent, and increases as the emission peak moves toward longer wavelengths. The EQE of green LEDs typically peaks at a low current density below $20 \, A/cm^2$, and rolls off rapidly at elevated currents, as shown in Fig. 4.2. It has been found that the droop is not a thermal issue, but solely induced by current injection [8]. Despite recent intense studies on this issue, the exact mechanism of the droop problem is not completely understood [6,7].

Both the "green gap" and efficiency droop are believed to be fundamental problems associated with the InGaN materials and their heterostructures. One of the most defining features of the III-nitride materials is the lack of high-quality native substrates [9, 10]. To date, all commercially available green LEDs are grown heteroepitaxially on c-plane sapphire and SiC substrates. The mismatches in lattice constant

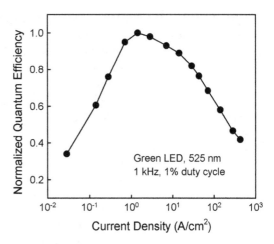

Fig. 4.2 Injection-current dependent EQE of a typical commercially available InGaN-based green LED grown on c-plane sapphire

and thermal expansion coefficient between the epilayer and substrate give rise to a high density of threading dislocations and large residual strain in the LED structures [9–11]. The contrasting thermodynamic and structural properties of InN and GaN lead to low miscibility, making it difficult to grow high-quality green-emitting InGaN/GaN multiple quantum wells (MQWs) [5]. Large compressive strain induced in InGaN QWs grown on polar planes causes strong piezoelectricity [11], which along with high dislocation density and large indium compositional inhomogeneity, gives rise to many unique electrical and optical characteristics of InGaN-based LEDs [9, 10]. Many of these effects become stronger as the indium content in the active region increases, and become major performance-limiting factors in green LEDs. Further improvement of the green LED performance must be made on the basis of a thorough understanding of these important material and structural issues.

In this chapter, we will review recent advances in the epitaxial growth of InGaN-based green LEDs on c-plane substrates as well as nonpolar and semipolar substrates. The piezoelectric polarization and carrier localization effects in InGaN QWs will be described and their influences on the performance of green LEDs will be addressed. Finally, our studies and other reported work on efficiency droop in green LEDs will be reviewed, and the possible mechanisms and mitigation strategies will be discussed.

4.2   Advances in Growth of c-Plane Green LEDs

4.2.1   Green LEDs on (0001) sapphire

To date, much more knowledge and experience have been accumulated in the technology of III-nitride growth on (0001) sapphire than on other substrates [9]. It is thus no surprise that most commercially available green LEDs are built on the c-plane of sapphire substrates. Large-area, high-quality sapphire is widely available in large quantities, and is fairly inexpensive. In addition, sapphire is transparent to UV and visible light; so LEDs can be flip-chip packaged for efficient light extraction [12]. However, LEDs on sapphire suffer from a critical disadvantage, namely, large mismatches in lattice constant and thermal expansion coefficient between nitride materials and sapphire. As a result, the crystalline quality of epilayers grown directly on sapphire is quite poor, containing a high density of threading dislocations and large biaxial strain. Many efforts have been devoted to developing high-quality buffer layers to accommodate the mismatches. A thin (~ 50 nm) low-temperature ($\sim 500°C$) GaN or AlN buffer layer is now widely adopted in nitride growth on sapphire. It has been proven that such a buffer layer can greatly reduce the dislocation density and improve the surface morphology of overlayers [13, 14].

Metal organic chemical vapor deposition (MOCVD) has evolved as the dominant technique for growing III-nitride LEDs [15]. The growth is usually conducted at low pressure. The common precursors include trimethylalluminum (TMAl), trimethylgallium (TMGa), and trimethylindium (TMIn) as the metal sources, and ammonia as the N source. Silane and bis-cyclopentadienyl-magnesium (Cp_2Mg) are commonly used for n- and p-type doping, respectively. Hydrogen or nitrogen is used as the carrier gas. A typical layer structure of green LEDs consists of an $In_xGa_{1-x}N(x > 0.2)$/GaN MQW active region, a n-GaN lower cladding layer, a p-AlGaN upper cladding layer, and a p-GaN top contact layer. The p-AlGaN is used as an electron-blocking layer (EBL) which is necessary to prevent over-injection of electrons. The contact layer is usually doped heavily with Mg to reduce the p-contact resistance. The ideal growth temperatures for different layers of the LED structure are different – GaN is typically grown at $1,000$–$1,100°C$, AlGaN requires a slightly higher temperature, and InGaN is grown at a much lower temperature ~ 700–$800°C$.

Despite the use of a low-temperature buffer, green LED structures grown on sapphire still contain a high density of threading dislocations in the range of 10^8–10^{10} cm^{-2} [9, 10]. Most dislocations traverse vertically from the substrate surface to the top layer and, depending on the growth condition of the capping layer, may terminate by forming various types of surface defects [10]. It has been found that the intersection of InGaN QWs with dislocations may promote the formation of large hexagonal pits called V-defects [16, 17]. Although these defects may create shunt leakage paths and degrade device reliability, they seem to have a limited adverse impact on the optical characteristics of green LEDs [9, 10]. Indeed, green LEDs grown on high-quality laterally overgrown GaN templates and free-standing GaN substrates showed quantum efficiencies similar to green LEDs grown on sapphire [18, 19]. It is generally accepted that localization effects in InGaN QWs arising from fluctuations in the indium composition and QW thickness prevent carriers from reaching the dislocation cores and greatly enhance the radiative efficiency of green LEDs [9].

While the entire LED structure is nearly fully relaxed with respect to the sapphire substrate during growth, residual compressive stress develops upon cooling to room temperature due to the larger thermal expansion coefficient of sapphire [20]. It has been found that excessive Si doping may change the stress from compressive to tensile [21], leading to wafer bowing and film cracking. In addition, in LEDs grown on polar substrates, the large lattice mismatch between InGaN and GaN leads to biaxial compressive strain in the InGaN QWs. The strain induces strong piezoelectric polarization and the quantum confined Stark effect (QCSE), which spatially separate electrons and holes, leading to a reduced radiative recombination rate and redshift of the emission peak [11]. Upon strain relaxation, misfit defects may form at the InGaN/GaN interfaces, acting as efficient nonradiative recombination centers. Apparently, the strain increases as more indium is incorporated into the QWs. Therefore, it has a greater adverse impact on the performance of green LEDs than on blue and UV LEDs.

In addition to the presence of a high dislocation density and large biaxial strain, poor crystal quality of InGaN alloys with high indium contents is another important limiting factor which hampers the further improvement of green LED performance [19]. The difficulty of growing InGaN alloys arises mainly from the contrasting thermodynamical properties of InN and GaN. Due to the high volatility of InN, InGaN QWs are typically grown at 700–800°C to ensure incorporation of high mole fractions of indium for green light emission. The temperature is much lower than the optimal growth temperatures of GaN and AlGaN. As a result, during the subsequent growth of p-type layers, the InGaN QWs may suffer from indium re-evaporation and inter-diffusion [22]. The low miscibility of InN in GaN imposes a restricted growth window for high-In content InGaN QWs and leads to significant compositional disorder, which is often manifested by the broadening of luminescent spectra [9]. In InGaN QWs with over 20% indium, phase separation may occur [23], leading to the formation of interfacial misfit defects driven by lattice mismatch.

Since the dislocation density in growth templates appears not to be a major performance-limiting factor, recent efforts to optimize the epitaxy of green LEDs on c-plane sapphire have focused on two aspects (1) Control of the QCSE in InGaN QWs by strain engineering and doping [24–31], and (2) optimization of growth temperatures to improve the quality of InGaN/GaN MQWs [22, 32–34].

The QCSE in InGaN QWs arising from piezoelectric polarization can be reduced by reducing strain in the QWs. A popular way is to insert a strain-relief layer between InGaN/GaN MQWs and the underlying n-GaN cladding layer. Huang et al. [24] reported on prestrained growth of InGaN/GaN MQW green LEDs by employing an underlying $In_{0.07}Ga_{0.93}N$/GaN QW for strain control. The LEDs showed a reduced electroluminescence (EL) peak blueshift with increasing injection current and 182% higher EL intensity at 20 mA. The effects of alternative strain–relief layers such as a thick InGaN layer [25, 26] and InGaN/GaN short-period superlattice [27], on the performance of green LEDs have also been studied. Such buried layers shift the lattice constant of the template close to the average value of the QWs and quantum barriers (QBs), and thus partly compensates the strain in the QWs. They may also benefit the LEDs by blocking dislocations and smoothing the surface morphology. Zhao et al. [28, 29] proposed unique two-layer and three-layer staggered InGaN QWs to overcome the strong QCSE in green LEDs. Theoretical study has showed that the staggered QW structures can enhance the overlap between

Fig. 4.3 (a) EL spectra and
(b) light output power versus
current density for
conventional InGaN QW and
three-layer staggered InGaN
QW LEDs on (0001) sapphire
emitting at 520–525 nm. The
inset shows the band lineups
of a three-layer staggered
InGaN QW (From Zhao et al.
[28])

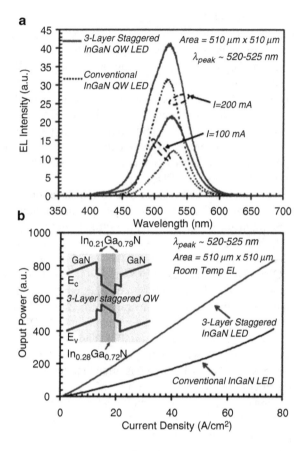

electron-hole wave functions, leading to increased radiative recombination rates. As
seen in Fig. 4.3, three-layer staggered InGaN QW LEDs emitting at 520–525 nm
have an output power 2–3.5 × higher than that of LEDs with a conventional InGaN
QW [29].

Intentional Si doping in the QBs and/or QWs have been reported to be effective
in screening the piezoelectric field in blue-emitting InGaN/GaN MQWs [30]. Free
carriers released from the doped layers may neutralize the polarization charge and
thus partially screen the piezoelectric field. However, in green LEDs where the QWs
are deeper, GaN QBs doped with Si ($\sim 5 \times 10^{17}$–4×10^{18} cm^{-3}) were found to block
hole transport, leading to increased forward voltage and reduced EL intensity [31].
Therefore, the use of Si doping to reduce the QCSE must be carefully optimized in
the design and growth of green LED structures.

In a typical epitaxial process of InGaN/GaN MQW LEDs, the entire MQW
structure is grown at the optimal temperature for InGaN, ~ 700–800°C, which is
much lower than the temperature required for growing high-quality GaN. As a
result, the crystal quality of the GaN QBs is poor, leading to increased carrier
leakage and nonradiative recombination. This problem is more severe for green

LEDs since lower growth temperatures are used to facilitate indium incorporation. Wu et al. [32] reported a study of green LEDs with the InGaN QWs and GaN QBs grown at 700°C and 950°C, respectively. The LEDs were 65% more efficient than LEDs with the MQWs grown at constant low temperature. Such improvement was attributed to the better quality of GaN QBs grown at 950°C.

Another problem encountered during the growth of green LEDs is the degradation of InGaN QWs as they are exposed to higher temperatures due to the low thermal stability of InGaN. Ju et al. [33] have found that a significant amount of indium was re-evaporated during a dual-temperature MQW growth, as revealed by a large redshift of PL. They found that the InGaN QWs can be well preserved with a 1 nm thick GaN coated on each QW at the QW growth temperature. The re-evaporation and inter-diffusion of indium may also occur during the subsequent high-temperature deposition of p-type AlGaN/GaN layers. Oh et al. [22] have characterized $In_{0.25}Ga_{0.75}N/GaN$ MQWs with a p-GaN capping layer deposited at 900°C and 1050°C. The former showed much more intense and well-defined X-ray

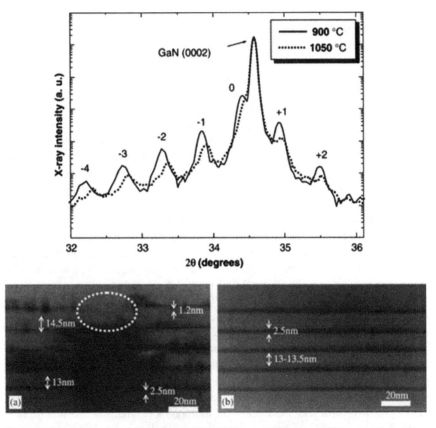

Fig. 4.4 (top) XRD spectra of InGaN/GaN MQW green LEDs on (0001) sapphire with p-GaN grown at different temperatures. (*bottom*) Cross-sectional TEM images of green-emitting InGaN/GaN MQWs with the p-GaN overlayer grown at (a) 1050°C and (b) 900°C (From Oh et al. [22])

diffraction (XRD) peaks as well as shaper and smooth InGaN/GaN interfaces, as illustrated in Fig. 4.4. The average indium compositions in the MQWs are indicated by the zero-order peak positions of the XRD spectra, and can be calculated using Vegard's law. It was found that the indium content was reduced to 18.2% in the latter, and its EL peak shifted to 460 nm. These results suggest that growing the p-type layers at 900°C can effectively suppress the re-evaporation and inter-diffusion of indium in green LEDs. In order to avoid the thermal damage to the active region of green LEDs during the high-temperature growth of the AlGaN EBL, Kim et al. [34] employed an alternative $In_{0.18}Al_{0.92}N$ EBL, which can be grown at a lower temperature ($\sim 780°C$), produces a large conduction band offset, and is lattice-matched to GaN. The InAlN EBL was found to be effective in preserving the quality of InGaN QWs and improving the EL efficiency of green LEDs.

4.2.2   Green LEDs on Free-Standing (0001) GaN

The growth of III-nitride device structures on bulk GaN substrates is considered as homoepitaxy due to the identical crystallography as well as similar lattice constants of GaN and its alloys with Al and In. It is expected that homoepitaxial growth can significantly reduce defect density and strain, and offer a better doping and impurity control, leading to improved device performance and yields. The growth procedure is also simplified as additional steps needed for heteroepitaxy, such as surface nitridation and low-temperature buffer layer growth, are not required. These incentives are the driving forces behind recent progress toward producing bulk GaN crystals by high-pressure solution growth [35–37] and hydride vapor phase epitaxy (HVPE) [38–40]. 2" free-standing HVPE GaN substrates are now commercially available, and the crystalline quality improves steadily as the technology advances. Even though sapphire and SiC will remain the substrates of choice for nitride growth before high-quality and low-cost GaN wafers become available, homoepitaxy provides a valuable avenue to explore the potential of high-quality devices.

Typical commercially available HVPE GaN wafers have dislocation densities in the range of 10^4–10^7 cm^{-2} [39]. Undoped materials have n-type conductivity due to the presence of a high density of point defects including N vacancies and O impurities, which may cause a redshift of PL spectra. Homoepitaxy on such a GaN substrate is relatively straightforward. The substrate is heated directly to the growth temperature in an ammonia-rich environment. 2D growth mode can be achieved on epi-ready (0001)GaN without using a buffer layer, leading to smooth surfaces with terrace steps [18, 41], which are in contrast to rough surfaces with a swirled step structure for typical GaN on sapphire. It has been found that vicinal surfaces with an offcut of 1–2 degrees may provide pre-defined atomic steps for 2D step-flow growth and yield smoother morphologies than the nominal c-plane [39]. Homoepitaxial GaN layers, in most cases, replicate the defect structure in the bulk GaN substrates [42]. Figure 4.5 shows XRD rocking curves of the (0002) reflection of a HVPE GaN substrate and a GaN epilayer grown atop, as compared

Fig. 4.5 XRD rocking
curves of the (0002)
reflection from a HVPE GaN
substrate (*solid line*), and
GaN epilayers grown on
c-plane bulk GaN (*dashed
line*) and sapphire (*dotted
line*) substrates

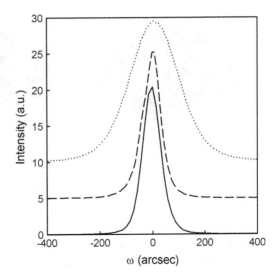

to a similar epilayer grown on sapphire. The FWHM of the homoepitaxial GaN is
79 arcsec, which is comparable to that of the substrate (85 arcsec), but much smaller
than for the heteroepitaxial GaN (230 aresec). Since the FWHM of the (0002) peak
reflects the degree of lattice distortion from dislocations, the smaller FWHM of the
homoepitaxial GaN confirms that the threading dislocation density is substantially
reduced. Our study also showed that impurity incorporation in homoepitaxial GaN
was significantly reduced compared to GaN grown on sapphire [43]. This may result
from dislocation reduction in homoepitxial GaN as impurities tend to aggregate
around microstructural defects and create localized states in the bandgap [44].

We have successfully grown InGaN/GaN MQW green LEDs on free-standing
HVPE (0001) GaN substrates using low-pressure MOCVD [18]. The GaN substrates, offcut by 1–2°C in the $< 11\underline{2}0 >$ direction, were unintentionally doped
and had a free carrier concentration $\sim 7 \times 10^{17}\,cm^{-3}$. Atomic force microscopy
(AFM) measurements showed that threading dislocations were decorated by surface
pits after the chemical mechanical polish and the density was determined to be
~ 1–$2 \times 10^7\,cm^{-2}$. The substrates were heated directly to $1,050°C$ in a steady
ammonia flow, followed by the growth of the green LED structure on the Ga-face.
The growth recipe for each layer was similar to that used for heteroepitaxial LEDs.
The structure consisted of a $2\,\mu m$ n-GaN layer, a 10-period InGaN/GaN MQW
active region with 3 nm QWs and 12 nm QBs, a $0.05\,\mu m$ p-type AlGaN cladding
layer and a $0.3\,\mu m$ p-GaN contact layer. The nominal indium mole fraction in the
active region was ~ 0.27.

The AFM image of the LED epilayer over a $5 \times 5\,\mu m$ area is illustrated in
Fig. 4.6b. The surface is free of surface pits, and a terrace-step structure is clearly
seen, suggesting that smooth 2D step-flow growth was achieved despite high indium
content in the QWs which may cause significant compositional disorder. In contrast,
the structure grown in the same epitaxy run on (0001) sapphire with a previously

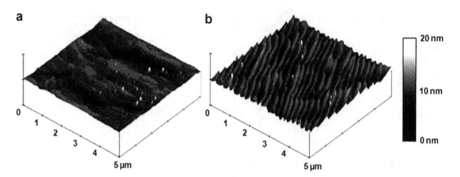

Fig. 4.6 $5 \times 5\,\mu m$ AFM images of InGaN/GaN MQW green LEDs grown on c-plane (**a**) sapphire and (**b**) GaN substrates

grown buffer exhibits a rougher morphology and V-defects with a size of 50–200 nm and a density of $\sim 1 \times 10^8\,cm^{-2}$ (see Fig. 4.6a). Most of the V-defects originate from the intersection of dislocations with InGaN QW layers, where large strain promotes their formation [16, 17]. The root-mean-square roughness values of the LEDs on sapphire and GaN are 1.4 and 0.5 nm, respectively. The dislocation density in the LED on GaN is expected to be similar to that in the GaN substrates [42], significantly lower than that in the LED on sapphire ($\sim 10^9\,cm^{-2}$).

One additional benefit of using bulk GaN substrates is that simple vertically structured LEDs can be fabricated. The vertical geometry, in combination with good thermal conductivity of GaN (5× higher than sapphire), allows the LEDs to operate at much higher current densities [45]. Top-emitting green LEDs with a chip size of $300 \times 300\,\mu m$ were fabricated using the standard LED fabrication process. A thin NiO_x/Au was used as the semitransparent p-type ohmic contact, whereas Ti/Al/Ti/Au was used as the n-type metallization. Figure 4.7 compares the current–voltage (I–V) characteristics of a vertical LED on GaN and a lateral LED on sapphire. The junction leakage currents in the LED on GaN at reverse bias and low forward bias are reduced by more than two orders of magnitude compared to those in the LED on sapphire. This correlates well with a reduced dislocation density in the former. An ideality factor of 2.1 can be extracted from the intermediate section of the forward I–V curve of the LED on GaN, suggesting that the injection current in this regime is mainly recombination current. This is in contrast to the dominant defect-assisted tunneling current in the LED on sapphire [46].

Figure 4.8 shows the EL spectra of the LEDs at increasing pulsed currents (1 kHz and 1% duty cycle) from 0.1 to 300 mA. The presence of Fabry–Perot interference fringes is a characteristic feature for LEDs grown on sapphire. As expected, they are absent in the LED on GaN. The peak wavelengths of the two LEDs at 20 mA are almost identical ~ 520 nm. The spectra exhibit a monotonic blueshift with increasing injection current. Up to 600 mA ($850\,A/cm^2$), the maximum current applied to the LEDs, the blueshift is ~ 107 meV. The current-induced blue shift, accompanied by a significant spectral broadening, occurs due to screening of the

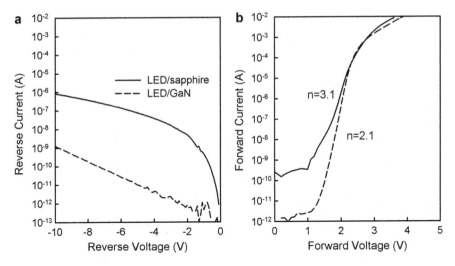

Fig. 4.7 (**a**) Reverse and (**b**) forward I–V characteristics of green LEDs grown on c-plane sapphire and GaN substrates

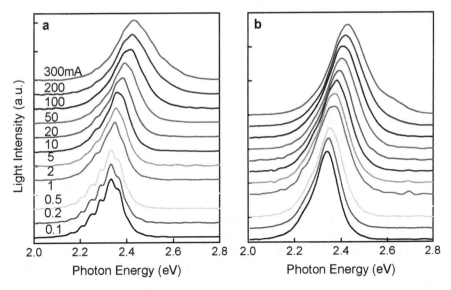

Fig. 4.8 EL spectra of green LEDs on c-plane (**a**) sapphire and (**b**) GaN substrates with increasing pulsed current. The peak intensity is normalized to unity and the spectra are shifted in the vertical direction for clarity

piezoelectric field and band filling of localized states [9]. The similar amounts of peak shift and spectral broadening suggest approximately the same degrees of strain and compositional fluctuation in the active regions despite very different growth modes on the GaN and sapphire substrates.

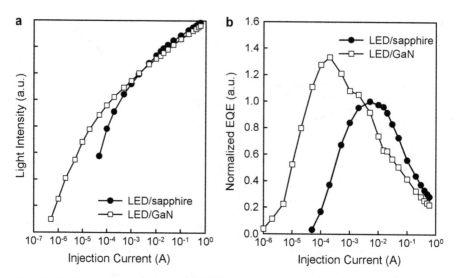

Fig. 4.9 (**a**) EL intensity and (**b**) normalized EQE of green LEDs grown on c-plane sapphire and GaN substrates as a function of pulsed injection current

Figure 4.9a compares the light output power–current ($L–I$) characteristics of the LEDs. Green light emission can be easily detected from the LED on GaN at currents below 1 μA due to low junction leakage. Below 0.1 mA, it significantly outperforms the LED on sapphire. However, the dependence of light intensity on current is superlinear, suggesting a significant influence of nonradiative recombination of the injected carriers. Above 0.1 mA, the $L–I$ curve is nearly linear on a log–log scale as the injection efficiency approaches unity and the radiative recombination process becomes dominant. The light intensity of the LED on sapphire increases rapidly and reaches the linear regime at ∼ 2 mA. At 10 mA, the LED on sapphire is actually 30% brighter than the LED on GaN. Ray-tracing simulation showed that the chip-to-air extraction efficiency of the former is 28% higher due to the smaller refractive index and higher transparency of sapphire compared to the HVPE GaN [41]. The internal quantum efficiencies (IQE) of the two LEDs at 10 mA are therefore comparable. This finding further corroborates the conclusion that dislocations play a minor role in the EL of green LEDs due to strong carrier localization effects [9].

Figure 4.9b compares the normalized EQE of the LEDs as a function of pulsed current. The EQE of both LEDs rises rapidly with increasing current and rolls off as current exceeds a characteristic current. The peak EQE of the LED on GaN is ∼ 33% higher. Using calculated light extraction efficiencies, the IQE is found to be ∼ 65%, which is 71% higher than that of the LED on sapphire. It is striking that the EQE of the LED on GaN peaks at a very low current ∼ 0.2 mA(0.3 A/cm^2) and a small bias ∼ 2.4 V. The peak-efficiency current density is one order of magnitude lower than that of typical green LEDs on sapphire. Above 0.2 mA, the EQE decreases rapidly and monotonically with increasing current, by 50% at 10 mA. Despite better material quality, the LED on GaN exhibits even more pronounced efficiency droop. This suggests that nonradiative recombination at dislocations is not the physical origin of the droop behavior [8].

Our study shows that, even though expitaxy on (0001) GaN substrates can substantially reduce the dislocation density in green LEDs, leading to reduced junction leakage current, they may not solve the well-recognized "green gap" and efficiency droop problems. Our results agree well with the report by Razeghi et al. [19], who observed no significant differences in terms of the overall performance between conventional green LEDs grown on sapphire and green LEDs on high-quality laterally-overgrown GaN templates. This finding reinforces that the performance of green LEDs is largely determined by the quality of the InGaN/GaN MQWs. The microstructural properties of the InGaN QWs grown on sapphire and bulk GaN are expected to be similar except for the difference in threading dislocation density. In both cases, the QWs are highly strained and probably very defective, largely due to the lattice mismatch between the InGaN QWs and GaN QBs.

4.3   Piezoelectric Polarization in c-Plane Green LEDs

III-Nitrides are polar crystals which typically crystallize in the wurtzite structure. Their noncentrosymmetry leads to strong spontaneous polarization [47]. In nitride heterostructures grown on the c-plane of sapphire, strain arising from lattice mismatches between AlN, GaN, and InN will cause additional piezoelectric polarization [11, 47]. Discontinuities in the net polarization at the heterointerfaces give rise to fixed space charges and thus large internal electric fields perpendicular to the growth plane. In an InGaN/GaN QW, the polarization-induced electric field bends the energy bands and leads to a spatial separation of the electron and hole wave functions [47], as illustrated in Fig. 4.10a. This property will result in an effective bandgap narrowing and a Stark shift in the excitonic absorption, an effect referred to as QCSE [48] The reduction in electron-hole overlap leads to a decrease in the transition probability for both absorption and recombination, and thus an increased probability of nonradiative recombination and less efficient light emission from the QW [9]. The QCSE increases as the mole fraction of indium in the InGaN QW increases. Therefore, green LEDs in particular are strongly impacted by the piezoelectric polarization effect.

The difference in spontaneous polarization between GaN and InN is small, whereas their lattice mismatch is quite large. Therefore, the spontaneous polarization contribution to the total polarization is negligible compared to the piezoelectric polarization in InGaN QWs [49]. In typical green LEDs, InGaN/GaN MQWs are grown on a thick layer of GaN. The QW layers are under compressive biaxial strain, which results in a polarization pointing toward [0001] direction, and an electric field pointing in the opposite direction. Since most LEDs have a p-GaN-up structure, the piezoelectric field in the InGaN QWs is opposite to the built-in electric field in the p–n junction. The influence of piezoelectric polarization on the band structure of a

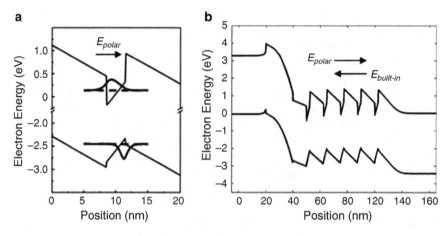

Fig. 4.10 Energy band diagrams of (**a**) an InGaN/GaN QW and (**b**) a green InGaN/GaN MQW LED structure grown on *c*-plane sapphire. The polarization-induced electric field in the QWs separates the electron and hole wave functions and bends the energy bands

c-plane green LED can be seen from Fig. 4.10b. When a reverse bias is applied, the electric field tends to compensate the piezoelectric field. So the band structure of the QWs bent by the piezoelectric field approaches the flat band condition. A blueshift of the PL peak with increasing reverse bias is thus expected. Masui et al. [50] performed PL measurements of green and amber InGaN/GaN MQW LEDs under electrical biases. Figure 4.11 shows the peak energy and PL intensity as a function of the applied voltage along with the $I-V$ characteristics. As expected, with increasing reverse bias, the emission peak constantly shifts toward the higher energy in both samples. It is noticed that the peak blueshift is accompanied by a reduction in PL intensity as the reverse bias is above a certain value. This is because photo-generated carriers may escape from the QWs via field emission as the energy barrier becomes sufficiently thin, and contribute to the reverse leakage current. In the amber LED under small biases, the PL intensity decreases significantly whereas its peak shifts toward lower energies as the voltage becomes less negative toward positive. This trend continues until the forward current starts flowing above the turn on voltage. These changes signal that the internal electric field is strengthened in the QWs, so the overlap between the electron and hole wavefunctions is further reduced. The total peak energy shifts between -10 to $+2$ V are 126 meV and 77 meV for the amber and green LEDs, respectively, suggesting that the amber LED experiences a greater QCSE than the green LED due to a higher indium content in the active region. The strength of the piezoelectric fields in green-emitting InGaN QWs could be as high as several MV/cm [51].

Under current injection at forward bias, the light intensity increases as EL turns on, and the peak shifts toward the higher energy with increasing current. Similar blueshifts have commonly been observed in the EL of blue and green InGaN/GaN MQW LEDs built on the *c*-plane [52, 53]. The shift may be attributed to the

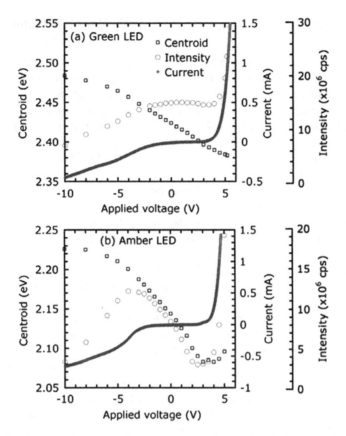

Fig. 4.11 Luminescence intensity and peak energy (shown as Centroid) of (**a**) green (510 nm) and (**b**) amber (585 nm) InGaN/GaN MQW LEDs grown on c-plane sapphire under different applied biases. The $I-V$ characteristics of the LEDs are also shown (From Masui et al. [50])

screening of the piezoelectric field by free carriers injected into the QWs. The band bending and bandgap narrowing are therefore mitigated. Note that band-filling at elevated currents is another mechanism which may cause EL blueshift in green LEDs [11], as detailed in Sect. 4.5. While these two factors cannot be easily deconvoluted under forward injection, the blueshifts under reverse bias seen in Fig. 4.11 are exclusively caused by the piezoelectric effect.

As discussed in Sect. 4.2.1, different strategies, including strain engineering and intentional doping, have been developed to reduce polarization and the resulting QCSE in blue and green LEDs. Thin InGaN QWs (2–3 nm) are generally employed to enhance carrier confinement, and thus the spatial overlap between electrons and holes. While the disadvantage of the QCSE may partially be overcome, thin QWs make carrier injection into the active region inefficient. To eliminate the polarization effects, LED structures must be grown on nonpolar substrates [47]. This is expected to be particularly beneficial in achieving high-efficiency green light emission.

4.4 Green LEDs on Nonpolar and Semipolar Substrates

The adverse effects of polarization are evitable as long as strained InGaN QWs are grown on a polar orientation. This motivates the growth of InGaN-based LEDs on nonpolar and semipolar substrates [47]. It is a convention to use hexagonal coordinates in nitride crystals and define the z axis to be parallel to the polar c axis [0001]. Planes which are parallel to the c axis are electrically nonpolar, and are called nonpolar planes. $\{11\bar{2}0\}$ and $\{10\bar{1}0\}$ are two important low-index nonpolar planes for growth of InGaN LEDs, called the a plane and m plane, respectively. The inclined planes in the structure, including $\{10\bar{1}1\}$, $\{11\bar{2}2\}$, and $\{10\bar{1}3\}$ are called semipolar planes. In the GaN crystal, they make angles of 28°, 32°, and 58° with respect to the c axis, respectively.

LEDs grown in nonpolar orientations have polarization vectors lying in the growth plane and thus parallel to heterointerfaces in device structures. Therefore, in nonpolar LEDs, there is no strain-induced electric field existing in the active region along the current-flowing direction. Given that the piezoelectric effect is eliminated, thicker InGaN QWs can be used. A larger active region volume with flat conduction and valence bands would reduce losses arising from current overflow and Auger recombination, leading to more efficient LEDs. Since green LEDs suffer more from the QCSE than blue LEDs, nonpolar growth provides a critical route to improve the efficiency of green LEDs and close the "green gap". It is expected that semipolar planes may also effectively reduce piezoelectric polarization. Calculations have suggested that the piezoelectric field may be reduced and even eliminated in semiploar planes with a titling angle of $\sim 45°$ with respect to the c-plane [54]. $\{10\bar{1}1\}$ and $\{11\bar{2}2\}$ are among such types of semipolar planes, and have been used as the growth planes for InGaN-based LEDs.

Early work on developing nonpolar and semipolar LEDs focused on the epitaxial growth of thick nonpolar and semipolar GaN templates on foreign substrates [47, 55–63]. Planar growth of nonpolar GaN was first achieved on (100) $\gamma - LiAlO_2$ by molecular beam epitaxy in 2000 [56] Since then, growth of smooth GaN templates along different nonpolar and semipolar orientations has been demonstrated. r-Plane sapphire and a-plane SiC were used for a-oriented GaN growth [57, 58], whereas m-plane SiC and (100) $\gamma - LiAlO_2$ were found to be suitable substrates for m-oriented GaN [59, 60]. In the meanwhile, several groups developed semipolar GaN templates on spinel, sapphire, and Si [61–63]. However, InGaN LEDs developed on those templates had very poor performance, largely due to the presence of high densities of threading dislocations and basal plane stacking faults in the templates [47]. There have been a few successful demonstrations of green LEDs on nonpolar and semipolar GaN templates [64–66]. At 20 mA, the optical output powers were in the microwatt range, and the maximum EQE was $\sim 0.1\%$.

Recently, the performance of nonpolar and semipolar LEDs was greatly improved using high-quality free-standing bulk-GaN substrates [67–73]. The substrates were obtained by slicing centimeter-thick c-axis GaN boules grown by HVPE along a-, m-, or semipolar planes. Despite small sizes, the free-standing

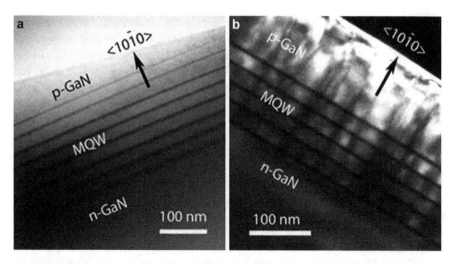

Fig. 4.12 Cross-sectional TEM micrographs of (**a**) *blue* (483 nm) and (**b**) *green* (516 nm) InGaN/GaN MQW LEDs grown on *m*-plane GaN substrates (From Detchprohm et al. [68])

substrates were large enough to conduct early research to investigate the potential of nonpolar and semipolar LEDs. Detchprohm et al. [67] have reported their studies on green LEDs on *m*- and *a*-plane GaN substrates. They compared 520–540 nm InGaN/GaN MQW LEDs grown on free-standing *a*-plane GaN and 6 μm-thick *a*-plane GaN templates deposited on *r*-plane sapphire. The former showed three times higher light output power. The results correlated well with better material quality of the LEDs on free-standing GaN, as revealed by TEM and high-resolution XRD. The same group also compared the performance of 516 nm green LEDs and 483 nm blue LEDs grown on free-standing *m*-plane GaN [68]. The green LED was found to be much less efficient. Cross-sectional TEM analysis, as seen in Fig. 4.12, shows that both device structures are free from any additional threading dislocations besides those protruding from the bulk substrate (nominally $< 5 \times 10^6 \, \text{cm}^{-2}$). In the green LED, a large number of misfit dislocations (10^9–$10^{10} \, \text{cm}^{-2}$) are created in the InGaN QWs and propagate along the growth direction. This is in contrast to the blue LED, which is essentially free of misfit dislocations. Apparently, as with *c*-plane green LEDs, large strain at the InGaN/GaN interfaces induces a high density of misfit dislocations, which act as nonradiative recombination centers and limit the efficiency of nonpolar green LEDs.

Two research groups have developed much more efficient green LEDs on the semipolar (11$\bar{2}$2) plane of free-standing GaN [69–72]. Sato et al. [69] reported green semipolar LEDs emitting at 516 nm with EQE of 10.5% and 6.3% at 20 mA and 200 mA, respectively. The same group has extended the wavelength to the yellow spectral region, and demonstrated 563 nm yellow LEDs with a similar EQE [71]. The InGaN yellow LEDs appeared to be more stable against changes in ambient temperature than AlInGaP yellow LEDs. These results suggest that the

($11\bar{2}2$) plane seems to facilitate high indium incorporation and has advantages over other nonpolar/semipolar planes for green LED growth. Despite these recent improvements, the efficiencies of the reported nonpolar and semipolar green LEDs are still lower than those of the state-of-the-art polar green LEDs. It is expected that their performance will be further improved through optimization of the LED structure and growth conditions.

Nonpolar green LEDs on a- and m-planes exhibited small or negligible peak blueshifts as the injection current was increased, indicating the suppression of the QCSE [66–68]. However, significant current-induced blueshifts have been observed in semipolar green LEDs [64, 69–71], which were often attributed to band-filling of localized states. Detchprohm et al. [67] have found that the indium mole fractions in nonpolar InGaN QWs must be significantly higher than those in c-plane InGaN QWs in order to achieve similar green emission. This is another indication of elimination of the QCSE in nonpolar QWs, since the QCSE would cause a large Stokes-like redshift of the emission wavelength. The absence of the QCSE in nonpolar LEDs was also confirmed by electrical-bias-applied PL measurements, which showed that the PL peak positions of m-plane InGaN/GaN MQWs were nearly independent of the external bias [74].

It has been found that nonpolar and semipolar LEDs exhibit anisotropic optical gain, leading to polarized light emission [64, 66, 72, 73]. As the InGaN active layer is under compressive strain, the electric field along the c-axis causes valence band splitting. The transition between the conduction band and $|X\rangle$ character valence band is predominant due to a higher hole occupation probability compared to the $|Z\rangle$ character band [73]. The EL thus linearly polarizes perpendicular to the c-axis. The polarization ratio is defined as:

$$\rho = \frac{I_\perp - I_{//}}{I_\perp + I_{//}}, \tag{4.1}$$

where $I_\perp$ and $I_{//}$ are the intensities of light with polarization perpendicular and parallel to the c-axis, respectively. As an example, Fig. 4.13 shows the variation of the EL intensity of an a-plane green LED (506 nm) measured by rotating a polarizer from $0°$ to $360°$. The polarization ratio was determined to be 67.4% [66]. By suppressing the influence of light scattering, Brinkley et al. [73] were able to measure ρ as high as 96% for a nonpolar green LED (513 nm) grown on m-plane GaN. Semipolar green LEDs on $\{11\bar{2}2\}$ and $\{10\bar{1}3\}$ orientations showed partially polarized emission with ρ in the range of 0.2–0.6 [64, 72]. On both nonpolar and semipolar planes, green LEDs have showed higher degrees of polarization than blue and near-UV LEDs [66, 72, 73]. This has been explained by a larger strain-induced valence band splitting in green LEDs as a result of higher indium incorporation. Employment of polarized LEDs as the backlighting sources in liquid crystal displays (LCDs) will promise further energy savings. A preliminary demonstration of light intensity modulations from an LCD unit combined with nonpolar InGaN-based LEDs has been performed by Masui et al. [75].

Fig. 4.13 Variation of the EL intensity of an *a*-plane green LED at 20 mA with the angular orientation of a polarizer (From Ling et al. [66])

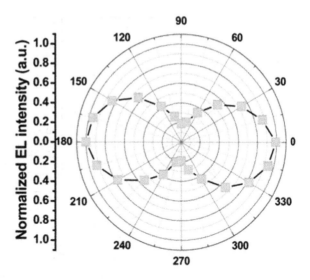

4.5   Carrier Localization in Green LEDs

In contrast to light-emitting devices based on conventional III-V semiconductors, InGaN-based LEDs on c-plane sapphire are surprisingly efficient despite the presence of a high density of threading dislocations. It is generally accepted that carrier localization effects arising from spatially inhomogeneous indium distribution play an important role in spontaneous emission from InGaN QWs [9]. The compositional fluctuations occur due to large contrasts in the thermodynamic and structural properties of GaN and InN. As the indium content in InGaN increases, the fluctuations increase, leading to a strong tendency toward phase separation [23, 76]. Nanoscale quantum dot (QD)-like structures in InGaN alloys have been observed directly by high-resolution TEM [77, 78], and the mean size of the QDs was found to increase with increasing indium content [79]. Indium-rich regions have reduced bandgaps and thus create potential minima, which localize carriers and excitons. Carrier transport among localized states and exciton relaxation into lower-lying states give rise to characteristic photoluminescence (PL) peak shifts and stretched exponential decay behaviors [80]. Submicron emission inhomogeneities have been revealed in cathodeluminescence and PL images of green-emitting InGaN QWs [81, 82]. The localization effects were also observed in semipolar and nonpolar InGaN QWs [83–85]. On some semipolar planes, adatoms experience enhanced surface migration than on the c-plane, leading to more homogeneous growth of InGaN and thus weaker carrier localization [84, 85].

The localization effects improve the radiative efficiency of InGaN-based LEDs in two ways. First, QD-like indium-rich regions in the InGaN QWs trap carriers, forming localized excitons. This further enhances the overlap of electrons and holes, and their recombination rates. Second, localization effects prevent carriers from diffusing to defects, and thus reduce nonradiative recombination. This explains that, despite poor material quality of high-indium InGaN alloys, green LEDs

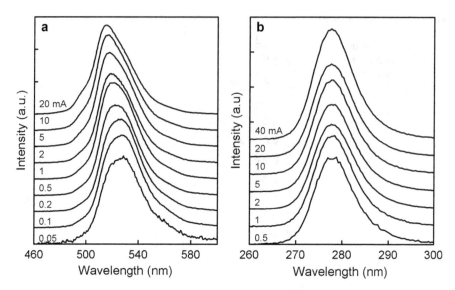

Fig. 4.14 EL spectra of (**a**) an InGaN-based green LED and (**b**) an AlGaN-based UV LED, both on (0001) sapphire, as a function of injection current. The peak intensities are normalized to unity and the spectra are shifted in the vertical direction for clarity

are relatively efficient. However, the localized state band can be filled up under high injection conditions. Delocalized carriers may be readily captured by defects, leading to degraded device performance.

One characteristic feature of InGaN-based LEDs is a pronounced blueshift of their EL peaks with increasing drive current. This may be attributed to the band-filling of localized states at potential minima in the InGaN QWs. Figure 4.14 compares the EL spectra of an InGaN/GaN MQW green LED and an AlGaN/AlGaN MQW UV LED, both on *c*-plane sapphire, at 300 K as a function of injection current. The spectra of the green LED are much broader due to strong alloy broadening, and show a large peak blueshift of 54 meV as the current is increased from 0.1 to 20 mA. In contrast, the emission peak of the UV LED has a small FWHM of ~ 10 nm and is almost independent of injection current. These results suggest much smaller compositional fluctuations and weaker localization effects in the AlGaN QWs [86, 87]. Carriers are, therefore, more likely to be trapped and recombine nonradiatively at defect states. This may be the main reason why the performance of the state-of-the-art AlGaN-based UV LEDs lags significantly behind that of InGaN-based LEDs. As discussed in Sect. 4.4.3, screening of polarization-induced electric field in polar LEDs by injected carriers may also cause EL blueshifts. In green LEDs, strong carrier localization in nanometer scale In-rich clusters considerably limits the effect of the piezoelectric polarization. Furthermore, polarization-induced charges at the heterointerfaces are partly screened by Si-doping in the QBs [30]. We therefore attribute the EL blueshift of the green LED seen in Fig. 4.14a largely to the band-filling of localized states.

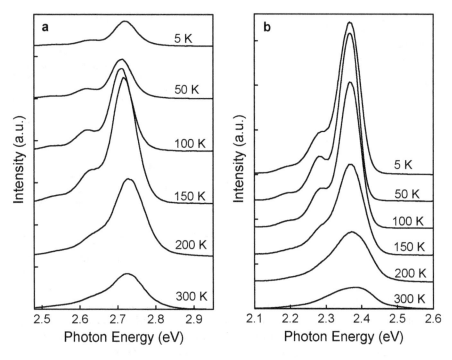

Fig. 4.15 EL spectra of (**a**) *blue* and (**b**) *green* InGaN/GaN MQW LEDs grown on (0001) sapphire substrates measured at 1 mA and different temperatures. The spectra are shifted in the vertical direction for clarity

The role of localized states in the radiative recombination process in InGaN-based LEDs can be better understood by means of temperature-dependent luminescence measurement [88]. We compared low-temperature (5–300 K) characteristics of green, blue, and UV LEDs with similar device structures grown on (0001) sapphire, but varying indium contents in the InGaN QWs. The respective nominal indium mole fractions in the QWs were 0.35, 0.20, and 0.04. Figure 4.15 illustrates the evolution of the EL spectra of the green and blue LEDs at 1 mA with decreasing temperature. For both LEDs, a series of LO-phonon replicas separated from each other by ∼ 90 meV appear at temperatures below 200 K. The LO-phonon energy appears to be weakly dependent on the indium concentration, and is reasonably close to that in GaN crystals (91 meV). The LO side-bands significantly broaden the low energy side of the main emission band, indicating carrier relaxation toward less energetic states via phonon emission. For the green LED, a monotonic increase in the emission intensity and a small shift of the peak energy with changing temperature are observed. In contrast, with decreasing temperature, the blue LED displays unique behaviors of increase and then decrease of the EL intensity, blueshift, and then redshift of the peak energy.

Fig. 4.16 EL peak energy shifts in InGaN/GaN MQW green, blue, and UV LEDs on (0001) sapphire substrates as a function of measurement temperature

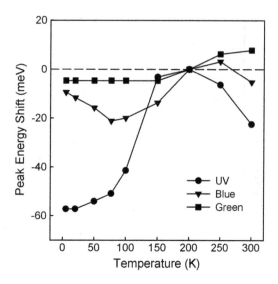

Figure 4.16 shows temperature-induced shifts of the EL peaks of the LEDs measured at 1 mA. With decreasing temperature, a blueshift and then a redshift are seen for all LEDs, although the magnitudes of the shifts are different. The redshift in the UV LED between 77 and 200 K is as large as 50 meV. Considering that the temperature-induced change of the GaN bandgap energy is ~26 meV, the actual displacement of the peak energy with respect to the band edge in this LED is −76 meV. The unusual redshift may be explained by carrier relaxation as follows. As the temperature is decreased, the carrier lifetime increases due to reduced nonradiative recombination rates, allowing more opportunity for carriers to relax into lower lying localized states. The total redshift, which is indicative of the degree of carrier relaxation, decreases as the indium content in the active region increases. In the green LED, the EL peak red shifts by only ~10 meV in the temperature range of 150–250 K and stabilizes at 2.37 eV at lower temperatures. This behavior can be interpreted as evidence of stronger localization effects in this LED, suggesting that a large number of localized states in the QWs are responsible for efficient carrier capture and recombination over the entire temperature range. In contrast, the UV LED demonstrates different emission mechanisms within different temperature ranges. While localized-state recombination appears to be important below 150 K when carriers are transferred to lower energy states, band-to-band transition becomes dominant at higher temperatures as the localized carriers are thermalized. Indeed, with increasing temperature above 200 K, the EL peak of the UV LED exhibits a redshift, which follows the characteristic temperature dependence of the GaN bandgap shrinkage [89]. The thermal energy at 150 K~13 meV may be estimated as the magnitude of the localization energy in the UV LED.

Figure 4.17 shows the *L–I* characteristics of the LEDs measured at 5 K. The light intensity is linear with current in the low injection regime for all the LEDs, indicating the dominant radiative recombination and constant quantum efficiency.

Fig. 4.17 *L–I*
characteristics of InGaN/GaN
MQW *green, blue,* and UV
LEDs on (0001) sapphire
substrates at 5 K. The arrows
indicate the onset of
transitions from linear to
sublinear current dependence

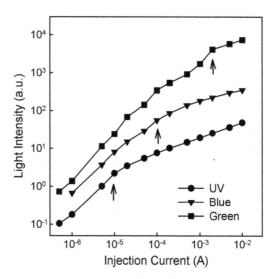

As current increases, the dependence of light intensity on forward current becomes
sublinear, following $L \sim I^m (m \sim 0.4)$. This is evidence that carrier capture becomes
inefficient at low temperatures. The point marking the transition from the linear to
sublinear dependence moves to higher injection currents in LEDs emitting at longer
wavelengths. In the UV LED, there are only a limited number of localized states.
Therefore, only a small portion of the injected carriers can transfer to localized
states and undergo radiative recombination. The density of localized states in the
green LED is much higher, allowing more carriers to be captured. The *L–I* data can
be fitted to the solution of a rate equation, which describes the carrier capture and
decapture processes in InGaN QWs at steady state:

$$\frac{dN}{dt} = (N_0 - N)j_c\sigma_{cap} - N\left(\frac{1}{\tau_{decap}} + \frac{1}{\tau_{rad}} + \frac{1}{\tau_{nonrad}}\right) = 0, \qquad (4.2)$$

where N_0 is the total number of localized states, N is the number of captured
carriers, j_c is the carrier injection flux, σ_{cap} is the effective cross-section for carrier
capture, and τ_{decap}, τ_{rad}, and τ_{nonrad} represent the lifetimes associated with carrier
decapture, radiative, and nonradiative recombination, respectively. Considering that
the nonradiative lifetime of localized carriers τ_{nonrad}, and the carrier decapture time
τ_{decap} at low temperatures (< 150 K) are quite long, we may obtain the EL intensity
L which is proportional to N/τ_{rad}:

$$L = \frac{N_0}{\frac{1}{j_c\sigma_{cap}} + \tau_{rad}} \qquad (4.3)$$

From the above equation, L is given by $L = N_0 j_c \sigma_{cap}$ in the limit of small j_c.
On the other hand, L saturates to the value of N_0/τ_{rad} at high currents. As seen from

Fig. 4.17, the output power of the green LED at currents > 10 mA is ∼200 times higher than that of the UV LED. Given that the radiative recombination lifetime $\tau_{rad}$is longer in the green LED due to the larger size of QD-like structures [90], we may deduce that the localized state density in the green LED is more than two orders of magnitude higher.

All these observations point to the presence of localized states in indium-rich nanoclusters. In addition, nano-scale variations in the thickness of InGaN QWs may also contribute to the localization effects [91, 92]. Nevertheless, it is worth noting that, some recent studies cast doubt on the hypothesis that the fluctuations in indium composition and QW width are the microstructural origins of carrier localization. It has been suggested that indium-rich regions observed by TEM may be an imaging artifact caused by electron beam damage rather than an as-grown feature of InGaN alloys [92–94]. Instead, different larger-scale structures on the order of tens of nanometers, like V-defects, have been suggested to cause the formation of thinner QWs around dislocations, providing a potential barrier to carrier diffusion to the dislocations [95, 96]. However, since the dimensions of these structures are significantly larger than typical localization length scales in InGaN [97], they likely play a minor role in carrier localization processes.

4.6   Efficiency Droop in Green LEDs

The efficiency of InGaN-based blue and green LEDs reaches its peak at a very low current density and decreases monotonically with increasing injection current. This phenomenon, known as efficiency droop [6, 7], is even seen in LEDs driven with short- and low-duty cycle pulses, when thermal effects due to self-heating are suppressed [98]. The droop problem is considered to be a critical restriction for using LEDs under high power conditions, which are desired for SSL applications. In an effort to investigate the underlying nonthermal mechanism, we studied and compared the optical characteristics of four commercially available LEDs, which had similar layer structures but different indium contents in the active regions. All the LEDs were grown on (0001) sapphire substrates by MOCVD. Three LEDs had an InGaN/GaN MQW active region with nominal indium mole fractions of 0.28, 0.17, and 0.09. The fourth LED had an In-free AlGaN-based MQW sandwiched between n-type and p-type AlGaN cladding layers. The room temperature peak energies of the LEDs at 20 mA were 2.39 eV (green), 2.63 eV (blue), 3.03 eV (near-UV), and 3.65 eV (UV). During EL measurements, self-heating was reduced by pulsing the LEDs with a small duty cycle of 1% at 1 kHz.

Figure 4.18 compares the normalized EQE-current ($\eta - I$) characteristics of the LEDs. The EQE of all the InGaN LEDs rises rapidly with current and rolls off as current exceeds a characteristic current. There are two important features in the data, evidencing the different degrees of nonthermal effects in these LEDs: (1) The characteristic current which marks the onset of the efficiency droop decreases with increasing indium content in the active region. It ranges from 1 mA for the green

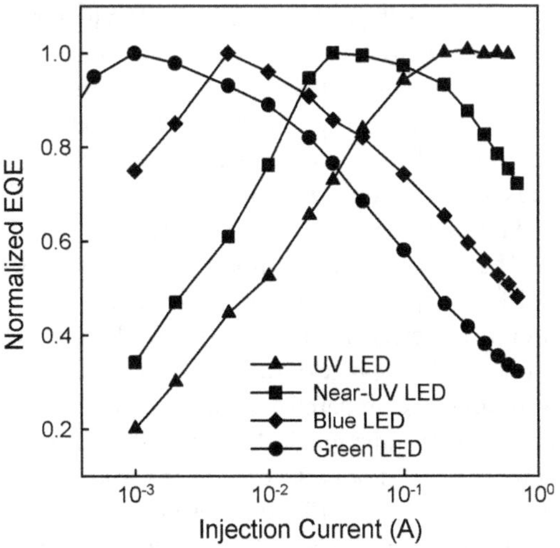

Fig. 4.18 Normalized EQE
of green, blue, near-UV, and
UV LEDs on (0001) sapphire
substrates as a function of
pulsed injection current

Fig. 4.19 Peak energy shifts
of the EL spectra of green,
blue, near-UV, and UV LEDs
on (0001) sapphire substrates
with respect to the peak
energies at 1 mA as a function
of pulsed injection current

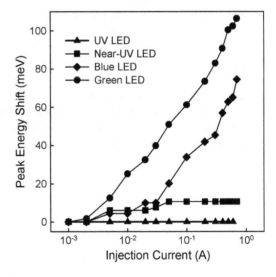

LED to 5 mA for the blue LED to 30 mA for the near-UV LED; (2) The efficiency droop is increasingly pronounced with increasing indium content in the active region. The EQE of the green LED shows the steepest drop, decreasing by 50% at 150 mA. In contrast, the EQE of the UV LED initially increases with increasing current and reaches the maximum at ∼ 200 mA. It remains nearly constant as current is further increased, implying a linear light output-current characteristic.

The current-dependent peak shifts of all the LEDs are summarized in Fig. 4.19. It is clear that the total peak shift is also a strong function of the indium content in the active region. Both the blue and green LEDs show a monotonic blue shift with increasing injection level over the entire current range, of a total amount of

∼ 75 and ∼ 106 meV, respectively. The total peak shift of the near-UV LED is only 11 meV, whereas the peak energy of the UV LED is constant. The large current-induced blueshift in the green LED, accompanying a significant spectral broadening, occurs due to the combined effects of band filling of localized states and partial screening of the piezoelectric field in the InGaN QWs. The absence of noticeable peak shift and spectral broadening in the UV LED is indicative of weak localization and piezoelectricity effects in the AlGaN QWs [86, 87].

The threading dislocation densities in all the LEDs are on the same order of ∼ 10^9 cm^{-2}, suggesting that dislocations may play a minor role in the efficiency droop. To verify this hypothesis, similar near-UV, blue and green LEDs were also grown on the c-plane of free-standing bulk GaN substrates with a dislocation density ∼ 10^7 cm^{-2}. Compared to their counterparts on sapphire, the LEDs on GaN showed much lower junction leakage and higher peak efficiencies [18, 41]. Figure 4.20 compares the normalized η–I characteristics of the near-UV and green LEDs. As seen, while the LEDs on GaN are of better quality, they display similar efficiency droop behaviors. This finding confirms that dislocations are not responsible for the decrease in EQE at elevated currents [8].

Different mechanisms have been suggested to explain the efficiency droop. Among them are (1) carrier delocalization from localized states followed by nonradiative recombination at defects [8, 99]; (2) electron leakage from InGaN QWs due to band structure modification by polarization [100, 101], or induced by impeded hole injection [7]; and (3) Auger nonradiative recombination which plays an important role at high carrier densities [102, 103]. However, to date, there is still a lack of consensus about the exact physical origin. There seems to be supporting evidence for all these models, which cannot be easily unified due to the large variety of sample materials studied by different groups.

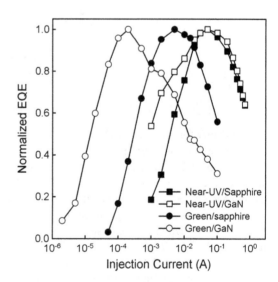

Fig. 4.20 Normalized EQE of near-UV and green LEDs on c-plane sapphire and bulk GaN substrates as a function of pulsed injection current

The wavelength-dependent efficiency droop seen in Fig. 4.18 may be interpreted as an evidence of carrier delocalization from indium-rich regions [8]. As current injection is increased, carriers are released to the conduction band due to filling-up of localized states. They can therefore diffuse to the defects nearby and recombine nonradiatively, resulting in a decreased EQE. This interpretation is supported by the fact that the efficiency droop in the green LED is accompanied by a monotonic peak blueshift and spectrum broadening up to 700 mA, both indicative of band-filling of localized states. The fact that the droop in green LEDs is more pronounced may be attributed to two factors: (1) With increasing indium content, the localization effects become stronger due to larger compositional fluctuations, so do the delocalization effects at high currents; (2) More misfit defects may form in the green-emitting QWs due to larger strain and more severe phase separation, acting as efficient nonradiative recombination centers. On the other hand, the radiative recombination in the UV LED is dominated by band-to-band transition due to the lack of localization effects [86]. Therefore, defects have a significant influence on its performance even at low injection levels. As defects are saturated at high currents, the EQE attains a saturation value.

Since a variety of material properties and effects in InGaN QWs may evolve with the indium content, the above results can also be explained by other aforementioned models. For instance, as more indium is incorporated into InGaN QWs, the piezoelectric polarization and QCSE also become stronger due to larger strain in the QWs. In the meanwhile, the Auger recombination rate may increase as the Auger recombination coefficient increases exponentially with decreasing bandgap and carriers are increasingly localized.

Several methods have been suggested to mitigate efficiency droop in c-plane LEDs, but experimental validation, mainly conducted with blue LEDs, only resulted in limited improvements [104–106]. If the Auger effect dominates in the efficiency droop, using wide InGaN QWs to reduce carrier densities may be effective in shifting the peak EQE to higher injection currents [105]. Currently, this method may not be feasible for green LEDs due to the difficulty of growing thick high-In content InGaN active layers. If droop is caused by polarization-enhanced carrier overflow, nonpolar growth should be a solution to the problem. However, contradictory results have been reported for LEDs grown on m-plane GaN substrates. Ling et al. [107] reported that m-plane blue-green (480 nm) LEDs exhibited much reduced efficiency droop compared to similar LEDs grown on c-plane sapphire. The experimental EQE-current curves, as illustrated in Fig. 4.21, are consistent with their simulations, which took into account of polarization-enhanced electron leakage and Auger recombination. In a later report by Lin et al. [108], significant efficiency droop was observed for blue-green LEDs with emission wavelengths of 490–506 nm. Since green LEDs suffer most from the droop problem, there is a compelling need to determine the underlying mechanism and develop an effective strategy for droop mitigation in green LEDs. This will be an essential part of the research effort to close the "green gap" in the LED technology.

Fig. 4.21 Experimental and
simulated EQE of *c*-plane
and *m*-plane blue–green
LEDs as a function of
injection current density
(From Ling et al. [107])

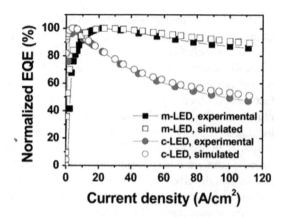

4.7 Conclusions

The EQE of the state-of-the-art green LEDs is about half of that of blue LEDs, and
lags further behind that of AlInGaP-based red LEDs. The "green gap" is a major
unsolved obstacle in current efforts to create high-efficiency and high-brightness
SSL solutions. The gap gets deeper at high injection currents, as green LEDs
exhibit more pronounced efficiency droop than other visible LEDs. These well-
recognized problems are believed to be associated with fundamental properties of
InGaN alloys and their heterostructures grown on polar planes of foreign substrates.
Green-emitting InGaN/GaN MQWs grown on (0001) sapphire are highly defective
and strained due to large differences in the structural and thermodynamic properties
of GaN and InN. The incorporation of high fractions of indium leads to phase
separation and introduces misfit defects. Biaxial strain induces strong piezoelectric
polarization and thus QCSE, which reduce the efficiencies of carrier injection and
radiative recombination. All these adverse effects evolve with increasing indium
incorporation, and thus present more significant challenges as emission moves
toward the green spectrum. Growth of green LEDs on high-quality (0001)GaN
templates and bulk substrates did not lead to more efficient devices, stressing
that poor quality of MQWs is the bottleneck for developing high-performance
green LEDs. Recent efforts have focused on strain engineering of device structures
and optimization of epitaxial conditions on sapphire, but only resulted in limited
performance improvements.

The control of polarization and QCSE appears to be a critical technical issue
in achieving high EQE of green LEDs. Nonpolar and semipolar growth provides
an important means to eliminate the QCSE in InGaN QWs. A recent breakthrough
has led to the demonstration of high-brightness green LEDs on semipolar ($11\bar{2}2$)
bulk GaN. However, the best numbers reported to date in terms of performance
are still lower than those of conventional *c*-plane green LEDs. This may be partly
attributed to the immature technologies due to the smaller and newer community
of nonpolar and semipolar LED research. While homoepitaxial growth of nonpolar

LEDs explores the ultimate performance of LEDs, the effort is restricted by the lack of low-cost and large-size GaN wafers. Nonpolar template growth may thus be an area of focus of further research. The well-established growth technologies and cost-effectiveness for LEDs on sapphire and SiC justify the enormous continuous efforts to improve the performance of current-generation green LEDs on these polar substrates in the future. The aforementioned obstacles must be overcome via effective strain control and further optimization of InGaN QW growth so as to close the "green gap" and enable high-efficiency high-quality SSL solutions.

Acknowledgements The author would like to thank his colleagues and collaborators: Dr. S. F. LeBoeuf, Dr. L. B. Rowland, Dr. W. Wang, Dr. C. H. Yan, and Dr. Y. Yang.

References

1. JY Tsao(ed) Light emitting diodes for general illumination, an OIDA Technology Roadmap. (Optoelectronics Industry Development Association Washington, DC, 2002)
2. Solid-State Lighting Research and Development Multi-Year Program Plan (2009) U.S. Department of Energy. http://apps1.eere.energy.gov/buildings/publications/pdfs/ssl/ssl_mypp2009_web.pdf
3. M.S. Shur, A. Zukauskas, Solid-State Lighting: Toward Superior Illumination, Proc. IEEE. **93**(10), 1691–1703 (2005)
4. S. Nakamura, S.J. Pearton, G. Fasol, *The Blue Laser Diode* (Springer, Heildelberg, Germany, 2000)
5. C. Wetzel, T. Detchprohm, MRS Internet J. Nitride Semicond. Res. **10**, 1–13 (2005)
6. J. Piprek, Phys. Stat. Sol. (a). **207**(10), 2217–2225 (2010)
7. J.H. Leach, X. Ni, J. Lee, U. Ozgur, A. Matulionis, H. Morkoc, Phys Stat, Sol (a). **207**(5), 1091–1100 (2010)
8. X.A. Cao, Y. Yang, H. Guo, J Appl Phys **104**, 093108 (2008)
9. S.F. Chichibu, Y. Kawakami, T. Sota, *Emission Mechanisms and Excitons in GaN and InGaN Bulk and QWs.* In: Nakamura S, Chichibu SF (ed), Introduction to Nitride Semi-conductor Blue Lasers and Light Emitting Diodes. (Taylor and Francis, New York, 2000)
10. X.A. Cao, *III-nitride Light-emitting diodes on novel substrates.* In: Neumark GF, Kuskovsky I, Jiang H (ed.) Wide Bandgap Light Emitting Materials and Devices, (Wiley-VCH, 2007)
11. J.H. Ryou P.D. Yoder, J. Liu, Z. Lochner, H. Kim, S. Choi, H.J. Kim, R.D. Dupuis, IEEE J Sel Top Quantum Electron **15**(4), 1080–1091 (2009)
12. J.J. Wierer, D.A. Steigerwald, M.R. Krames, J.J. O'Shea, M.J. Ludowise, G. Christenson, Y.C. Shen, C. Lowery, P.S. Martin, S. Subramanya, W. Gotz, N.F. Gardner, R.S. Kern, S.A. Stockman, Appl Phys Lett **78**, 3379–3381 (2001)
13. S.P. DenBaars, S. Keller, MOCVD of Group III-Nitrides In: Gallium Nitride (II) Pankove JI, Moustakas TD (ed), (Academic, San Diego, 1998)
14. H. Amano, N. Sawaki, I. Akasaki, Y. Toyoda, Appl Phys Lett **48**, 353–355 (1986)
15. S. Nakamura, Jpn J Appl Phys **30**, L1705–1707 (1991)
16. X.H. Wu, C.R. Elsass, A. Abare, M. Mack, S. Keller, P.M. Petroff, S.P. DenBaars, J.S. Speck, S.J. Rosner, Appl Phys Lett **72**, 92–94 (1998)
17. D.I. Florescu, S.M. Ting, J.C. Ramer, D.S. Lee, V.N. Merai, A. Parkeh, D. Lu, E.A. Armour, L. Chernyak, Appl Phys Lett **83**, 33–35 (2003)
18. Y. Yang, X.A. Cao, C.H. Yan, Appl Phys Lett **94**, 041117 (2009)
19. M. Razeghi, C. Bayram, R. McClintock, F.H. Teherani, D.J. Rogers, V.E. Sandana, J Light Emitting Diodes **2**(1), 1–32 (2010)

20. M. Leszczynski, T. Suski, H. Teisseyre, P. Perlin, I. Grzegory, J. Jun, S. Porowski, T.D. Moustakas, J Appl Phys **76**, 4909–4911 (1994)
21. L.T. Romano, C.G. Van de Walle, I.I.I.J.W. Ager, W. Götz, R.S. Kern, J Appl Phys **87**, 7745–7752 (2000)
22. M.S. Oh, M.K. Kwon, I.K. Park, S.H. Baek, S.J. Park, S.H. Lee, J.J. Jung, J Cryst Growth **289**, 107–112 (2006)
23. N.A. El-Masry, E.L. Piner, S.X. Liu, S.M. Bedair, Appl Phys Lett **72**, 40–42 (1998)
24. C.F. Huang, T.C. Liu, Y.C. Lu, W.Y. Shiao, Y.S. Chen, J.K. Wang, C.F. Lu, C.C. Yang, J Appl Phys **104**, 123106 (2008)
25. X.H. Wang, L.W. Guo, H.Q. Jia, Z.G. Xing, Y. Wang, X.J. Pei, J.M. Zhou, H. Chen, Appl Phys Lett **94**, 111913 (2009)
26. C.H. Chen, Y.K. Su, S.J. Chang, G.C. Chi, J.K. Sheu, J.F. Chen, C.H. Liu, Y.H. Liaw, IEEE Electron Dev Lett **23**(3), 130–132(2002)
27. W.V. Lundin, A.E. Nikolaev, A.V. Sakharov, E.E. Zavarin, G.A. Valkovskiy, M.A. Yagovkina, S.O. Usov, N.V. Kryzhanovskaya, V.S. Sizov, P.N. Brunkov, A.L. Zakgeim, A.E. Cherniakov, N.A. Cherkashin, M.J. Hytch, E.V. Yakovlev, D.S. Bazarevskiy, M.M. Rozhavskaya, A.F. Tsatsulnikov, J Cryst Growth **315**(1), 267–271 (2011)
28. H. Zhao, G. Liu, X.H. Li, G.S. Huang, J.D. Poplawsky, S.T. Penn, V. Dierolf, N. Tansu, Appl Phys Lett **95**, 061104 (2009)
29. H. Zhao, R.A. Arif, N. Tansu, IEEE J Sel Top Quantum Electron **15**, 1104–1114 (2009)
30. E.H. Park, D. Nicol, H. Kang, I.T. Ferguson, S.K. Jeon, J.S. Park, T.K. Yoo, Appl Phys Lett **90**, 031102 (2006)
31. J.H. Ryou, J. Limb, W. Lee, J. Liu, Z. Lochner D. Yoo, R.D. Dupuis, IEEE Photon Technol Lett **20**(21), 1769–1771 (2008)
32. L.W. Wu, S.J. Chang, Y.K. Su R.W. Chuang, T.C. Wen, C.H. Kuo, W.C. Lai, C.S. Chang, J.M. Tsai, J.K. Sheu, IEEE Trans Electron Dev **50**(8), 1766–1770 (2003)
33. J.W. Ju, E.S. Kang, H.S. Kim, L.W. Jang, H.K. Ahn, J.W. Jeon, I.H. Lee, J.H. Baek, J Appl Phys **102**, 053519 (2007)
34. H.J. Kim, S. Choi, S.S. Kim, J.H. Ryou, P.D. Yoder, R.D. Dupuis, A.M. Fischer, K. Sun, F.A. Ponce, Appl Phys Lett **96**, 101102 (2010)
35. M. Bockowski, P. Strak, I. Grzegory, B. Lucznik, S. Porowski, J Cryst Growth **310**(17), 3924–3933 (2008)
36. D. Ehrentraut, Z. Sitar, MRS Bull **34**(4), 259–265 (2009)
37. T. Hashimoto, E. Lettsa, M. Ikaria, Y. Nojima, J Cryst Growth **312**(18), 2503–2506 (2010)
38. D. Gogova, A. Kasic, H. Larsson, C. Hemmingsson, B. Monemar, F. Tuomisto, K. Saarinen, L. Dobos, B. Pecz, P. Gibart, B.Beaumont J Appl Phys **96**(1), 799–806 (2004)
39. X. Xu, R.P. Vaudo, J. Flynn, J. Dion, G.R. Brandes, Phys Stat Sol (a) **202**, 727–730 (2005)
40. K. Fujito, S. Kubo, H. Nagaoka, T. Mochizuki, H. Namita, S. Nagao J Cryst Growth **311**(10), 3011–3014 (2009)
41. X.A. Cao, S.F. LeBoeuf, M.P. D'Evelyn, S.D. Arthur, J. Kretchmer, C.H. Yan, Z.H. Yang, Appl Phys Lett **84**(21), 4313–4315 (2004)
42. C.R. Miskys, M.K. Kelly, O. Ambacher, G. Martinez-Criado, M. Stutzmann, Appl Phys Lett **77**, 1858–1860 (2000)
43. X.A. Cao, H. Lu, S.F. LeBoeuf, C. Cowen, S.D. Arthur, W. Wang, Appl Phys Lett **87**, 053503 (2005)
44. I. Arslan, N.D. Browning, Phys Rev Lett **91**, 165501 (2003)
45. X.A. Cao, S.D. Arthur, Appl Phys Lett **85**(18), 3971–3973 (2004)
46. X.A. Cao, J.M. Teetsov, M.P. D'Evelyn, D.W. Merfeld, C.H. Yan, Appl Phys Lett **85**(1), 7–9 (2004)
47. H. Masui, S. Nakamura, S.P. DenBaars, U.K. Mishra, IEEE Trans Electron Dev **57**(1), 88–100 (2010)
48. D.A.B. Miller, D.S. Chemla, T.C. Damen, A.C. Gossard, W. Wiegmann, T.H. Wood, C.A. Burrus, Phys Rev Lett **53**(22), 2173–2176 (1984)
49. F. Bernardini, V. Fiorentini, Appl Surf Sci. **166**(1–4), 23–29 (2000)

50. H. Masui, J. Sonoda, N. Pfaff, I. Koslow, S. Nakamura, S.P. DenBaars, J Phys D:Appl Phys **41,**165105 (2008)
51. Y.D. Jho, J.S. Yahng, E. Oh, D.S. Kim, Phys Rev B **66**, 1–11 (2002)
52. S.F. Chichibu, A.C. Abare, M.S. Minsky, S. Keller, S.B. Fleischer, J.E. Bowers, E. Hu, U.K. Mishra, L.A. Coldren, S.P. DenBaars, Appl Phys Lett **73**(14), 2006–2008 (1998)
53. P. Waltereit, O. Brandt, J. Ringling, K.H. Ploog, Phys Rev B **64**(24), 245305 (2001)
54. A.E. Romanov, T.J. Baker, S. Nakamura, J.S. Speck, J Appl Phys **100**(2), 023522 (2006)
55. B.A. Haskell, S. Nakamura, S.P. DenBaars, J.S. Speck, Phys. Stat. Sol (b) **244**(8), 2847–2858 (2007)
56. P. Waltereit, O. Brandt, A. Trampert, H.T. Grahn, J. Menniger, M. Ramsteiner, M. Reiche, K.H. Ploog Nature **406**, 865–868 (2000)
57. M.D. Craven, S.H. Lim, F. Wu, J.S. Speck, S.P. DenBaars, Phys Stat Sol (a), **194**(2), 541–544 (2002)
58. M.D. Craven, A. Chakraborty, B. Imer, F. Wu, S. Keller, U.K. Mishra, J.S. Speck, S.P. DenBaars, Phys Stat Sol (c) (7), 2132–2135 (2003)
59. Y.S. Cho, Q. Sun, I.H. Lee, T.S. Ko, C.D. Yerino, J. Han, B.H. Kong, H.K. Cho, S. Wang, Appl Phys Lett **93**, 111904 (2008)
60. P. Waltereit, O. Brandt, M. Ramsteiner, A. Trampert, H.T. Grahn, J. Menniger, M. Reiche, K.H. Ploog, J Cryst Growth 227–228, 437–441 (2001)
61. T.J. Baker, B.A. Haskell, F. Wu, P.T. Fini, J.S. Speck, S. Nakamura, Jpn J Appl Phys **44**(28–32), L920–L922 (2005)
62. T.J. Baker, B.A. Haskell, F. Wu, J.S. Speck, S. Nakamura, Jpn J Appl Phys **45**(4–7), L154–L157 (2006)
63. T. Hikosaka, T. Tanikawa, Y. Honda, M. Yamaguchi, N. Sawaki, Phys Stat Sol (c) **5**(6), 2234–2237 (2008)
64. R. Sharma, P.M. Pattison, H. Masui, R.M. Farrell, T.J. Baker, B.A. Haskell, F. Wu, S.P. DenBaars, J.S. Speck, S. Nakamura, Appl Phys Lett **87**, 231110 (2005)
65. N.F. Gardner, J.C. Kim, J.J. Wierer, Y.C. Shen, M.R. Krames, Appl Phys Lett **86**, 111101 (2005)
66. S.C. Ling, T.C. Wang, J.R. Chen, P.C. Liu, T.S. Ko, B.Y. Chang, T.C. Lu, H.C. Kuo, S.C. Wang, J.D. Tsay, IEEE Photon Technol Lett **21**(16), 1130–1132 (2009)
67. T. Detchprohm, M. Zhu, Y. Li, Y. Xia, C. Wetzel, E.A. Preble, L. Liu, T. Paskova, D. Hanser, Appl Phys Lett **92**, 241109 (2008)
68. T. Detchprohm, M. Zhu, Y. Li, L. Zhao, S. You, C. Wetzel, E.A. Preble, T. Paskova, D. Hanser, Appl Phys Lett **96**, 051101 (2010)
69. M. Funato, M. Ueda, Y. Kawakami, Y. Narukawa, T. Kosugi, M. Takahashi, T. Mukai, Jpn J Appl Phys **45**(24–28), L659–L662 (2006)
70. H. Sato, A. Tyagi, H. Zhong, N. Fellows, R.B. Chung, M. Saito, K. Fujito, J.S. Speck, S.P. DenBaars, S. Nakamura, Phys Stat Sol (RRL) **1**(4), 162–164 (2007)
71. H. Sato, R.B. Chung, H. Hirasawa, N. Fellows, H. Masui, F. Wu, M. Saito, K. Fujito, J.S. Speck, S.P. DenBaars, S. Nakamura, Appl Phys Lett **92**(22), 221110 (2008)
72. N. Fellows, H. Sato, H. Masui, S.P. DenBaars, S. Nakamura, Jpn J Appl Phys **47**(10), 7854–7856 (2008)
73. S.E. Brinkley, Y.D. Lin, A. Chakraborty, N. Pfaff, D. Cohen, J.S. Speck, S. Nakamura, S.P. DenBaars, Appl Phys Lett **98**, 011110 (2011)
74. T. Onuma, H. Amaike, M. Kubota, K. Okamoto, H. Ohta, J. Ichihara, H. Takasu, S.F. Chichibu, Appl Phys Lett **91**(18), 181903 (2007)
75. H. Masui, H. Yamada, K. Iso, J.S. Speck, S. Nakamura, S.P. DenBaars, J Soc Inf Displays **16**, 571–578 (2008)
76. M. Marques L.K. Teles, L.M.R. Scolfaro, L.G. Ferreira, J.R. Leite, Phys Rev B **70**, 073202 (2004)
77. L. Nistor, H. Bender, A. Vantomme, M.F. Wu, J.V. Lauduyt, K.P. O'Donnell, R. Martin, K. Jacobs, I. Moerman, Appl Phys Lett **77**, 507–509 (2000)

78. D. Gerthsen, E. Hahn, B. Neubauer, A. Rosenauer, H.M. Schon, A. Rizzi, Phys Stat Sol (a) **177**, 145–150 (2000)
79. K.P. O'Donnell, R.W. Martin, P.G. Middleton, Phys Rev Lett **82**, 237–240 (1999)
80. S.W. Feng, Y.C. Cheng, Y.Y. Chung, C.C. Yang, Y.S. Lin, C. Hsu, K.J. Ma, J.I. Chyi, J Appl Phys **92**, 4441–4448 (2002)
81. S. Chichibu, K. Wada, S. Nakamura, Appl Phys Lett **71**, 2346–2348 (1997)
82. K. Okamoto, A. Kaneta, Y. Kawakami, S. Fujita, J. Choi, M. Terazima, T. Mukai, J Appl Phys **98**, 064503 (2005)
83. Y.J. Sun, O. Brandt, S. Cronenberg, S. Dhar, H.T. Grahn, K.H. Ploog, P. Waltereit, J.S. Speck Phys Rev B **67**, 041306(R) (2003)
84. M. Funato, Y. Kawakami, J Appl Phys **103**, 093501 (2008)
85. M. Funato, A. Kaneta, Y. Kawakami, Y. Enya, K. Nishizuka, M. Ueno, T. Nakamura, Appl Phys Express **3**, 021002 (2010)
86. X.A. Cao, S.F. LeBoeuf, T.E. Stecher, IEEE Electron Dev Lett **27**, 329–331 (2006)
87. X.A. Cao, S.F. LeBoeuf, IEEE Electron Dev Lett **54**(12), 3114–3417 (2007)
88. X.A. Cao, S.F. LeBoeuf, L.B. Rowland, C. Yan, H. Liu, Appl Phys Lett **83**(21), 3614–3616
89. W. Shan, T.J. Schmidt, X.H. Yang, S.J. Hwang, J.J. Song, B. Goldenberg, Appl Phys Lett **66**, 985–987 (1995)
90. I.L. Krestnikov, N.N. Ledentsov, A. Hoffmann, D. Bimberg, A.V. Sakharov, W.V. Lundin, A.F. Tsatsulnikov, A.S. Usikov, Z.I. Alferov, Y.G. Musikhin, D. Gerthsen, Phys Rev B **66**, 155310 (2002)
91. D.M. Graham, A. Soltani-Vala, P. Dawson, M.J. Godfrey, T.M. Smeeton, J.S. Barnard, M.J. Kappers, C.J. Humphreys, E.J. Thrush, J Appl Phys **97**, 103508 (2005)
92. M.J. Galtrey, R.A. Oliver, M.J. Kappers, C.J. Humphreys, P.H. Clifton, D. Larson, D.W. Saxey, A. Cerezo, J Appl Phys **104**, 013524 (2008)
93. J.P. O'Neill, I.M. Ross, A.G. Cullis, T. Wang, P.J. Parbrook, Appl Phys Lett **83**, 1965–1967 (2003)
94. T.M. Smeeton, C.J. Humphreys, J.S. Barnard, M.J. Kappers, J Mater Sci **41**, 2729–2737 (2006)
95. A. Hangleiter, F. Hitzel, C. Netzel, D. Fuhrmann, U. Rossow, G. Ade, P. Hinze, Phys Rev Lett **95**, 127402 (2005)
96. N.K. van der Laak, R.A. Oliver, M.J. Kappers, C.J. Humphreys, J Appl Phys **102**, 013513 (2007)
97. D.M. Graham, P. Dawson, M.J. Godfrey, M.J. Kappers, C.J. Humphreys, Appl Phys Lett **89**, 211901 (2006)
98. Y. Yang, X.A. Cao, C.H. Yan, IEEE Trans Electron Dev **55**(7), 1771–1775 (2008)
99. B. Monemar, B.E. Sernelius, Appl Phys Lett **91**, 181103 (2007)
100. M.H. Kim, M.F. Schubert, Q. Dai, J.K. Kim, E.F. Schubert, J. Piprek, Y. Park, Appl Phys Lett **91**, 183507 (2007)
101. K.J. Vampola, M. Iza, S. Keller, S.P. DenBaar, S. Nakamura, Appl Phys Lett **94**(6), 061116 (2009)
102. Y.C. Shen, G.O. Mueller, S. Watanabe, N.F. Gardner, A. Munkholm, M.R. Krames, Appl Phys Lett **91**, 141101 (2007)
103. K.T. Delaney, P. Rinke, C.G. Van de Walle, Appl Phys Lett **94**(19), 191109 (2009)
104. J. Xu, M.F. Schubert, A.N. Noemaun, D. Zhu, J.K. Kim, E.F. Schubert, M.H. Kim, H.J. Chung, S. Yoon, C. Sone, Y. Park, Appl Phys Lett **94**, 011113 (2009)
105. N.F. Gardner, G.O. Müeller, Y.C. Shen, G. Chen S. Watanabe, Appl Phys Lett **91**, 243506 (2007)
106. Y.J. Lee, C.H. Chen, C.J. Lee, IEEE Photon Technol Lett **22**(20), 1506–1508 (2010)
107. S.C. Ling, T.C. Lu, S.P. Chang, J.R. Chen, H.C. Kuo, S.C. Wang, Appl Phys Lett **96**, 231101 (2010)
108. Y.D. Lin, A. Chakraborty, S. Brinkley, H.C. Kuo, T. Melo, K. Fujito, J.S. Speck, S.P. DenBaars, S. Nakamura, Appl Phys Lett **94**(26), 261–108 (2009)

Chapter 5
Improved Light Extraction Efficiency in GaN-Based Light Emitting Diodes

Jihyun Kim

Abstract The improvement in light extraction efficiency (LEE) of GaN-based LEDs is one of the most important areas for increasing the external quantum efficiency for solid-state lighting applications. We summarize the advances in this field.

GaN-based light-emitting diodes (LEDs) have made a lot of progress with regard to growth, doping, and p-ohmic contacts [1–3]. The external quantum efficiency (EQE) of LEDs is equal to IQE (internal quantum efficiency)×LEE (light extraction efficiency). Although the internal quantum efficiency (IQE) has been improved by advances in epitaxial growth techniques, the methods used to extract the photons from the quantum wells (QWs) still need to be improved. It has been reported that the LEE has to be over 90% to achieve 200 lm/W in white LEDs.

According to Snell's law, the light extraction angle (escape angle or escape cone) between GaN ($n \sim 2.5$) and air ($n \sim 1$) was about 23.5°. Snell's law states that the escape angle (θ) can be calculated by $\theta = \sin^{-1}(n_{air}/n_{GaN})$ where n_{air} and n_{GaN} are the index of air and GaN, respectively. Therefore, the total internal reflection (TIR) resulting from the narrow escape cone prevents the protons from escaping from the semiconductor. In other applications such as optical fibers and optical waveguides, TIR is required. If Indium tin oxide, which is a common transparent conducting oxide ($n \sim 2.19$), is used, the escape angle can be increased up to 27°, which is still narrow. The photons outside of the escape angle will be re-absorbed after multiple reflections or increase the operating temperature by dissipation (Fig. 5.1).

Several novel techniques have been developed to overcome the low escape angle or avoid the wave-guiding effect, including photonic crystal (PC) structures, photoelectrochemical (PEC) wet etching, natural lithography, graded-refractive index layer (GRIN), and patterned sapphire substrate (PSS). The basic idea of the surface texturing techniques is (1) to increase the effective surface area, which can

J. Kim (✉)

Department of Chemical and Biological Engineering, Korea University, Seoul 136-701, Korea

e-mail: hyunhyun7@korea.ac.kr

S. Pearton (ed.), *GaN and ZnO-based Materials and Devices*,
Springer Series in Materials Science 156, DOI 10.1007/978-3-642-23521-4_5,
© Springer-Verlag Berlin Heidelberg 2012

Fig. 5.1 Schematic of escape angle between air and GaN LED

Fig. 5.2 TIR inside GaN LED. TIR occurs in flat surface, but the photon can find the escape angle in textured surface

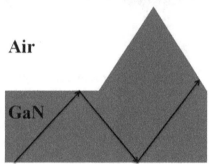

further increase the number of scattering events and (2) to randomize the path of the light rays so that the photons may eventually find the escape angle [4]. A lot of methods to make GaN LEDs brighter have been reported (Fig. 5.2).

5.1 PEC Etch: c-Plane vs. a-Plane

Among these methods, PEC was first employed to improve the light extraction efficiency (LEE) because this is cheap, simple, and easy to implement [5]. In addition, the chemical wet-etch method causes no damage to the crystal lattice, compared with the dry-etching method such as the reactive-ion etch and inductively coupled plasma etch. Since GaN is a compound semiconductor with a wurtzite structure, the etch mechanism depends on the growth direction and surface atom. There are polar, semi-polar, and non-polar GaN, which depend on the growth direction. Until now, c-plane growth with the Ga-face is the most popular method used to grow GaN-based LEDs.

1. c-plane with Ga-face

It has been shown that Ga-face is chemically inert because of the triple dangling bonds, which have negative charges. After the Ga atom is removed, the negatively charged triple dangling bonds repel KOH-ions, which also have negative charges. Since the wet etch process cannot proceed, the Ga-face becomes very stable in base solutions [6]. Therefore, most efforts have been focused on wet-etching the Nitrogen-face instead of the Ga-face (Fig. 5.3).

Fig. 5.3 Hexagonal GaN structure

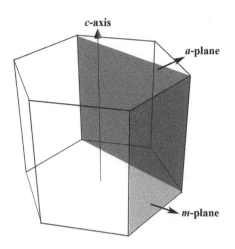

2. c-plane with N-face

There is a negatively charged single dangling bond on the N-face, which is more vulnerable to the base solution. The etch mechanism reported by Li et al. is as follows:

$$2GaN + 3H_2O \xrightarrow{\text{KOH}} Ga_2O_3 + 2NH_3,$$

where KOH acts as both the catalyst for the reaction and the solvent for the resulting Ga_2O_3 [7]. Once the wet etch begins on the N-face, typical pyramidal structures with six sidewall facets {10–1–1} are observed [8]. These pyramidal shapes are extremely effective in extracting photons inside the GaN layer because the photons will finally find the escape angle after multiple reflections inside the pyramid. Fujii et al. reported that the light intensity was improved by KOH^- treatment [5]. However, there are still difficulties associated with optimizing the wet-etch conditions and achieving uniformity across large wafers for further commercialization. Recently, in commercial LEDs, the pyramidal structure in N-face GaN is formed by wet etching after the substrate is lifted-off by a laser (Figs. 5.4 and 5.5).

3. Non-polar plane

Since the piezoelectric field makes the hole and electron wave functions misaligned, the low radiative recombination efficiency within QWs is unavoidable in the polar GaN structure. However, this problem can be solved using non-polar a- or m-plane GaN, where the m-plane is more stable than the a-plane in a base solution. In addition, the side wall of the a- and m-plane GaN can be textured using a base solution since one of the side walls will be N-face GaN. Recently, Jung et al. reported an increase in the LEE in an a-plane InGaN/GaN LED with KOH-based PEC etching [9]. Figure 5.6 shows that both the top and side wall were textured. Note that the hexagonal pyramids were formed in the $(-c)$-axis, which is N-polar (Fig. 5.6).

Fig. 5.4 Six sidewall facets
{10–1–1}

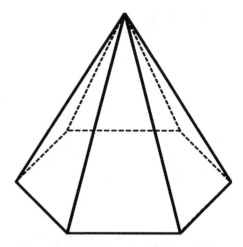

Fig. 5.5 Six sidewall facets
are exposed after PEC etch

Fig. 5.6 Top-view SEM
image of a-plane GaN after
PEC treatment

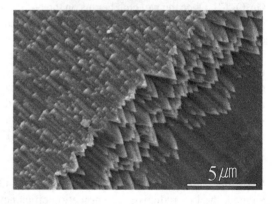

5.2   Natural Lithography

Although, the hexagonal pyramids formed by chemical wet etch are highly effective
in extracting photons, uniformity is still a problem. Alternatively, the dry-etching
method based on a self-assembled monolayer (so-called natural lithography) has
been intensively investigated using polymer- and SiO_2-based nanospheres [10, 11].
The Stöber method has been widely used to synthesize SiO_2 nanospheres of various

Fig. 5.7 Top-view SEM
image of ordered nanospheres

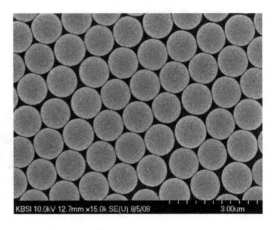

KBSI 10.0kV 12.7mm ×15.0k SE(U) 8/5/08   3.00um

sizes [12]. Various (pseudo) photonic crystal structures have been successfully fabricated by natural lithography using the spin-coating method, drop-coating method, Langmuir-Blodgget method, etc. [13, 14]. Either an additive or subtractive process can be used to texture the surface of optical devices such as solar cells and LEDs. There have been a few reports on the photonic crystal effects using natural lithography technique [15, 16].

The SEM image in Fig. 5.7 shows that well-ordered nanospheres can be fabricated using the spin-coating method, where the diameter of nanospheres was 300 nm.

1. Surface texturing by dry-etch with nanosphere masks

Spheres made of polymers or SiO_2 can also be used. Since SiO_2 is more resistant to the dry etch process, SiO_2-based nanospheres are more often used to roughen GaN. To remove SiO_2-based nanospheres, diluted HF solution (buffered oxide etchant) can be used. Since SiO_2-based nanospheres are heavier than common polymers, heating makes SiO_2-based nanospheres sink, which enables one to fabricate convex or concave structures. By applying a diluted HF solution, a concave structure can be created. Both concave and convex structures are very efficient in extracting photons from the semiconductor [17]. Jang et al. reported that the depth that the nanospheres sink can be controlled by the heating time [18]. Figure 5.8 depicts GaN coated with SiO_2-based nanospheres, followed by ICP dry etch.

A concave structure was achieved by dipping the GaN coated with nanospheres into a diluted HF solution (Fig. 5.9).

2. Graded-Refractive Index (GRIN) layer with SiO_2-nanosphere-based natural lithography

When light rays move along the orthogonal direction (90° to the surface), the transmission at the interface between the semiconductor and air can be expressed as $4n/(n+1)^2$, where n is the refractive index of a semiconductor. For GaN, this value will be 0.82 because $n = 2.5$ for the orthogonal incidence. If an anti-reflective

Fig. 5.8 SEM image after
ICP etch with SiO$_2$
nanosphere-based mask

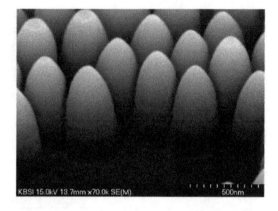

Fig. 5.9 Concave structure
after removing SiO$_2$
nanospheres by diluted HF
solution

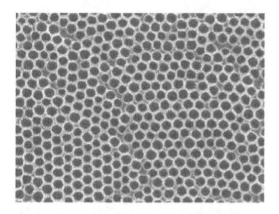

Fig. 5.10 LED with
anti-reflective coating

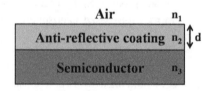

coating is deposited on top of GaN-based LEDs, the LEDs can become brighter by reducing reflection. SiN$_x$ has been commonly deposited to match the refractive index in optical devices. The reason for the anti-reflective coating is to achieve destructive interference, which would be most effective if the amplitude of the two wave functions is similar. Therefore, $\sqrt{2.5}(= 1.58)$ is preferred for a single anti-reflective layer, which is close to the refractive index of SiN$_x$. The optimal thickness of destructive interference has to be precisely designed for multi anti-reflective layers. (Fig. 5.10).

When the escape cone is considered, the $1/4n^2$ relationship allows only 4% of the generated photons inside the QWs to transmit to the air [19]. An alternative method is to vary the porosity. The refractive index can be adjusted by controlling the porosity of the materials, where an increase in porosity would lower the effective refractive index. This technique is very effective because it is difficult to find the adequate material with good electrical properties, high transparency, and thermal conductivity in terms of the refractive index. Multilayers with different refractive indexes can be deposited to match the indexes between GaN(2.5) and air(1). Kim et al. reported that the combination of natural lithography and wet etching can be used to generate the GRIN structure in GaN-based LEDs [17]. In addition, the deposition of ITO at oblique angles allowed researchers to control the refractive index of ITO from 2.19 to 1.17, which increased the LEE by 24.3% [20].

3. Nano-needles using natural lithography

Nanostructures such as nanorods, nanowires, and nanorings have been investigated for potential applications in LEDs and nano-optoelectronics. Most nanostructures have been fabricated by the bottom-up method, where the vapor–liquid–solid (VLS) method is one of the most common techniques used to grow GaN nanowires and nanorods. One of the issues with the bottom-up approach is that the quality, doping concentrations, diameter, and length are very difficult to control. These factors need to be tightly controlled if they are to be applied in a semiconductor fabrication facility. However, control over these factors can be possible using the top-down method. In this approach, the doping concentrations are uniform because the carrier concentrations will be same as the substrate. The length can controlled by the etch rate and time. The diameter can be decided by the size of the nanospheres. Some group have used reactive-ion etching to decrease the diameter of the nanospheres, followed by ICP dry etching to produce ultra-sharp nano-needles. Micro- and nano-needles are similar to fin, which can enhance thermal management. Therefore, the operating temperature can be lowered by employing micro- or nano-needle structures. In addition, the GRIN layer from 2.5 to 1 can be produced using a cone shape. Fresnel reflection can be minimized if the structure is fully optimized (Fig. 5.11).

These nano- or micro-needles can be used in drug-release applications and for the development of patch or painless syringes. In addition, the reflectance can be controlled using a needle-like structure for solar-cell applications (Fig. 5.12).

4. Backside roughening

The backside of GaN-based LEDs has been coated with Ag or Al to direct the photons to the front. Since the planar structure is good for wave-guiding effects, the roughed interface between the reflective metal and semiconductor is effective

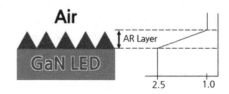

Fig. 5.11 Graded index by nanostructured surface

Fig. 5.12 Nano-needles
fabricated by ICP dry etch

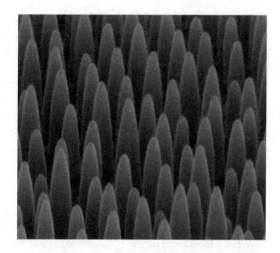

Fig. 5.13 Cross-sectional
SEM image with convex
reflectors

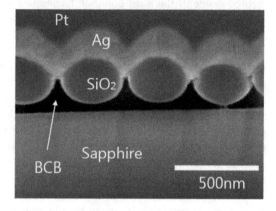

in randomizing the light rays, which can improve the LEE. Sapphire, which is the
most commonly used substrate for the growth of GaN-based LEDs, cannot be easily
shaped. Therefore, the additive layer can be deposited through the combination of
a soft material and SiO_2-based nanospheres. The cross-sectional image (Fig. 5.13)
shows the flat interface between sapphire and BCB (benzo-cyclo-butene), where
the refractive index of BCB was similar to that of sapphire. After SiO_2-based
nanospheres sank at elevated temperature, Ag was deposited by electron-beam
evaporation (Fig. 5.13).

In addition, a concave mirror was fabricated by removing the SiO_2-based
nanospheres using a diluted HF solution, followed by the deposition of Ag. Platinum
(Pt) was deposited to protect the surface layer from Ga-ions when a focused-ion
beam was used to cut the sample (Fig. 5.14).

Fig. 5.14 Cross-sectional
SEM image with concave
reflectors

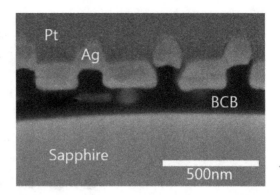

5.3 Photonic Crystal

Outside of the escape cone, the light would propagate in the guided modes, which
will be re-absorbed or escape from the edge of the LED. Two approaches have
been developed when the photonic crystal (PC) structure is used in LEDs. Since the
photonic bandgap can redirect the photons, the photonic bandgap can be employed.
However, there are many issues associated with growth and fabrications. Instead,
PC can be used to extract the photons in guided modes by diffraction. A LEE as
high as 73% has been achieved using PC structures, where the thickness of the
LED and the depth, geometry and size of the holes were shown to be the key
parameters [21, 22]. Electron-beam lithography and focused-ion beam techniques
have been conventionally employed to produce various geometries of the PC
designs. Recently, the stamping method, which is cheaper and simpler than the
E-beam lithography and FIB technique, has been used to create PC patterns on GaN-
based LEDs.

5.4 Plasmonics

Internal quantum efficiency (IQE) depends on defects and dislocations, which
mainly cause non-radiative recombination within QWs. By coupling the QWs and
surface plasmons (SPs), IQE can be improved. Silver and aluminium have been
widely used as plasmonic metal, where SP-QW coupling improved the spontaneous
recombination rate. Eventually, this will brighten GaN-based LEDs due to the
energy transfer between the surface Plasmon and QW. Biteen et al. reported fourfold
enhancement of PL intensity [23]. Okamoto et al. reported that InGaN/GaN GaN
coated with Ag produced a 14-fold increase in PL intensity compared with the
uncoated sample [24]. In addition, they found that the PL enhancement highly
depended on the distance between the QWs and plasmonic metal because the SP
exponentially decayed with the distance from the plasmonic metal. Therefore, the

Fig. 5.15 Schematic images
with plasmonic metals
embedded close to QWs

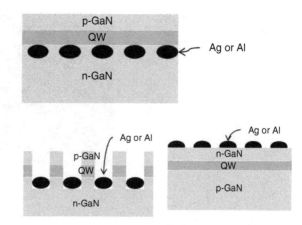

key parameter in this approach is how closely the plasmonic metal is located to the
QWs. Okamoto et al. proposed a possible mechanism for this phenomenon. Once
the excitons are generated within the near-field of the metal layer, QW energy can be
coupled to the SP. Therefore, the fast decay rate via QW-SP coupling will brighten
GaN-based LEDs [25]. Kwon et al. successfully inserted Ag nanoparticles between
the QWs and n-GaN layer, which improved the optical output power by 32.2% at a
100 mA current [26].

Since p-GaN cannot be highly doped compared with n-GaN, the SP-QW coupling will be very poor if the plasmonic metal is deposited on top of conventionally
thick p-GaN. Alternative methods have been developed to achieve good SP-QW
coupling (Fig. 5.15).

5.5 Chip Shaping

The thermal conductivity of Sapphire is very low compared to that of GaN.
Ohmic metallizations can be used as the reflectors in flip-chip (FC) configuration.
In addition, the refractive index of Sapphire is 1.76, which provides wider escape
cone compared to GaN. Multiple opportunities to find the escape cone can be
achieved by shaping the Sapphire (Fig. 5.16). Lee et al. used $H_2SO_4 : H_3PO_4$ (3:1)
based wet etching to shape the Sapphire and under these conditions the EL intensity
was increased by 55% at a current injection of 350 mA [27]. Kao et al. reported
that GaN-based LED with a 22° undercut side wall was enhanced by 170%. In this
study, ICP-RIE etching was used to shape the mesa side wall [28]. In addition, side
wall texturing, which resembled wave-like side walls, was shown to improve light
extraction in the horizontal direction [29].

Fig. 5.16 Schematic diagram of shaped sapphire

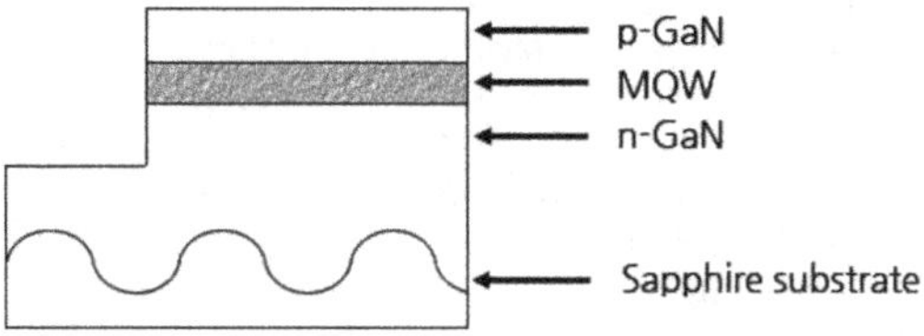

Fig. 5.17 Schematic diagram of GaN-based LED on PSS

5.6 Patterned Sapphire Substrate

The lack of the commercial cheap homoepitaxial substrates for the growth of GaN made Sapphire the most common substrate for the growth of a GaN epilayer. Huge lattice mismatch between Sapphhire and GaN made dislocation inevitable in the GaN layer. Therefore, several techniques such as lateral epitaxial overgrowth have been developed with the aim of reducing the number of growth defects. Since the PSS structure has many advantages with regard to reducing dislocations and increasing the number of scattering events, many commercial LED makers use the PSS techniques (Fig. 5.17)

Yamada et al. used the PSS technique to enhance the LEE [30]. Tadatomo et al. demonstrated GaN-based Ultraviolet LED by employing PSS structure [31]

Acknowledgements The research at Korea University was supported by the Carbon Dioxide Reduction and Sequestration Center, one of the twenty-first Century Frontier R&D Program funded by the Ministry of Education, Science and Technology of Korea.

References

1. S. Nakamura, G. Fasol, S. Pearton, *The Blue Laser Diode: The Complete Story/Edition 2*, (Springer Berlin, 2000)
2. E.F. Schubert, J.K. Kim, Science, **308**, 1274 (2005)
3. M. Fukuda, *Optical Semiconductor Devices*, (Wiley, New York, 1999)
4. I. Schnitzer, E. Yablonovitch, C. Caneau, T.J. Gmitter, A. Scherer, Appl. Phys. Lett. **63**(16), 2174 (1993)
5. T. Fujii, Y. Gao, R. Sharma, E.L. Hu, S.P DenBaars, S. Nakamura, Appl. Phys. Lett. **84**, 855 (2004)
6. Younghun Jung, Kwang Hyeon Baik, Fan Ren, Stephen J. Pearton, Jihyun Kim, J. Electrochem. Soc., **157**(6), H676–H678 (2010)

7. D. Li, M. Sumiya, S. Fuke, D. Yang, D. Que, Y. Suzuki, Y. Fukuda, J. Appl. Phys. **90**, 4219 (2001)

8. H.M. Ng, N.G. Weimann, A. Chowdhury, J. Appl. Phys. **94**, 650 (2003)

9. Younghun Jung, Jihyun Kim, Soohwan Jang, Kwang Hyeon Baik, Yong Gon Seo, Sung-Min Hwang, Opt. Exp. Vol. **18**, No.9, pp. 9728–9732 (2010)

10. C.H. Chiu, T.C. Lu, H.W. Huang, C.F. Lai, C.C. Kao, J.T. Chu, C.C. Yu, H.C. Kuo, S.C. Wang, C.F. Lin, T.H. Hsueh, Nanotechnology **18**, 445201 (2007)

11. X. Qian, J. Li, D. Wasserman, W.D. Goodhue, Appl. Phys. Lett. **93**, 231907 (2008)

12. W. Stöber, A. Fink, E.J. Bohn, Colloid Interface Sci., **26**,62 (1968)

13. Y. Li, W. Cai, G. Duan, Chem. Mater., **20**, 615 (2008)

14. Ching-Mei Hsu, Stephen T. Connor, Mary X. Tang, Yi Cui, Appl. Phys. Lett. **93**, 133109 (2008)

15. Wai Yuen Fu, Kenneth Kin-Yip Wong, H.W. Choi, Appl Phys Lett **95**, 133125 (2009)

16. Yan-Kuin Su, Jian-Jhong Chen, Chuing-Liang Lin, Shi-Ming Chen Wen-Liang Li, Chien-Chih Kao, Jpn J Appl Phys., **47**(8), 6706 (2008)

17. B.J. Kim, H. Jung, S.H. Kim, J. Bang, J. Kim, IEEE Photonics Technol. Lett., Vol. **21**, pp.700–7002 (2009)

18. Se Gyu Jang, Dae-Geun Choi, Chul-Joon Heo, Su Yeon Lee, Seung-Man Yang, Adv. Mater. **20**, 4862 (2008)

19. B.E.A. Saleh, M.C. Teich, *Fundamentals of Photonics*, 2nd (Wiley-Interscience, 2006)

20. J.K. Kim, S. Chhajed, M.F. Schubert, E.F. Schubert, A.J. Fischer, M.H. Crawford, J. Cho, H. Kim, C. Sone, Adv. Mater., **20**, 801 (2008)

21. J.J. Wierer, A. David, M.M. Megens, Nat. Photonics, **3**, 163 (2009)

22. A. David, H. Benisty, C. Weisbuch, J. Display Tech., **3**, 133 (2007)

23. J.S. Biteen, D. Pacifici, N.S. Lewis, H.A. Atwaer, Nano Lett. **5**, 1768 (2005)

24. K. Okamoto, I. Niki, A. Shvartser, Y. Narukawa, T. Mukai, A. Scherer, Nat. Mater. **3**, 601 (2004)

25. K. Okamoto, I. Niki, A. Scherer, Y. Narukawa, T. Mukai, Y. Kawakami, Appl. Phys. Lett., **87**, 071102 (2005)

26. M.-K. Kwon, J.-Y. Kim, B.-K. Kim, I.-K. Park, C.-Y. Cho, C.C. Byeon, S.-J. Park, Adv. Mater., **20**, 1253 (2008)

27. C.E. Lee, H.C. Kuo, Y.C. Lee, M.R. Tsai, T.C. Lu, S.C. Wang, C.T. Kuo, IEEE Photon. Technol. Lett., **20** (3), 184 (2008)

28. C.-C. Kao, H.-C. Kuo, H.-W. Huang, J.-T. Chu, Y.-C. Peng, Y.-L. Hsieh, C.Y. Luo, S.-C. Wang, C.-C. Yu, C.-F. Lin, IEEE Photon. Technol. Lett., **17** (1), 19, (2005)

29. C.S. Chang, S.J. Chang, Y.K. Su, C.T. Lee, Y.C. Lin, W.C. Lai, S.C. Shei, J.C. Ke, H.M. Lo, IEEE Photon. Technol. Lett., **16** (3), 750, (2004)

30. M. Yamada, T. Mitani, Y. Narukawa, S. Shioji, I. Niki, S. Sonobe, K. Deguchi, M. Sano T. Mukai, Jpn. J. Appl. Phys., **41**, L1431 (2002)

31. K. Tadatomo, H. Okagawa, Y. Ohuchi, T. Tsunekawa, T. Jyouichi, Y. Imada, M. Kato, H. Kudo, T. Taguchi, Phys. Stat. Sol. (a) **188**(1), 121–125 (2001)

Chapter 6
GaN-Based Sensors

F. Ren, B.H. Chu, K.H. Chen, C.Y. Chang, Victor Chen, and S.J. Pearton

Abstract Recent progress in the use of surface-functionalized GaN for sensing of gases, heavy metals, UV photons and biological molecules is reviewed. The use of such sensors for monitoring nerve cells is also explored. Additionally, we briefly review progress with InN-based chemical sensors. For the detection of gases such as hydrogen, the GaN is coated with a catalyst metal such as Pd or Pt to increase the detection sensitivity at room temperature. Functionalizing the surface with oxides, polymers, and nitrides is also useful in enhancing the detection sensitivity for gases and ionic solutions. The use of enzymes or adsorbed antibody layers on the semiconductor surface leads to highly specific detection of a broad range of antigens of interest in the medical and homeland security fields. We give examples of recent work showing sensitive detection of glucose, lactic acid, prostate cancer, and breast cancer markers. Nerve cell-GaN surface coupling allowed for the analysis of cell reactions to different neuroinhibitors. InN chemical sensors have also been shown to be viable for sensing anions, pH, and polarity.

6.1 Introduction

Chemical sensors are needed for applications including homeland security, medical, and environmental monitoring and also food safety. A desirable goal is the ability to simultaneously analyze a wide variety of environmental and biological gases

F. Ren (✉) · B.H. Chu · K.H. Chen · C.Y. Chang · V. Chen
Department of Chemical Engineering, University of Florida, Gainesville, FL 32611, USA
e-mail: ren@che.ufl.edu

S.J. Pearton
Department of Materials Science and Engineering, University of Florida, Gainesville, FL 32611, USA
e-mail: spear@mse.ufl.edu

S. Pearton (ed.), *GaN and ZnO-based Materials and Devices*,
Springer Series in Materials Science 156, DOI 10.1007/978-3-642-23521-4_6,
© Springer-Verlag Berlin Heidelberg 2012

and liquids in the field and to be able to selectively detect a target analyte with high specificity and sensitivity. In the area of detection of medical biomarkers, many different methods have been employed [1–9]. Depending on the sample condition, these methods may show variable results in terms of sensitivity for some applications and may not meet the requirements for a handheld biosensor.

For biological and medical sensing applications, disease diagnosis by detecting specific biomarkers (functional or structural abnormal enzymes, low molecular weight proteins, or antigen) in blood, urine, saliva, or tissue samples has been established. Many techniques possess a major limitation in that only one analyte is measured at a time. Examples of detection of biomarkers using electrical measurements with semiconductor devices include carbon nanotubes for lupus erythematosus antigen detection [4], compound semiconductor nanowires and In_2O_3 nanowires for prostate-specific antigen detection [5], and silicon nanowire arrays for detecting prostate-specific antigen [9]. In clinical settings, biomarkers for a particular disease state can be used to determine the presence of disease as well as its progress.

Silicon-based sensors are still the dominant component of the semiconductor segment due to their low cost, reproducible and controllable electronic response. However, these sensors are not suited for operation in harsh environments, for instance, high temperature, high pressure, or corrosive ambients. Si will be etched by some of the acidic or basic aqueous solutions encountered in biological sensing. By sharp contrast, GaN is not etched by any acid or base at temperatures below a few hundred degrees. Therefore, III nitrides are alternative options to supplement silicon in these applications because of their chemical resistance, high temperature/high power capability, high electron saturation velocity, and simple integration with the existing GaN-based UV light-emitting diode, UV detectors, and wireless communication chips.

The high electron sheet carrier concentration of nitride HEMTs provides an increased sensitivity relative to simple Schottky diodes fabricated on GaN layers or field effect transistors (FETs) fabricated on the AlGaN/GaN HEMT structure. The gate region of the HEMT can be used to modulate the drain current in the FET mode or use as the electrode for the Schottky diode. A variety of gas, chemical, and health-related sensors based on HEMTs have been demonstrated with proper surface functionalization of the gate area of the HEMTs, including the detection of hydrogen, mercury ion, prostate-specific antigen (PSA), DNA, and glucose [10–14, 16–52].

In this chapter, we discuss recent progress in the functionalization of GaN for applications in detection of gases, pH measurement, biotoxins and other biologically important chemicals and the monitoring of nerve cell responses to outside inhibitors. We also look at fabrication and functionalization of InN sensors for sensing anions, pH, and polarity.

6.2 Gas Sensing

6.2.1 H_2 Sensing

There is interest in the detection of hydrogen sensors for use in hydrogen-fuelled automobiles and with fuel cells for space craft and other long-term sensing applications. These sensors are required to detect hydrogen near room temperature with minimal power consumption and weight and with a low rate of false alarms. Unlike conventional sensors where the changes of the sensing material's conductivity or resistivity are used to detect the gas concentration, by integrating the gas sensing material such as Pd or Pt metal on the gate electrode of the HEMTs, the change of the sensing material's conductivity can be amplified through Schottky diode or FET operation. H_2 is dissociated when adsorbed on Pt and Pd at room temperature. The reaction is as follows:

$$H_{2(ads)} \rightarrow 2H^+ + e^-$$

Dissociated hydrogen causes a change in the channel and conductance. This makes the integrated semiconductor device based sensors extremely sensitive and the sensors have a broad dynamic range of the sensing concentration. A surface covered reference semiconductor device can also be easily fabricated side-by-side with sensing device to eliminate the temperature variation of the ambient and fluctuation of the supplied voltage to the sensors.

Figure 6.1 shows a schematic of a Schottky diode hydrogen sensor on AlGaN/GaN HEMT layer structure and a photograph of packaged sensors. The devices were exposed to either 100% pure N_2 or 1% H_2 in nitrogen. For these diodes, the current increases upon introduction of H_2, through a lowering of the effective barrier height. The data were fit to the relations for thermionic emission and showed decreases in Schottky barrier height Φ_B of 30–50 mV at 50°C and larger changes at higher temperatures. The decrease in barrier height is completely reversible upon removing the H_2 from the ambient and results from diffusion of atomic hydrogen to the metal/GaN interface, altering the interfacial charge.

Figure 6.2 shows time response of diode forward current at a fixed bias of 0.6 V when switching back and forth between the ambient from N_2 to 1% H_2 balanced with nitrogen. The change in forward current for the diode was in the micro-amp range with a bias voltage of 0.6 V, which was corresponding to a power consumption of 3.6 μW. The response time was less than 1 s. Using the same layer structure, the sensor can also be fabricated as a field effect transistor. Excellent response time and repeatability were also achieved, as illustrated in Fig 6.2 (bottom).

To achieve the goal of detecting reactions due to hydrogen only and excluding other changes caused by variables such as temperature and moisture, a differential detection interface was used. Several kinds of differential devices have been fabricated and each of its performance has been evaluated to select the most effective solution. These differential devices have two sensors integrated on the same chip. The two sensors are identical except one is designed to react to hydrogen whereas

Fig. 6.1 Cross-sectional schematic of completed Schottky diode on AlGaN/GaN HEMT layer structure (*top*) and plan-view photograph of device (*bottom*)

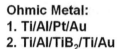

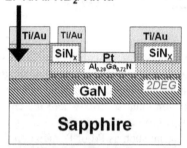

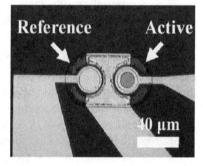

the other one is covered by dielectric protection layer and not exposed to ambient gas. One sensor reacted promptly with the exposure of hydrogen while the other, the reference diode, had no significant response as expected, proving the functionality of the differential sensor.

We have also found that nanostructured wide bandgap materials functionalized with Pd or Pt are sensitive than their thin film counterparts because of the large surface-to-volume ratio. Impressive results have been demonstrated with GaN, InN, and ZnO nanowires or nanobelts that are sensitive to hydrogen down to approximately 20 ppm at room temperature. As an example, Fig 6.3 (top) shows scanning electron microscopy (SEM) micrographs of as-grown nanowires. A layer of 10-nm-thick Pd was deposited by sputtering onto the nanowires to verify the effect of catalyst on gas sensitivity. The bottom of Fig. 6.3 shows the measured resistance at a bias of 0.5 V as a function of time from Pd-coated and uncoated multiple GaN nanowires exposed to a series of H_2 concentrations (200–1,500 ppm) in N_2 for 10 min at room temperature. Pd-coating of the nanowires improved the sensitivity to ppm level H_2 by a factor of up to 11. The addition of Pd appears to be effective in catalytic dissociation of molecular hydrogen. Diffusion of atomic hydrogen to the metal/GaN interface alters the surface depletion of the wires and hence the resistance at fixed bias voltage. The resistance change depended on the gas concentration but the variations were small at H_2 concentration above 1,000 ppm. The resistance after exposure to air was restored to 90% of initial level within 2 min.

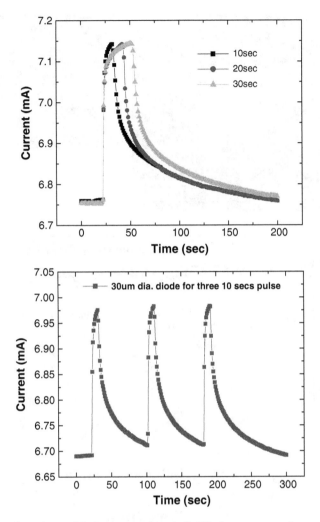

Fig. 6.2 Time dependence of drain-source current in GaN hydrogen sensor when switching from N_2 to $1\%H_2/99\%$ N_2 ambient and back again for periods of 10–30 s (*top*) or sequential 10 s pulses

6.2.2 O_2 Sensing

The current technology for O_2 measurement, referred to as oximetry, is small and convenient to use. However, it does not provide a complete measure of respiratory sufficiency. A patient suffering from hypoventilation (poor gas exchange in the lungs) given 100% oxygen can have excellent blood oxygen levels while still suffering from respiratory acidosis due to excessive CO_2. The O_2 measurement is also not a complete measure of circulatory sufficiency. If there is insufficient blood flow or insufficient hemoglobin in the blood (anemia), tissues can suffer hypoxia

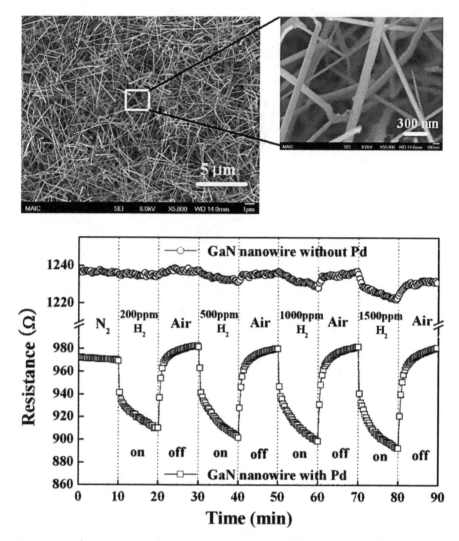

Fig. 6.3 SEM images of as-grown GaN nanowires (*top*) and measured resistance at an applied bias of 0.5 V as a function of time from Pd-coated and uncoated multiple GaN nanowires exposed to a series of H_2 concentrations (200–1,500 ppm) in N_2 for 10 min at room temperature

despite high oxygen saturation in the blood that does arrive. The current oxide-based O_2 sensors can operate at very high temperatures, e.g., commercialized solid electrolyte ZrO_2 (700°C) or the semiconductor metal oxides such as TiO_2, Nb_2O_5, $SrTiO_3$, and CeO_2 (> 400°C). However, it is important to develop a low operation temperature and high sensitivity O_2 sensor to build a small, portable, and low-cost O_2 sensor system for biomedical applications.

The conductivity mechanism of most metal oxides based semiconductors results from electron hopping from intrinsic defects in the oxide film and these defects

Fig. 6.4 Schematic of AlGaN/GaN HEMT based O_2 sensor (*top*) and drain current of IZO functionalized HEMT sensor measured at fixed source-drain during exposure to different O_2 concentration ambients. The drain bias voltage was 0.5 V and measurements were conducted at 117°C

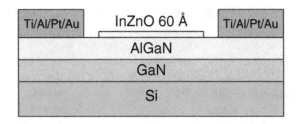

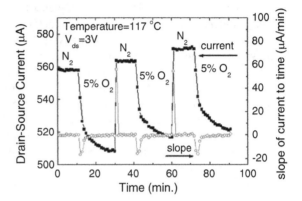

are related to the oxygen vacancies generated during oxide growth. Typically, the higher the concentration of oxygen vacancies in the oxide film, the more conductive is the film. InZnO (IZO) films have been used in fabricating thin film transistors and the conductivity of the IZO is also found to depend on the oxygen partial pressure during oxide growth. IZO is a good candidate for O_2 sensing applications [53–59].

The schematic of the oxygen sensor is shown at the top of Fig 6.4. The bottom part of the figure shows the device had a strong response when it was tested at 120°C in pure nitrogen and pure oxygen alternately at $V_{ds} = 3$ V. When the device was exposed to oxygen, the drain-source current decreased, whereas when the device was exposed to nitrogen, the current increased. The IZO film provides a high oxygen vacancy concentration, which makes the film readily sense oxygen and create a potential on the gate area of the AlGaN/GaN HEMT. This oxygen sensor can operate with a high sensitivity at a relatively low temperature compared to many oxide-based oxygen sensors which operate from 400 to 700°C.

6.2.3 CO_2 Sensing

The detection of carbon dioxide (CO_2) gas has attracted attention in the context of global warming, biological and health-related applications such as indoor air quality control, process control in fermentation, and in the measurement of CO_2 concentrations in the exhaled breath of patients with lung and stomach diseases.

In medical applications, it can be critical to monitor CO_2 and O_2 concentrations in the circulatory systems for patients with lung diseases in the hospital [16,61]. The current technology for CO_2 measurement typically uses IR instruments, which can be very expensive and bulky.

The most common approach for CO_2 detection is based on non-dispersive infrared (NDIR) sensors, which are the simplest of the spectroscopic sensors. The best detection limits for the NDIR sensors are currently in the range of 20–10,000 ppm. In the recent years, monomers or polymers containing amino groups, such as tetrakis (hydroxyethyl) ethylenediamine, tetraethylene-pentamine, and polyethyleneimine (PEI) have been used for CO_2 sensors to overcome the power consumption and size issues found in the NDIR approach Most of the monomers or polymers are utilized as coatings of surface acoustic wave transducers. The polymers are capable of adsorbing CO_2 and facilitating a carbamate reaction. PEI has also been used as a coating on carbon nanotubes for CO_2 sensing by measuring the conductivity of nanotubes upon exposing to the CO_2 gas For example, CO_2 adsorbed by a PEI coated nanotube portion of a NTFET (nanotube field effect transistor) sensor lowers the total pH of the polymer layer and alters the charge transfer to the semiconducting nanotube channel, resulting in the change of NTFET electronic characteristics.

A schematic cross-section of the device is shown in Fig 6.5. The interaction between CO_2 and amino group containing compounds with the influence of water molecules is based on an acid–base reaction. The purpose of adding starch into the PEI in our experiment was to enhance the absorption of the water molecules into the PEI/starch thin film. Several possible reaction mechanisms have been suggested. The key reaction was that the primary amine groups, $-NH_2$, on the PEI main chain reacted with CO_2 and water forming $-NH_3^+$ ions and the CO_2 molecule became $OCOOH^-$ ions. Thus, the charges, or the polarity, on the PEI main chain were changed. The PEI/starch was coated on the gate region of the HEMT. The charges of the PEI changed through the reactions between $-NH_2$ and CO_2 as well as water molecules. These are then transduced into a change in the concentration of the 2DEG in the AlGaN/GaN HEMTs [62].

Figure 6.5 (bottom) shows the drain current of PEI/starch functionalized HEMT sensors measured exposed to different CO_2 concentration ambients. The measurements were conducted at 108°C and a fixed source-drain bias voltage of 0.5 V. The current increased with the introduction of CO_2 gas. This was due to the increase in the net positive charges on the gate area, thus inducing electrons in the 2DEG channel. The response to CO_2 gas had a wide dynamic range from 0.9 to 50%. Higher CO_2 concentrations were not tested because there is little interest in these for medical related applications. The response times were on the order of 100 s. The signal decay time was slower than the rise time and was due to the longer time required to purge CO_2 out from the test chamber.

The effect of ambient temperature on CO_2 detection sensitivity was investigated. The drain current changes were linearly proportional to the CO_2 concentration for all the tested temperatures. However, the HEMT sensors showed higher sensitivity for the higher testing temperatures. There was a noticeable change of the sensitivity

Fig. 6.5 Schematic of
AlGaN/GaN HEMT based
CO_2 sensor (*top*) and drain
current of PEI/starch
functionalized HEMT sensor
measured at fixed
source-drain during exposure
to different CO_2
concentration ambients. The
drain bias voltage was 0.5 V
and measurements were
conducted at 108°C

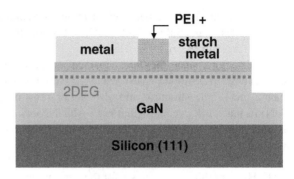

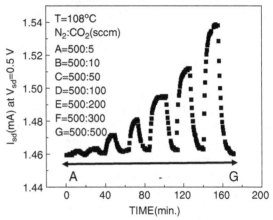

from the sensors tested at 61°C to those tested at 108°C. This difference is likely due
to higher ambient temperature increasing the reaction rate between amine groups
and CO_2 as well as the diffusion of CO_2 molecules into the PEI thin film. The
sensors exhibited reversible and reproducible characteristics.

In conclusion, PEI/starch functionalized HEMT sensors for CO_2 detection with
a wide dynamic range from 0.9 to 50%. The sensors were operated at low bias
voltage (0.5 V) for low power consumption applications. The sensors showed higher
sensitivity at the testing temperature higher than ~100°C. The sensors showed good
repeatability. This electronic detection of CO_2 gas is a significant step towards a
compact sensor chip, which can be integrated with a commercial available hand-
held wireless transmitter to realize a portable, fast and high sensitive CO_2 sensor.

6.2.4   CH$_4$ Sensing

Of particular interest in developing wide bandgap sensors are methods for detecting
ethylene (C_2H_4), which offers problems because of its strong double bonds and
hence the difficulty in dissociating it at modest temperatures. Ideal sensors have

the ability to discriminate between different gases and arrays that contain different metal oxides (e.g., SnO_2, ZnO, CuO, WO_3) on the same chip can be used to obtain this result [63, 67–83]. Another prime focus should be the thermal stability of the detectors, since they are expected to operate for long periods at elevated temperature. MOS diode-based sensors have significantly better thermal stability than a metal-gate structure and also better sensitivity than Schottky diodes on GaN. In this work, we show that both AlGaN/GaN MOS diodes and Pt/ZnO bulk Schottky diodes are capable of detection of low concentrations (10%) of ethylene at temperatures between 50 and 300°C (ZnO) or 25 and 400°C (GaN).

Pt/Sc_2O_3/AlGaN/GaN MOS-HEMT diodes were found to be capable of detecting C_2H_4. A possible mechanism for the current increases involves atomic hydrogen which is either decomposed from C_2H_4 in the gas phase or chemisorbed on the Pt Schottky contacts than catalytically decomposed to release atomic hydrogen. The hydrogen can then diffuse rapidly through the Pt metallization and the underlying oxide to the interface where it forms a dipole layer and lowers the effective barrier height. We emphasize that other mechanisms could be present; however, the activation energy for the current recovery is $\sim$1 eV, similar to the value for atomic hydrogen diffusion in GaN, which suggests that this is at least a plausible mechanism. As the detection temperature is increased, the response of the MOS-HEMT diodes increases due to more efficient cracking of the hydrogen on the metal contact. Note that the changes in both current and voltage are quite large and readily detected. In analogy with results for MOS gas sensors in other materials systems, the effect of the introduction of the atomic hydrogen into the oxide is to create a dipole layer at the oxide/semiconductor interface that will screen some of the piezo-induced charge in the HEMT channel.

6.3   Sensor Functionalization

Specific and selective molecular functionalization of the semiconductor surface is necessary to achieve specificity in chemical and biological detection. Devices such as FETs can readily discriminate between adsorption of oxidizing and reducing gas molecules from the changes in the channel conductance. However, precise identification of a specific type of molecule requires functionalization of the surface with specific molecules or catalysts. Effective biosensing requires coupling of the unique functional properties of proteins, nucleic acids (DNA, RNA), and other biological molecules with the solid-state "chip" platforms. These devices take advantage of the specific, complementary interactions between biological molecules that are fundamental aspects of biological function. Specific, complementary interactions are what permit antibodies to recognize antigens in the immune response, enzymes to recognize their target substrates, and the motor proteins of muscle to shorten during muscular contraction. The ability of biological molecules, such as proteins, to bind other molecules in a highly specific manner is the underlying principle of the "sensors" to detect the presence (or absence) of target molecules – just as it is in the biological senses of smell and taste.

In our HEMT devices, the surface is generally functionalized with an antibody or enzyme layer. The success of the functionalization is monitored by a number of methods. The first test is a change in surface tension when the functional layer is in place and the change in surface bonding can in some cases be seen by X-ray photoelectron spectroscopy. Typically, a layer of Au is deposited on the gate region of the HEMT as a platform to attach a chemical such as thioglycolic acid, whose S-bonds readily attach to the Au. The antibody layer can then be attached to the thioglycolic acid. When the surface is completely covered by these functional layers, the HEMT will not be sensitive to buffer solutions or water that does not contain the antigen of interest, as shown in Fig 6.6. For detecting hydrogen, the gate region is functionalized with a catalyst metal such as Pt or Pd. In other cases, we immobilize an enzyme to catalyze reactions, as is used for the detection of glucose. In the presence of the enzyme glucose oxidase, glucose will react with oxygen to produce gluconic acid and hydrogen peroxide. Table 6.1 shows a summary of the surface functionalization layers we have employed for HEMT sensors to date. There are many additional options for the detection of biotoxins and biological molecules of interest using different protein or antibody layers. The advantage of this approach is that large arrays of HEMTs can be produced on a single chip and functionalized with different layers to allow for the detection of a broad range of chemicals or gases.

6.4   pH Measurement

The measurement of pH is needed in many different applications, including medicine, biology, chemistry, food science, environmental science, and oceanography. Solutions with a pH less than 7 are acidic and solutions with a pH greater than 7 are basic or alkaline.

Ungated AlGaN/GaN HEMTs exhibit large changes in current upon exposing the gate region to polar liquids. The polar nature of the electrolyte introduced led to a change of surface charges, producing a change in surface potential at the semiconductor/liquid interface. The use of Sc_2O_3 gate dielectric produced superior results to either a native oxide or UV ozone-induced oxide in the gate region. The ungated HEMTs with Sc_2O_3 in the gate region exhibited a linear change in current between pH 3 and 10 of 37 $\mu A/pH$. The HEMT pH sensors show stable operation with a resolution of <0.1 pH over the entire pH range. 100 Å Sc_2O_3 was deposited as a gate dielectric through a contact window of SiN_x layer. For comparison, we also fabricated devices with just the native oxide present in the gate region and also with the UV ozone induced oxide. Figure 6.7 shows a scanning electron microscopy (SEM) image (top) and a cross-sectional schematic (bottom) of the completed device. The gate dimension of the device is $2 \times 50 \ \mu m^2$. The pH solution was applied using a syringe autopipette (2–20 μl).

Fig. 6.6 Example of
successful functionalization
of HEMT surface; the device
is no longer sensitive to water
when the surface is
completely covered with the
functional layer

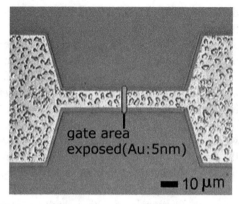

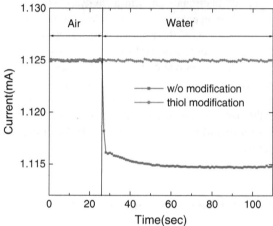

Table 6.1 Summary of surface functional layers used with HEMT sensors

Detection	Mechanism	Surface functionalization
H_2	Catalytic dissociation	Pd, Pt
Pressure change	Polarization	Polyvinylidene difluoride
Botulinum toxin	Antibody	Thioglycolic acid/antibody
Proteins	Conjugation/hybridization	Aminopropylsilane/biotin
pH	Adsorption of polar molecules	Sc_2O_3, ZnO
Hg^{2+}	Chelation	Thioglycolic acid/Au
KIM-1	Antibody	KIM-1 antibody
Glucose	GO_X immobilization	ZnO nanorods
Prostate Specific Antigen	PSA antibody	Carboxylate succimdyl ester/ PSA antibody
Lactic acid	LO_X immobilization	ZnO nanorods
Chloride ions	Anodization	Ag/AgCl electrodes; InN
Breast Cancer	antibody	Thyioglycolic acid/c-erbB antibody
CO_2	Absorption of water/charge	Polyethylenimine/starch
DNA	Hybridization	$3'$-thiol-modified oligonucleotides
O_2	Oxidation	InGaZnO

Fig. 6.7 SEM and schematic
of gateless HEMT with oxide
in the gate region

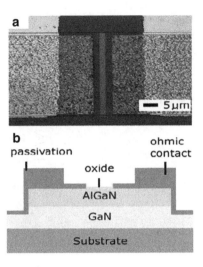

The adsorption of polar molecules on the surface of the HEMT affected the surface potential and device characteristics. Figure 6.8 shows the current at a bias of 0.25 V as a function of time from HEMTs with Sc_2O_3 in the gate region exposed for 150 s to a series of solutions whose pH was varied from 3 to 10. The current is significantly increased upon exposure to these polar liquids as the pH is decreased. By comparison, devices with the native oxide in the gate region showed a higher sensitivity of $\sim$70 μA/pA but a much poorer resolution of $\sim$0.4 pH and evidence of delays in response of 10–15 s. The latter may result from deep traps at the interface between the semiconductor and native oxide, whose density is much higher than at the Sc_2O_3–nitride interface. The devices with UV-ozone oxide in the gate region did not show these incubation times for detection of pH changes and showed similar sensitivities of gate source current as the Sc_2O_3 gate devices ($\sim$40 μA/pH) but with poorer resolution ($\sim$0.25 pH). As mentioned earlier, the pH range of interest for human blood is 7–8. Figure 6.8 (bottom) shows the current change in the HEMTs with Sc_2O_3 at a bias of 0.25 V for different pH values in this range. Note that the resolution of the measurement is <0.1 pH.

6.5   Exhaled Breath Condensate

There is significant interest in developing rapid diagnostic approaches and improved sensors for determining early signs of medical problems in humans. Exhaled breath is a unique bodily fluid that can be utilized in this regard [84–97]. Exhaled breath condensate pH is a robust and reproducible assay of airway acidity. Humans, even when they are extremely ill, will not have blood (or interstitial space between cells in tissue) pH below 7. When they do drift below this value, it almost invariably equals mortality.

Fig. 6.8 Change in current in gateless HEMT at fixed source-drain bias of 0.25 V with pH from 3–10 (*top*) and change in current in gateless HEMT at fixed source-drain bias of 0.25 V with pH from 7–8 (*bottom*)

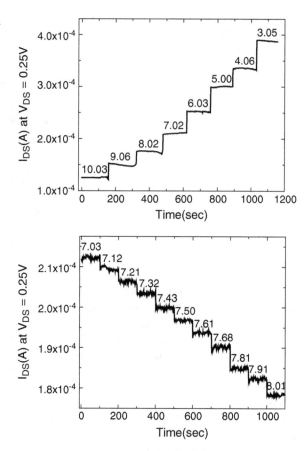

While most applications will detect substances or diseases in the breath as a gas or aerosol, breath can also be analyzed in the liquid phase as exhaled breath condensate (EBC). Analytes contained in the breath originating from deep within the lungs (alveolar gas) equilibrate with the blood, and therefore the concentration of molecules present in the breath is closely correlated with those found in the blood at any given time. EBC contains dozens of different biomarkers, such as adenosine, ammonia, hydrogen peroxide, isoprostanes, leukotrienes, peptide, cytokines, and nitrogen oxide. Analysis of molecules in EBC is noninvasive and can provide a window on the metabolic state of the human body, including certain signs of cancer, respiratory diseases, liver, and kidney functions [98–105].

The glucose oxidase enzyme (GOx) is commonly used in biosensors to detect the levels of glucose for diabetics. By keeping track of the number of electrons passed through the enzyme, the concentration of glucose can be measured. Due to the importance and difficulty of glucose immobilization, numerous studies have been focused on the techniques of immobilization of glucose with carbon nanotubes, ZnO nano-materials, and gold particles. ZnO-based nano-materials are especially interesting due to their non-toxic properties, low cost of fabrication, and favorable electrostatic interaction between ZnO and the GO_x lever. However, the activity of GO_x is highly dependent on the pH value of the solution [105]. The pH value of a typical healthy person can vary significantly depending on the health condition of each individual, e.g., the pH value for patients with acute asthma was reported as low as 5.23 ± 0.21 ($n = 22$) as compared to 7.65 ± 0.20 ($n = 19$) for the control subjects [104]. To achieve accurate glucose concentration measurement with immobilized GO_x, it is necessary to determine the pH value and glucose concentration with an integrated pH and glucose sensor.

6.6 Heavy Metal Detection

The detection of Superfund contaminants in ground water, along with testing of their effects on the environment and aquatic wildlife, can greatly improve environmental monitoring and management. While techniques for detection of hazardous environmental chemicals are readily available, they require the transportation of samples to a laboratory for analysis. Data analysis and collection require skilled expertise, is expensive, and require prolonged amounts of time. Thus, current detection techniques do not allow real time monitoring of environmental toxicants.

Arsenic and mercury are two of the most serious contaminants in the USA. Through different human activities, these contaminants discharge into soil and water resources, which impair the ecological and economic value of those resources.

Mercury is another serious contaminant in the USA. The majority (65%) of mercury enters into the environment from stationary combustion, of which coal-fired power plants are the largest aggregate source (40% of US mercury emissions in 1999).

A schematic cross-section of the device with Hg^{2+} ions bound to thioglycolic acid functionalized on the gold gate region and plan view photomicrograph of a completed device is shown in Fig 6.9. In some cases, we utilized sensors functionalized with thioglycolic acid, $HSCH_2COOH$, which is an organic compound that contains both a thiol (mercaptan) and a carboxylic acid functional group. A self assembled monolayer of thioglycolic acid molecule was adsorbed onto the gold gate due to strong interaction between gold and the thiol group. The extra thioglycolic acid molecules were rinsed off with DI water. An increase in the hydrophilicity of the treated surface by thioglycolic acid functionalization was confirmed by contact angle measurements which showed a change in contact angle from $58.4°$ to $16.2°$ after the surface treatment. The drain current of both sensors (i.e., those with only Au gate and those with the thioglycolic acid functionalized onto the Au layer) further reduced after exposure to different concentrations of Hg^{2+} ion solutions. The drain current was reduced $\sim$60% for the thioglycolic acid functionalized AlGaN/GaN HEMT sensors while bare-Au-gate sensors had less than a 3% change in drain current. The mechanism of the drain current reduction for bare Au gate and thioglycolic acid functionalized AlGaN/GaN HEMT sensors was

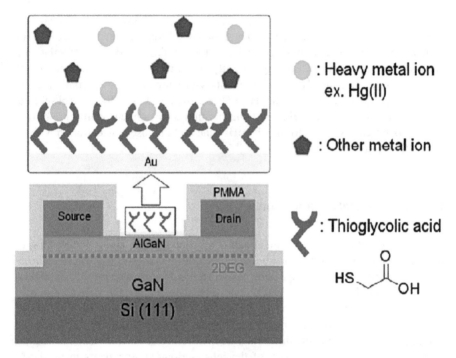

Fig. 6.9 Schematic of AlGaN/GaN HEMT. The Au-coated gate area was functionalized with thioglycolic acid

quite different. For the bare Au-gate devices, Au–mercury amalgam formed on the surface of the bare Au-gates when the Au-gate electrode was exposed to Hg^{2+} ion solutions. The formation rate of the Au–mercury amalgam depended on the solution temperature and the concentration of the Hg^{2+} ion solution. For the higher Hg^{2+} ion concentration solution, 10^{-5} M, the bare Au-gate based sensor took less than 15 s for the drain to reach steady state. However, the drain current required 30–55 s to reach steady state, when the sensor was exposed to the less concentrated Hg^{2+} ion solutions.

A response time of <5 s was obtained for the thioglycolic acid fictionalized AlGaN/GaN HEMT sensors, when the sensor was exposed to 10^{-5} M of the Hg^{2+} ion solution. This is the shortest response time of Hg^{2+} ion detection ever reported. For the thioglycolic acid functionalized AlGaN/GaN HEMT, the thioglycolic acid molecules on the Au surface align vertically with carboxylic acid functional group toward the solution. The carboxylic acid functional group of the adjacent thioglycolic acid molecules form chelates of $R - COO^-(Hg^{2+})^-OOC - R$ with Hg^{2+} ions when the sensors are exposed to the Hg^{2+} ion solution. The charges of trapped Hg^{2+} ion in the $R - COO^-(Hg^{2+})^-OOC - R$ chelates changed the polarity of the thioglycolic acid molecules, which were bonded to the Au-gate through –S–Au bonds. This is why the drain current changes in response to mercury ions. The difference in drain current for the device exposed to different Hg^{2+} ion

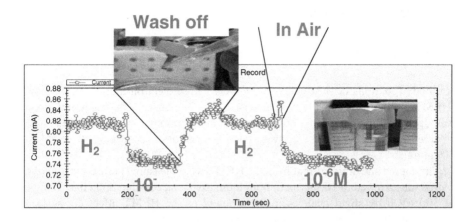

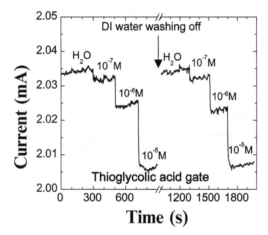

Fig. 6.10 Time-dependent response of the drain current for detecting Na^+, Mg^{2+}, or Hg^{2+} with a thioglycolic acid functionalized Au-gate HEMT sensor before and after washing the sensor in water

concentration to the DI water is illustrated in Fig 6.10. The Hg^{2+} ion concentration detection limit for the thioglycolic acid functionalized sensor is 10^{-7} M, which is equivalent to 2 ppb (parts per billion). The thioglycolic acid fucntionalized sensor also showed excellent sensing selectivity (over 100 times higher selectivity) over Na^+ and Mg^{2+} ions. The sensors can also be rinsed in DI water after a detection event and re-used, as shown in Fig 6.10. This is very convenient when making sure that false positives will not affect the ultimate judgment as to the presence of Hg in the test sample. Since our sensor chip is very compact (1 × 5 mm) and operates at extremely low power (8 μW based on 0.3 V of drain voltage and 80 μA of drain current operated at 11 Hz), it can be integrated with a commercial available hand-held wireless transmitter to realize a portable, fast response, and high sensitive Hg^{2+} ion detector.

6.7 Biotoxin Sensors

6.7.1 Botulinum

Antibody-functionalized Au-gated AlGaN/GaN high electron mobility transistors (HEMTs) show great sensitivity in detecting botulinum toxin. The botulinum toxin was specifically recognized through botulinum antibody, anchored to the gate area, as shown in Fig 6.11. We investigated a range of concentrations from 0.1 to 100 ng/ml.

The source and drain current from the HEMT were measured before and after the sensor was exposed to 100 ng/ml of botulinum toxin at a constant drain bias voltage of 500 mV. Any slight changes in the ambient of the HEMT affect the surface charges on the AlGaN/GaN. These changes in the surface charge are transduced into a change in the concentration of the 2DEG in the AlGaN/GaN HEMTs, leading

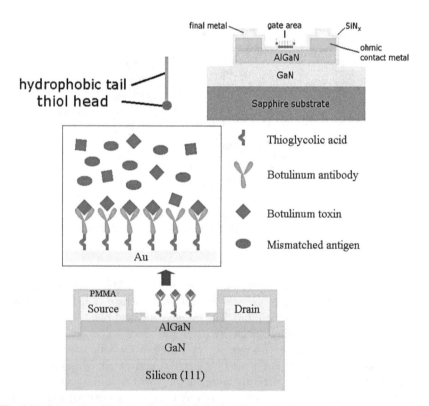

Fig. 6.11 Schematic of functionalized HEMT for botulinum detection

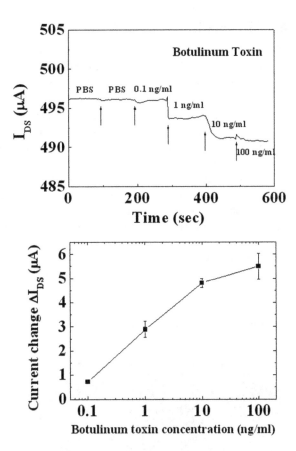

Fig. 6.12 Drain current of an AlGaN/GaN HEMT versus time for botulinum toxin from 0.1 to 100 ng/ml (*top*) and change of drain current versus different concentrations from 0.1 to 100 ng/ml of botulinum toxin (*bottom*)

to a decrease in the conductance for the device after exposure to botulinum toxin. Figure 6.12 (top) shows a real time botulinum toxin detection in PBS buffer solution using the source and drain current change with constant bias of 500 mV. No current change can be seen with the addition of buffer solution for around 100 s, showing the specificity and stability of the device. In clear contrast, the current change showed rapid response in less than 5 s when target 1 ng/ml botulinum toxin was added to the surface. The abrupt current change due to the exposure of botulinum toxin in a buffer solution was stabilized after the botulinum toxin thoroughly diffused into the buffer solution. Different concentrations (from 0.1 to 100 ng/ml) of the exposed target botulinum toxin in a buffer solution were detected. The sensor saturates above 10 ng/ml of the toxin. The limit of detection of this device was below 1 ng/ml of botulinum toxin in PBS buffer solution. The source–drain current change was nonlinearly proportional to botulinum toxin concentration, as shown in Fig 6.12 (bottom).

6.8 Biomedical Applications

AlGaN/GaN HEMTs can be used for measurements of pH in EBC and glucose, through integration of the pH and glucose sensors onto a single chip and with additional integration of the sensors into a portable, wireless package for remote monitoring applications [16, 107]. Figure 6.13 shows an optical microscopy image of an integrated pH and glucose sensor chip and cross-sectional schematics of the completed pH and glucose device. The gate dimension of the pH sensor device and glucose sensors was $20 \times 50 \, \mu m^2$.

For the glucose detection, an array of 20–30 nm diameter and $2 \, \mu m$ tall ZnO nanorods were grown on the $20 \times 50 \, \mu m^2$ gate area. The lower right inset in Fig 6.13 shows closer view of the ZnO nanorod arrays grown on the gate area. The total area of the ZnO was increased significantly with the ZnO nanorods. The ZnO nanorod matrix provides a microenvironment for immobilizing negatively charged GO_x while retaining its bioactivity, and passes charges produced during the GO_x and glucose interaction to the HEMT. The GO_x solution was prepared with a concentration of 10 mg/mL in 10 mM phosphate buffer saline (pH value of 7.4). After fabricating the device, $5 \, \mu l$ GO_x ($\sim$100 U/mg) solution was precisely introduced to the surface of the HEMT using a pico-liter plotter. The sensor chip

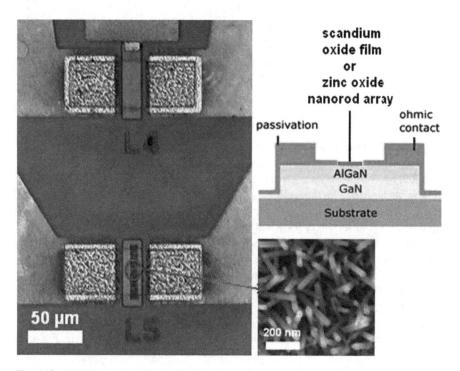

Fig. 6.13 SEM image of an integrated pH and glucose sensor. The insets show a schematic cross-section of the pH sensor and also an SEM of the ZnO nanorods grown in the gate region of the glucose sensor

was kept at $4°C$ in the solution for 48 h for GO_x immobilization on the ZnO nanorod arrays followed by extensively washing to remove the un-immobilized GO_x.

To take advantage of quick response (less than 1 s) of the HEMT sensor, a real-time EBC collector is needed [16, 109]. The amount of EBC required to cover the HEMT sensing area is very small. Each tidal breath contains around 3 μl of EBC. The contact angle of EBC on Sc_2O_3 has been measured to be less than $45°$, and it is reasonable to assume a perfect half sphere of EBC droplet formed to cover the sensing area of $4 \times 50\,\mu m^2$ gate area. The volume of a half sphere with a diameter of 50 μm is around 3×10^{-11} L. Therefore, 100,000 of 50 μm diameter droplets of EBC can be formed from each tidal breath.

To condense entire 3 μl of water vapor, only ~7 J of energy is needed to be removed for each tidal breath, which can be easily achieved with a thermal electric module, a Peltier device. The AlGaN/GaN HEMT sensor is directly mounted on top of the Peltier unit (TB-8–0.45–1.3 HT 232, Kryotherm), which can be cooled to precise temperatures by applying known voltages and currents to the unit. During our measurements, the hotter plate of the Peltier unit was kept at $21°C$, and the colder plate was kept at $7°C$ by applying bias of 0.7 V at 0.2 A. The sensor takes less than 2 s to reach thermal equilibrium with the Peltier unit. This allows the exhaled breath to immediately condense on the gate region of the HEMT sensor.

The glucose was sensed by ZnO nanorod functionalized HEMTs with glucose oxidase enzyme localized on the nanorods as shown in Fig. 6.14. This catalyzes the reaction of glucose and oxygen to form gluconic acid and hydrogen peroxide. Glucose detection using Au nano particles, ZnO nanorods, or carbon nanotubes with GO_x immobilization is based on electrochemical measurement [102–105]. Since

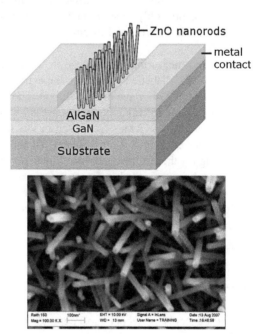

Fig. 6.14 Schematic of ZnO nanorod functionalized HEMT (*top*) and SEM of nanorods on gate area (*bottom*)

Fig. 6.15 Change in
drain-source current in
HEMT glucose sensors both
with and without localized
enzyme

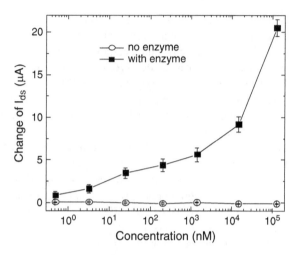

a reference electrode is required in the solution, the volume of sample cannot be easily minimized. The current density is measured when a fixed potential is applied between nanomaterials and the reference electrode. This is a first-order detection and the range of detection limit of these sensors is $0.5–70\,\mu M$. Even though the AlGaN/GaN HEMT-based sensor used the same GOx immobilization, the ZnO nanorods were used as the gate of the HEMT. Although the response of the HEMT-based sensor is similar to that of an electrochemical-based sensor, a much lower detection limit of $0.5\,nM$ was achieved for the HEMT due to this amplification effect. Since no reference electrode is required for the HEMT-based sensor, the amount of sample only depends on the area of gate dimension and can be minimized. The sensors do not respond to glucose unless the enzyme is present, as shown in Fig 6.15.

Although measuring the glucose in the EBC is a noninvasive and convenient method for the diabetic application, the activity of the immobilized GO_x is highly dependent on the pH value of the solution. The GOx activity can be reduced to 80% for pH 5–6. If the pH value of the glucose solution is larger than 8, the activity drops off very quickly. When the glucose sensor was used in a pH-controlled environment, the drain current stayed fairly constant. The human pH value can vary significantly depending on the health condition. Since we cannot control the pH value of the EBC samples, we needed to measure the pH value while determining the glucose concentration in EBC. With the fast response time and low volume of EBC required for HEMT-based sensor, a hand-held and real-time glucose sensing technology can be realized.

6.8.1 Prostate Cancer Detection

Prostate cancer is the second most common cause of cancer death among men in the USA. The most commonly used serum marker for the diagnosis of prostate cancer is prostate-specific antigen (PSA). The market size for prostate cancer testing is enormous. According to the American Cancer Society, prostate cancer is the most common form of cancer among men, next to skin cancer. It is estimated that during 2007, in the USA alone, 218,890 new cases of prostate cancer will be diagnosed and 1 in 6 men will be diagnosed with prostate cancer during his lifetime.

Prostate-specific antigen (PSA) is made by cells in the prostate gland and although PSA is mostly found in semen, a certain amount is found in the blood as well. Most men have PSA levels under 4 ng/ml blood. When prostate cancer develops, the PSA level usually goes up above 4 ng/ml; however, about 15% of men with a PSA below four will have prostate cancer on biopsy. If the patient's PSA level is between four and ten, their chance of having prostate cancer is about 25%. If the patient's PSA level is above ten, there is more than a 50% chance they have prostate cancer, which increases as the PSA level goes up. If the patient's PSA level is high, the doctor may advise a prostate biopsy to find out if they have cancer.

Generally PSA testing approaches are expensive, time-consuming, and need sample transportation. A number of different electrical measurements have been used for rapid detection of PSA. For example, electrochemical measurements based on impedance and capacitance are simple and inexpensive but need improved sensitivities for use with clinical samples. Resonant frequency changes of an anti-PSA antibody coated microcantilever enable a detection sensitivity of $\sim$10 pg/ml but this micro-balance approach has issues with the effect of the solution on resonant frequency and cantilever damping. Antibody-functionalized nanowire FETs coated with antibody provide for low detection levels of PSA, but the scale-up potential is limited by the expensive e-beam lithography requirements.

Antibody functionalized Au-gated AlGaN/GaN HEMTs shown schematically in Fig 6.16 were found to be effective for detecting PSA at low concentration levels. The PSA antibody was anchored to the gate area through the formation of carboxylate succinimdyl ester bonds with immobilized thioglycolic acid. The devices could detect a range of concentrations from 1 μg/ml to 10 pg/ml. The lowest detectable concentration was two orders of magnitude lower than the cut-off value of PSA measurements for clinical detection of prostate cancer. Figure 6.17 shows the real time PSA detection in PBS buffer solution using the source and drain current change with a constant bias of 0.5 V [42]. No current change can be seen with the addition of buffer solution or nonspecific bovine serum albumin (BSA), but there was a rapid change when 10 ng/ml PSA was added to the surface. The abrupt current change due to exposure of PSA in a buffer solution could be stabilized after the PSA diffused into the buffer solution. The ultimate detection limit appears to be a few pg/ml.

6.8.2 Kidney Injury Molecule Detection

Problems such as acute kidney injury (AKI) or acute renal failure (ARF) are unfortunately still associated with a high mortality rate. An important biomarker for

Fig. 6.16 Schematic of HEMT sensor functionalized for PSA detection

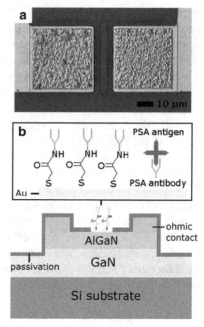

Fig. 6.17 Drain current versus time for PSA detection when sequentially exposed to PBS, BSA, and PSA

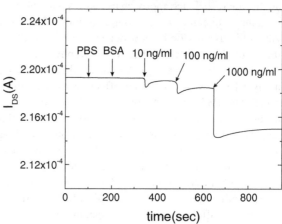

early detection of AKI is the urinary antigen known as kidney injury molecule-1 or KIM-1 and this is generally carried out with the ELISA technique. The biomarker can also be detected with particle-based flow cytometric assay, but the cycle time is several hours. Electrical measurement approaches based on carbon nanotubes, nanowires of In_2O_3 or Si, or Si or GaN FETs look promising for fast and sensitive detection of antibodies and potentially for molecules such as KIM-1.

The functionalization scheme in the gate region began with thioglycolic acid followed by KIM-1 antibody coating. The gate region was deposited with a 5 nm thick Au film. Then the Au was conjugated to specific KIM-1 antibodies with a

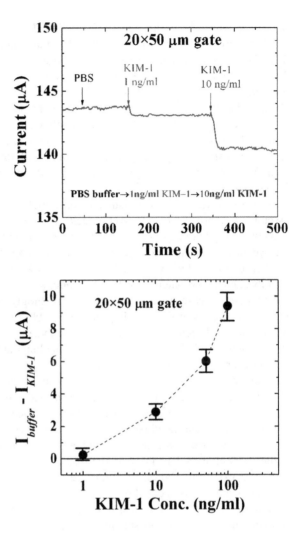

Fig. 6.18 Time-dependent current signal when exposing the HEMT to 1 ng/ml and 10 ng/ml KIM-1 in PBS buffer

self-assembled monolayer of thioglycolic acid. The HEMT source-drain current showed a clear dependence on the KIM-1 concentration in phosphate-buffered saline (PBS) buffer solution, as shown in Fig 6.18 where the time-dependent source-drain current at a bias of 0.5 V is plotted for KIM-1 detection in PBS buffer solution. The limit of detection (LOD) was 1 ng/ml using a $20 \times 50\,\mu$m gate sensing area.

6.8.3 Breast Cancer

When breast cancer is discovered early on, there is a much better chance of successful treatment. Therefore it is highly recommended that women check their breasts monthly from the age of 20. Clinical breast examinations should be conducted every 3 years from ages 20–39 and an annual mammogram for women 50 and older. Studies indicate a 96% survival rate if patients could be screened every 3 months. Thus, mortality in breast cancer patients could be reduced by increasing the frequency of screening. However, this is not presently feasible due to the lack of cheap and reliable technologies that can screen breast cancer non-invasively.

Antibody-functionalized Au-gated AlGaN/GaN high electron mobility transistors (HEMTs) show promise for detecting c-erbB-2 antigen. The c-erbB-2 antigen was specifically recognized through c-erbB antibody, anchored to the gate area. We investigated a range of clinically relevant concentrations from 16.7 to $0.25\,\mu g/ml$.

The Au surface was functionalized with thioglycolic acid. We anchored a self-assembled monolayer of thioglycolic acid, $HSCH_2COOH$, an organic compound and containing both a thiol (mercaptan) and a carboxylic acid functional group, on the Au surface in the gate area through strong interaction between gold and the thiol group of the thioglycolic acid. The devices were first placed in the ozone/UV chamber and then submerged in 1 mM aqueous solution of thioglycolic acid at room temperature. This resulted in the binding of thioglycolic acid to the Au surface in the gate area with the COOH groups available for further chemical linking of other functional groups. The device was incubated in a phosphate buffered saline (PBS) solution of $500\,\mu g/ml$ c-erbB-2 monoclonal antibody for 18 h before real time measurement of c-erbB-2 antigen.

After incubation with a PBS buffered solution containing c-erbB-2 antibody at a concentration of $1\,\mu g/ml$, the device surface was thoroughly rinsed off with deionized water and dried by a nitrogen blower. The source and drain current from the HEMT were measured before and after the sensor was exposed to $0.25\,\mu g/ml$ of c-erbB-2 antigen at a constant drain bias voltage of 500 mV. Any slight changes in the ambient of the HEMT affect the surface charges on the AlGaN/GaN.

Figure 6.19 (top) shows real time c-erbB-2 antigen detection in PBS buffer solution using the source and drain current change with constant bias of 500 mV. No current change can be seen with the addition of buffer solution for around 50 s, showing the specificity and stability of the device. In clear contrast, the current change showed a rapid response in less than 5 s when target $0.25\,\mu g/ml$ c-erbB-2 antigen was added to the surface. The abrupt current change due to the exposure of c-erbB-2 antigen in a buffer solution was stabilized after the c-erbB-2 antigen thoroughly diffused into the buffer solution. Three different concentrations (from 0.25 to $16.7\,\mu g/ml$) of the exposed target c-erbB-2 antigen in a buffer solution were detected. The limit of detection of this device was $0.25\,\mu g/ml$ c-erbB-2 antigen in PBS buffer solution. The source-drain current change was nonlinearly proportional to c-erbB-2 antigen concentration, as shown in Fig 6.19 (bottom). Between each test, the device was rinsed with a wash buffer of pH 6.0 phosphate buffer solution containing KCl to strip the antibody from the antigen.

Clinically relevant concentrations of the c-erbB-2 antigen in the saliva and serum of normal patients are 4–$6\,\mu g/ml$ and 60–$90\,\mu g/ml$, respectively. For breast cancer patients, the c-erbB-2 antigen concentrations in the saliva and serum are

Fig. 6.19 Drain current of an AlGaN/GaN HEMT over time for c-erbB-2 antigen from 0.25 to 17 μg/ml (*top*) and change of drain current versus different concentrations from 0.25 to 17 μg/ml of c-erbB-2 antigen

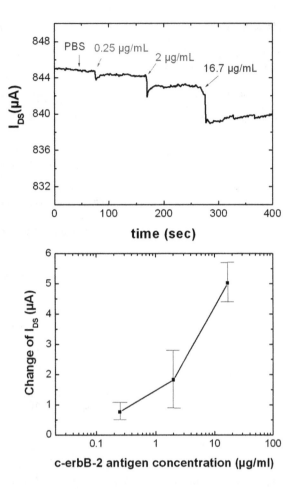

9–13 μg/ml and 140–210 μg/ml, respectively. Our detection limit suggests that HEMTs can be easily used for the detection of clinically relevant concentrations of biomarkers.

6.8.4   *Lactic Acid*

Lactic acid can also be detected with ZnO nanorod-gated AlGaN/GaN HEMTs. Interest in developing improved methods for detecting lactate acid has been increasing due to its importance in areas such as clinical diagnostics, sports medicine, and food analysis.

A ZnO nanorod array, which was used to immobilize lactate oxidase (LO_x), was selectively grown on the gate area using low temperature hydrothermal decomposition (Fig 6.20, top). The array of one-dimensional ZnO nanorods provided a large

Fig. 6.20 Schematic
cross-sectional view of the
ZnO nanorod gated HEMT
for lactic acid detection (*top*)
and plot of drain current
versus time with successive
exposure to lactate acid from
167 nM to 139 µM (*bottom*)

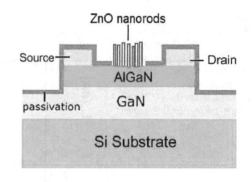

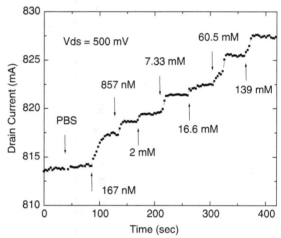

effective surface area with high surface-to-volume ratio and a favorable environment
for the immobilization of LO_x. The AlGaN/GaN HEMT drain-source current
showed a rapid response when various concentrations of lactate acid solutions were
introduced to the gate area of the HEMT sensor. The HEMT could detect lactate
acid concentrations from 167 nM to 139 µM. Figure 6.20 (bottom) shows a real time
detection of lactate acid by measuring the HEMT drain current at a constant drain-
source bias voltage of 500 mV during exposure of HEMT sensor to solutions with
different concentrations of lactate acid. The sensor was first exposed to 20 µl of
10 mM PBS and no current change could be detected with the addition of 10 µl
of PBS at approximately 40 s, showing the specificity and stability of the device.
By contrast, a rapid increase in the drain current was observed when target lactate
acid was introduced to the device surface. The sensor was continuously exposed to
lactate acid concentrations from 167 nM to 139 µM.

As compared with the amperometric measurement based lactate acid sensors, our
HEMT sensors do not require a fixed reference electrode in the solution to measure
the potential applied between the nano-materials and the reference electrode [110–
115]. The lactate acid sensing with the HEMT sensor was measured through the

drain current of HEMT with a change of the charges on the ZnO nanorods and the detection signal was amplified through the HEMT. Although the time response of the HEMT sensors is similar to that of electrochemical based sensors, a significant change of drain current was observed for exposing the HEMT to the lactate acid at a low concentration of 167 nM due to this amplification effect. In addition, the amount of sample, which is dependent on the area of gate dimension, can be minimized for the HEMT sensor due to fact no reference electrode is required. Thus, measuring lactate acid in the exhaled breath condensate (EBC) can be achieved as a noninvasive method.

6.8.5 Chloride Ion Detection

Chlorine is widely used in the manufacture of many products and items directly or indirectly, e.g., in paper product production, antiseptic, dye-stuffs, food, insecticides, paints, petroleum products, plastics, medicines, textiles, solvents, and many other consumer products. It is used to kill bacteria and other microbes in drinking water supplies and waste water treatment.

HEMTs with a Ag/AgCl gate are found to exhibit significant changes in channel conductance upon exposing the gate region to various concentrations of chorine ion solutions. The Ag/AgCl gate electrode, prepared by potentiostatic anodization, changes electrical potential when it encounters chorine ions [Fig 6.21 (top)]. This gate potential changes lead to a change of surface charge in the gate region, inducing a higher positive charge on the AlGaN surface and increasing the piezo-induced charge density in the HEMT channel. These anions create an image positive charge on the Ag gate metal for the required neutrality, thus increasing the drain current of the HEMT.

Figure 6.21 (bottom) shows the time dependence of Ag/AgCl HEMT drain current at a constant drain bias voltage of 500 mV during exposure to solutions with different chlorine ion concentrations. The HEMT sensor was first exposed to DI water and no change of the drain current was detected with the addition of DI water at 100 s. This stability was important to exclude possible noise from the mechanical change of the NaCl solution. By sharp contrast, there was a rapid response of HEMT drain current observed in less than 30 s when the target of 1×10^{-8} M NaCl solution was switched to the surface at 175 s. 1×10^{-7} M of NaCl solution was then applied at 382 s and it was accompanied with a larger signal corresponding to the higher chlorine concentration. Further real time tests were carried out to explore the detection of higher Cl^- ion concentrations. The sensors were exposed to 10^{-8}, 10^{-7}, 10^{-6}, 10^{-5}, and 10^{-4} M solutions continuously and repeated five times to obtain the standard deviation of source-drain current response for each concentration. The limit of detection of this device was 1×10^{-8} M chlorine in DI water. Between each test, the device was rinsed with DI water. These results suggest that our HEMT sensors are recyclable with simple DI water rinse.

Fig. 6.21 Schematic
cross-sectional view of a
Ag/AgCl gated HEMT (*top*)
and time-dependent drain
current of a Ag/AgCl gated
AlGaN/GaN HEMT exposed
to different concentrations of
NaCl solutions (*bottom*)

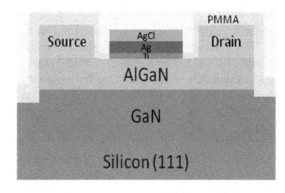

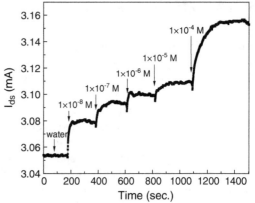

Real time detection of chloride ion detection with HEMTs with an InN thin film
in the gate region has also been demonstrated. The InN, deposited by Molecular
Beam Epitaxy, provided fixed surface sites for reversible anion coordination. The
potential change in the gate area induced a change of the piezo-induced charge
density in the electron channel in the HEMT.

6.8.6 Pressure Sensing

Piezoelectric materials are used widely as sensitive pressure sensors and piezoelectric gauges are typically fabricated with materials such as PZT, lithium niobate, and
quartz. There is a very high piezoelectric effect in polarized polyvinylidene fluoride
(PVDF). Polarized PVDF has also become an important piezoelectric material due
to its flexibility, low density, low mechanical impedance, and easy fabrication as
a ferroelectric. Because of its versatility, PVDF has many applications in low-cost
and disposable pressure sensors. HEMTs with a polarized polyvinylidene difluoride
(PVDF) film coated on the gate area exhibited significant changes in channel
conductance upon exposure to different ambient pressures. The PVDF thin film

Fig. 6.22 Schematic of
HEMT sensor coated with
polarized PVDF (*top*) and
drain current of a PVDF
gated AlGaN/GaN HEMT as
a function of pressure. The
PVDF was polarized by
grounding the copper chuck
holding the sample and the
copper wire electrode was
applied with 10 kV

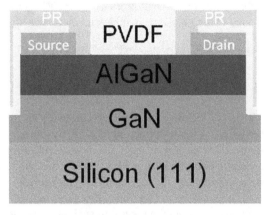

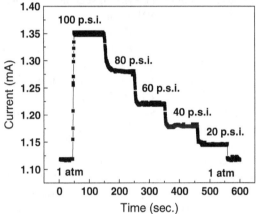

was deposited on the gate region with an inkjet plotter. Next, the PDVF film was
polarized with an electrode located 2 mm above the PVDF film at a bias voltage
of 10 kV and 70°C. A schematic of the HEMT is shown in Fig 6.22. Variations in
ambient pressure induced changes in the charge in the polarized PVDF, leading to
a change of surface charges on the gate region of the HEMT. Changes in the gate
charge were amplified through the modulation of the drain current in the HEMT. By
reversing the polarity of the polarized PVDF film, the drain current dependence on
the pressure could be reversed. Our results indicate that HEMTs have potential for
use as pressure sensors. For the pressure sensing measurement, the HEMTs sensor
were mounted on a carrier and put in a pressure chamber. Figure 6.22 (bottom) also
shows real time pressure detection with the polarized PVDF gated HEMT. The drain
current of HEMT sensor showed a rapid decrease in less than 5 s when the ambient
pressure was changed to 20 psig. A further decrease of the drain current for the
HEMT sensor was observed when the chamber pressure increased to 40 psig. These
abrupt drain current decreases were due to the change of charges in the PVDF film

upon a shift of ambient pressure. A HEMT sensor without the PDVF coating was loaded in the pressure chamber and there was no change of drain current observed.

6.8.7 Traumatic Brain Injury

TBI is one of the most frequent causes of morbidity and mortality in the modern battlefield. The development of a fast response and portable TBI sensor can have tremendous impact in early diagnosis, and proper management of TBI. Accurate and early diagnosis of a soldier's health in acute care environments can significantly simplify decisions about situation management. For example, decisions need to be made about whether to admit or discharge injured soldiers or to transfer other facility with advanced diagonal system, such as computer tomography (CT) and magnetic resonance imaging (MRI) scans.

TBI antibody can be functionalized on the HEMT surface and fast response of TBI antigen was achieved. The detection of limit of detection (LOD) was in the one-tenth of μg/ml range; however, this is not low enough for practical use. The typical TBI antigen concentration in the TBI patient's serum is in the range of ng/ml. We have used HEMT sensors to detect the kidney injury molecules and prostate-specific antigen and achieved the LOD in the range of 1–10 pg/ml range. The reason for higher LOD for the TBI antigen detection was due to the much smaller size of the TBI antigen. Smaller antigens carry less charges, thus provide less effect on the drain current of the HEMT sensor. Based on the promising biomarker and device data, we have recently used HEMTs for detecting a biomarker UCH-L1 (BA0127) antigen involved in Traumatic Injury Molecule. The gate region was functionalized with a specific antibody to traumatic brain injury antigen. The HEMT current showed a decrease as a function of TBI antigen concentration in PBS buffer (Fig 6.23). This shows time-dependent current change in BA0127 (UCH-L1) antibody modified HEMTs upon exposure to 2 μg/ml, and then to 16.9, 80, and 188 μg/ml of BA0127 (UCH-L1) in PBS buffer. The response time is around 6 s. The preliminary limit of detection (LOD) was found to be 20 μg/ml,

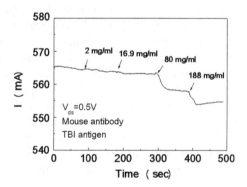

Fig. 6.23 Time-dependent current signal when exposing the HEMT to 2 to 188 μg/ml BA0127 TBI antigen in PBS buffer

demonstrating the potential for TBI detection with accurate, rapid, noninvasive, and high-throughput capabilities.

6.9   Nerve Cell Monitoring

Mammalian nerve cells are one of the most sophisticated signaling mechanisms in living organisms. The study of the physiological response mechanisms in these cells can pave the way for better treatment of neurological diseases such as Alzheimer's or Parkinson's disease, which are rooted in disrupted cell processes. By studying nerve cells' response to neuroinhibitors (which induce the disrupted cell processes), we can gain a better understanding of these diseases. Additionally, in the field of pharmacology, it is important to study the effects of chemical compounds on different cells in the body [16, 117]. AlGaN/GaN biosensors offer a viable method of monitoring and studying human nerve cell response to inhibitors.

Because of the excellent chemical stability and biocompatibility of the AlGaN/GaN sensor surface, no surface functionalization is necessary to couple the cell lines to the sensor [118]. However, to ensure stable measurement conditions, it was necessary to use a buffered cell media and a covered measurement setup. NG 108-15 cells were proliferated on the sensor over a period of 3 days. Between the compact cell layer and the sensor surface, an ion accumulation channel (called the cleft) was formed [119]; cell activity can be monitored by probing the ion concentration or potential changes (that result from various neuroinhibitors acting on the cells) in the cleft. Before monitoring the cells, it was confirmed that the sensor was sensitive to changes in ion concentrations of Na^+, one of the primary ions that the neuroinhibitors PMSF (phenylmethanesulphonylfluoride), DFP (diisoproppylfluorophosphate), and amiloride cause a change in the ionic concentration (Fig 6.24).

The AlGaN/GaN sensor was used to monitor alkalization and breathing of the nerve cells, as well as the cells' responses to being dosed with different

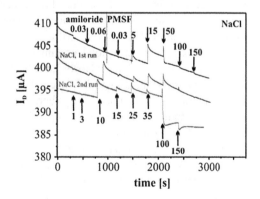

Fig. 6.24 Sensor response upon dosing of Na^+ ions in SCZ (substituted cell media with zero ions) solution. Amiloride and PMSF are supplied prior to the Na^+ dosing. *Numbers* refer to end concentrations in the solutions in units of mM (after [118])

experiment	cell reaction	sensor response	underlying effect
alkalization	consumption of H^3O^+ and diffusion of CO_2 into the atmosphere	I_D decreases ($\downarrow$)	increasing of OH⁻ ions
breathing	decreasing CO_2 concentration in the cell medium, creating non-equilibrium of membrane potential	I_D oscillates ($\updownarrow$)	alternating HCO_3-concentration in the cell medium
dosing of DFP	inhibiting of AChE, thus, keeping the Na^+ channels open	I_D decreases ($\downarrow$)	depletion of positive ions in the cleft (Na^+) due to equilibration of the concentration gradient
dosing of amiloride	blocking of Na^+ channels	I_D increases ($\uparrow$)	accumulation of positive ions in the cleft (Na^+, K^+)
dosing of PMSF	inhibiting of AChE, thus, keeping the Na^+ channels open	I_D decreases ($\downarrow$)	depletion of positive ions in the cleft (Na^+) due to equilibration of the concentration gradient

Fig. 6.25 Summary of the observed sensor responses to cell reactions caused by different treatments (after [118])

neuroinhibitors. Figure 6.25 (table) gives a summary of the results: alkalization caused the drain current to decrease as a result of an increase in OH⁻ ions. Cell breathing resulted in the drain current oscillating because of alternating HCO_3^- concentrations in the cell medium. Both dosing with DFP and PMSF resulted in a decrease in drain current; this was caused by a depletion of positive ions in the cleft. Dosing with amiloride caused a blocking of Na^+ channels, resulting in an increase in drain current (because of the accumulation of positive ions in the cleft). Furthermore, cellular response to multiple neurotoxins within the same treatment was measured. Figure 6.26 shows cell reaction to alternating dosings of PMSF and amiloride. Recording the sensor signal in cycles, it was first observed that the initial dosing with 10 μm PMSF caused a ~10 uA drop in I_D, while subsequent doses (even after a 2 h break time) produced no discernable effect on I_D. However, even after the first dose, amiloride continued to cause decreases in drain current. In the subsequent treatment, when the PMSF was split into several 3 μm doses, I_D still stopped decreasing after the total amount of PMSF approached ~12 μm (suggesting a saturation I_D of ~12 μm). However, after the 3 h break, the cells had a partial recovery, as a first dose of PMSF elicited a decrease in I_D. Amiloride continued decreasing the I_D. When acting together, the neuroinhibitors had a different effect on the cells than when they were applied individually, and this discrepancy was readily detected by the sensors.

Finally, ion flux calculations/modeling provided strong evidence that the Na^+ flux plays a major role in signal generation in the sensor.

The viability of AlGaN/GaN sensors for the quantitative analysis of cell reactions to different neuroinhibitors has been demonstrated. The sensors show excellent

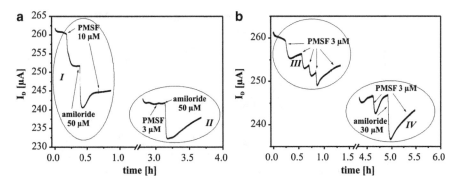

Fig. 6.26 I_{DS} versus time as reaction of the cells to different inhibitors at (**a**) treatment 1 with a break of 2 h and (**b**) treatment 2 with a break of 3 h (after [118])

biocompatibility with the nerve cells, stable operation under various physiological stresses, and strong signal resolution. These conditions allow for long-duration measurements and enable easy measurement procedures. Although it cannot be fully determined what ion(s) is involved in sensor signal generation, quantitative modeling suggests $Na^{+'}$s role in creating signals. However, such models are not perfect and need further refinement. With further improvements, these sensors and sensing procedures show large potential for use in pharmaceutical screening, drug detection, tumor analysis, and other biomedical applications.

6.10   InN Sensors

InN has become an attractive material for chemical and biological sensing applications because of its unusually strong surface electron accumulation layer [120–125]. This accumulation layer forms a natural 2DEG at the InN surface (unlike the 2DEG which occurs at the buried interface of the heterostructure of a HEMT). Adsorption on the InN surface modulates the electron density in the accumulation layer, giving rise to significant changes in surface conductance [126]. Work demonstrating the use of InN-based ISFETs for applications in sensing anions, pH, and polarity is presented.

N-face InN film grown on a silicon substrate was shown to be an excellent sensing material that can selectively respond to anions in aqueous solutions [126]. Using an InN-based ISE (ion-selective electrode), potentiometric measurements were conducted to assess the change in potential in the InN-electrode interface when exposed to IA/IIA chloride and hydroxyl solutions of various concentrations. Results indicated that the ISE selectively responds to chlorine ions (anions) instead of cations. Additionally, the ISE's dynamic response to concentration changes of KCl was measured. The measurements showed that the InN-based ISE has a fast response time and long-term stability and yields results with high repeatability.

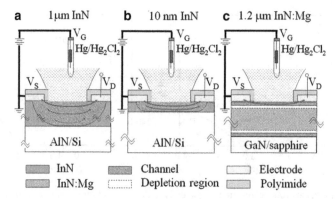

Fig. 6.27 Schematic diagrams of three investigated ISFET structures, including $1\,\mu\text{m}$-thick -c-InN, 10-nm-thick -c-InN, and $1.2 - \mu\text{m}$-thick a-InN:Mg epilayers (after [126])

Fig. 6.28 Dynamic response of the ultrathin InN-based ISFET in a KCl solution with varying KCl concentrations and exposed to a constant V_{DS} of 0.2 V (after [126])

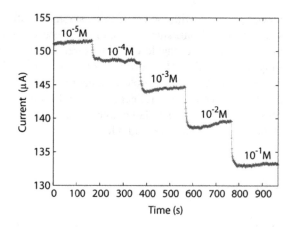

Three InN-based ISFETs were fabricated. Initial I_{DS}–V_{GS} characterization (Fig 6.27) showed that the 10 nm thick –c-InN and the $1.2\,\mu\text{m}$ thick a-InN:Mg ISFETs showed relatively higher current variation ratios than the $1\,\mu\text{m}$ thick –c-InN one. As a result, these two were used for the sensing studies. Compared with ISFETs based on other III-nitrides, InN-based ISFETs exhibit higher current ratio upon the same magnitude of gate voltage change [16, 128].

Ultrathin (10 nm thick) InN-based ISFETs were used to perform chlorine anion sensing. The ISFET was exposed to varying concentrations of KCl ranging from 10^{-5} to 10^{-1} M and given a constant V_{DS} of 0.2 V. The ISFET showed clear changes in current upon exposure to different ion concentrations (see Fig 6.28). The change in current with respect to each order of magnitude of change in chlorine ion concentration was found to be $5\,\mu\text{A}$, indicating a sensitivity of $5\,\mu\text{A}/\text{pCl}$ (pCl denotes $-\log[\text{Cl}^-]$). In addition, the response time was shown to be less than 10 s. Figure 6.29 shows the semi-log plot of I_{DS} versus KCl concentration; I_{DS} decreases linearly as concentration increases logarithmically.

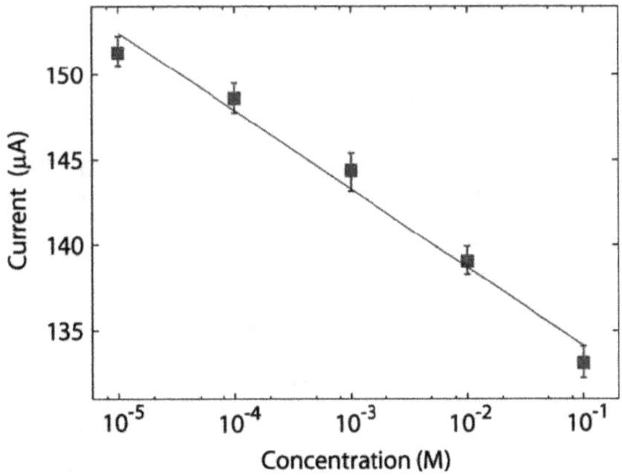

Fig. 6.29 Current response of the InN ISFET vs. the log of the concentration of KCl; the tested device was shown to have a sensitivity of 5 pCl (after [126])

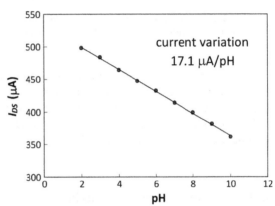

Fig. 6.30 Current response of an ultrathin ISFET to various pH buffers ranging from pH 2 to pH 10 at a constant V_{DS} of 0.2 V (after [126])

Ultrathin (10 nm) and 1.2 μm thick a-InN:Mg ISFETs were used to analyze the response of the InN and InN:Mg surfaces to changes in pH. The ultrathin ISFET was shown to have an average current variation of 17.1 μA/pH (corresponding to a current variation ratio of 4.12%) when exposed to different pH levels (see Fig 6.30). It was also shown to have a gate sensitivity of ~58.3 mV/pH, very close to the theoretical Nernst response of 59 mV/pH. The a-InN:Mg ISFET was shown to have a gate sensitivity of 56.5 mV/pH. Figure 6.31 shows the dynamic response of an a-InN:Mg ISFET to different pH levels while being given a constant V_{DS} of 0.1 V. This figure reveals the distinct and abrupt changes when the pH changed, with an estimated average response time of 10 s. Later measurements of the ISFETs in AC mode instead of DC mode yielded signal resolutions that were improved by nearly one order of magnitude. This is most likely because the AC mode reduced the effects of environmental signal noise.

The chemical response of the InN surface to polar liquids (IPA, acetone, methanol) was assessed using the ultrathin ISFETs. For three samples given a fixed V_{DS} of 0.25 V, I_{DS} was measured as each sample was exposed to the polar liquids. Figure 6.16 shows that the average I_{DS} of methanol, IPA, and acetone are 466, 462,

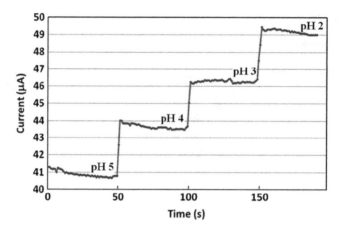

Fig. 6.31 Dynamic response of the a-InN:Mg ISFET to pH levels ranging from 5 to 2 (pH changes done with titration with HCl solution) while given a constant V_{DS} of 0.1 V. (after [126])

and $405\,\mu A$ respectively. The current variation of the ISFETs displayed a linear relation to the value of p/ε of the polar liquid, which is attributed to the potential drop at the liquid–solid interface due to the molecular dipole of the polar liquid.

There is great interest in InN for their unique electronic properties. InN has great potential for applications in chemical and biological sensing.

6.11 Summary and Conclusions

We have summarized recent progress in GaN-based sensors. These devices can take advantage of the advantages of microelectronics, including high sensitivity, possibility of high-density integration, and mass manufacturability. Frequent screening can catch the early development of diseases, reduce the suffering of the patients due to late diagnoses, and lower the medical cost. This frequency cannot be achieved with the current methods of mammography due to high cost to the patient and invasiveness (radiation). We have also reviewed process with InN-based sensors; these sensors display similar qualities to GaN sensors and have similar potential sensing applications.

There are still some critical issues. First, the sensitivity for certain antigens (such as prostate or breast cancer) needs to be improved further to allow sensing in body fluids other than blood (urine, saliva). Second, a sandwich assay allowing the detection of the same antigen using two different antibodies (similar to ELISA) needs to be tested. Third, integrating multiple sensors on a single chip with automated fluid handling and algorithms to analyze multiple detection signals, and fourth, a package that will result in a cheap final product are needed. Fourth, the stability of surface functionalization layers in some cases is not conducive to long-term storage and this will limit the applicability of those sensors outside of clinics.

Acknowledgements This work is supported by the ONR funded Center for Sensor Materials and Technologies, NSF and ARO. The collaborations with T. Lele, B.P. Gila, W. Johnson and A. Dabiran are greatly appreciated.

References

1. A.L. Burlingame, R.K. Boyd, S.J. Gaskell, Mass spectrometry. Anal. Chem. **68**, 599–611 (1996)
2. K.W. Jackson, G. Chen, Anal. Chem. **68**, 231–242 (1996)
3. J.L. Anderson, E.F. Bowden, P.G. Pickup, Anal. Chem. **68**, 379–401 (1996)
4. R.J. Chen, S. Bangsaruntip, K.A. Drouvalakis, N.W.S. Kam, M. Shim, Y. Li, W. Kim, P.J. Utz, H. Dai, Proc. Natl. Acad. Sci. USA **100**, 4984–4990 (2003)
5. C. Li, M. Curreli, H. Lin, B. Lei, F.N. Ishikawa, R. Datar, R.J. Cote, M.E. Thompson, C. Zhou, J. Am. Chem. Soc. **127**, 12484–12498 (2005)
6. J. Zhang, H.P. Lang, F. Huber, A. Bietsch, W. Grange, U. Certa, R. Mckendry, H.-J. Güntherodt, M. Hegner, Ch. Gerber, Nat. Nanotechnol. **1**, 214–220 (2006)
7. F. Huber, H.P. Lang, C. Gerber, Nat. Nanotechnol. **3**, 645–646 (2008)
8. A. Sandu, Nat. Nanotechnol. **2**, 746–748 (2007)
9. G. Zheng, F. Patolsky, Y. Cui, W.U. Wang, C.M. Lieber, Nat. Biotechnol. **23**, 1294–1296 (2005)
10. J. Jun, B. Chou, J. Lin, A. Phipps, S. Xu, K. Ngo, D. Johnson, A. Kasyap, T. Nishida, H.T. Wang, B.S. Kang, T. Anderson, F. Ren, L.C. Tien, P.W. Sadik, D.P. Norton, L.F. Voss, S.J. Pearton, Solid State Electron. **51**, 1018–1022 (2007)
11. X. Yu, C. Li, Z.N. Low, J. Lin, T.J. Anderson, H.T. Wang, F. Ren, Y.L. Wang, C.Y. Chang, S.J. Pearton, C.H. Hsu, A. Osinsky, A. Dabiran, P. Chow, C. Balaban, J. Painter, Sens. Actuat. B **135**, 188–194 (2008)
12. H.T. Wang, T.J. Anderson, F. Ren, C. Li, Z.N. Low, J. Lin, B.P. Gila, S.J. Pearton, A. Osinsky, A. Dabiran, Appl. Phys. Lett. **89**, 242111–242114 (2006)
13. H.T. Wang, T.J. Anderson, B.S. Kang, F. Ren, C. Li, Z.N. Low, J. Lin, B.P. Gila, S.J. Pearton, A. Osinsky, A. Dabiran, Appl. Phys. Lett. **90**, 252109–252111 (2007)
14. T.J. Anderson, H.T. Wang, B.S. Kang, F. Ren, S.J. Pearton, A. Osinsky, Amir Dabiran, P.P. Chow, Appl. Surf. Sci. **255**, 2524–2526 (2008)
15. J Kim, B.P. Gila, G.Y. Chung, C.R. Abernathy, S.J. Pearton, F. Ren, Solid-State Electron. **47**, 1487–1490 (2003)
16. H.T. Wang, T.J. Anderson, F. Ren, C. Li, Z.N. Low, J. Lin, B.P. Gila, S.J. Pearton, A. Osinsky, A. Dabiran, Appl. Phys. Lett. **89**, 242111–242113 (2006)
17. H.T. Wang, B.S. Kang, F. Ren, R.C. Fitch, J.K. Gillespie, N. Moser, G. Jessen, T. Jenkins, R. Dettmer, D. Via, A. Crespo, B.P. Gila, C.R. Abernathy, S.J. Pearton, Appl. Phys. Lett. **87**, 172105–172107 (2005)
18. J. Schalwig, G. Muller, U. Karrer, M. Eickhoff, O. Ambacher, M. Stutzmann, L. Gogens, G. Dollinger, Appl. Phys. Lett. **80**, 1222–1224 (2002)
19. B.P. Luther, S.D. Wolter, S.E. Mohney, Sens. Actuat. B **56**, (1999) 164–168
20. B.S. Kang, R. Mehandru, S. Kim, F. Ren, R.C. Fitch, J.K. Gillespie, N. Moser, G. Jessen, T. Jenkins, R. Dettmer, D. Via, A. Crespo, K.H. Baik, B.P. Gila, C.R. Abernathy, S.J. Pearton, Phys. Status Solidi C **2**, 2672–2674 (2005)
21. H.T. Wang, B.S. Kang, F. Ren, L.C. Tien, P.W. Sadik, D.P. Norton, S.J. Pearton, J. Lin, Appl. Phys. A Mater. Sci. Proc. **81**, 1117–1120 (2005)
22. J.S. Wright, W. Lim, B.P. Gila, S.J. Pearton, F. Ren, W. Lai, L.C. Chen, M. Hu, K.H. Chen, J. Vac. Sci. Technol. B **27**, L8–L10 (2009)
23. J.L. Johnson, Y. Choi, A. Ural, W. Lim, J.S. Wright, B.P. Gila, F. Ren, S.J. Pearton, J. Electron. Mater. **38**, 490–494 (2009)

24. W. Lim, J.S. Wright, B.P. Gila, J.L. Johnson, A. Ural, T. Anderson, F. Ren, S.J. Pearton, Appl. Phys. Lett. **93**, 072110–072112 (2008)

25. L. Tien, P. Sadik, D.P. Norton, L. Voss, S.J. Pearton, H.T. Wang, B.S. Kang, F. Ren, J. Jun, J. Lin, Appl. Phys. Lett. **87**, 222106–222108 (2005)

26. O. Kryliouk, H.J. Park, H.T. Wang, B.S. Kang, T.J. Anderson, F. Ren, S.J. Pearton, J. Vac. Sci. Technol. B **23**(1891), 1891–1894 (2005)

27. L. Tien, H.T. Wang, B.S. Kang, F. Ren, P.W. Sadik, D.P. Norton, S.J. Pearton, J. Lin, Electrochem. Solid-State Lett. **8**, G239–241 (2005)

28. H.T. Wang, B.S. Kang, F. Ren, L.C. Tien, P.W. Sadik, D.P. Norton, S.J. Pearton, J. Lin, Appl. Phys. Lett. **86**, 243503–243505 (2005)

29. M. Eickhoff, J. Schalwig, G. Steinhoff, O. Weidemann, L. Görgens, R. Neuberger, M. Hermann, B. Baur, G. Müller, O. Ambacher, M. Stutzmann, Phys. Status Solidi C **0**(6), 1908–1918 (2003)

30. R. Mehandru, B. Luo, B.S. Kang, J. Kim, F. Ren, S.J. Pearton, C.-C. Pan, G.-T. Chen, J.-I. Chyi, Solid-State Electron. **48**, 351–353 (2004)

31. R. Neuberger, G. Muller, O. Ambacher, M. Stutzmann, Phys. Stat. Sol. **183**(2), R10–R12 (2001)

32. G. Parvesh, P. Sujata, H. Subhasis, M. Gupta, R.S. Gupta, Solid-State Electron. **51**, 130–135 (2007)

33. L. Shen, R. Coffie, D. Buttari, S. Heikman, A. Chakraborty, A. Chini, S. Keller, S.P. DenBaars, U.K. Mishra, IEEE Electron Dev. Lett. **25**, 7–9.(2004)

34. B.S. Kang, H.T. Wang, T.P. Lele, F. Ren, S.J. Pearton, J.W. Johnson, P. Rajagopal, J.C. Roberts, E.L. Piner, K.J. Linthicum, Appl. Phys Lett. **91**, 112106–112108 (2007)

35. A. El Kouche, J. Lin, M.E. Law, S. Kim, B.S. Kim, F. Ren, S.J. Pearton, Sens. Actuat. B Chem. **105**, 329–333 (2005)

36. H.T. Wang, B.S. Kang, F. Ren, S.J. Pearton, J.W. Johnson, P. Rajagopal, J.C. Roberts, E.L. Piner, K.J. Linthicum, Appl. Phys. Lett. **91**, 222101–222103 (2007)

37. B.S. Kang, S. Kim, F. Ren, B.P. Gila, C.R. Abernathy, S.J. Pearton, Sens. Actuat. B Chem. **104**, 232–236 (2005)

38. H.T. Wang, B.S. Kang, T.F. Chancellor Jr., T.P. Lele, Y. Tseng, F. Ren, S.J. Pearton, A. Dabiran, A. Osinsky, P.P. Chow, Electrochem. Solid-State Lett. **10**, J150–152 (2007)

39. K.H. Chen, H.W. Wang, B.S. Kang, C.Y. Chang, Y.L. Wang, T.P. Lele, F. Ren, S.J. Pearton, A. Dabiran, A. Osinsky, P.P. Chow, Sens. Actuat. B Chem. **134**, 386–389 (2008)

40. S.J. Pearton, T. Lele, Y. Tseng, F. Ren, Trend Biotechnol. **25**, 481–482 (2007)

41. H.T. Wang, B.S. Kang, T.F. Chancellor Jr., T.P. Lele, Y. Tseng, F. Ren, S.J. Pearton, J.W. Johnson, P. Rajagopal, J.C. Roberts, E.L. Piner, K.J. Linthicum, Appl. Phys. Lett. **91**(042114), 042114–042116 (2007)

42. B.S. Kang, H.T. Wang, F. Ren, B.P. Gila, C.R. Abernathy, S.J. Pearton, D.M. Dennis, J.W. Johnson, P. Rajagopal, J.C. Roberts, E.L. Piner, K.J. Linthicum, Electrochem. Solid-State Lett. **11**, J19–J21 (2008)

43. B.S. Kang, H.T. Wang, F. Ren, B.P. Gila, C.R. Abernathy, S.J. Pearton, J.W. Johnson, P. Rajagopal, J.C. Roberts, E.L. Piner, K.J. Linthicum, Appl. Phys. Lett. **91**, 012110–012112 (2007)

44. B.S. Kang, G. Louche, R.S. Duran, Y. Gnanou, S.J. Pearton, F. Ren, Solid-State Electron. **48**, 851–854 (2004)

45. J.R. Lothian, J.M. Kuo, F. Ren, S.J. Pearton, J. Electron. Mater. **21**, 441–445 (1992)

46. J.W. Johnson, B. Luo, F. Ren, B.P. Gila, W. Krishnamoorthy, C.R. Abernathy, S.J. Pearton, J.I. Chyi, T.E. Nee, C.M. Lee, C.C. Chuo, Appl. Phys. Lett. **77**, 3230 (2000)

47. B.S. Kang, S.J. Pearton, J.J. Chen, F. Ren, J.W. Johnson, R.J. Therrien, P. Rajagopal, J.C. Roberts, E.L. Piner, K.J. Linthicum, Appl. Phys. Lett. **89**, 122102–122104 (2006)

48. B.S. Kang, F. Ren, L. Wang, C. Lofton, W. Tan, S.J. Pearton, A. Dabiran, A. Osinsky, P.P. Chow, Appl. Phys. Lett. **87**, 023508–023510 (2005)

49. B.S. Kang, H. Wang, F. Ren, S.J. Pearton, T. Morey, D. Dennis, J. Johnson, P. Rajagopal, J.C. Roberts, E.L. Piner, K.J. Linthicum, Appl. Phys. Lett. **91**, 252103–252105 (2007)

50. B.S. Kang, S. Kim, F. Ren, J.W. Johnson, R. Therrien, P. Rajagopal, J. Roberts, E. Piner, K.J. Linthicum, S.N.G. Chu, K. Baik, B.P. Gila, C.R. Abernathy, S.J. Pearton, Appl. Phys. Lett. **85**, 2962–2964 (2004)
51. S.J. Pearton, B.S. Kang, S. Kim, F. Ren, B.P. Gila, C.R. Abernathy, J. Lin, S.N.G. Chu, J. Phys. Condens. Mat. **16**, R961–985 (2004)
52. L. Voss, B.P. Gila, S.J. Pearton, H. Wang, F. Ren, J. Vac. Sci. Technol. **B23**, 6–10 (2005)
53. W. Lim, J.S. Wright, B.P. Gila, S.J. Pearton, F. Ren, W. Lai, L.C. Chen, M. Hu, K.H. Chen, Appl. Phys. Lett. **93**, 202109–202111 (2008)
54. K.D. Mitzner, J. Sternhagen, D.W. Galipeau, Sens. Actuat. **B93**, 92 (2003)
55. P. Mitra, A.P. Chatterjee, H.S. Maiti, Mater. Lett. **35**, 35 (1998)
56. E.M. Logothetis, Automotive oxygen sensors, in *Chemical Sensor Technology*, vol. 3, ed. by N. Yamazoe (Elsevier, Amsterdam, 1991)
57. Y. Xu, X. Zhou, O.T. Sorensen, Sens. Actuat. B **65**, 2–9 (2000)
58. L. Castañeda, Mater. Sci. Eng. B **139**, 149–157 (2007)
59. J. Gerblinger, W. Lohwasser, U. Lampe, H. Meixner, Sens. Actuat. B **26**(93), 93–98 (1995)
60. R. Yakimova, G. Steinhoff, R.M. Petoral Jr., C. Vahlberg, V. Khranovskyy, G.R. Yazdi, K. Uvdal, A. Lloyd Spetz, Biosens. Bioelectron. **22**, 2780–2785 (2007)
61. A. Trinchi, Y.X. Li, W. Wlodarski, S. Kaciulis, L. Pandolfi, S.P. Russo, J. Duplessis, S. Viticoli, Sens. Actuat. A **108**, 263–270 (2003)
62. Y-L Wang, L.N. Covert, T.J. Anderson, W Lim, J. Lin, S.J. Pearton, D.P. Norton, J.M. Zavada, F. Ren, Electrochem Solid-State Lett **11**(3) H60–H62 (2007)
63. M.J. Thorpe, K.D. Moll, R.J. Jones, B. Safdi J. Ye Science **311** 1595–1598 (2006)
64. M.J. Thorpe, D. Balslev-Clausen, M.S. Kirchner J. Ye, Optics Express **16**, 2387–2393 (2008)
65. J. Kong, H. Dai, J. Phys. Chem. B **105** 2890–2895 (2001)
66. S. Satyapal, T. Filburn, J. Trela, J. Strange, Energ Fuel **15**(250) 250–254 (2001)
67. D.B. Dell'Amico, F. Calderazzo, L. Labella, F. Marchetti, G. Pampaloni, Chem. Rev. **103**, 3857–3897 (2003)
68. S.M. Savage, A. Konstantinov, A.M. Saroukan, C. Harris, Proc. ICSCRM'99. 511–515 (2000)
69. K.D. Mitzner, J. Sternhagen, D.W. Galipeau, Sens. Actuat. B **9**, 92–97 (2003)
70. S.J. Pearton, D.P. Norton, K. Ip, Y.W. Heo, T. Steiner, Prog. Mater. Sci. **50**, 293–344 (2005)
71. J.R. LaRoche, Y.W. Heo, B.S. Kang, L. Tien, Y. Kwon, D.P. Norton, B.P. Gila, F. Ren, S.J. Pearton, J. Electron. Mater. **34**, 404–409 (2005)
72. Q. Wan, Q.H. Li, Y.J. Chen, T.H. Wang, X.L. He, J.P. Li, C.L. Lin, Appl. Phys. Lett. **84**, 3654–3656 (2004)
73. F. Tsao, J.Y. Chen, C.H. Kuo, G.C. Chi, C.J. Pan, P.J. Huang, C.J. Tun, B.J. Pong, T.H. Hsueh, C.Y. Chang, S.J. Pearton, F. Ren, Appl. Phys. Lett. **92**, 203110–203112 (2008)
74. K. Keem, H. Kim, G.T. Kim, J.S. Lee, B. Min, K. Cho, M.Y. Sung, S. Kim, Appl. Phys. Lett. **84**, 4376–4378 (2004)
75. Y.W. Heo, V. Varadarjan, M. Kaufman, K. Kim, D.P. Norton, F. Ren, P.H. Fleming, Appl. Phys. Lett. **81**(3046), 3046–3048 (2002)
76. D.P. Norton, Y.W. Heo, M.P. Ivill, K. Ip, S.J. Pearton, M.F. Chisholm, T. Steiner, Mater. Today 34–40, June (2004)
77. B. Gou, Z.R. Qiu, K.S. Wong, Appl. Phys. Lett. **82**, 2290–2292 (2003)
78. T. Koida, S.F. Chichibu, A. Uedono, A. Tsukazaki, M. Kawasaki, T. Sota, Y. Segewa, H. Koinuma, Appl. Phys. Lett. **82**, 532–537 (2003)
79. O. Lopatiuk, W. Burdett, L. Chernyak, K.P. Ip, Y.W. Heo, D.P. Norton, S.J. Pearton, B. Hertog, P.P. Chow, A. Osinsky, Appl. Phys. Lett. **86**, 012105–012107 (2005)
80. B.S. Kang, Y.W. Heo, L.C. Tien, D.P. Norton, F. Ren, B.P. Gila, S.J. Pearton, Appl. Phys. A **80**, 1029–1033 (2005)
81. B.S. Kang, F. Ren, Y.W. Heo, L.C. Tien, D.P. Norton, S.J. Pearton, Appl. Phys. Lett. **86**, 112105–112107 (2005)
82. I. Horvath, J. Hunt, P.J. Barnes, Eur. Respir. **26**, 523–529 (2005)
83. K. Namjou, C.B. Roller, P.J. McCann, IEEE Circ. Dev. Mag. **22**, 22–28 (2006)

84. R.F. Machado, D. Laskowski, O. Deffenderfer, T. Burch, S. Zheng, P.J. Mazzone, T. Mekhail, C. Jennings, J.K. Stoller, J. Pyle, J. Duncan, R.A. Dweik, S.C. Erzurum, Am. J. Respir. Crit. Care. Med. **171**, 1286–1295 (2005)
85. T. Kullmann, I. Barta, Z. Lazar, B. Szili, E. Barat, M. Valyon, M. Kollai, I. Horvath, Eur. Respir. **29**, 496–502 (2007)
86. J. Vaughan, L. Ngamtrakulparit, T.N. Pajewski, R. Turner, T.A. Nguyen, A. Smith, P. Urban, S. Hom, B. Gaston, J. Hunt, Eur. Respir. J. **22**, 889–895 (2003)
87. J.F. Hunt, K. Fang, R. Malik, et al., Am. J. Respir. Crit. Care. Med. **171**, 1286–1292 (2005)
88. K. Kostikas, G. Papatheodorou, K. Ganas, K. Psathakis, P. Panagou, S. Loukides, Am. J. Respir. Cirt. Care. Med. **165**, 1364–1369 (2002)
89. I. Horvath, J. Hunt, P.J. Barnes, Eur. Respir. J. **26**(9), 523–548 (2005)
90. K. Czebe, I. Barta, B. Antus, M. Valyon, I. Horváth, T. Kullmann, Resp. Med. **102**(5), 720–725 (2008)
91. K. Bloemen, G. Lissens, K. Desager, G. Schoeters, Resp. Med. **101**(6), 1331–1337 (2007)
92. S. Park, H. Boo, T.D. Chung, Anal. Chi. Act. **46**, 556–560 (2006)
93. P. Pandey, S.P. Singh, S.K. Arya, V. Gupta, M. Datta, S. Singh, B.D. Malhotra, Langmuir **23**, 3333–3339 (2007)
94. J.F. Hunt, K. Fang, R. Malik, A. Snyder, N. Malhotra, T.A.E. Platts-Mills, B. Gaston, Am. J. Respir. Crit Care Med. **161**, 694–696 (2000)
95. T. Balaji, M. Sasidharan H. Matsunaga, Analyst **130**, 1162–1167 (2005)
96. G.Q. Shi G. Jiang, Anal Sci **18**, 1215–1219 (2002)
97. A. Caballero, R. Martínez, V. Lloveras, I. Ratera, J. Vidal-Gancedo, K. Wurst, A. Tárraga, P. Molina, J. Veciana, J. Am. Chem. Soc. **127**, 15666–15672 (2005)
98. E. Coronado, J.R. Galán-Mascarós, C. Martí-Gastaldo, E. Palomares, J.R. Durrant, R. Vilar, M. Gratzel, Md.K. Nazeeruddin, J. Am. Chem. Soc. **127**, 12351–12356 (2005)
99. Y.K. Yang, K.J. Yook, J. Tae, J. Am. Chem. Soc. **127**, 16760–16765 (2005)
100. M. Matsushita, M.M. Meijler, P. Wirsching, R.A. Lerner, K.D. Janda, Org. Lett. **7**, 4943–4948 (2005)
101. C.C. Huang, H.T. Chang, Anal. Chem. **78**, 8332–8843 (2006)
102. S.S. Arnon, R. Schechter, T.V. Inglesby. JAMA **285**(8), 256–265 (2001)
103. R.A. Greenfield, B.R. Brown, J.B. Hutchins J.J. Iandolo, R. Jackson, L.N. LSlater, M.S. Bronze, Am. J. Med. Sci. **323**, 326–334 (2002)
104. R. McIntyre, L. Bigler, T. Dellinger, M. Pfeifer, T. Mannery, C. Streckfus, Oral Surg. Oral Med. Oral Pathol. Oral Radiol. Endod. **88**(6), 687–693 (1999)
105. C. Streckfus, L. Bigelr, T. Dellinger, M. Pfeifer, A. Rose, J.T. Thigpen, Clin. Oral Investig. **3**(3), 138–144 (1999)
106. Y.-L. Wang, B.H. Chu, K.H. Chen, C.Y. Chang, T.P. Lele, G. Papadi, J.K. Coleman, B.J. Sheppard, C.F. Dungen, S.J. Pearton, J.W. Johnson, P. Rajagopal, J.C. Roberts, E.L. Piner, K.J. Linthicum, Appl. Phys. Lett. **94**, 243901–243903 (2009)
107. T. Anderson, F. Ren, S.J. Pearton, B.S. Kang, H.-T. Wang, C.-Y. Chang, J. Lin, Sensors. **9**(6), 4669–4702 (2009)
108. G. Shekhawat, S.H. Tark, V.P. Dravid, Science **311**, 1592–1597 (2006)
109. H.T. Wang, B.S. Kang, F. Ren, S.J. Pearton, J.W. Johnson, P. Rajagopal, J.C. Roberts, E.L. Piner, K.J. Linthicum, Appl. Phys. Lett. **91**, 222101–222103 (2007)
110. S.C. Hung, B.H. Chou, C.Y. Chang, K.H. Chen, Y.L. Wang, S.J. Pearton, A. Dabiran, P.P. Chow, G.C. Chi, F. Ren, Appl. Phys. Lett. **94**, 043903–043905 (2009)
111. Y.L. Wang, B.H. Chu, K.H. Chen, C.Y. Chang, T.P. Lele, Y. Tseng, S.J. Pearton, J. Ramage, D. Hooten, A. Dabiran, P.P. Chow, F. Ren, Appl. Phys. Lett. **93**, 262101–262103 (2008)
112. B.H. Chu, B.S. Kang, F. Ren, C.Y. Chang, Y.L. Wang, S.J. Pearton, A.V. Glushakov, D.M. Dennis, J.W. Johnson, P. Rajagopal, J.C. Roberts, E.L. Piner, K.J. Linthicum, Appl. Phys. Lett. **93**, 042114–042116 (2008)
113. S.C. Hung, Y.L. Wang, B. Hicks, S.J. Pearton, F. Ren, J.W. Johnson, P. Rajagopal, J.C. Roberts, E. Piner, K. Linthicum, G.C. Chi, Electrochem. Solid. State Lett. **11**, H241–243 (2008)

114. C.Y. Chang, B.S. Kang, H.T. Wang, F. Ren, Y.L. Wang, S.J. Pearton, D.M. Dennis, J.W. Johnson, P. Rajagopal, J.C. Roberts, E.L. Piner, K.J. Linthicum, Appl. Phys. Lett. **92**, 232102–232104 (2008)
115. S.C. Hung, B. Hicks, Y.L. Wang, S.J. Pearton, D.M. Dennis, F. Ren, J.W. Johnson, P. Rajagopal, J.C. Roberts, E.L. Piner, K.J. Linthicum, G.C. Chi, Appl. Phys. Lett. **92**, 193903–193905 (2008)
116. D.K. Jorkasky, Toxicol. Lett. **539**, 102–103 (1998)
117. L.B. Kinter, J.P. Valentin, Fundam. Clin. Pharmacol. **16**, 175 (2002)
118. I. Cimalla, M. Gebinoga, A. Schober, V. Polyakov, V. Lebedev, V. Cimalla, AlGaN/GaN sensors for direct monitoring of nerve cell response to inhibitors (in press 2010), 14–15
119. D. Braun, P. Fromherz, Phys. Rev. Lett. **81**, 5241 (1998)
120. H. Lu, W.J. Schaff, L.F. Eastman, C.E. Stutz, Appl. Phys. Lett. **82**, 1736–1738 (2003)
121. K.A. Rickert, A.B. Ellis, F.J. Himpsel, H. Lu, W. Schaff, J.M. Redwing, F. Dwikusuma, T.F. Kuech, Appl. Phys. Lett. **82**, 3254–3256 (2003)
122. T.D. Veal, I. Mahboob, L.F.J. Piper C.F. McConville, H. Lu W.J Schaff J. Vac. Sci. Technol. B **22**(4), 2175–2178 (2004)
123. I. Mahboob, T.D. Veal, L.F.J. Piper, C.F. McConville, H. Lu, W.J. Schaff, J. Furthmüller, F. Bechstedt. Phys. Rev. B **69**, 201307 (2004)
124. I. Mahboob, T.D. Veal, C.F. McConville, H. Lu, W.J. Schaff, Phys. Rev. Lett. **92**, 036804 (2004)
125. S.X. Li, K.M. Yu, J. Wu, R.E. Jones, W. Walukiewicz, J.W. Ager III, W. Shan, E.E. Haller, H. Lu, W.J. Schaff, Phys. Rev. B **71**, 161201 (2005)
126. Y.W. Chang, Y.S. Lu, J.A. Yeh, Y.L. Hong, H.M. Lee, S. Gwo. InN-based Chemical sensors (in press, 2010)
127. G. Steinhoff, M. Hermann, W.J. Schaff, L.F. Eastman, M. Stutzmann, M. Eickhoff, Appl. Phys. Lett. **83**, 177–179 (2003)
128. J.L. Chiang, Y.C. Chen, J.C. Chou C.-C Chen, Jpn. J. Appl. Phys **41**, 541–545 (2002)

Chapter 7
GaN HEMT Technology

Wayne Johnson and Edwin L. Piner

Abstract A review of nitride-based HEMT technology, beginning with substrate considerations and moving to crystal growth, device processing, packaging, and finally to products. The final section introduces important areas anticipated for future GaN HEMT development: InAlN HEMTs, GaN-on-diamond, and heterointegration.

7.1 Introduction

Since the first III-nitride electronic device fabricated in 1993 [1], GaN HEMT technology has progressed from fundamental materials research to application as a platform technology for product lines spanning multiple end-use markets. Although there have been numerous reports in the literature of MESFETs and MISFETs with reasonable performance, the most noteworthy advantage of III-nitride materials over other wide bandgap semiconductors, such as SiC, is the availability of heterostructures. For the most widely investigated AlGaN/GaN heterostructures, the sheet charge is a result of the large spontaneous and piezoelectric polarization-induced field and large conduction band offset. Even in the absence of modulation doping, this built-in field induces a two-dimensional electron gas (2DEG) that is linearly proportional to the Al-mole fraction across typical Al contents of <30%. A sheet electron density above $10^{13} \mathrm{cm}^{-2}$ can be realized, representing a factor of

W. Johnson (✉)
Kopin Corporation, Taunton, MA 02780, USA
e-mail: wjohnson@nitronex.com

E.L. Piner
Texas State University – San Marcos, San Marcos, TX 78666, USA
e-mail: epiner@txstate.edu

S. Pearton (ed.), *GaN and ZnO-based Materials and Devices*,
Springer Series in Materials Science 156, DOI 10.1007/978-3-642-23521-4_7,
© Springer-Verlag Berlin Heidelberg 2012

5–10× above typical GaAs or InP pHEMTs. The associated mobility at this high current density is typically in the range of 1,300–2,000 cm^2/Vs.

Remarkable progress has been made in recent years in high-performance GaN HEMTs grown on a variety of substrates including sapphire, SiC, and Si. This chapter is structured to provide an overview of GaN HEMT technology in much the same order that the technology has matured: from substrate considerations and crystal growth to device processing, packaging, and products. The final section introduces important areas anticipated for future GaN HEMT development: InAlN HEMTs, GaN-on-diamond, and heterointegration.

7.2 Substrate Considerations

The GaN epitaxial process has benefited greatly from the viability of heteroepitaxy. In fact, the commercial success of the GaN HEMT is directly attributable to the ability to deposit high-quality epitaxy on non-GaN substrates and the ability to integrate GaN HEMT device fabrication in standard GaAs processing lines. None of the current commercial success would have been economically viable had the GaN HEMT required a native-GaN substrate that, while improving, is limited today to 2-in. wafers and was essentially non-existent in 2000, when the viability of integration with existing GaAs fabrication processes was beginning to emerge [2].

There are three substrates that have been utilized to achieve commercial success with heteroepitaxial GaN-based devices: sapphire (Al_2O_3), silicon carbide (SiC), and silicon (Si). In addition, long-lifetime GaN laser diodes (LDs) have matured via homoepitaxy on GaN substrates [3]. While many other materials have been used as substrates to deposit GaN, many with good crystalline quality, the vast majority of research and development efforts throughout the world have focused on the aforementioned four key materials.

Table 7.1 Properties of key substrate materials

	Al_2O_3	SiC	Si	GaN
Lattice (mismatch, %)	14	3.5	17	–
TCE (mismatch, %)	−26	25	56	–
Thermal Conductivity (W/cm-k at 25°C)	0.4	3.0–3.8	1.5	1.5
Electrical Resistivity (Ω − cm) commercial high-purity	>1E14	>1E5	1E4–1E5	>1E5
Thermal stability (relative)-at Tg	Very high	Very high	Moderate	–
Energy gap, Eg (eV)	9.9	4H-SiC:3.26, 6H-SiC:3.03	1.12	3.4
Substrate size (mm)	150	150	400	50
Substrate cost (relative)	High	Very high	Low	Very high
"Integratability"	Low	Moderate	Very high	–

GaN HEMT technology poses several key challenges to the selection of the substrate material. Table 7.1 summarizes those challenges and compares the properties of Al_2O_3, SiC, Si, and GaN. The net outcome of comparing the various substrate options is that there is not, currently, one perfect material choice. Therefore, various tradeoffs must be considered during the material and device design process. The differing requirements and tolerances between applications (or markets) typically drive the decision regarding the most applicable substrate. While commercial GaN HEMT products and their markets will be discussed later in this chapter, there are general substrate-related themes that can be analyzed.

7.2.1 Silicon

When it comes to substrate size and cost, clearly silicon is the winner. This is due to the long history of development and refinement of this critical semiconductor material, which GaN-on-Si heteroepitaxy has been able to leverage. The silicon industry has also established the commercial viability of a wide range of substrate electrical resistivity for both n- and p-type conduction. While the high-quality, low-cost, and large-diameter of Si substrates are well understood, other inherent advantages of GaN-on-Si include the ability to leverage established Si processes for wafer grinding and polishing, via-hole formation, and AuSi eutectic die attach [4]. Applications for GaN-on-Si HEMTs have employed substrates across this wide range of resistivity; particularly p-type for power switching and intrinsic (float-zone, highly resistive) silicon for RF power amplifiers.

Silicon has a reasonable thermal conductivity that is similar to bulk crystalline GaN, but inferior to the chief competing substrate material; SiC. This is a key design consideration for maximum exploitation of the exceptional power density offered by GaN HEMTs. The inferior Si substrate thermal conductivity can limit the scaling of critical device dimensions, which can be a factor in applications where the junction temperature rise is exceptionally limited. Furthermore, compared to SiC, the lattice and coefficient of thermal expansion mismatches create a challenge to grow epitaxial GaN. These mismatches have been researched extensively and several techniques to overcome their impact on epitaxial crystal quality have been developed, and are discussed in the next section. Silicon's "thermal stability," meaning the relative inertness of the material at typical GaN growth temperature, is also inferior to SiC and Al_2O_3. This phenomenon manifests itself typically through auto-doping of the Si substrate during nucleation of the GaN epitaxy [5]. In particular, the Al and Ga, whether introduced in the growth chamber intentionally or unintentionally, can diffuse into the silicon from the gas phase creating a p-doped layer at the surface of the substrate. While considerable work has been done to mitigate the auto-doping effect, for RF amplifier applications at higher frequencies (above ∼6 GHz) the capacitive coupling of this conductive layer can limit the dc-to-RF conversion efficiency of the device [6].

7.2.2   SiC

Interest in SiC as a substrate for GaN HEMT applications has been driven predominately by intense military R&D funding since the mid-1990s. The CW power density demonstrated by GaN HEMTs first exceeded the best GaAs performance ($\sim$2 W/mm) in 1997 and has since astonishingly increased by an additional 20× [7]. Virtually every military application using GaAs devices – those requiring high efficiency, high power and bandwidth and/or high frequency – can benefit from GaN HEMT technology. RF amplifiers for these applications require the best heat dissipation and electrical resistivity characteristics; both are provided by semi-insulating SiC.

The substrate size of SI–SiC has matured from 25 mm a decade ago, to 150 mm at present, making it viable for device fabrication in standard compound semiconductor processing equipment and facilities. While still very expensive compared to Si or even Al_2O_3, and therefore limiting for certain applications, economies of scale have driven the cost of SI–SiC substrates to a fraction of the price paid for smaller substrates only a few years ago. Furthermore, the mismatch of the key crystalline properties between SiC and GaN epitaxy is sufficiently minor to afford a wide window for optimal crystal quality. Stated simply, it is easier to nucleate and grow high-quality GaN-on-SiC than on any other non-GaN substrate; in particular, GaN-on-Si and GaN-on-Al_2O_3.

7.2.3   Sapphire

GaN optoelectronic devices (blue, green, and white LEDs) have driven the development and production of Al_2O_3 as a substrate material since 1993 [8]. GaN LEDs have created a multi-billion US dollar worldwide industry that predominately leverages Al_2O_3 substrates. The industry has concurrently pushed demand for Al_2O_3 wafer scaling up to 100 mm (and 150 mm) and the price down. The approaches originally developed for nucleating and growing GaN LED structures have been applied to the growth of GaN HEMT structures [9]. These nucleation schemes very effectively mitigate the lattice incongruity between GaN and Al_2O_3 while the CTE mismatch favors compression of the epi, in contrast with the tension typically exhibited by GaN-on-SiC and GaN-on-Si. While compressive stress may manifest itself in wafer bow, it does not result in epilayer cracking as may occur when growing on SiC or Si.

While much of the early GaN HEMT research utilized Al_2O_3 substrates, the low thermal conductivity of the material has largely relegated its use to research today. The very high-power density observed with GaN HEMTs is achieved with exceptionally high current that produces significant heating within the 2DEG channel of the HEMT. Even during I–V sweeps, this self-heating will result in reduced current at high voltage. The recent rapid advent of cost-effective GaN-on-Si and the military interest in cutting-edge performance GaN-on-SiC has left little viable interest in HEMTs built on GaN-on-Al_2O_3.

7.2.4 GaN

A wide range of devices based on heteroepitaxial GaN have been researched and many commercialized, even though the dislocation density in the device structure region of the epi typically ranges from 2E8 to 2E9 cm^{-2}. By comparison, a typical Si or GaAs-based semiconductor device will have fewer than 1E4 cm^{-2} dislocations. Any more and the device lifetime is significantly degraded. Why GaN-based devices can endure such high dislocation densities is not completely understood, but the performance is unquestionable and has allowed the commercial success of GaN-on-Al$_2$O$_3$, -on-SiC, and -on-Si. Because of this success, the effort to develop homoepitaxial GaN and, therefore, bulk GaN substrates (as is needed in the Si and GaAs industries) is secondary to heteroepitaxial GaN.

Interestingly, there is one exception: GaN-based LDs. Early GaN LDs experienced significant degradation [10] that has been attributed to threading screw dislocations acting to conduct current, ultimately resulting in a device short and burnout [11]. Because of the commercial interest in 405-nm GaN LDs for high-density data storage, GaN substrates have been developed in spite of the difficulty associated with growth of GaN boules. While the interest in GaN-on-GaN for HEMTs is currently limited to academic pursuits, should a device reliability issue be encountered that is attributed to threading dislocations, GaN substrates will be the likely pathway to resolve the issue.

7.3 Epitaxy and Device Structures

7.3.1 GaN Epitaxy

The epitaxial growth of GaN HEMT structures has become standard for both metalorganic chemical vapor deposition (MOCVD) and molecular beam epitaxy (MBE) systems. The majority of HEMT growth worldwide is conducted by MOCVD. This is due to the historical chronology of GaN epitaxial development (MOCVD was first to produce commercial device quality material) and the lower cost structure (both process and ownership) inherent in MOCVD. Whereas in other material systems, most notably GaAs, MBE typically provides higher purity sources resulting in "cleaner" epi, in GaN the MBE process often uses metalorganic sources for the column III elements and/or gas source (i.e., ammonia) for the nitrogen (column V) element. The use of these source materials substantially reduces the higher purity advantage in the as-grown film realized in other (again, notably GaAs) material systems. The use of metalorganic sources in MBE is termed MO-MBE (for metalorganic MBE) [12] and the use of ammonia for the nitrogen source is GSMBE (for gas source MBE) [13] or NH$_3$-source MBE. For MBE processes that use N$_2$ as the nitrogen source, due to the extremely high binding energy of the N$_2$ molecule, it must be split with an RF-plasma. This process is termed plasma-assisted MBE (PA-MBE) [14].

Another traditional advantage of MBE is the precise control of switching source materials thereby producing atomically abrupt interfaces. An abrupt interface is critical to the performance of HEMTs due to the proximity of the 2DEG sheet charge within a few angstroms of the barrier layer/channel layer interface (most typically AlGaN/GaN). The interface abruptness has a dramatic impact on the mobility of the 2DEG, in particular. Advances in the process of switching gases in the MOCVD reactor have largely mitigated this MBE advantage as evidenced by approximately equal mobilities for similar GaN HEMT structures grown by MBE and MOCVD.

Recent GaN HEMT devices produced by MOCVD and MBE demonstrate statistically equivalent RF performance. Historically, this has not been so. All of the early record power densities and epi device structure breakthroughs occurred on MOCVD-grown material. During this same time period MBE researchers struggled with source material experimentation, eventually leading to MOMBE and NH_3-source MBE, and the establishment of the proper nucleation conditions to produce high crystalline quality films. Through the focused effort of several research teams, and in particular the DARPA-sponsored Wide Band Gap Semiconductor for RF Applications (WBGS-RF) Program, the MBE-to-MOCVD performance gap was closed [15].

Heterogeneous GaN epitaxy for HEMTs, specifically GaN-on-SiC and GaN-on-Si for reasons discussed in the previous section, encompasses three key elements: nucleation of the film, buffer layer structure, and device layer structure. While, generally, the MBE and MOCVD processes are similar, each element will be discussed in the context of the MOCVD growth environment that is the current mainstay of modern GaN industries.

7.3.2 Nucleation

To facilitate subsequent growth of high crystalline quality GaN epi, a template must first be established that accommodates the lattice mismatch between the substrate and GaN [16]. Various schemes have been developed and reported in the literature [17–21]. A typical process utilizes an AlN layer most often deposited at high temperature, nominally 1,100°C, a V–III gas phase ratio of ~1,000, and growth pressure tending lower than that of Ga-containing III–N film growth, ~30–100 mbar. The thickness range is ~20–500 nm, with growth on SiC typically on the thinner end (20–100 nm) while growth on Si is thicker (200–500 nm). The process parameters are selected to optimize the density of the AlN crystallite nuclei that form on the substrate surface.

Considering the additional complexities of lower thermal stability and higher CTE mismatch inherent in Si substrates, the nucleation layer must also exhibit proper compensating properties to ensure good GaN-on-Si crystal quality [22–25]. While the impurities from the reactor (gas lines and/or coating of the chamber walls and ceiling) and premature gas-phase reactions of the Al- and Ga-containing species can be nearly eliminated with proper reactor design and attention to laminar flow dynamics, respectively [5], a source of Al necessarily must be introduced to precipitate growth of the AlN nucleation layer. At the nominal AlN nucleation layer growth temperature, $\sim 1,100°C$, the activation energy for diffusion, Q, of Al in Si is 3.2 eV. Using the standard Arrhenius expression,

$$D = D_o \exp(-Q/kT),$$

the diffusion coefficient, D, of Al in Si at $\sim 1,100°C$ is 0.19 $\mu m^2/min$.

A complex competition occurs for the Al adatoms at the Si substrate surface to either diffuse into the substrate, incorporate onto the growing AlN nuclei, or desorb back into the gas phase. Fortunately, in this competition the binding of the Al adatoms to N creates a relatively strong bond. Furthermore, diffusion of Al within AlN is very difficult. These two physical phenomena inherently lead to the approach to coalesce the AlN nuclei very quickly as one method to minimize the concentration of Al that may diffusion into the Si substrate. Nuclei with spacing on the order of less than 10 nm are typical, resulting in coalescence within the first 10 nm thickness of the growing AlN nucleation layer. In terms of rate, this typically occurs within one or 2 min after growth is initiated.

A second method that can compliment the first is to modify the condition of the Si substrate surface. One successful approach in this regard is to nitridate the silicon surface to form SiN_x, then grow AlN on the SiN_x [26]. The SiN_x aids in reducing the concentration of Al that diffuses into the substrate by acting as a diffusion barrier. The nitrogen source for the SiN_x is typically, and conveniently, the ammonia used also as the nitrogen source to grow the AlN nucleation layer. The challenge with this approach is to control the thickness and structure of the SiN_x so that AlN nuclei with sufficient density can form and coalesce. Figure 7.1 is a high-resolution STEM image of the AlN nucleation layer/Si substrate interface in between which an optimal SiN_x layer has been formed by nitridation of the silicon. Registry of the AlN crystalline lattice to the silicon lattice occurs through the amorphous SiN_x layer resulting in a sufficiently high AlN nuclei density ensuring complete coalescence. (The registry of the crystals through the amorphous layer is most likely a result of

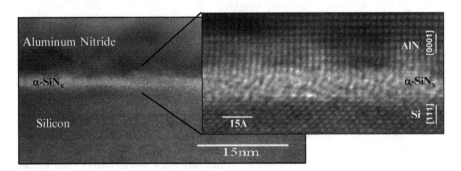

Fig. 7.1 High resolution STEM image of the AlN nucleation layer/Si substrate interface illustrating amorphous SiN_x interfacial layer

the sense of the Si lattice strain field through the amorphous SiN_x. It has not been observed to be due to pinholes within the SiN_x in which AlN nuclei formed. In fact, such a nucleation mechanism suffers from low nuclei density and poly-crystallinity of the subsequent III–N film.)

While overcoming the lattice mismatch issue is the primary function of the AlN nucleation layer, it also provides a highly electrically insulating film. This is important for RF power device operation to avoid capacitive coupling to the RF energy that would otherwise negatively impact the device dc-RF conversion efficiency.

Furthermore, and more critical to the integrity of the GaN epitaxy, the AlN nucleation layer can also assist in the stress management due to the CTE mismatch in three ways. First, the CTE mismatch is greatest between GaN and silicon, hence the significant propensity for this materials system to crack either during epi growth or cool-down from the growth temperature [27]. However, the mismatch between GaN and SiC is sufficient to also cause epilayer cracking under the right conditions. The CTE mismatch between AlN and silicon is substantially less and, in fact, is approximately zero between AlN and SiC. Second, the AlN nucleation layer threading dislocation density is two to three *orders of magnitude* larger than the typical density in the near-device region of the epi, previously discussed. This very high density is due to one or both of misalignment of the coalescing AlN nuclei resulting in, essentially, grain boundary dislocations and misfit dislocations at the substrate/AlN interface that turn up into the AlN layer during growth due to strain and dislocation defect energetics. These defects may contribute to alleviating the stress due to CTE mismatch by accommodating the strain energy by their orientation and mutual interaction. Third, the morphology of the AlN nucleation layer can be engineered to spread the strain field at the AlN/subsequent layer interface. Just as the MOCVD growth environment can be tailored to allow for near-atomically smooth, planar interfaces, it can also be tailored toward the opposite; near-three-dimensional growth and highly textured, rough, interfaces. Figure 7.2 is a cross-sectional STEM image of a GaN-on-Si epiwafer where the roughness of the AlN/AlGaN interface has been designed to be intentionally rough. The peak-to-valley height variation is on the order of 100 nm. This roughness spreads the stress over the thickness of the height variation that would otherwise be concentrated at the AlN/AlGaN interface.

7.3.3 Buffer Layer Structure

Between the nucleation layer and the device layer(s) resides, typically, a sequence of layers generally referred to as the buffer layer structure. The principle functions of this structure are: CTE stress mitigation, threading dislocation density reduction, and electrical isolation. A typical buffer structure comprised of two AlGaN layers and one GaN layer is shown in Fig. 7.2.

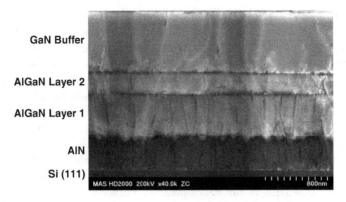

Fig. 7.2 XSTEM image of a GaN-on-Si epiwafer illustrating roughness of lower AlN/AlGaN interface and progressive smoothing for subsequent layers

Mitigating the effects of the CTE mismatch is important to ensure crack-free epitaxy on a wafer that has wafer bow sufficiently low to allow for device fabrication on standard processing equipment. As Table 7.1 indicates, the relative magnitude of the CTE mismatch is much less with GaN-on-SiC compared to GaN-on-Si. For this reason, $\sim$3 μm of crack-free low-bow GaN-on-SiC can be grown without additional stress mitigating layers (e.g., the two AlGaN layers in Fig. 7.2 for GaN-on-Si) while $\sim$0.5 μm is the limit for GaN-on-Si. On the other hand, with a proper CTE mismatch mitigation buffer layer structure, an (Al)GaN epilayer thickness of 14 μm has been demonstrated on silicon. The specific process employed to achieve this result is insertion of multiple low-temperature AlN interlayers within the GaN buffer. The interlayers are typically 10–20 nm thick [28, 29] and are designed to decouple the thermal stress in the GaN above and below each LT–AlN layer [30]. Epilayer thicknesses of this magnitude are equivalent to that grown on SiC, also with CTE mismatch mitigating layers, and are important because they can provide sufficient III–N material to block high voltage (at least 1,200 V) for power switching devices.

While threading dislocations may benefit the energetics of stress mitigation, as discussed previously in regards to the function of the nucleation layer, they pose a concern for 2DEG characteristics and device degradation. Dislocations threading through the 2DEG impact charge mobility/velocity due to lattice perturbation and electron scattering. Dislocations, especially of the screw type, can also trap charge and act as vertical charge transport sites that can manifest as device leakage or short- or long-term device degradation and, ultimately, device failure. While it is feasible, and energetically favorable from the perspective of minimum free energy of the system, to have no dislocations in a crystal, heterogeneous epitaxy inherently produces such defects. Fortunately, the same energy reduction consideration will drive the system toward the tendency to annihilate threading dislocations in the buffer layer structure. This occurs through a process of dislocation bending wherein two separate dislocations with separate lattice strain fields will combine to form a loop thereby eliminating the lattice strain associated with the defects. With proper growth processes, dislocation defect densities on the order of 2E8 to 2E9 cm^{-2} in the near-device region of the buffer layer structure are obtained. GaN-on-SiC tends toward the lower end of the range while GaN-on-Si tends toward the higher end. Both commercial discrete transistors and MMICs tolerate this dislocation density to produce power density superior to any material system with reliability exceeding system requirements [15].

GaN HEMTs sustain both high voltage and high current making it imperative to provide a highly electrically resistive (semi-insulating) buffer layer structure to minimize device leakage. Point defects such as impurity elements substituting for the Ga or N sites and lattice vacancies predominately determine the conductivity, and inversely resistivity, of the buffer material. Point defects, while energetically stable and, therefore, impossible to eliminate completely, are, however, afforded a level of control in the growth environment.

Various impurities form acceptor and donor states in the band structure and can be compensated to a net neutral charge with the intentional addition of certain dopants. Two commonly employed compensating impurity dopants are carbon and iron. Carbon is amphoteric, but because of the relative atomic size difference between Ga and N, carbon most often sits on the N site and produces an acceptor state. (The location of the state within the bandgap is disputed. Recent results give strong indication it is a relatively shallow acceptor state [31].) Fe-doped GaN has been shown experimentally to produce a deep acceptor state [32]. Because the usual impurities associated with GaN growth produce an n-type background, acceptor states (p-type), with the appropriate matching concentration, will yield semi-insulating GaN buffer material. (A note of caution is worthwhile here. Experience with other III–V material systems has shown that unintended consequences can arise when compensation doping is employed. One key example: Vanadium doping in early GaAs material to produce SI–GaAs substrates later proved to be a device reliability issue. Much time, effort, and expense ensued to remedy the problem.). The growth environment, ultimately manifested in the boundary layer and the surface of the growing epitaxial film, offers multiple factors for minimizing and controlling the relative abundance of the incorporated impurities. Generally, the factors are grouped as temperature, pressure, and V–III gas phase ratio. Minimizing unintentional impurity incorporation involves optimizing these factors simultaneously. This optimization is also conditioned on the specific design of the reactor so there is not one specific recipe. However, when coupled with utilizing the highest purity source gases (metalorganics and ammonia), this approach has proven to yield high crystal quality, semi-insulating GaN buffer material.

7.4   Device Layer Structure

The HEMT device advantage, as the acronym implies, is predicated on realizing a high electron mobility region in the epitaxy. The layers designed for the realization, and optimization, of the high electron mobility region comprise the device layer structure. Certain heterojunctions in the GaN material system produce a 2DEG with high mobility and exceptionally high charge due to band alignment that results in a dip of the conduction band below the Fermi energy level. The band alignment is a product of the combined effects of spontaneous and piezoelectric polarization resulting from the heterojunction and strain in, particularly, the [1] crystallographic orientation. The first heterojunction to demonstrate the 2DEG was AlGaN/GaN [1]. The exceptionally promising results from this early device quickly yielded, first, RF performance and, later, power switching performance far exceeding that of incumbent technologies. Key developments in the HEMT device layer structure have largely driven this rapid rate of performance improvement.

The most commonly employed, and only commercialized, HEMT 2DEG heterojunction is AlGaN/GaN. The AlGaN barrier layer Al content typically is in the range of 15–30% mole fraction with a thickness of nominally 15–30 nm. The precise values of each are specific to each fabrication facility. Generally, they depend on the target device performance, but are also balanced by the ability to control the charge within the channel (i.e., minimal leakage and ability of the gate to modulate the channel).

The 2DEG characteristics follow the trends of (1) higher Al content → higher charge density, and (2) thicker barrier → higher (absolute value) pinch-off voltage. There are limits, however. There is a lower limit on thickness below which the 2DEG will not form due to the above mentioned conduction band bending polarization effects being insufficient to create the well. There is an upper limit on thickness above which the lattice mismatch stress will exceed the maximum yield strength causing the AlGaN to relax through the formation of cracks. Both limits are functions of the Al content. With regards to Al content, while a higher mole percentage produces more charge, this typically comes at the expense of device leakage issues and crystal quality, especially above ∼30%. However, in the ranges noted, the 2DEG will have representative sheet charge density of $0.8–1.1 \times 10^{13} \, \text{cm}^{-2}$ with mobility of $1,300–2,000 \, \text{cm}^2/\text{Vs}$.

Various design improvements to the basic AlGaN/GaN device layer structure have been developed. Three, in particular, are noteworthy: GaN cap, AlN interlayer, and back barrier structures.

7.4.1 GaN Cap

A GaN capping layer above the AlGaN barrier layer is utilized for a variety of purposes. First, the binary GaN material is uniform in alloy composition and the nature of the growth tends to planarize, both of which can effectively "heal" surface inhomogeneities that may be present in the AlGaN layer. The improved surface can result in more uniform device performance across the wafer. Second, the presence of a GaN cap above typical AlGaN barrier layer has been demonstrated to increase the effective Schottky barrier height due to piezoelectric polarization effects. Corresponding reductions in reverse leakage current have been shown [33].

Third, the GaN cap layer has been doped n-type, typically with silicon, to improve Ohmic contact resistance. Last, a relatively thick GaN cap layer has been shown to obviate the need to passivate surface states that would otherwise act as charge trapping centers [34]. Doped and/or thick GaN cap layers are typically recess etched to form the gate.

7.4.2 AlN Interlayer

Both the 2DEG sheet charge and mobility have demonstrated improved performance with the inclusion of a thin AlN interlayer between the AlGaN and GaN layers [35]. AlN provides a larger band offset to GaN and, hence, more channel charge. The binary AlN is not susceptible to the "compositional pulling effect" resulting in an AlN/GaN heterointerface that can be sharper than AlGaN/GaN thereby minimizing alloy scattering of the channel electrons and, thus, better mobility. The thickness of the AlN interlayer ranges from 0 to 2 nm (A thickness of 0 describes the situation of intentionally inverting the profile of the AlGaN ternary to the composition that would otherwise be realized due to the compositional pulling effect to enhance interface abruptness and thereby realize the aforementioned benefit of reduced alloy scatter.). While the AlN/GaN heterojunction should have the same benefits and even better performance, research has shown the crystal quality and AlN thickness limitation due to stress relaxation cracking to be a significant hinderance.

7.4.3 Back Barrier Structures

The high 2DEG sheet charge density can readily produce $>1\,A/mm$ of drain current in the HEMT. Such large current being forced through such a narrow 2DEG region can lead to insufficient carrier confinement to prevent soft pinchoff, high sub-threshold current, and high output conductance which can result in reduced power performance, particularly for devices with short gate lengths ($<0.3\,\mu m$) [36]. Conversely, inherent charge in the buffer due to impurities can lead to device leakage. A back barrier structure beneath the 2DEG channel can mitigate this leakage and two such structures have been developed: AlGaN and InGaN. In the case of AlGaN back barrier, the GaN buffer is replaced with a low Al content, but still larger band gap AlGaN buffer, on which a thin GaN channel is grown to maximize the 2DEG sheet charge mobility [37]. Alternatively, a properly designed InGaN layer inserted in the GaN buffer, also just below the 2DEG region, will have a band offset that also produces a conduction band barrier [38]. Both structures have proven to be effective in minimizing buffer leakage and keeping the electrons confined to the channel region of the device layer structure.

7.5 HEMT Device Processing

Borrowing from GaAs pHEMT technology, conventional GaN HEMT processing schemes often proceed through a sequence of (1) mesa isolation, (2) ohmic metalization and anneal, (3) gate and contact pad metalization, and (4) passivation. However, in this sequence, the channel region of the HEMT is exposed during all pre-passivation process steps to various photoresist coats, developer solutions, solvents, and plasma descums, as well as effects from the ambient. Since the surface of GaN-based devices – HEMTs in particular – has been shown to be extremely sensitive [39], passivating as early as possible during the device fabrication sequence is an attractive approach. This practice effectively "locks-in" the qualities of the surface/passivant interface, eliminating unwanted contributions from subsequent process steps. Process sequences with early passivation and dielectrically defined T-gates, as depicted in Fig. 7.3, have become increasingly popular among GaN HEMT device manufacturers, borrowing from similar progression of GaAs pHEMT processing to protect the surface and improve device performance [40]. The process flow of Fig. 7.3 proceeds through ohmic metalization and anneal, passivation, implant isolation, gate and contact pad etching and metallization. For a typical device, subsequent fabrication steps (not shown) include formation of a source-connected field plate over an additional encapsulant dielectric, frontside electroplating and/or interconnect, and backside processing (e.g., thinning, via etch, backside metal). A typical plan view of a multi-finger GaN RF power device is given in Fig. 7.4.

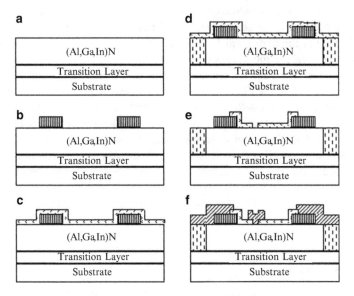

Fig. 7.3 Process sequence for an early passivation GaN HEMT illustrating (**a**) starting substrate; (**b**) ohmic metalization and anneal; (**c**) passivation; (**d**) ion implant interdevice isolation; (**e**) passivation etch to define gate and contact pad openings; (**f**) gate and contact pad metalization

Fig. 7.4 Top-down view of
typical multi-finger GaN
HEMT power device showing
gate (G), source (S), and
drain (D) pad areas. Note
airbridge interconnection of
source fingers

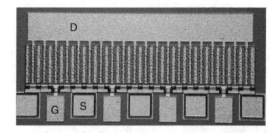

7.5.1 Metalization

7.5.1.1 Ohmic

Most modern GaN HEMT ohmic contact schemes employ Ti/Al/x/Au, where x
may be Ti, Ni, Pt, Mo, or Pd. Despite many studies, the explicit roles of the
individual constituent elements are still not well agreed in the community. Cross-
sectional transmission electron microscopy (XTEM) has shown the existence of Au
at the GaN surface of Ti/Al/Pt/Au contacts, suggesting significant Au diffusion into
the contact during annealing. In fact, Ni, Pd, and Pt have all been shown to be
permeable to Au during typical contact processing [41]. One issue unique to GaN
HEMT contact formation is the extremely high annealing temperatures required to
minimize contact resistance. Typical annealing conditions are ∼850°C, 30 s. in a
nitrogen ambient. This very high annealing temperature is well above the melting
point of Al (661°C), and typically leads to rough ohmic contact morphology. The
edge definition of the patterned contact is also affected by the roughening, and edge
acuity may become an issue for very small device features, such as the channel
of an HEMT. For the traditional Ti/Al-based scheme, a tradeoff between electrical
properties and surface morphology seems to exist, with smoother morphologies
available at lower annealing temperatures while sacrificing Rc. To combat this problem for deep submicron scaling, silicon ions have been implanted into source-drain
regions to achieve a non-alloyed ohmic. Contact resistance as low as 0.2 ohm-
mm has been measured with this approach [42]. Other approaches utilizing Si
implantation include the use of surface-protective cap layers, enabling post-implant
high temperature annealing (to ∼1, 200°C) and very low contact resistance without
damaging the surface and/or channel [43, 44]. Additional advances in GaN ohmic
contacts are expected due to significant interest in increasing the high-frequency
performance of GaN HEMTs, for which minimizing source resistance will play a
key role.

7.5.1.2 Schottky

Historically, most AlGaN/GaN HEMTs have employed Ni/Au as the gate contact
due to comparable electrical properties and improved adhesion to the (Al)GaN

surface relative to Pt. For Ni gates, it has been shown that an interfacial layer of NiO_x can lead to increased HEMT parametric drift but can be partially mitigated by thermal treatment [45]. Recently, the use of Pt has become more prevalent in GaN HEMTs and state-of-the-art enhancement and depletion mode devices have been demonstrated using a Pt/Au or Pt/Ti/Au Schottky contact [46]. Very recent studies have shown that Pt may be preferable to Ni for reduced leakage and improved device stability under reliability stress [47]. At present, the large barrier height of Pt, combined by the propensity of Ni to oxidize, seem to favor Pt as the Schottky barrier metal of choice. Many future devices, however, are expected to employ thin gate insulator layers and optimization of metal scheme for these structures has yet to emerge.

7.5.2   Isolation

7.5.2.1   Dry Etching

High-density, inductively-coupled chlorine-based plasmas have become the most common method to etch GaN due to high plasma uniformity over large areas, lower ion energy (and thus lattice damage) relative to RIE, and lack of electromagnets and waveguides necessary with ECR. Other techniques, including chemically assisted ion beam etching, reactive ion beam etching, and laser ablation have been applied to the III-nitrides, with varying levels of success [48–50]. In the nearterm, ICP etching will likely remain the enabling technology for applications where etch isolation is required or preferred. However, despite the success of high-density plasma etch techniques, dry etch-induced lattice damage can severely degrade material properties and HEMT performance. The damage may occur as lattice defects, unintentional passivation, or preferential N_2 loss producing surface non-stoichiometry and/or conductivity. Such damage can be manifested as extrinsic leakage currents in isolated areas of GaN HEMTs, decreasing breakdown voltage and limiting power performance. For non-HEMT electronic devices, plasma damage is an especially troublesome issue for III–N bipolar electronic devices, such as BJTs and HBTs, and has been a fundamental limitation to their development. For an npn HBT where a dry etch step is required to uncover the p-base, etch damage can decrease the hole concentration or completely reverse the surface conductivity type, depending on the extent of the damage.

7.5.2.2   Ion Implantation

Ion implantation can be used to produce interdevice isolation without disrupting device planarity. This may be advantageous in certain processing sequences where etching or step coverage of metalization layers off mesa sidewalls is undesirable. Implantation into semiconductors creates damage-related energy states that trap free

Fig. 7.5 Simulated vacancy distribution from nitrogen ion isolation implant showing SiN_x and (Al,Ga)N layers

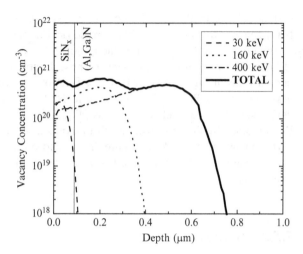

carriers or create recombination sites, leading to high-resistivity material. Isolation implants into GaN have been demonstrated with numerous species including H^+, He^+, N^+, O^+, F^+, P^+/He^+, Ti^+, Cr^+, and Fe^+ [51–56] and have been reviewed by several authors [57]. Although most energy levels created by implant damage in (Al,Ga)N are not near midgap, they are still typically sufficient to produce insulating material with sheet resistances of $\sim 10^9$–10^{12} Ω/sq.

An implant suitable for HEMT device isolation is given in Fig. 7.5, illustrating vacancy concentration of $>10^{20} cm^{-3}$ from the AlGaN surface to a depth of $\sim 0.5\,\mu m$. Such implants can be designed using Monte Carlo simulations and can be performed through SiN_x passivant, if present. In many cases, it may be desirable to implant through a dielectric layer, such as SiN_x, as this can help increase the vacancy concentration in the important near-surface region. Energy/dose conditions for the implant of Fig. 7.5 were $30\,keV/6\times10^{12}\,cm^{-2}$, $160\,keV/1.8\times10^{13}\,cm^{-2}$, and (doubly charged) $400\,keV/2.5\times10^{13}\,cm^{-2}$. Typical sheet resistance of implanted material was $10^{11}\,\Omega/sq$ and the implant remained thermally stable to 600°C. A model using a combination of resistive buffer layer current and Poole-Frenkel current through the implanted region has been used to fit experimental data of HEMT isolation leakage vs. applied voltage [58].

7.5.3 Passivation and Field Plating

Well-known polarization effects in GaN HEMTs lead to surface states that can have a significant detrimental impact on device performance. When unpassivated, these positively charged surface donor states can trap negative charge and lead to a "virtual gate" phenomenon that depletes the 2DEG of carriers, reducing drain current and power performance [59]. In 2000, Green et al. reported the first GaN

HEMT passivated with SiN_x [60] and this is still by far the most widely employed material for terminating the surface of traditional AlGaN/GaN heterostructures. Such SiN_x layers are typically deposited by PECVD at a baseplate temperature of $\sim300°C$ using SiH_4 and NH_3 precursors combined with inerts such as Ar, N_2, and/or He. Stress and fixed charge in these SiN_x films can be tailored by modifying growth parameters and can have an impact on device performance. The primary results of appropriate passivation are (1) an increase in sheet electron concentration due to termination of surface states and (2) a reduction in trap-related dispersion in fabricated devices. Both serve to greatly increase drain current and power density of RF GaN HEMTs relative to unpassivated structures. The SiN_x passivant can also be used to create dielectrically defined T-gates. For such a process, as illustrated in Fig. 7.3, a low-damage SiN_x etch recipe is required for accurate pattern transfer and critical dimension control of gate openings. For reproducible control of gate lengths <1 μm, a dry etch is required. For SiN_x etching, Fl-containing chemistries such as CF_4 or SF_6 can be used, often in a high density, low power plasma. The dielectrically defined T-gate geometry has both advantages and disadvantages at the device level. Overlap of gate metalization onto the surface of the passivant can be optimized to create a gate field plate and reduce peak electric field in the semiconductor, increasing breakdown voltage. Unfortunately, the presence of high-k SiN_x surrounding the gate also leads to parasitic loading and increases the fringing capacitance, degrading gain of RF GaN HEMTs. Bilayer dielectric films consisting of thin SiN_x and an overlayer of a lower dielectric constant material such as SiO_2 have been explored with good results [61].

Additionally, for aggressively scaled HEMTs with gate length <100 nm, the aspect ratio of SiN_x thickness to gate length may complicate accurate pattern transfer required for gate length definition [62]. For such devices, it is advantageous to make the dielectric very thin. For deposition of thin passivant films, atomic layer deposition (ALD) has received considerable recent attention, borrowing from techniques developed to support growth of high-k dielectrics for sub-65 nm Si CMOS. Interestingly, many of these high-k passivant films can be optimized as surface passivation and for use as gate insulators for high-frequency GaN MOS-HEMTs, leading to reduced gate capacitance and leakage current relative to traditional Schottky gate HEMTs. Solutions receiving considerable recent attention include Al_2O_3 passivant/insulator films [63]. Other approaches include AlN films as passivant/insulator and heat spreading layer [64].

In addition to the field plate formed in the gate electrode, most modern GaN HEMTs employ an additional dielectric encapsulation layer over the drift region that serves to protect the device and act as the capacitor dielectric for deposition of a second field plate. This field plate is typically tied to the grounded source electrode (source field plate, or SFP) and is located above the high field region at the drain edge of the gate as shown in Fig. 7.6. Optimization of SFP geometry can lead to simultaneous reduction in peak field and feedback capacitance. Reduction in peak field has been demonstrated to increase breakdown voltage [65] and also increase output power density of RF HEMTs. Reduction in feedback capacitance increases microwave gain relative to devices without such field plate(s).

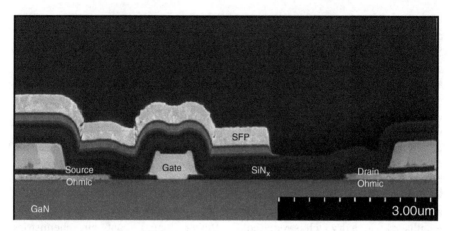

Fig. 7.6 Cross-sectional STEM of GaN-on-Si HEMT showing epilayers and device construction

7.6 HEMT Packaging and Products

Packaging technologies for GaN HEMTs have received surprisingly little attention in the technical literature, despite the critical importance of appropriate packaging to the realization of high-performance devices. The high-power density of GaN can lead to dissipated power densities at levels significantly higher than competitive technologies such as Si or GaAs. For such devices, a primary challenge of any packaging solution is heat sinking to minimize thermal resistance. Without appropriate thermal management at both the chip and package level, the high-power density of GaN can cause performance degradation and/or fundamental limitations in device safe operating area.

7.6.1 Air Cavity Packaging

Most RF GaN HEMTs are assembled in air cavity packages with excellent high-frequency properties and heat sinking directly through the package flange. Air cavity assembly of high-power GaN-on-SiC HEMT chips typically involves eutectic die attach at ~300°C with a ~1 mil thick AuSn preform. GaN-on-Si is the only GaN platform able to leverage the well-established AuSi eutectic process used by other power FET technologies such as Si LDMOS. For AuSi eutectic attach, the GaN-on-Si chip with backside Au electroplating is scrubbed onto the flange under force at a temperature of ~400°C. The eutectic reaction proceeds very quickly and, when optimized, can produce <5% total voiding with a thin (few μm) bondline and excellent thermal properties. Typical air cavity packaging of a GaN active chip with associated MOS capacitor matching components is shown in Fig. 7.7.

Fig. 7.7 Example of ceramic air cavity packaging for high power GaN RF power transistors. The power transistors are shown both unlidded (*top*) and lidded (*bottom*). In the unlidded picture, the GaN chips as well as input and output matching MOS capacitors are shown

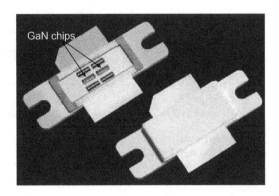

Optimization of packaging materials is also critical for thermal design. Thermal impedance from substrate to package can be comparable to that from junction to substrate, underscoring the importance of the package for any high-power device technology. For GaN HEMTs, novel Cu-flanged air cavity packages are an active area of development. Integration of Cu as a heat sink presents two main challenges: (1) the CTE of Cu precludes the use of conventional Al_2O_3 ceramic materials as the windowframe dielectric; and (2) the CTE mismatch between Cu and the GaN chip – combined with the ductility of Cu – can cause package bowing during assembly. Item #1 is currently being addressed by package suppliers through development of organic materials with suitable dielectric properties and thermal expansion closely matched to Cu. Most of these organic materials cannot withstand AuSi eutectic die attach temperatures, so AuSn eutectic attach is currently used for all devices packaged in Cu. Item #2 can be mitigated by optimizing thickness and/or prebow (intentionally bowing the package prior to assembly so it becomes flat post-assembly). Other heatsink materials under investigation for high-power GaN packaging include AlSiC and Cu-diamond.

7.6.2 Plastic Overmold

Some plastic packaging schemes for GaN HEMTs have emerged, enabling GaN performance attributes to be combined with the assembly scale and cost structure of overmold packages [66, 67]. Such an approach is attractive for pre-driver and driver applications, areas not often considered for GaN devices. A typical plastic packaging process for GaN HEMTs utilizes a copper heatsink, thermally conductive and electrically conductive epoxy die attach, and RoHS-compliant plastic overmolding. SOIC-based products are available with peak power levels ranging from 5 W to nearly 50 W. Additional details of the SOIC packaging approach are illustrated in Fig. 7.8 [4].

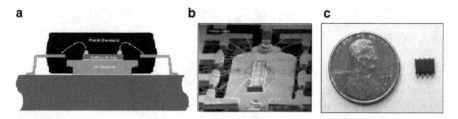

Fig. 7.8 (**a**) Cross-sectional schematic of SOIC package showing input and output leads, Cu heatsink, GaN-on-Si chip, and plastic overmold. (**b**) GaN-on-Si device in SOIC package immediately prior to overmolding. The chip dimensions are 1.65×0.55 mm. (**c**) GaN-on-Si product in 8-pin SOIC package with size reference

7.6.3   GaN HEMT Products

The first commercially available GaN HEMT products were launched by Eudyna Devices (now Sumitomo Electric Device Innovations, Inc.) in February 2006. These early discrete transistor products operated from a supply voltage of 50 V and ranged in power from 10 to 180 W across W-CDMA and WiMAX frequency bands. The Sumitomo product portfolio has broadened over the past 5 years and now includes ~30 products at frequencies up to 3.5 GHz and with power levels as high as 400 W at S-band.

In the U.S., GaN products for RF applications are currently available from TriQuint, Nitronex, Cree, and RFMD. Foundry services have recently been announced for most U.S. GaN HEMT manufacturers, primarily for GaN-on-SiC. In many cases, these foundry offerings are accompanied by process design kits, non-linear scalable models, and support across multiple microwave design software platforms, clearly attesting to the maturation of GaN HEMTs over the past decade. Commercial products are largely discrete RF power transistors operating at 28 and 48 V, with MMICs becoming available in the past 1–2 years for applications such as broadband communications and jamming. Such devices facilitate improved communications transmit distance and extend the umbrella size of electronic protection units.

GaN HEMTs also hold great promise for high-efficiency power conversion and power management applications. Here, GaN technology is expected to deliver performance figures-of-merit ten times greater than those attainable from modern Si-based power electronics. As an example, the GaN-enabled capability to increase microprocessor power conversion frequency and efficiency is expected to result in smaller converter footprint, reduced external component count and parasitic losses, and could save billions of dollars of energy annually.

Within the past 1–2 years, the first GaN HEMTs for power management/power conversion applications have been commercially launched. GaN HEMTs in this space are expected to make significant design-ins over the coming years, with the GaN power device opportunity recently forecast to reach a $350 M market size in 2015 [68]. The first products by International Rectifier were low-voltage power stages in LGA packages built on a GaN-on-Si platform. These products contain a driver IC matched to a GaN power device offering switching speeds up to 3–5 MHz. End-use applications for these early products include multiphase and point-of-load applications such as servers and switches. The first commercial enhancement mode devices were introduced in 2009 from Efficient Power Conversion Corp. and today extend to a voltage rating of 200 V. GaN devices for high-voltage power applications today lag considerably behind the maturity level of RF GaN HEMTs. However, with a power electronics market size of over \$16 billion US dollars in 2010 and tremendous promise for GaN materials in this space, this is envisioned as an extremely exciting area for GaN product development over the next decade.

7.7 Future Directions for GaN HEMTs

7.7.1 GaN-on-Diamond

GaN HEMT technology has demonstrated record power density compared to all other material systems. The power density is so great ($40 + \text{W/mm}$ [7]) that the limiting factor for scaling the device gate periphery is heat generation in the channel. Currently, commercial GaN HEMTs operate at power densities of $\sim 7\,\text{W/mm}$ [15]. To close the gap between the research and production power densities, better heat dissipation away from the channel is required. The ideal material for this purpose is diamond due to its exceptionally high thermal conductivity.

Direct growth of GaN on diamond substrates has been demonstrated [69, 70]. However, the size of the single-crystal diamond is small and scaling is a challenge due to the inherent kinetic limitations for growing the crystal larger. An alternative approach that has received recent attention involves replacing the original substrate (Si, SiC, or Al_2O_3) with poly-crystalline diamond. Two pathways for this approach are being pursued. The first involves forming a poly-crystalline diamond layer 50+ microns thick on a sacrificial Si wafer and through etching, or other means, removing the original substrate on which the GaN HEMT structure has been grown, then bonding the diamond layer to the GaN and removing the sacrificial Si wafer [71]. The second involves growing 10–25 μm of diamond directly on top of the GaN HEMT structure (along with a sacrificial handle wafer on the diamond) then flipping the wafer and removing the original substrate and the nucleation and buffer layers [72]. The two process routes are shown schematically in Fig. 7.9. Based on thermal simulations, a 2× increase (minimum) of the production-capable power density while maintaining a constant junction temperature should be readily realized with either approach. Initial device results have been published [73] and further materials optimization is underway with full performance capability demonstrations anticipated in the near future.

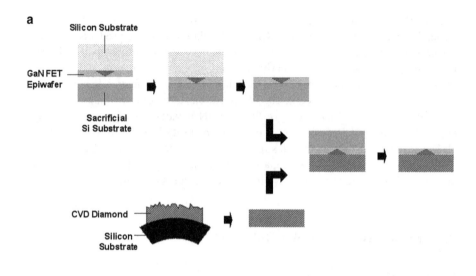

Fig. 7.9 (**a**) Process route for forming a GaN HEMT on poly-crystalline diamond by depositing diamond on sacrificial Si substrate. (**b**) Process route for forming a GaN HEMT on poly-crystalline diamond by direct deposition of diamond on GaN

7.7.2 Heterointegration

Today's circuit designer is forced to choose the proper material/device type based on performance requirements and other circuit-level considerations. Circuit boards are often comprised of chips made from several different materials systems (e.g., Si, GaAs, InP, GaN). If different materials could be embedded within one chip, circuit design and, hence, circuit performance would realize a dramatic paradigm shift.

GaN-on-Si is a primary candidate material system to realize multiple functionality on a single chip. Such a chip could leverage the logic capability of silicon with the power performance of GaN, for example. The key consideration for optimizing these two device structures on the same chip is reconciling the disparities in the device fabrication processes. While much work is left to be done, a promising break-through has recently been made in which heterointegration of GaN HEMT and Si MOS device operation was demonstrated in the research lab environment [74] where the two devices were only one micron apart. Thus, the basic feasibility has been proven and, with further research, the capability will become commercially viable.

7.7.3 InAlN HEMTs

Sheet charge density and electron transport properties of AlGaN/GaN heterostruc-tures limit the minimum sheet resistance of AlGaN/GaN HEMTs to $\sim$250 ohms/sq [75]. To fabricate AlGaN/GaN HEMTs with improved frequency response, the thickness of the AlGaN barrier layer should be reduced as the gate length is scaled [76,77]. However, at barrier layer thicknesses below $\sim$15 nm, surface depletion can significantly degrade the sheet charge density, thereby increasing sheet resistance and reducing HEMT current density [78]. The InAlN/GaN material system offers an attractive alternative for such high-power, high-frequency applications. The InAlN alloy can be synthesized lattice matched to GaN with large refractive index contrast and sheet charge density roughly twice that of typical AlGaN/GaN HEMTs. This sheet charge density is due to the more than 4$\times$ increase in spontaneous polarization of $In_{0.17}Al_{0.83}N$/GaN as compared to a traditional $Al_{0.2}Ga_{0.8}N$/GaN HEMT structure as shown in Table 7.1. Although there is no strain-induced piezoelectric component to the overall polarization charge of the lattice-matched InAlN heterostructures, the spontaneous component dominates and leads to a total polarization charge more than 2$\times$ that of $Al_{0.2}Ga_{0.8}N$/GaN. This enables extremely high sheet charge density even in thin ($\sim$10 nm) barrier HEMT structures. Gonshchorek et al. have calculated InAlN 2DEG densities ranging from 2.2×10^{13} for $In_{0.21}Al_{0.79}N$ to 3.5×10^{13} cm^{-2} for $In_{0.07}Al_{0.93}N$ in 14-nm thick layers [79]. These properties can be exploited to fabricate InAlN/GaN HEMTs with very high current density, low access resistance, aggressive scaling, and monolithic integration of normally on and normally off operation [80–82].

Despite these advantages, differences in fundamental properties of the InN and AlN binary components dictates remarkably different growth conditions and has limited progress in the growth of high quality InAlN alloys. Typical MOCVD growth conditions for AlN are $>1,200°C$ with low V/III ratio of $<1,000$. Due to the much lower decomposition temperature, typical MOCVD InN growth takes place at $<800°C$ with V/III ratio that can be two orders of magnitude higher than for AlN. The compromise in choice of ternary InAlN alloy growth conditions typically results in poor material quality, often manifested in rough or pitted surface morphol-ogy and compositional inhomogeneity. Furthermore, many early films have been characterized by significant amounts (several %) of unintentionally incorporated

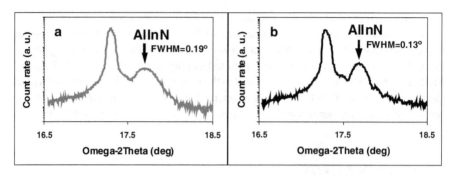

Fig. 7.10 X-ray scans from 30 nm thick InAlN layers grown at (**a**) 100 torr and (**b**) 50 torr reactor pressure. The In composition in these samples was 12–14% (slightly tensile strain conditions)

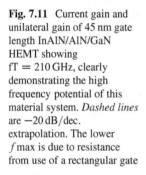

Fig. 7.11 Current gain and unilateral gain of 45 nm gate length InAlN/AlN/GaN HEMT showing fT = 210 GHz, clearly demonstrating the high frequency potential of this material system. *Dashed lines* are −20 dB/dec. extrapolation. The lower f max is due to resistance from use of a rectangular gate

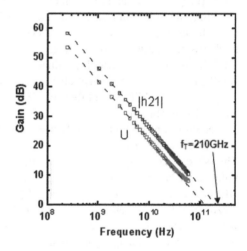

Ga [83] and compositional variations and indium segregation, particularly around V-shaped defects [84, 85].

To improve InAlN material quality, lower MOCVD growth pressure is favored. At 50 torr, AFM $1 \times 1\,\mu m^2$ root mean square surface roughness and X-ray peak full width at half maximum (FWHM) have been shown to be significantly improved relative to growth at 100 torr [86] as shown in Fig. 7.10

Additionally, a thin AlN spacer layer has been utilized between the GaN buffer and AlInN barrier layer to decrease allow scattering and improve electron mobility [79, 87, 88]. Growth conditions for this spacer are highly important and can lead to significant differences in surface morphology and 2DEG sheet resistance, with values as low as 167 ohm/sq demonstrated in [86]. In this work it was also shown that InAlN thickness in the range of 10–60 nm and In composition variation from 12 to 18% had nearly no effect on the sheet resistance of InAlN/GaN HEMT structures employing a 1 nm AlN barrier (Fig. 7.11 and Table 7.2).

Table 7.2 Polarization properties of traditional AlGaN/GaN HEMT heterostructure and lattice-matched InAlN/GaN heterostructure. Note absence of piezoelectric polarization ΔP_{piezo} but extremely large spontaneous polarization charge ΔP_o for the InAlN barrier From [95]

Structure	$\Delta P_o (10^{12} \, cm^{-2})$	$P_{piezo} (10^{12} \, cm^{-2})$	$P_{total} (10^{12} \, cm^{-2})$	$\Delta E_c (eV)$
$Al_{0.2}Ga_{0.8}N$/GaN	6.5	5.32	11.8	0.3
$Al_{0.83}In_{0.17}N$/GaN	27.3	0	27.3	0.68

At the time of this writing, current state-of-the-art InAlN HEMT device performance includes current density of $>2.5 \, A/mm$ [46], transconductance $>1 \, S/mm$ [89], power density of $>10 \, W/mm$ at $10 \, GHz$ [90], and cutoff frequency of $>200 \, GHz$ [91, 92]. It is still too early to assess the impact of the lattice-matched InAlN alloy on HEMT reliability performance, as there is not yet sufficient data to draw conclusions. Initial studies are beginning to emerge [93, 94].

Abbreviations

ΔP_o	Spontaneous polarization charge
P_{piezo}	Piezoelectric polarization charge
ΔE_c	Conduction band discontinuity
2DEG	Two-dimensional electron gas
ALD	Atomic layer deposition
AlGaN	Aluminum gallium nitride
AlN	Aluminum nitride
BJT	Bipolar junction transistor
CTE	Coefficient of thermal expansion
ECR	Electron cyclotron resonance
GaN	Gallium nitride
GSMBE	Gas source molecular beam epitaxy
HBT	Heterojunction bipolar transistor
HEMT	High electron mobility transistor
ICP	Inductively coupled plasma
InAlN	Indium aluminum nitride
LD	Laser diode
LED	Light-emitting diode
MESFET	Metal semiconductor field effect transistor
MISFET	Metal insulator semiconductor field effect transistor
MBE	Molecular beam epitaxy
MOCVD	Metalorganic chemical vapor deposition
MO-MBE	Metalorganic molecular beam epitaxy
PA	Power amplifier
PA-MBE	Plasma-assisted molecular beam epitaxy

P_{total} Total polarization charge
RF   Radio frequency
RIE   Reactive ion etch
SiN_x Silicon nitride

References

1. M.A. Khan, T.N. Kuznia, A.R. Bhattaraia, D.T. Olson, Appl. Phys. Lett. **62**, 1786 (1993)
2. N. Maeda, T. Saitoh, K. Tsubaki, T. Nishida, N. Kobayashi, Phys. Stat. Sol. (b) **216**, 727 (1999)
3. S. Nagahama, N. Iwasa, M. Senoh, T. Matsushita, Y. Sugimoto, H. Kiyoku, T. Kozaki, M. Sano, H. Matsumura, H. Umemoto, K. Chocho, T. Mukai, Jpn. J. Appl. Phys. **39**, L647–L650 (2000)
4. J.W. Johnson, S. Singhal, A.W. Hanson, R.J. Therrien, A. Chaudhari, W. Nagy, P. Rajagopal, Q. Martin, T. Nichols, J.C. Roberts, E.L. Piner, I.C. Kizilyalli, K.J. Linthicum, Mater. Res. Soc. Symp. Proc. **1068**, 3–12 (2008)
5. P. Rajagopal, J.C. Roberts, J.W. Cook, Jr., J.D. Brown, E.L. Piner, K.J. Linthicum, Mater. Res. Soc. Symp. Proc. **798**, 61–66 (2004)
6. D.M. Fanning, L.C. Witkowski, C. Lee, D.C. Dumka, H.Q. Tseng, P. Saunier, W. Gajewski, E.L. Piner, K.J. Linthicum, J.W. Johnson, International Conf. on Compound Semi. Manufacturing Technologies (GaAs MANTECH) Digest, 2005
7. Y.-F. Wu, M. Moore, A. Saxler, T. Wisleder, P. Parikh, In: *IEEE Device Research Conf.*, State College, PA, USA, (2006) pp. 151–152
8. S. Nakamura, G. Fasol, *"The blue laser diode, GaN based light emitters and lasers"* (Springer, Berlin, 1997)
9. B.P. Keller, S. Keller, D. Kapolnek, W.-N. Jiang, Y.-F. Ww, H. Masui, X. Wu, B. Heying, J.S. Speck, U.K. Mishra, S.P. Denbaars J. Electron. Mater. **24**(11), 1707 (1995)
10. S. Nakamura, M. Senoh, S. Nagahama, N. Iwasa, T. Yamada, T. Matsushita, Y. Sugimoto, H. Kiyoku, Appl. Phys. Lett. **70**, 868 (1997)
11. L. Marona, P. Wisniewski, P. Prystawko, I. Grzegory, T. Suski, S. Porowski, P. Perlin, R. Czernecki, M. Leszczyski, Appl. Phys. Lett. **88**, 201111 (2006)
12. T. Li, R.P. Campion, C.T. Foxon, S. Rushworth, L. Smith, In: International Conference on Molecular Beam Epitaxy, 143–144 (2002) 10.1109/MBE.2002.1037800
13. J. Bardwell, Y. Liu, H. Tang, J. Webb, S. Rolfe, J. Lapointe, Electron. Lett. **39**, 564 (2003)
14. M. Micovic, A. Kurdoghlian, P. Janke, P.H.D. Wong, J. Moon, L. McCray, C. Nguyen, IEEE Trnas. Eletron. Dev. **48**, 591 (2001)
15. M. Rosker, C. Bozada, H. Dietrich, A. Hung, D. Via, S. Binari, E. Vivierios, E. Cohen, J. Hodiak, CS MANTECH Conference Digest, May 18–21st, (Tampa, Florida, USA, 2009)
16. T. Sasaki, J. Cryst. Growth **129**, 81 (1993)
17. S. Yoshida, S. Misawa, S. Gonda, Appl. Phys. Lett. **42**, 427 (1983)
18. H. Amano, N. Sawaki, I. Akasaki, Y. Toyoda, Appl. Phys. Lett. **48**, 353 (1986)
19. S. Nakamura, Jpn. J. Appl. Phys. **30**, L1705 (1991)
20. T.W. Weeks Jr., M.D. Bremser, K.S. Ailey, E.P. Carlson, W.G. Perry, E.L. Piner, N.A. El-Masry, R.F. Davis, J. Mater. Res. **11**(4), 1081 (1996)
21. E.L Piner, Y.W. He, K.S. Boutros, F.G. McIntosh, J.C. Roberts, S.M. Bedair, N.A. El-Masry, Mater. Res. Soc. Symp. Proc. **395**, 307 (1996)
22. M.-H. Kim, Y.-G. Do, H.C. Kang, D.Y. Noh, S.-J. Park, Appl. Phys. Lett. **79**, 2713 (2001)
23. E. Feltin, B. Beaumont, M. Laugt, P. de Mierry, P. Vennéguès, H. Lahrechè, M. Leroux, P. Gibart, Appl. Phys. Lett. **79**, 3230 (2001)
24. A. Dadgar, M. Poschenrieder, J. Bläsing, K. Fehse, A. Diez, A. Krost, Appl. Phys. Lett. **80**, 3670 (2002)

25. Y. Fu, D.A. Gulino, J. Vac. Sci. Technol. A. **18**(3), 965 (2000)
26. J.C. Roberts, J.W. Cook, Jr., P. Rajagopal, E.L. Piner, K.J. Linthicum, Mater. Res. Soc. Symp. Proc. **1068**, 147 (2008)
27. P. Rajagopal, T. Gehrke, J.C. Roberts, J.D. Brown, T.W. Weeks, E.L. Piner, K.J. Linthicum, Mater. Res. Soc. Symp. Proc. **743**, 3 (2003)
28. A. Krost, A. Dadgar, Mater. Sci. Eng. **B93**, 77 (2002)
29. A. Dadgar, J. Christen, S. Richter, F. Bertram, A. Diez, J. Bläsing, A. Krost, A. Strittmatter, D. Bimberg, A. Alam, M. Heuken, IPAP Conference Series **1**, 845 (2000)
30. A. Reiher, J. Bläsing, A. Dadgar, A. Diez, A. Krost, J. Crys. Growth **248**, 563 (2003)
31. J.L. Lyons, A Janotti, C.G. Van de Walle, Appl. Phys. Lett. **97**, 152108 (2010)
32. S. Heikman, S. Keller, S.P. DenBaars, U.K. Mishra, Appl. Phys. Lett. **81**, 439 (2002)
33. E.T. Yu, X.Z. Dang, L.S. Yu, D. Qiao, P.M. Asbeck, S.S. Lau, G.J. Sullivan, K.S. Boutros, J.M. Redwing, Appl. Phys. Lett. **73**, 1880 (1998)
34. L. Shen, R. Coffie, D. Buttari, S. Heikman, A. Chakraborty, A. Chini, S. Keller, S.P. DenBaars, U.K. Mishra, IEEE Electron Dev. Lett. **25**(1), 7 (2004)
35. L. Shen, S. Heikman, B. Moran, R. Coffie, N.-Q. Zhang, D. Buttari, I.P Smorchkova, S. Keller, S.P. DenBaars, U.K. Mishra, IEEE Electron Dev. Lett. **22**(10), 457 (2001)
36. B. Heying, W.-B. Luo, I. Smorchkova, S. Din, M. Wojtowicz, 2010 IEEE MTT-S International Microwave Symposium Digest (MTT), 23–28 May 2010, Anaheim, CA, pp. 1218–1221, DOI: 10.1109/MWSYM.2010.5517568
37. M. Micovic, M.P. Hashimoto, M. Hu, I. Milosavljevic, J. Duvall, P.J. Willadsen, W. Wong, A. Conway, A. Kurdoghlian, P. Dellman, J. Moon, A. Schmitz, M. Delaney 2004 IEEE IEDM Techncial Digest. **27**(1), 807 (2004)
38. T. Palacios, A. Chakraborty, S. Heikman, S. Keller, S. P. DenBaars, U. K. Mishra, IEEE Electron Dev. Lett. **27**(1), 13 (2006)
39. R. Vetury, N.Q. Zhang, S. Keller, U. Mishra, IEEE Transact. Electron Dev. **48**, 560 (2001)
40. D. Fanning, L. Witkowski, J. Stidham, H.-Q. Tserng, M. Muir, P. Saunier, 2002 GaAs MANTECH, Paper 5a, San Diego, CA (2002).
41. Q.Z. Liu, S.S. Lau, Solid-State Electron **42**, 677 (1998)
42. Y. Pei, F. Recht, N. Fichtenbaum, S. Keller, S.P. DenBaars, U.K. Mishra, Electron Lett. **43**, 1466 (2007)
43. C. Nguyen, P. Shah, E. Leong, M. Derenge, K. Jones, Semiconductor Device Research Symposium, 2009, (College Park, MD, Dec. 9–11, 2009)
44. M. Placidi, A. Perez-Tomas, A. Constant, G. Rius, N. Mestres, J. Millan, P. Godignon, Spanish Conference on Electron Devices 2009, Santiago de Compostela, Spain, Feb 11–13. 2009
45. S. Singhal, J.C. Roberts, P. Rajagopal, T. Li, A.W. Hanson, R. Therrien, J.W. Johnson, I.C. Kizilyalli, K.J. Linthicum, In: IEEE International Reliability Physics Symposium, San Jose, CA, March 26–30, 2006
46. P. Saunier, Intl. Workshop on Nitride Semiconductors, I1.1, Tampa, FL, Sept. 2010
47. C.-F. Lo, L. Liu, T.S. Kang, R. Davies, B.P. Gila, S.J. Pearton, I.I. Kravchenko, O. Laboutin, Y. Cao, J.W Johnson, F. Ren, Electrochem. Solid-State Lett. **14**, H264 (2011).
48. W.-J. Lee, H.-S. Kim, J.-W. Lee, T.-I. Kim, G.-Y. Yeom, Jpn. J. Appl. Phys. **37**, 7006 (1998)
49. M. Endo, Z. Jin, S. Kasai, H. Hasegawa, Jpn. J. Appl. Phys. **41**(4B), 2689 (2002)
50. T. Anderson, F. Ren, S.J. Pearton, M.A. Mastro, R.T. Holm, R.L. Henry, C.R. Eddy, J.Y. Lee, K.-Y. Lee, J. Kim, J. Vac. Sci. Technol. B. **24**, 2246 (2006)
51. S.C. Binari, H.B. Dietrich, G. Kelner, L.B. Rowland, K. Doverspike, K.D. Wickenden, J. Appl. Phys. **78**, 3008 (1995)
52. S.J. Pearton, C.R. Abernathy, P.W. Wisk, W.S. Hobson, F. Ren, Appl. Phys. Lett. **63**, 1143 (1993)
53. S.J. Pearton, C.R. Abernathy, C.B. Vartuli, J.C. Zolper, C. Yuan, R.A. Stall, Appl. Phys. Lett. **67**, 1435 (1995)
54. J.W. Johnson, B.P. Gila, B. Luo, K.P. Lee, C.R. Abernathy, S.J. Pearton, J.I. Chyi, T.E. Nee, C.M. Lee, C.C. Chuo, T.J. Anderson, F. Ren, J. Electrochem. Soc. **148**, G303 (2001)
55. G. Harrington, Y. Hsin, Q.Z. Liu, P.M. Asbeck, S.S. Lau, M.A. Khan, J.W. Yang, Q. Chen, Electron. Lett. **34**, 193 (1998)

56. X.A. Cao, S.J. Pearton, G.T. Dang, A.P. Zhang, F. Ren, R.G. Wilson, J.M. Van Hove, J. Appl. Phys. **87**, 1091 (2000)
57. S.O. Kucheyev, J.S. Williams, S.J. Pearton, Mater. Sci. Eng. R. **33**, 51 (2001) and references therein
58. C.F. Lo, T.S. Kang, L. Liu, C.Y. Chang, S.J. Pearton, I.I. Kravchenko, O. Laboutin, J.W. Johnson, F. Ren, Appl. Phys. Lett. **97**, 262116 (2010).
59. R. Vetury, N. Q. Zhang, S. Keller, U. K. Mishra, IEEE Trans. Electron Dev. **48**(3), 560–566 (2001)
60. B.M. Green, K.K. Chu, E.M. Chumbes, J.A. Smart, J.R. Shealy, L.F. Eastman, IEEE Electron Dev. Lett. **21**(6), 268–270 (2000)
61. S. Rajan; Yi Pei; Zhen Cheng; S.P. DenBaars; U.K. Mishra, Device Research Conference, 2008, 131–132
62. T. Palacios, A. Chakraborty, S. Heikman, S. Keller, S.P. DenBaars, U.K. Mishra, IEEE Electron Dev. Lett. **27**(1), 13–15 (2006)
63. B. Lu, O.I. Saadat, T. Palacios, IEEE Electron Dev. Lett. **31**, 990–992 (2010)
64. N. Tsurumi, H. Ueno, T. Murata, H. Ishida, Y.Uemoto, T. Ueda, K. Inoue, T. Tanaka, IEEE Transact Electron Dev. **57**(5), (2010)
65. W. Saito, T. Nitta, Y. Kakiuchi, Y. Saito, K. Tsuda, I. Omura, A M. Yamaguchi, IEEE Trans. Electron Dev. **54**, 1825 (2007)
66. R. Therrien, W. Nagy, I. Kizilyalli, RF Design, Feb. 2007, p. 26
67. U.H. Andre, R.S. Pengelly, A.R. Prejs, S.M. Wood, E.J. Crescenzi, High Frequency Electronics, June 2007, p. 16
68. P. Roussel, "Power GaN 2010" Market Report from Yole Development, Nov. 2010
69. C.R. Miskys, J.A. Garrido, C.E. Nebel, M. Hermann, O. Ambacher, M. Eickhoff, M. Stutzmann, Appl. Phys. Lett. **82**, 290 (2003)
70. A. Dussaigne, M. Malinverni, D. Martin, A. Castiglia, N. Grandjean, J. Cryst. Growth. **311**(21), 4539 (2009)
71. D. Francis, J. Wasserbauer, F. Faili, D. Babic, F. Ejeckam, W. Hong, P. Specht, E. R. Weber, Proceedings of the CS MANTECH, Austin, TX, May 14–17, (2007), pp. 133–136
72. E.L. Piner, J.C. Roberts, IEEE Microwave Theory and Techniques Society International Microwave Symposium, High-Power-Density Packaging of Gallium Nitride Workshop (IEEE MTT-S 2010 Workshop WME IMS), May 23, 2010, Anaheim, CA, USA
73. J.G. Felbinger, M.V.S. Chandra, Y. Sun, L.F. Eastman, J. Wasserbauer, F. Faili, D. Babic, D. Francis, F. Ejeckam, IEEE Electron Dev. Lett. **28**, 948 (2007)
74. J.W. Chung, J. Lee, E.L. Piner, T. Palacios, IEEE Electron Dev. Lett., **30**(10), 1015 (2009)
75. U.K. Mishra, P. Parikh, Y.F. Wu, Proc. IEEE, **90**(6), 1022–1031 (2002)
76. Y. Awano, M. Kosugi, K. Kosemura, T. Mimura, M. Abe, IEEE Trans. Electron Dev. **36**(10), 2260–2266 (1989)
77. G.H. Jessen, R.C. Fitch, J.K. Gillespie, G. Via, A. Crespo, D. Langley, D.J. Denninghoff, M. Trejo, E.R. Heller, IEEE Trans. Electron Dev. **54**(10), 2589–2597 (2007)
78. J.P. Ibbetson, P.T. Fini, K.D. Ness, S.P. DenBaars, J.S. Speck, U.K. Mishra, Appl. Phys. Lett. **77**(2) 10 (2000)
79. M. Gonschorek, J.-F. Carlin, E. Feltin, M.A. Py, N. Grandjean, V. Darakchieva, B. Monemar, M. Lorenz, G. Ramm, J. Appl. Phys. **103**, 093714 (2008)
80. J. Kuzmik, IEEE Electron Dev. Lett. **22**(11), 510–512, (2001)
81. M. Gonshorek, J.-F. Carlin, E. Feltin, M.A. Py, N. Grandjean, Appl. Phys. Lett. **89**(6) (2006)
82. C. Ostermaier, G. Pozzovivo, J.-F. Carlin, B. Basnar, W. Schrenk, Y. Douvry, C. Gaquière, J. -C. DeJaeger, K. Cico, K. Fröhlich, M. Gonschorek, N. Grandjean, G. Strasser, D. Pogany, J. Kuzmik, IEEE Elect. Dev. Lett. **30**(10) (2009)
83. M. Trejo, G.H. Jessen, A. Crespo, J.K. Gillespie, D. Langley, D. Denninghoff, G.D. Via, J. Carlin, D. Tomich, J. Grant, H. Smith, 2008 CS MANTECH Conference, Chicago, Illinois
84. S.-L. Sahonta, G.P. Dimitrakopoulos, T. Kehagias, J. Kioseoglou, A. Adikimenakis, E. Iliopoulos, A. Georgakilas, H. Kirmse, W. Neumann, P.H. Komninou, Appl. Phys. Lett. **95**, 021913 (2009)

85. T. Kehagias, G.P. Dimitrakopulos, J. Kioseoglou, H. Kirmse, C. Giesen, M. Heuken, A. Georgakilas, W. Neumann, T.H. Karakostas, P.H. Komninou, Appl. Phys. Lett. **95**, 071905 (2009)
86. O. Laboutin, C.-F. Lo, L. Liu, F. Ren, W. Johnson, Intl. Workshop on Nitride Semiconductors, A4.6, (Tampa, FL,2010)
87. J. Xie, X. Ni, M. Wu, J.H. Leach, Ü. Özgür, H. Morkoc, Appl. Phys. Lett. **91**, 132116 (2007)
88. M. Gonschorek, J.-F. Carlin, E. Feltin, M.A. Py, N. Grandjean, Appl. Phys. Lett. **89**, 062106 (2006)
89. P. Saunier, Intl. Workshop on Nitride Semiconductors, I1.1, Tampa, FL, Sept. 2010
90. J.C. De Jaeger, C. Gaquiere, Y. Douvry, N. Defrance, V. Hoel, S Delage, N. Sarazin, E. Morvan, M. Alomari, E. Kohn, A. Dussaigne, J.F. Carlin, J. Kusmik, C. Ostermaier, D. Pogany, 18th International Conference on Microwave Radar and Wireless Communications, Vilnius, Lithuania, June 2010. p. 1–4
91. H. Sun, A.R. Alt, H. Benedickter, E. Feltin, J.-F. Carlin, M. Gonschorek, N. Grandjean, C.R. Bolognesi, IEEE Electron Dev. Lett. **31**, 9 (2010)
92. R. Wang. G. Li, O. Laboutin, Y. Cao, W. Johnson, G. Snider, P. Fay, D. Jena, H. Xing, IEEE Electron Dev. Lett. **32**, 892 (2011).
93. J. Kuzmík, G. Pozzovivo, C. Ostermaier, G. Strasser, D. Pogany, E. Gornik, J.-F. Carlin, M. Gonschorek, E. Feltin, N. Grandjean, J. Appl. Phys. **106**(12), 124503, (2009)
94. J.H. Leach, M. Wu, X. Ni1, X. Li, Ü. Özgür, H Morkoç, J. Liberis, E. Šermukšnis, A. Matulionis, H. Cheng, C. Kurdak, Y.-T. Moon, Physica Status Solidi (a). **207**(6) 1345–1347 (2010)
95. J. Kumzik, IEEE Electron Dev. Lett. **22**, 510 (2001).

Chapter 8
Recent Advances in High-Voltage GaN MOS-Gated Transistors for Power Electronics Applications

T. Paul Chow and Z. Li

Abstract The recent progress and present status of the development of high-voltage lateral power GaN field-effect transistors for power switching applications are reviewed and discussed. The basic device structures explored and the performance improvement in blocking voltage and specific on-resistance over the last few years are presented. The technical challenges and reliability issues that still need to be addressed are pointed out.

8.1 Introduction

Similar to SiC, GaN has many attractive material properties, such as high breakdown field, good bulk electron mobility, and thermal conductivity, which make it suitable for high-power switching as well as high-frequency amplifying electronic applications [1] (see Table 8.1). On the other hand, thin layers of GaN have been successfully heteroepitaxially grown, with suitable AlN/GaN buffer layers, on silicon, sapphire, and SiC substrates. In addition, GaN has demonstrated two-dimensional electron gas (2DEG) at heterojunction interfaces (AlGaN/GaN, AlInN/GaN) that has very high electron mobilities ($\sim 2,000\,\mathrm{cm}^2/(\mathrm{V\,s})$). Furthermore, the fabrication processing steps for GaN devices have thermal budgets that are compatible with state-of-art silicon nanometer CMOS foundries and can exploit these infrastructures for cost-effective, large-scale GaN power device commercialization and manufacturing.

T. Paul Chow (✉) · Z. Li
Department of Electrical, Computer and Systems Engineering, Rensselaer Polytechnic Institute,
Troy, NY 12180, USA
e-mail: chowt@rpi.edu

S. Pearton (ed.), *GaN and ZnO-based Materials and Devices*,
Springer Series in Materials Science 156, DOI 10.1007/978-3-642-23521-4_8,
© Springer-Verlag Berlin Heidelberg 2012

Table 8.1 Semiconductor properties

Material	E_g (eV)	Direct/ Indirect	n_i (cm^{-3})	ε_r	μ_n (cm^2 /(V s))	E_c (cm^2 /(V s))	v_{sat} (10^7 cm/s)	λ (W/cm − K)
Si	1.1	I	1.5×10^{10}	11.8	1,350	0.3	1.0	1.5
2H-Ga/n	3.39	D	1.9×10^{-10}	9.9	1,000^a 2,000**	3.3*	2.5	1.3
3C-GaN	3.27	D	8×10^{-9}	9.9	1,000	1*	2.5	1.3
3C-SiC	2.2	I	6.9	9.6	900	1.2*	2.0	4.5
4H-SiC	3.26	I	8.2×10^{-9}	10	720^a 650^c	2.0	2.0	4.5
6H-SiC	3.0	I	2.3×10^{-6}	9.7	370^a 50^c	2.4	2.0	4.5

Note: a – mobility along a-axis, c-mobility along c-axis, *Estimated value, **2DEG

A few years ago, discrete RF GaN power HEMTs on silicon substrates as well as high-voltage discrete power GaN Schottky diodes (600 V, 4–8 A) on sapphire substrates have been commercialized [2–4]. More recently, high-voltage (200 V, 12 A) GaN power transistors on silicon substrates are becoming commercially available [5,6].

In this paper, we will present the recent advances in high-voltage GaN power transistors. These transistors can function both as discrete power switches or integrated with other logic or analog device elements to form power ICs.

8.2 Device Structures and Design

Most of the high-voltage GaN power transistors reported are lateral in structure because the GaN/AlGaN epitaxial layers are grown on insulating or high-resistivity substrates. The schematic cross-sections of several high-voltage lateral GaN transistor structures are shown in Fig. 8.1a–d. Power GaN transistors are interesting in that they can either resemble that of lateral silicon power MOSFETs or AlGaAs/GaAs Schottky-gate HEMTs. Besides a MOS channel, lateral high-voltage GaN MOSFETs have a lightly doped drain region to support high-drain voltages and ion-implanted source/drain regions. By contrast, power AlGaN/GaN HEMTs have Schottky-gate-controlled heterojunction 2DEG channel, a heterojunction drift region and alloyed source/drain contacts. In either type of transistors, the surface electric field must be suppressed so as to maximum the blocking voltage. The MOS Channel-HEMT (MOSC-HEMT, previously called hybrid MOS-HEMT [7]) combines the best features of the MOSFET and HEMT.

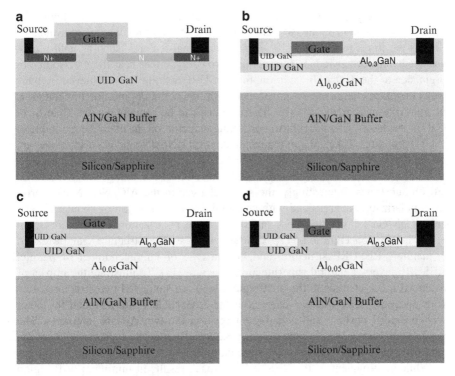

Fig. 8.1 Schematic cross-sections of high voltage lateral GaN: (**a**) MOSFET, (**b**) HEMT, (**c**) MOS-HEMT, and (**d**) MOS-channel HEMT (MOSC-HEMT)

8.3 Device Design

The REshaping of SUrface Field (RESURF) principle [8] is employed to maximize the breakdown voltage (BV) of lateral high-voltage power devices since it can suppress the surface electric field and force the avalanche process to be initiated in the bulk junction region. The key design parameter is the space charge per unit area (N_{RESURF})($= \int N_{epi} \, dx$) in the drift region. Adoption of an optimum N_{RESURF} leads to inherent lateral and vertical electric field shaping due to non-planar, two- or three-dimensional depletion action originating from interactions of lateral and vertical pn or MIS junctions [9–11].

To determine the optimum RESURF charge density (N_{RESURF}), we use Gauss' Law to yield

$$\int \xi \times dA \approx \xi \times A = - \int \rho dV/\varepsilon_s = -Q/\varepsilon_s = -qN/\varepsilon_s,$$

where ξ is the electric field in the depleted drift region, A the area, Q the total amount of space charges enclosed, and ε_s the semiconductor permittivity. Since

$N_{RESURF} = Q_{RESURF}/qA \approx \xi\varepsilon_s/q$ and $\xi \approx 1.5 \times 10^5 V/cm$ in silicon, the optimum N_{RESURF} is about $10^{12}/cm^2$, but, for GaN and SiC, where the avalanche field is ten times higher, it is $10^{13}/cm^2$. Additional features, such as field plates, are needed to further reduce surface field crowding, either near the gate/source or drain side of the device. In addition, the RESURF action can be achieved in many ways (a) space charge can be introduced with in situ epi doping, ion implantation, or polarization charges, and (b) the substrates can be semiconductors (like Si or GaN), insulators (like sapphire), or semiconductor on insulator/semiconductor (SOI) structures (like silicon/oxide/silicon or GaN/AlN/silicon). Here, we are focusing on lateral GaN power devices on SOI substrates consisting of AlGaN/GaN- or GaN-active epitaxial layers/AlN, GaN- or AlGaN-insulating buffer layers on silicon substrates. Interestingly, the space charges in the AlGaN/GaN case arise from polarization charges and with an aluminum mole fraction of about 22% and an AlGaN/GaN layer thickness of 20 nm; these polarization charges are about $10^{13}/cm^2$ in concentration [12]. Consequently, such a structure can be called a "natural" RESURF structure, and, in the case of infinite number of parallel RESURF layers, a natural superjunction structure [13]. To illustrate the impact of balanced space charges in the drift region, we have simulated two structures, one with uncompensated positive charges at the AlGaN/GaN interface and the other with balanced positive and negative charges on the two AlGaN interfaces. The equipotential line profiles, shown in Fig. 8.2, clearly indicate field crowding at the gate corner in the former but much more uniform electric field in the latter [14].

While employment of the RESURF approach results in minimum drift region length for a given BV, it also leads to a minimum specific on-resistance ($R_{on,sp}$). In particular, using an one-dimensional pn junction breakdown theory, one deduce a $R_{on,sp} \propto (BV)^{2.5}$ relationship for unipolar rectifiers or transistors and we almost always find this relationship to be applicable in vertical discrete devices. By contrast, lateral RESURF-type unipolar devices have a $R_{on,sp} \propto (BV)^n$, where n is between 1.1 and 1.5, significantly smaller than that of vertical devices. In addition,

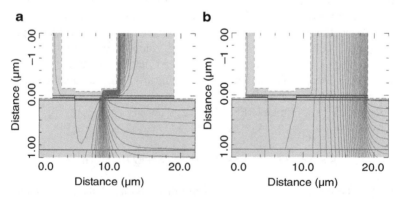

Fig. 8.2 Simulated equivalent potential line distribution for 1-zone RESURF MOSC-HEMT: (**a**) without GaN cap at V_{DS} of 40 V and (**b**) with GaN cap at V_{DS} of 646 V

for power FET structures, we also need to consider the on-resistance component (R_{ch}) contributed from the channel region and this resistance is directly related to the field-effect carrier mobility achievable from a particular FET process. The relative on-resistance contributions from the channel and drift regions are strongly dependent on the blocking or BV desired. In 600-V lateral silicon RESURF devices, the drift region resistance dominates. However, in lateral GaN RESURF devices of the same blocking voltage, the on-resistances from the MOS channel, relative to the resistance of the drift region, is becoming increasingly dependent on the channel length and field-effect mobility. Consequently, the choice of channel region design and technology in GaN power FETs plays a major role in ultimate device performance and proper channel scaling is necessary and critical with blocking voltages below 600 V. To illustrate these considerations, the specific on-resistance versus blocking voltage for GaN MOSFETs, using the set of device and material parameters assumed in Table 8.2, is shown in Fig. 8.3 [15]. Also, included for comparison is the one-dimensional bulk vertical GaN device performance projection.

For power MOSC-HEMTs, the channel resistance has become so dominant that scaling it to below submicron dimensions is necessary. We have performed a systematic channel scaling study and found that scaling the channel length to below 1 μm is needed to achieve specific on-resistance below 10 mΩ − cm² [16], as shown in Fig. 8.4. It can also be noted that the $R_{on,sp}$ of the conventional HEMTs is less sensitive to the channel length because of a higher channel mobility, but it is more difficult to achieve a sufficiently large positive threshold voltage for enhancement mode operation.

Table 8.2 Device parameters for lateral GaN MOSFETs in calculating the breakdown voltage/on-resistance tradeoff

$L_G(\mu m)$	$N_{s,CH}(cm^{-2})$	$\mu_{CH}(cm^2/(V\,s))$	$E_{RES}(V/\mu m)$	$\Phi_{RES}(cm^{-2})$	$\mu_{RES}(cm^2/(V\,s))$
3	1×10^{13}	540	100	1×10^{13}	900

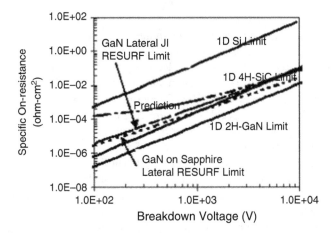

Fig. 8.3 Projections of the performance of GaN devices

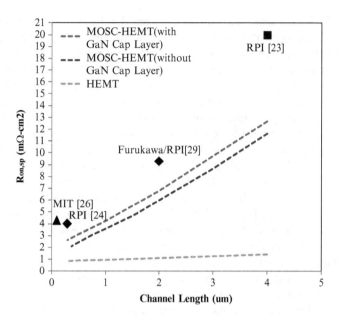

Fig. 8.4 $R_{on,sp}$ versus MOS channel length from experiments and simulations for HEMT and MOSC-HEMT without or with GaN cap layer

8.4 Experimental Results

The first high-voltage GaN power transistor is a MOS-HEMT on sapphire, with a $R_{on,sp}$ of $1.7\,m\Omega - cm^2$ and a BV of $1.3\,kV$ [17]. Subsequently, there are many attempts to refine the device structures and process to improve its characteristics. The main structural variants include insertion of a better dielectric between the Schottky metal and AlGaN layer, varying Al mole fraction in the AlGaN layer, and better surface passivation layer. The device parameters optimized are: BV, off-state leakage current, gate leakage, and frequency-dependent current dispersion. However, most of the off-state leakage current reported are often excessive (usually, the BV was measured at 1 mA/mm) [18]. Recently, the situation has been improved, and, in one case, accomplished with an in situ grown surface SiN passivation layer [19]. The highest BV for a GaN HEMT reported is $10.4\,kV$, with a specific on-resistance of $186\,m\Omega - cm^2$ [20]. It is interesting to note that all the high-voltage commercial GaN power transistors so far base on the basic HEMT device structure approach, though mostly on silicon substrates [5, 6].

While most of the GaN transistors demonstrated utilized a heterojunction channel and Schottky (sometimes in conjunction with a dielectric) gate (see Table 8.3), we have focused on developing an enhancement mode MOS channel technology. Previously, we have successfully optimized the GaN/SiO_2 interface using a PECVD oxide followed with a high-temperature $(1,000°C)$ nitrogen annealing. We first characterize the MOS capacitors and found the extracted interface state densities to

Table 8.3 Summary of GaN high-voltage transistors

Device	BV (V)	Ron, sp (m$\Omega \times$ cm^2)/Imax(A)	Other features	Research group/company	References
HEMT	1,300	1.7	SiO$_2$/Si$_3$N$_4$ as gate insulator L_{GD} = 20 µm	UCSB	N.Q. Zhang et al., IEDM (2001)
HEMT	8,300	186	L_{GD} = 125 µm, field plate	Panasonic	Y. Uemoto et al., IEDM (2007)
HEMT (GIT)	800	2.6	L_{GD} = 7.5, V_{th} = 1 V, p-AlGaN/AlGaN under gate	Panasonic	Y. Uemoto et al., TED (2007)
HEMT	1,800	7, 120 A	L_{GD} = 15 µm, Si substrate	Furukawa	N. Ikeda et al., ISPSD (2008)
HEMT	1,580	4	L_{GD} = 16 µm	Cornell	J. Shi et al., PSS(C) (2008)
HEMT	100	300 mA/mm	V_{th} = 0.75 V, Fluoride plasma treatment	HKUST	D. Song et al., EDL (2007)
HEMT	1,500	5.3	L_{GD} = 20 µm, Si substrate transfer	MIT	B. Lu et al., EDL (2010)
HEMT	2,200	18	L_{GD} = 20 µm, Si substrate removal	IEMC	P. Srivastava et al., EDL (2011)
HEMT	1,100	6.1	L_{GD} = 22 µm	U. of Sheffield/Powdec	A. Nakajima et al., EDL (2011)

(continued)

Table 8.3 continued

Device	BV (V)	Ron, sp (mΩ × cm²)/Imax(A)	Other features	Research group/company	References
HEMT	10,400	186	$L_{GD} = 125\,\mu m$, thick poly-AlN passivation	Panasonic	M. Yanagihara et al., PSS(A) (2009)
MOSFET	1,570	2.5 A	$V_{th} = 3\,V$, implanted RESURF region	RPI	W. Huang et al., ISPSD (2008)
MOSFET	770 forward	80 mA/mm	$V_{th} = 5.1\,V$	Rohm	H. Otake et al., JJAP (2007)
MOSFET	1,050 reverse		Monolithic Schottky pair, reverse blocking, $L_{GD} = 18\,\mu m$	RPI	W. Huang et al., ISPSD (2007)
MOSFET	2,500		$L_{GD} = 30\,\mu m$, field plates, lower temperature anneal	RPI	W. Huang et al., ISPSD (2008)
		30	$L_{GD} = 20\,um$, $V_{th} = 2\,V$, higher temperature anneal		
MOSFET	730	34	$L_{GD} = 16\,\mu m$	RPI	W. Huang et al., ISPSD (2009)
MOS-HEMT	1,000	5	$L_{GD} = 15\,\mu m$, $V_{th} = 0.5\,V$, Si substrate	NEC	K. Ota et al., IEDM (2009)

Device				Institution	Reference
MOS-HEMT	320	800 mA/mm	L_{GD} = 5 μm, V_{th} = 3 V. L_{ch} = 0.8 μm, GaN/AlN/GaN GaN cap	Fujits	M. Kanamurua et al., EDL (2010)
MOS-HEMT	1,200	9	L_{GD} = 5 μm, V_{th} = 0.64 V, plasma treated gate	HRL	R. Chu et al., EDL (2011)
MOS-HEMT	640	3.8	L_{GD} = 12 μm, V_{th} = 1 V	Toshiba	M. Kuraguchi et al., PSS(c) 2007
MOS-HEMT		20	L_{GD} = 20 μm, V_{th} = 2.5 V	RPI	W. Huang et al., ISPSD (2008)
MOS-HEMT	1,300		L_{GD} = 24 μm, field plate	RPI	K. Tang et al=., ISPSD (2009)
MOS-HEMT	650	9.3, 100A	L_{GD} = 15 μm, V_{th} = 2.7 V, Si substrate	Furukawa/RPI	Kambayashi et. Al. and T.P Chow, SSE (2010)
MOS-HEMT	400	200 mA/mm	L_{GD} = 6.5 μm, V_{th} = 5.2 V, Si substrate	Sharp	T. Oka et al., EDL (2008)
MOS-HEMT	643	4.3	L_{GD} = 18 μm, V_{th} = 2.9 V, L_{ch} = 0.1 μm, Si substrate	MIT	B. Lu et al., EDL (2010)

Fig. 8.5 Output $I-V$
characteristics of GaN
MOSC-HEMT with 3-μm
channel length and 20-μm
RESURF length

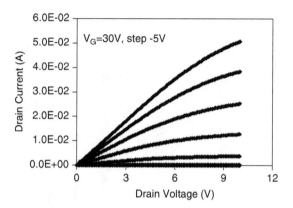

be very low and approaching that in Si MOS [21]. Subsequently, we have, using this GaN MOS process, experimentally demonstrated a record high field-effect inversion electron mobility of $167\,\mathrm{cm}^2/(\mathrm{V\,s})$ in GaN MOSFETs with ion-implanted source/drain regions. In addition, since then, we have demonstrated several high-voltage enhancement-mode GaN lateral power MOS-gated FET structures (MOS-FETs and MOSC-HEMTs) on sapphire substrates [7, 15, 22–25]. These prototype transistors have specific on-resistances up to more than 50× better than the silicon counterparts at the same blocking voltage range (600–2.5 kV) and off-state leakage current density more than 100× less than that in conventional HEMTs. The main advantages of the MOSC-HEMT is the low off-state leakage and low gate leakage current, often several orders of magnitude less than that of conventional HEMTs. On the other hand, the poorer channel mobility in the MOS channel requires the downscaling of the channel length, down to submicron dimensions, to minimize the dominance of the channel on-resistance [24, 26].

The output $I-V$ characteristics of a 3-μm GaN MOSC-HEMT on sapphire substrate with specific on-resistance of $20\,\mathrm{m\Omega}$-cm^2 are shown in Fig. 8.5 [23]. We have recently demonstrated a 0.3-μm channel GaN MOSC-HEMT on silicon substrate with specific on-resistance of $4\,\mathrm{m\Omega}$-cm^2 and a BV over 350 V [24]. Hence, the on-state performance of the HEMT and MOSC-HEMT is now quite similar. In addition, we have found that under avalanche conditions, the MOSC-HEMT is survivable [24] when the drain current is limited but we have not found a report that a HEMT or MOS-HEMT has survived such an avalanche test.

8.5 Technical Challenges and Reliability

The key technical challenges for GaN power FET development and commercialization include current collapse, avalanche capabilities, device ruggedness, threshold voltage control, and long-term reliability.

Drain current collapse under ac frequencies is a well-known phenomenon and there are probably many causes. Minimization of surface electric field with field plates and reduced polarization charges are among the many ways to control it (see, for example, [24, 27]). We would like to point out that our GaN MOS-gated FETs do not exhibit, while MOSC-HEMTs do exhibit, the current collapse behavior commonly seen in AlGaN/GaN HEMTs at room temperature. The temperature dependence of the current collapse behavior is also quite different, indicating the complexities of this degradation phenomenon [28]. Many ways have been suggested to control or increase the threshold voltage control in either MOS-HEMT or MOSC-HEMT (see, for example, [27]) but more efforts are needed to make it as controllable as that in state-of-art Si nanometer MOSFETs.

8.6 Summary

We have reviewed the recent advances of high-voltage lateral GaN power FETs. Its successful commercialization will impact many consumer and industrial power electronic applications, resulting in significant energy efficiency and conservation efforts.

Acknowledgement This work was supported by SRC (Task 1961.001) and the NSF ERC on Smart Lighting (Award #EEC-0812056).

References

1. T.P. Chow, R. Tyagi, IEEE Trans. Electron Dev. **41**, 1481 (1994)
2. http://www.irf.com/product-info/ganpowir/
3. http://www.nitronex.com/pdfs/AN-010%20LDMOS.pdf
4. http://www.veloxsemi.com/pdfs/Advan_of_GaN_Schottky.PDF
5. http://www.irf.com/pressroom/articles/560PEE0811.pdf
6. http://epc-co.com/epc/documents/product-training/Characterization_guide.pdf
7. W. Huang, Z. Li, T.P. Chow, Y. Niiyama, T. Nomura, S. Yoshida, Proc. Int. Symp. Power Semicond. Dev. and ICs, pp. 295–298, (2008)
8. J.A. Appels, H.M.J. Vaes, IEDM Tech. Dig., pp. 238–241, (1979)
9. H. Vaes, J. Appels, IEDM Tech. Dig. 87–90, (1980)
10. E.J. Wildi, P.V. Gray, T.P. Chow, H.R. Chang, M. Cornell, IEDM Tech. Dig. 268–271, (1982)
11. S. Merchant, E. Arnold, H. Baumgart, S. Mukherjee, H. Pein, R. Pinker, Proc. Int. Symp. Power Semicond. Dev. ICs. 31–35, (1991)
12. O. Ambacher, J. Smart, J.R. Shealy, N.G. Weimann, K. Chu, M. Murphy, W.J. Schaff, L.F. Eastman, R. Dimitrov, L. Wittmer, M. Stutzmann, W. Rieger, J. Hilsenbeck, J. Appl. Phys., **85**(6), 3222 (1999)
13. H. Ishida, D. Shibata, M. Yanagihara, Y. Uemoto, H. Matsuo, T. Ueda, T. Tanaka, D. Ueda, Electron Dev. Lett. **29**, 1087 (2008)
14. Z. Li, T.P. Chow, Int. Conf. on Nitride Semiconductors, Tampa, FL, Oct., 2010

15. W. Huang, T. Khan, T.P. Chow, Proc. Int. Symp. Power Semicond. Dev. and ICs, 796–798, (2006)
16. Z. Li, T.P. Chow, Solid-State Electronics, **56**(1), 111 (2011)
17. N.-Q. Zhang, B. Moran, S.P. DenBaars, U.K. Mishra, X.W. Wang, T.P. Ma, Tech. Dig., 589–592 (2001)
18. See, for example, S. Yagi, M. Shimizu, M. Inada, Solid-State Electronics, **50**(6), 1057 (2006)
19. J. Derluyn, M. Van Hove, D. Visalli, A. Lorenz, D. Marcon, P. Srivastava, K. Geens, B. Sijmus, J. Viaene, X. Kang, J. Das, F. Medjdoub, K. Cheng, S. Degroote, M. Leys, G. Borghs, M. Germain, IEDM Tech. Dig., 157–160 (2009)
20. M. Yanigihara, Y. Uemoto, T. Ueda, T. Tanaka, D. Ueda, Phys. Status Solidi A, **206**(6), 1221 (2009)
21. W. Huang, T. Khan, T.P. Chow, J. Electronic Mater. **35**, 726–732 (2006)
22. W. Huang, T. Khan, T.P. Chow, Electron Dev. Lett. **27**, 796–798 (2006)
23. W. Huang, T.P. Chow, Y. Niiyama, T. Nomura, S. Yoshida, Proc. Int. Symp. Power Semicond. Dev. and ICs, 291–294, (2008)
24. Z. Li, T.P. Chow, to be published
25. K. Tang, Z. Li, T.P. Chow, Y. Niiyama, T. Nomura, S. Yoshida, Proc. Int. Symp. Power Semicond. Dev. and ICs, 279–282, (2009)
26. B. Lu, T. Palacios, Electron Dev. Lett. **31**(9), 990 (2010)
27. R. Chu, A. Corrion, M. Chen, R. Li, D. Wong, D. Zehner, B. Hughes, K. Boutros, Electron Dev. Lett. in press (2011)
28. Z. Li, T. Marron, H. Naik, T.P. Chow, Proc. Int. Symp. Power Semicond. Dev. and ICs, 225–228 (2010)
29. H. Kambayashi, Y. Satoh, S. Ootomo, T. Kokawa, T. Nomura, S. Kato, T. P. Chow, Solid-State Electronics **54**, 660 (2010)

Chapter 9
Radiation Effects in GaN

Alexander Y. Polyakov

Abstract A review of electron, proton, and neutron damage in GaN and AlGaN is presented. A comparison of theoretical and experimental threshold displacement energies is given, along with a summary of energy levels introduced by different forms of radiation, carrier removal rates and role of existing defects.

9.1 Introduction

In recent years, group III-nitrides have become third in importance after Si and GaAs among semiconductor materials systems and are rapidly moving to overtaking GaAs. GaN-based light emitting diodes (LEDs) and laser diodes (LDs) dominate in the field of semiconductor light sources for UV/visible range. High electron mobility AlGaN/GaN transistors HEMTs will be the backbone of high-power/high-frequency/high-temperature electronics in the near future. GaN-based power devices are seriously considered as competitors for SiC devices. UV GaN photodetectors find wider and wider applications. Many of these devices will make their way into satellite and military systems where issues of radiation tolerance are critical. Hence, the practical need of radiation effects studies for GaN devices. In addition, as for other semiconductor materials, irradiation provides a versatile instrument to controllably change the spectrum of defects in GaN and thus to understand the nature of grown-in and radiation defects. There is also, again as for other semiconductors, considerable interest to using radiation to modify characteristics of GaN and related devices. Finally, it appears that GaN shows good promise as a material for radiation detectors for specialized applications. Naturally,

A.Y. Polyakov (✉)
Institute of Rare Metals, B. Tolmachevsky, 5, Moscow 119017, Russia
e-mail: aypolyakov@gmail.com

S. Pearton (ed.), *GaN and ZnO-based Materials and Devices*,
Springer Series in Materials Science 156, DOI 10.1007/978-3-642-23521-4_9,
© Springer-Verlag Berlin Heidelberg 2012

understanding of the nature of processes occurring under and after irradiation is very important for proper design of such detectors.

All these considerations make radiation effects studies in GaN and related materials and devices of much practical and scientific importance. What follows is a survey showing the current state of affairs in this field. To summarize it briefly, some preliminary results for chosen types of materials, chosen kinds of devices, and chosen radiation sources have been obtained. Among the studied materials absolutely dominate measurements on undoped n-GaN, for devices the main bulk of research was performed for GaN LEDs and AlGaN/GaN HEMTs, the most popular radiation sources were protons and electrons with some work also reported for γ-irradiation and neutron irradiation. In most cases the results are somewhat fragmentary, the choice of materials, devices, radiation sources was to a large extent dictated by their availability rather than actual needs. Once the practical applications of GaN in special systems will really be considered we will undoubtedly see a strong revival of interest to radiation effects studies of GaN, this time in order to assess the reliability of devices in question and to produce the data necessary for intelligent modeling, design, and optimization. However, even the published preliminary results unambiguously indicate a much higher radiation hardness of GaN compared to Si and GaAs and even these initial efforts have uncovered the essential properties of radiation defects in GaN so that these preliminary results will serve as a very useful foundation for future work.

9.2   Fundamental Studies of Radiation Defects in GaN and Related Materials

9.2.1   Threshold Displacement Energy: Theory and Experiment

Perhaps the first experimental study of displacement threshold energy for GaN was published in [1] where the authors monitored the changes in luminescence spectra of GaN LEDs as a function of electron irradiation with energies 300–1,400 keV. They observed that generation of the band attributed to the formation of Ga vacancy started from electron energy of 440 keV which corresponds to the Ga displacement energy of 19 ± 2 eV. Comparison of defect production efficiency for 2 MeV protons and 2.5 MeV electrons showed that the protons were 1,000 times more effective, whereas calculations of the number of displaced atoms based on the measured threshold energy predicted the ratio of 250. The discrepancy was attributed to the difference in defects self-annealing rates depending on irradiating particles type and energy. It was noted that the measured displacement threshold in GaN was much higher than in Si and GaAs, similar to SiC, but lower than in diamond which reflects the difference in the bond strength and can serve as an indication of relative radiation hardness of respective materials. The authors of [1] failed to find a clearcut threshold energy for the nitrogen displacement.

However, measurements of the changes in electrical properties of undoped n-GaN films irradiated with electrons with energies (700–1,000 keV) persuaded Look et al. [2] that nitrogen vacancies were introduced with the rate close to 1 cm^{-1} at such electron energies. Molecular dynamics calculations of displacement effects in GaN performed in [3] might explain why certain discrepancies could exist when comparing effectiveness of radiation defects introduction by various particles based on singular values of threshold defect formation. It was found that (a) there exists a wide distribution of threshold energies for both Ga and N sublattices and (b) that effects of recombination induced by self-annealing caused by athermal local energy transfer are very important. The minimal energies of defects formation were found to be 18 ± 1 eV for Ga and 22 ± 1 for nitrogen, but the average displacement energy was much higher, 45 ± 1 eV (Ga) and 109 ± 2 (N). This modeling predicts about five times higher radiation tolerance of GaN compared to GaAs.

The actual experimental measurement of defects accumulation in GaN produced by various high-energy ion species at different temperatures were performed by means of Rutherford backscattering and transmission electron microscopy in [4]. The authors observed that radiation defects in GaN were mobile even at low temperatures, that the doses of ions producing amorphization were more than an order of magnitude higher than for GaAs. For light ions trapping of primary radiation defects by the implanted species was observed (this was particularly the case for C ions). For hydrogen irradiation it was found that, for doses exceeding $\sim10^{16}$ cm^{-2}, formation of hydrogen bubbles became an issue. For heavy ions (starting with Si) the main factor in defect production was the ion mass and energy and hence the number of primary displacements initiated by such ions. Recombination of radiation defects during irradiation was very prominent even at low temperature and the rate of recombination increased with irradiation temperature. At room temperature the doses of irradiation producing amorphization increased from some 10^{14} cm^{-2} for heavy ions to $\sim10^{16}$ cm^{-2} for light ions, but increasing the implantation temperature increased the dose of amorphization by several orders of magnitude. For heavily amorphized material the formation of nitrogen bubbles was observed. These experimental observations are in good agreement with the results of theoretical modeling performed in [3] which shows that the main features of radiation defects formation process in GaN are reasonably well understood.

9.2.2 Radiation Defects in GaN: Defects Levels, Effects on Charge Carriers Concentration, Mobility, Lifetime of Charge Carriers, Thermal Stability of Defects

9.2.2.1 Levels of Radiation Defects in GaN

The most basic primary defects produced in GaN by any type of irradiation are Frenkel pairs in the Ga and N sublattices. Theoretical estimates that have become by now reasonably reliable predict that nitrogen vacancies in GaN are electronic resonances with levels in conduction band. By capturing electrons they are turned into effective-mass-like (EM) shallow donors [5] and measurements of electrical properties of undoped n-AlGaN films as a function of Al composition suggest that such EM V_N donors could have the ionization energy of 40–60 meV in GaN and that the Al concentration at which respective resonance levels emerge in the forbidden gap is close to 20% [6]. For nitrogen interstitials N_i theoretical calculations predict the existence of a deep acceptor near 1 eV from the conduction band edge [7, 8]. Ga vacancies V_{Ga} in the triply charged state are supposed to produce an acceptor state near $E_v + 1$ eV, whilst Ga interstitials form negative-U type donors whose $+/0$ transition level is close to the conduction band edge and $3 + /2+$ charge transition level is located near $E_v + 2.6$ eV (E_c–0.8 eV) [7, 8]. Strangely enough, most of these states seem to be indeed observed in radiation effects experiments performed for GaN.

One of the first systematic studies of radiation defects in n-GaN was the pioneering work of Look et al. [2] already mentioned earlier. The authors showed that electron irradiation with energies $0.7 - 1$ MeV introduced new donors with ionization energy close to 0.06 eV with introduction rate of 1 cm^{-1}. The net electron concentration hardly changed at all while the mobility of electrons decreased with increasing dose. The analysis of the temperature dependence of mobility taking into account the existence of a heavily n-type interfacial layer near the substrate suggested that acceptor centers were introduced at a rate similar to the rate of the 0.06 eV donors. The observed effects were explained by the formation of Frenkel pairs in the nitrogen sublattice, with the nitrogen vacancies V_N responsible for the 0.06 eV donors and nitrogen interstitials N_i associated with the compensating acceptors. Deep levels transient spectroscopy (DLTS) measurements performed on 1 MeV electron-irradiated n-GaN indeed showed the presence of deep electron traps with the activation energy 0.9 eV that could be attributed to N_i acceptors [9]. DLTS spectra of n-GaN films irradiated with ^{60}Co γ-rays also showed the presence of electron traps G1 with activation energy 80 meV [10] close to the one observed for V_N donors in [2]. In undoped n-GaN samples irradiated with 10 MeV electrons, we observed deep electron traps with activation energy 1 eV that were shown to be acceptors and that we attributed to N_i acceptors [11]. At the same time, the authors of [12] observed that 2.5 MeV electron irradiation at 4.2 K produced in GaN a strong defect photoluminescence PL band centered near 0.95 eV, for which optically detected electron paramagnetic resonance (ODEPR) spectra could be obtained. The PL line in question was attributed to Ga vacancy V_{Ga} with a level near $E_v + 1$ eV and the ODEPR process was interpreted as interaction with two different Ga_i interstitial centers with levels close to $E_v + 2.6$ eV. The quenching of the ODEPR signal for annealing to room temperature was then attributed to moving of the Ga_i defect away from V_{Ga}.

If, however, only Frenkel pairs were produced by irradiation of GaN one would expect that, for n-GaN, the carrier removal rate at the initial stage of irradiation, when the density of radiation defects is lower than the concentration of dopant donors, will be equal to the production rate of V_{Ga} and N_i, whilst for high irradiation doses the Fermi level will be pinned by the shallower of the two EM-like native donors, V_N or V_{Ga}^{1+}. In p-type GaN the initial carrier removal rate should be close to that in n-type until the aggregate density of V_N and Ga_i donors exceeds the density of acceptor dopants after which the Fermi level should jump to the level of V_{Ga} near $E_v + 1\,eV$ and get pinned there. This generally is not the case because primary defects recombine, form complexes with each other, with dopants and with extended defects and because at high energy of irradiating particles the energy of the primary recoils becomes so high that they produce collision cascades and form heavily disordered regions (DRs) with a very high defect density in the core. DLTS measurements on n-GaN films irradiated with electrons, γ-rays, and protons with energies of some MeV bound to produce predominantly point radiation defects showed effective formation of electron traps with apparent activation energies 0.13 eV, 0.16 eV, and 0.18–0.2 eV [10, 13–18](ER1, ER2, ER3 according to the defect nomenclature introduced in [15]). These traps tend to produce a broad feature in DLTS due to the proximity of emission rates of components, but this broad peak can be deconvoluted into separate defects contribution due to the large difference in electron capture cross-sections [9, 10, 13–16]. It was shown in [13] that for the ER3 trap the apparent activation energy is the sum of the trap ionization energy which is 0.06 eV, i.e., close to that of V_N and the capture activation energy of about 0.14 eV. Based on this observation it was suggested that the ER3 traps are complexes of nitrogen vacancies with some other native defects, such as V_N–N_i or V_N–N_{Ga}–$Ga_N - N_i$ [9]. The shallower defects ER1 and ER2 have been observed in a variety of surface treatments bound to produce nitrogen vacancies and are also thought to be V_N-related [9]. It was assumed that the 0.13 and 0.2 eV electron traps were donors as well as the 0.06 eV V_N donors [9]. By implication, the carrier removal in irradiated n-GaN should then be regulated by the balance between relatively shallow V_N-related donors, deep N_i acceptors, deep Ga_i donors, and V_{Ga} acceptors. However, this nice picture is considerably spoiled by experimental measurements of the ER3 traps ionization energy dependence on applied electric field that seems to indicate that these traps are acceptors [14]. An alternative suggested identity of these traps offered in [14–16] associates them with $V_{Ga}N_i^{2-}$ complexes. In addition, it should be noted that DLTS spectra of n-GaN films irradiated with light particles (γ-rays, electrons with several MeV energy, protons with several MeV energy) are dominated by the ER1–ER3 relatively shallow electron traps. With rare exceptions, only for higher electron energies [11], heavier ions (He, N) [15, 19], neutron irradiation [20], or higher density of defects in proton implanted samples (as for 150 keV protons with the dose over $5 \times 10^{14}\,cm^{-2}$ [21]) are deeper traps commonly detected. Electron traps that are generally observed in these cases show the activation energies of 0.75–0.8 eV, 0.95–1.2 eV [11, 15, 19–21]. The 0.8 eV traps show measurable decrease of the apparent ionization energy with increasing electric field, whereas the 1 eV traps energy does not vary with electric field. The first type of behavior is expected for donors because of the Poole-Frenkel effect [22], whilst the second type of behavior is characteristic of acceptors that are neutral when they emit an electron [22]. The attribution based mostly on theoretical calculations tends to ascribe the first to Ga_i^{2+}

deep donors and the second to N_i^{2-} deep acceptors [9,11,15,20]. In addition to these deep centers we reported that implantation of n-GaN with 150 keV protons to doses higher than 5×10^{14} cm^{-2} created deep electron traps whose energies tractably increased from 0.2 eV at low dose to 0.25 eV, 0.32 eV, and 0.45 eV at higher dose suggesting that these centers could be larger complexes produced by addition of new radiation defects to the more simple radiation defects formed at low doses [21]. Measurements of deep traps spectra on neutron transmutation doped and annealed n-GaN reported by us [23] indicate that the 0.2 eV and the 0.45 eV traps could incorporate donor doping atoms.

The knowledge of hole traps behavior in irradiated n-GaN is less satisfactory compared to electron traps. Part of the problem is that DLTS spectroscopy studies on n-GaN have mostly been done on Schottky diodes where probing of the traps in the lower half of the bandgap can only be achieved by optical injection of holes (techniques such as deep levels optical spectroscopy (DLOS) [24] or optical deep levels transient spectroscopy (ODLTS) [25]). The results of these measurements critically depend on one's ability to provide complete recharging of deep traps within the space charge region. When the lifetime strongly decreases after irradiation, the ability to fully recharge hole traps becomes an issue and could lead to erroneous conclusions in some cases. A good example is the measurements of concentration for the 0.9–1 eV hole trap attributed to V_{Ga} acceptors. These traps produce a prominent hole traps peak in ODLTS with activation energy close to 0.9 eV (see e.g. [26]) as shown in Fig. 9.1. The 0.95 eV defect PL band in irradiated GaN can be reasonably assigned to recombination on these hole traps [12, 27–29]. Also, these traps are believed to be related to the famous yellow recombination band in GaN that is attributed to a donor–acceptor pairs (DAP) transition involving a shallow donor and the $E_v + 0.9$ eV hole trap (see e.g. [6, 30–33]). We have seen already that electron irradiation at 4.2 K strongly increases the intensity of the 0.95 eV PL band [12]. Detailed studies seem to indicate that the defects in question are stable up to annealing temperatures of 500°C, even though the ODEPR signal vanishes because of the increased spatial separation of V_{Ga} and Ga_i [12]. Thus,

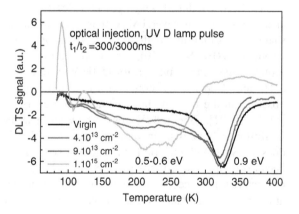

Fig. 9.1 ODLTS spectra measured for undoped n-GaN before and after irradiation with fast reactor neutrons; measurement conditions: reverse bias −1 V, optical injection by 5-s-long deuterium UV lamp pulse, time window 300/3,000 ms

Fig. 9.2 Changes in room temperature MCL spectra measured before and after irradiation of undoped n-GaN with fast neutrons

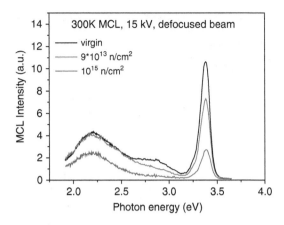

Fig. 9.3 The dependence of the 0.9 eV hole trap ODLTS peak amplitude on the driving current, i.e., intensity of the 365-nm-wavelength GaN LED used for excitation Fig. 9.3. $E_v + 0.9$ eV saturation with light

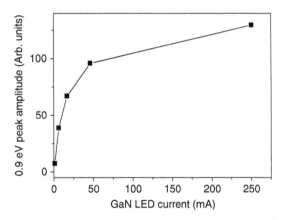

one would expect that irradiation will increase the magnitude of the ODLTS peak $E_v + 0.9$ eV and enhance the yellow band intensity. The former is seldom the case, however. Figure 9.1 shows the change in ODLTS spectra of undoped ELOG n-GaN sample irradiated with fast reactor neutrons (the behavior for 10 MeV electron irradiation is similar). It can be seen that the respective peak magnitude actually decreases after irradiation. At the same time, in our experiments with luminescence spectra measurements on neutron, electron, or proton-irradiated samples we see that at the initial stages the relative intensity of the V_{Ga}-related DAP yellow band increases (Fig. 9.2). A partial explanation of the observed controversy is provided by Figs. 9.3 and 9.4. Figure 9.3 demonstrates that the magnitude of the $E_v + 0.9$ eV peak needs a high light intensity to be saturated. In irradiated samples the actual injection level could considerably decrease compared to virgin samples because of reduced lifetime. Hence, one can move from the saturation regime in Fig. 9.3 to the low injection level regime and hence a lower ODLTS signal simply because of the lifetime decrease. Figure 9.4 illustrates how the apparent activation energy of the peak changes with changing the excitation intensity measured as the forward

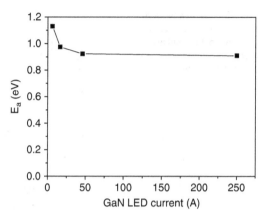

Fig. 9.4 Changes in apparent activation energy of the 0.9 eV hole traps with increasing driving current of the GaN LED

current through the 365 nm-wavelength GaN LED (peak power of 3 W). It can be seen that the energy measurably increases with decreasing the excitation level. Detailed understanding is lacking at the moment, but the most obvious way to explain the results would be to assume that the peak observed is a superposition of one or several positive and negative peaks due, respectively, to electron and hole traps. The effective activation energy deduced from the Arrhenius plot of $1/\tau T^2$ versus $1,000/T$ for the peak will depend on contributions from both types of traps (τ above is the characteristic emission time of the trap, T is the temperature position of the peak with the chosen value of spectrometer time window [25]). In Fig. 9.1 we show the progress of ODLTS spectra in undoped n-GaN with increasing fluence. It can be seen that after irradiation with high fluence in place of the 0.9 eV hole trap one observes the 0.9 eV electron trap due to Ga$_i$. This would suggest that filling of the latter with electrons is more effective than filling the former with holes. For lower fluences the contribution from these traps will interfere with measurements decreasing the 0.9 eV hole peak amplitude. Complete analysis has not been completed, but preliminary results do suggest that this effect could be partly responsible for the apparent behavior of the 0.9 eV hole trap. The concentration of the electron traps can be independently determined from usual DLTS and thus deconvolution of the ODLTS signal performed. For 10 MeV electron irradiation the electron trap interfering with ODLTS measurements of 0.9 eV hole traps seems to be the 1 eV N$_i$ acceptor rather than the 0.9 eV Ga$_i$ donor. These results indicate that, for 10 MeV electron irradiation, the introduction rate for the 0.9 eV hole traps is close to, but lower than the introduction rate of the 1 eV N$_i$ electron trap.

Moreover, there seem to exist several hole traps with a level near $E_v + 0.8$–0.9 eV, but with vastly different hole capture cross-section (see Fig. 9.5 comparing Arrhenius plots for major hole traps observed in two undoped n-GaN samples cut from the center and the periphery of the wafer). In addition, the figure shows the plot for the sample that was irradiated with neutrons and annealed to 1,000°C. One can note that even after annealing to 1,000°C a hole trap with the activation energy close to that of V$_{Ga}$, but a lower hole capture cross-section is present. It is tempting to

Fig. 9.5 Arrhenius dependences of the dominant hole traps in undoped n-GaN measured by ODLTS: measurements in the center of the wafer, at the periphery of the wafer, in the center of the wafer after irradiation with $1.5 10^{17}\,\mathrm{cm}^{-2}$ fast neutrons and annealing at $1,000^\circ C$

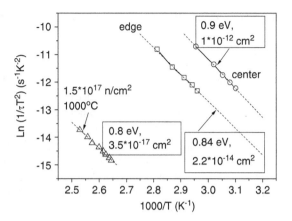

attribute the new trap to the same family of defects as V_{Ga}, which would mean that, even after very high temperature annealing such traps can survive. The difference between the hole traps in question could be, e.g., due to some of them being trapped at dislocations, but this needs further research. Photoluminescence studies on yellow band-related defects in GaN also suggest the presence of several types of such bands [32] which was attributed, e.g., to complexing of Ga vacancies with impurities, such as hydrogen. Those are most likely to be the members of the 0.9 eV hole traps family annealed at moderate temperatures.

Measurements of deep traps spectra for irradiated p-GaN are very scarce. Part of the problem is that standard DLTS measurements on Schottky diodes on p-GaN are very difficult because of the problems with high series resistance of the films and strong freeze-out of relatively deep Mg acceptors even at moderately low temperatures. To our knowledge the actual studies of radiation defects in p-GaN were carried out for 100 keV protons in [34] and for fast reactor neutrons in [35]. For proton irradiation conductivity versus temperature, low-frequency capacitance–voltage $C-V$ measurements, admittance spectra measurements, photoinduced transient current PICTS [36] and current deep traps transient CDLTS [37] spectroscopy indicated that deep electron traps near $E_c - 0.5 - 0.6\,\mathrm{eV}$ and deep hole traps with levels near $E_v + 0.3\,\mathrm{eV}$ and $E_v + 0.85\,\mathrm{eV}$ were introduced. The electron traps at E_c–0.5–0.6 eV are believed to be complexes of Mg acceptors with native defects and were shown to also give rise to intense blue defect luminescence band centered at 2.9 eV [38, 39]. The $E_v + 0.85\,\mathrm{eV}$ traps are probably the same as the V_{Ga}-related hole traps in n-type films. Their in-grown concentration in p-type GaN should be very low because of the high formation energy [7]. However, irradiation being a very nonequilibrium process, it can introduce such defects and as a result produce the yellow luminescence band in heavily irradiated p-GaN films, while yellow luminescence is manifestly absent in virgin p-GaN [34].

Neutron irradiation was performed for two types of p-GaN films, one grown by hydride vapor phase epitaxy (HVPE) and the other by MBE [35]. HVPE films differed from MBE films by a very slight temperature dependence of conductivity, much lower mobility, and the presence of additional acceptors with activation energy of 0.12 eV, more shallow than the ordinary Mg acceptors with activation energy of 0.15 eV. These two types of samples also differed by the luminescence spectra and by photosensitivity. In HVPE samples the luminescence was dominated by blue luminescence band due to transitions involving the E_c–0.5–0.6 eV Mg-related centers discussed above. In MBE films the concentration of these centers was much lower. Accordingly, the photosensitivity of HVPE samples was higher due to trapping of electrons on the E_c–0.5–0.6 eV level. Irradiation of both types of samples compensated acceptors and slightly but measurably increased the activation energy of major acceptors from 0.15 to 0.18 eV. It stronger decreased the concentration of shallow 0.12 eV acceptors compared to the 0.15 eV acceptors in the HVPE samples. In luminescence spectra of the MBE samples it increased the contribution of the blue luminescence band. Both types of samples remained p-type up to the dose of 1.7×10^{17} cm^{-2} and in both we observed the emergence of the yellow luminescence band due to the formation of the $E_v + (0.8$–$0.9)$ eV V_{Ga} related acceptors after high doses. At the fluence of 10^{18} cm^{-2} both samples converted to high-resistivity n-type with the Fermi level pinned near E_c–$(0.9$–$1)$ eV. Clearly, save for the E_c–$(0.5$–$0.6)$ eV Mg-related centers, the radiation defects introduced in p-type GaN are the same as in n-type, but the carrier removal rate was about 20 times higher than in n-GaN [35].

Obviously, much has still to be done to uncover the origin of the main radiation-induced centers in GaN. Table 9.1 summarizes the published data on properties of radiation defects introduced by different types of irradiation.

9.2.2.2 Carrier Removal and Deep Traps Introduction Rates; Disordered Regions

Some of the most basic questions of radiation physics are what determines the changes of electrical properties of irradiated material and where the Fermi level is stabilized after high doses of radiation. In Fig. 9.6 we present the carrier concentration changes for n-GaN samples irradiated with 10 MeV electrons. Figure 9.7 compares these changes to the defect introduction rates measured for the major deep traps detected by DLTS, the 0.18 eV electron traps and the 1 eV electron traps. The carrier removal rate estimated from Fig. 9.6 is 0.4 cm^{-1}, the 0.18 eV traps and the 1 eV traps introduction rates are, respectively, 0.2 cm^{-1} and 0.8 cm^{-1}. As discussed above, the contribution from the other acceptors, the V_{Ga}-related $E_v + 0.9$ eV hole traps in ODLTS spectra of irradiated samples, is underestimated by the interference of the 1 eV electron traps due to N_i acceptors. Deconvolution of the 0.9 eV hole trap ODLTS feature gives the upper limit of the V_{Ga} introduction rate as about 0.4 cm^{-1}. Thus, if we assume that the initial carrier removal rate deduced from Fig. 9.6, 0.4 cm^{-1}, comes from the difference in introduction rates of all these acceptors and the introduction rate of the 0.06 eV V_N donors the latter should be close to 1.4 cm^{-1}, i.e., $\sim$7 times that of the 0.18 eV traps. A similar relation between the concentrations of these traps was determined from electron concentration and mobility fitting and

Table 9.1 Defects introduced by various types of irradiation in GaN

Material type, donor or acceptor concentration (cm^{-3})	Irradiation type	Carrier removal rate (cm^{-1})	Defect levels (activation energy, eV from which band edge (conduction C or valence V), defect type (donor D, acceptor A)	Defect production rate (cm^{-1})	Proposed identity	References
N-GaN, 10^{17}	^{60}Co γ-rays, 21 MRad	–	0,09, C, ?	9.1	G1=ER1	[10]
			0.11, C, ?	7.3	G2=ER2	
			0.14, C, ?	4.8	G3=ER3	
N-GaN, 10^{17}	0.7–1 MeV electrons	–	0.06, C, D	1	V_N	[2]
			?, ?, A	1	N_i	
N-GaN,10^{17}	1 MeV electrons	–	0.13, C, D,	~0.1	ED1=ER2	[9, 40]
			0.2, C, D	~0.1	ED2=ER3	
			0.9,C,?	~0.1	AD2=Ni	
N-GaN, 10^{17}	1 MeV electrons	–	0.18, C, D	0.2	V_N–N_i	[9, 40]
			0.06	1	V_N	
N-GaN, 10^{16}	10 MeV electrons	0.4	0.18, C, D	0.2	ED2=ER3	[11]
			1, C, A	0.8	N_i	
N-GaN, 10^{16}	0.2–2.5 MeV	–	0.13, C, ?	–	ER1	[15]
			0.16, C, ?		ER2	
			0.2, C, A		ER3= $V_{Ga}N_i^{2-}$	
N-GaN,10^{16}	Protons, 2 MeV	260	0.13, C, ?	30	ER1	[15]
			0.16, C, ?	400	ER2	
			0.2, C, A	600	ER2	
N-GaN,10^{16}	Protons, 150 keV	100	0.2, C, ?		=ER3	[21]
			0.25, C, ?		Complex of ER3	
			0.32, C, ?		Complex of ER3	
			0.45, C, ?		Complex of ER3	
			0.6, C, A			
			0.8, C, D		Ga$_i$	

(Continued)

Table 9.1 (Continued)

Material type, donor or acceptor concentration (cm^{-3})	Irradiation type	Carrier removal rate (cm^{-1})	Defect levels (activation energy, eV from which band edge (conduction C or valence V), defect type (donor D, acceptor A)	Defect production rate (cm^{-1})	Proposed identity	References
N-GaN, 2×10^{18}	Protons, 150 keV	10,000				[47]
N-GaN, 4×10^{17}	Protons, 150 keV	3,500				[47]
N-GaN, 10^{16}	5.4 MeV He	–	0.2, C, A	3,300		[15]
			0.78, C, D	1,500	ER3	
			0.95, C, A	3,300	N_i	
N-GaN, 10^{16}	N ions, 300 keV	–	0.7, C, ?			[19]
N-GaN, 10^{16},	Fast reactor neutrons	5	0.2, C, D		ER3	[20]
			0.8, C, D		Ga_i	
N-GaN, 1e17	2 MeV electrons, 4.2 K		0.9, V, A		V_{Ga}	[12]
p-GaN, 2×10^{17} (MBE)	Fast reactor neutrons	100	0.8, C, D		Ga_i	[35]
			0.5, C, ?		Mg-related	
			0.85, V, A		V_{Ga}	
			0.9, C, ?		Between Ga_i and N_i	
p-GaN, 1e18 (HVPE)	Fast reactor neutrons	100	0.5, C, ?		Mg-related	[35]
			0.85, V, A		V_{Ga}	
			0.9, C, ?		Between Ga_i and N_i	
p-GaN, 1e18	Protons, 100 keV	10,000	0.5, C, ?		Mg-related	[34]
			0.3, V, A			
			0.85, V, A		V_{Ga}	

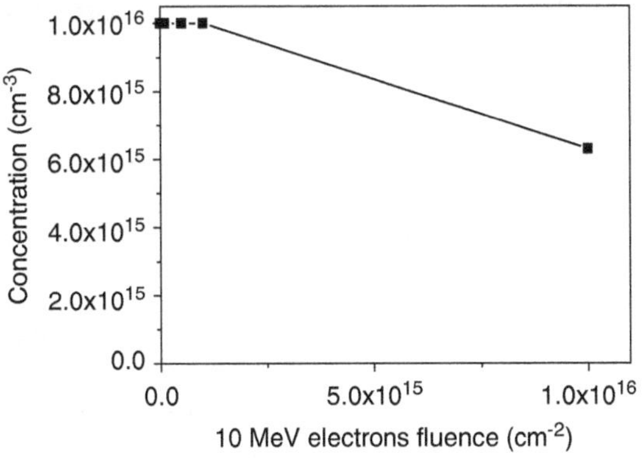

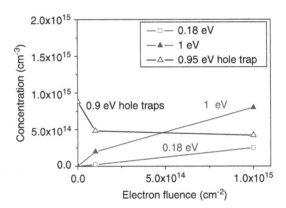

Fig. 9.6 Electron concentration changes induced in undoped n-GaN films by room temperature irradiation with 10 MeV electrons

Fig. 9.7 Changes in concentration of major electron and hole traps induced in undoped n-GaN by room temperature 10 MeV electron irradiation

DLTS measurements for n-GaN irradiated with 1 MeV electrons in [40] (1 cm^{-1} for the V_N centers versus 0.2 cm^{-1} for the 0.18 eV centers). Hence, the data on electron removal by relatively low-energy electrons can be consistently explained by the introduction rates of well-documented radiation point defects.

This apparently is not the case for fast neutron-irradiated GaN. Figure 9.8 shows the evolution of DLTS spectra of undoped n-GaN samples with donor concentration of 1.6×10^{16} cm^{-3}. The main deep traps that can be associated with such irradiation are the 0.18 eV ER3 electron traps and the 0.8 eV Ga$_i$ electron traps. However, as clearly seen from Fig. 9.9, the introduction rate of the shallower ER3 traps is very low, much lower than for electron irradiation, the introduction rate of the 0.8 eV traps is below 1 cm^{-1} and is much lower than the electron removal rate of 5 cm^{-1}. Besides, since these traps are deep donors they cannot contribute to carriers' removal. The studies of hole traps are summarized by Fig. 9.10. It shows ODLTS spectra of n-GaN samples before and after irradiation with fast reactor neutrons. The 0.9 eV hole traps signal can be clearly seen for the virgin sample. It decreased slightly after irradiation, most likely again due to the contribution from the 0.8 eV electron traps peak (the apparent changes of the 0.9 eV hole traps concentration with

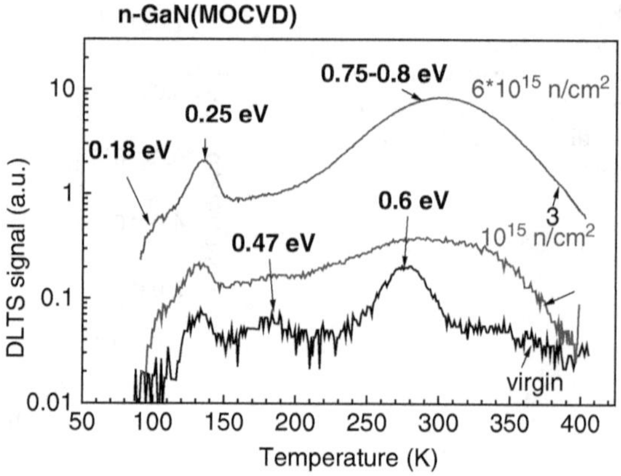

Fig. 9.8 Evolution of DLTS spectra of undoped n-GaN films as a result of fast reactor neutron irradiation; measurements conditions: reverse bias of –1 V, forward bias pulse of 1 V, 2-s-long, time windows 100/1,000 ms

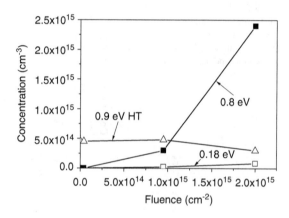

Fig. 9.9 Changes in the concentration of major electron and hole traps in undoped n-GaN induced by fast reactor neutron irradiation

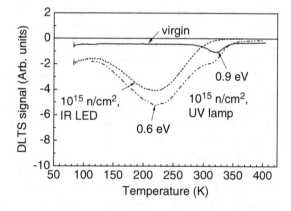

Fig. 9.10 ODLTS spectra measured on the virgin and neutron-irradiated undoped n-GaN; reverse bias –1 V, injection with a pulse of UV deuterium lamp excitation, 5-s-long, time window 300/3,000 ms; also shown is the spectrum taken with 1.4 eV LED excitation

dose are also shown in Fig. 9.9). As discussed above, the introduction rate for these deep acceptors should not be higher than the introduction rate of the 0.8 eV donors. Hence, these hole traps also cannot account for the observed carrier removal rate.

In [20] we suggested that the removal rate observed in neutron-irradiated n-GaN can be explained by the formation of DRs of the type described by Gossick [41]. Some idea of the properties of such DRs in GaN can be obtained from Fig. 9.10. It can be seen that irradiation produces a broad hole-trap-like feature at temperatures 100–300 K. The apparent activation energy of this peak is 0.6–0.7 eV. If this were a true hole trap or band of hole traps they would be located near $E_v + 0.6$ eV and their ODLTS signal should not be produced by optical excitation with the photon energy lower than 2.8 eV. However, as seen from Fig. 9.10, the signal in question is effectively generated even for photon energy of 1.4 eV (mind that the true hole trap peak of the $E_v + 0.9$ eV center is not observed with this excitation). At the same time, we note that for neutron-irradiated samples we see a strong persistent photocapacitance/photoconductivity PPC signal with optical threshold close to 1 eV. As argued in our paper [20] both the PPC phenomena and the appearance of the quasi-hole-trap signal in ODLTS stem from the existence of regions in which the bands are bent upwards by about 1 eV so that electrons released from deep centers inside these regions are swept out by the built-in electric field of the region and have to overcome the barrier of about 1 eV to be recaptured by their host traps. (For intrinsic excitation one simply deals with accumulation of holes in the well and increased concentration of persistent free electrons.) It is quite natural to associate the regions in question with Gossick-like DRs [41], i.e. heavily disordered core regions surrounded by the space charge region with a strong band bending. For very high neutron irradiation doses the outer regions of the DRs overlap and the Fermi level pinning position in such material gives some idea of the Fermi level position in the core of the DR. Our measurements on heavily neutron-irradiated GaN show that, irrespective of the starting conductivity type and doping, the Fermi level in such samples is pinned near $E_c-(0.9-1)$ eV [42]. We pointed to an obvious correlation between this Fermi level pinning position and the Fermi level pinning at the surface of n-GaN Schottky barriers. This is closely linked to the Fermi level stabilization F_s or charge neutrality (CNL) concept introduced long time ago to explain similar correlation in many other III–V materials (see e.g., [43,44]). Several theoretical models have been offered to account for the observed F_s position in various materials. In one class of models the Fermi level is believed to be trapped between the levels of the major native defects (see e.g. [43] and references therein). Mind that in GaN it is located very comfortably between the levels of N_i acceptors and of Ga_i donors. Moreover, lattice parameter measurements in heavily neutron-irradiated GaN show a measurable increase indicating that the dominant defects could be interstitials [20,35,42]. Rutherford backscattering experiments on neutron-irradiated GaN also point to a very high density of interstitials, predominantly Ga_i (see e.g., [45]).

In the other set of models the F_s position is determined by the gap-induced states and can be estimated based on the known band structure [44]. Theoretical estimates made along these lines place the F_s in GaN near $E_c-0.8$ eV [46].

Although any of these models will qualitatively explain the behavior of neutron-irradiated n- and p-GaN, much needs to be done to build the quantitative model. For example, the dependence of the carrier removal rate on starting donor density

Fig. 9.11 Changes in electron concentration induced in undoped n-GaN films with different starting concentration by fast reactor neutron irradiation; also shown are corresponding changes for undoped and lightly doped ELOG GaN

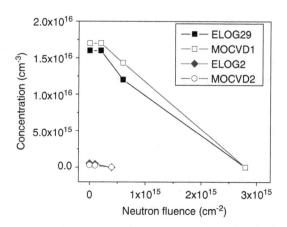

is not easily explained by the classical Gossick model. Figure 9.11 shows such a dependence for n-GaN films with donor concentration of 10^{15} cm^{-3} and 1.6×10^{16} cm^{-3}. It can be seen that the carrier removal rate decreases by about five times when the donor concentration thus decreases. We also note that the carrier removal rate in neutron-irradiated p-GaN is about 20 times higher than for n-GaN despite the much higher concentration of acceptors in p-GaN than donors in n-GaN that should have overpowered the effect of increased barrier height in p-GaN DRs [20, 35]. The same asymmetry of carrier removal rates was observed for proton-implanted p-GaN and n-GaN samples [21, 34] and suggests that interaction of primary defects with Mg ions present in very high concentrations could be an additional factor in both cases.

The dependence of electron removal rate was also observed for variously doped n-GaN films irradiated with 100 keV protons [47] and with ^{60}Co γ-rays [48]. Clearly, a lot remains to be done to get to the stage where accurate predictions of the changes of electrical properties of GaN with irradiation will become possible. However, the already existing data undoubtedly presents a useful guide for extrapolations and preliminary modeling.

9.2.2.3 Effect of Irradiation on Lifetime of Nonequilibrium Charge Carriers

It is a common practice in radiation effects community to characterize the radiation damage for lifetime by lifetime degradation constant K_τ (see e.g. [49]):

$$\tau_0/\tau = 1 + K_\tau F \tag{9.1}$$

where τ_0 is the lifetime value before irradiation, τ is the lifetime after irradiation with the fluence F. Direct measurements of lifetimes in GaN of either type before and after irradiation are rather scarce. In our paper [21] we measured the diffusion length in n-GaN before and after proton irradiation with 150 keV protons. For that electron beam induced current (EBIC) profiling technique was used. Before irradiation the diffusion length was 1.2 μm which converts to the lifetime of 10 ns if we assume the hole mobility of 100 cm²/Vs. Proton irradiation with the dose of 5×10^{14} cm⁻² decreased the diffusion length to 0.6 μm (estimated lifetime of 1.4 ns, the hole mobility was also assumed to be 100 cm²/Vs). From (9.1) the lifetime degradation coefficient K_τ (protons) is found to be 1.6×10^{-14} cm² for such irradiation.

Systematic data of this kind has not as yet been collected for GaN. Some idea of the lifetime damage constant could be obtained from photoluminescence intensity measurements because the PL (or in our case microcathodoluminescence (MCL)) intensity is also related to lifetime. In these measurements the lifetime degradation constant can be obtained from the well-known expression (see e.g. [49]):

$$I_0/I = 1 + K_\tau F \qquad (9.2)$$

where I_0 is the intensity before irradiation and I is the intensity after irradiation with fluence F. Measurements of the bandedge MCL intensity in neutron-irradiated undoped n-GaN were performed in [20]. It was found that strong changes in intensity occurred after the neutron fluence of 2×10^{15} cm⁻². Calculations according to (9.2) yielded in this case the K_τ(neutrons) value of about 3×10^{-16} cm².

Similar estimates for HVPE p-GaN and MBE p-GaN samples irradiated with fast neutrons yield for lifetime damage constants the values of:
K_τ(neutrons, HVPE p-GaN) = (7–8) $\times 10^{-17}$ cm²,
K_τ(neutrons, MBE p-GaN)= 4×10^{-16} cm².

The nature of defects responsible for the lifetime degradation is not quite clear at the moment. For proton implanted n-GaN we noted that strong decrease in the measured lifetime and in the bandedge luminescence intensity started after doses at which deep electron traps 0.8 eV presumably associated with Ga_i (see previous two sections) were introduced [21]. However, this problem needs much more study. We would also like to point out that, in GaN, where the dislocation density is fairly high, one has to consider possible effects of dislocations on diffusion length or lifetime measurements (see next section).

9.2.2.4 Effects of Dislocation Density, Comparison of Irradiation of Standard GaN and Low-Dislocation-Density Material Grown by Epitaxial Lateral Overgrowth

As is well known, GaN films are mainly grown on lattice-mismatched substrates: sapphire, SiC, and therefore contain a very high density of dislocations. In standard GaN films grown on sapphire using low temperature GaN or AlN buffers the dislocation density is on the order of 10^9 cm⁻². By very careful optimization of the growth conditions, the density of dislocations can be reduced to some 10^8 cm⁻² for growth on sapphire and to low 10^8 cm⁻² for growth on SiC. The dislocation density can be decreased to about 10^6 cm^{-2} for epitaxially laterally overgrown ELOG GaN films prepared on sapphire by overgrowing a regular pattern prepared on a GaN template by deposition of a dielectric mask (usually, SiO$_2$ or Si$_3$N$_4$) (see e.g., [50]). A simplest example is a pattern consisting of SiO$_2$ stripes. Growth over the stripes proceeds predominantly in lateral direction which leads to an efficient filtering out of dislocations in these ELOG wing regions, while in between the stripes growth proceeds mostly in the vertical direction and inherits the dislocation density of the starting GaN template, see e.g., [50]. Typically, ELOG films are prepared by metalorganic chemical vapor deposition (MOCVD) and the dislocation density in the ELOG wing regions is about 10^6–10^7 cm^{-2}, whereas in the ELOG window regions it is on the order of some 10^8 cm^{-2} (see e.g., [51]). Also, recently it was shown that thick (over 200 μm) GaN films grown on sapphire by HVPE have the dislocation density below 10^7 cm^{-2} and can be separated from sapphire by various means to form an improved GaN substrate for epitaxial growth of GaN-based device structures (see e.g., [52]). Therefore, interaction of radiation defects with dislocations is something to be considered for structures prepared by ordinary epitaxy. At the same time, advanced devices, such as high-power LEDs and LDs tend to be grown either on low-dislocation-density GaN substrates or using some version of ELOG technique. This could have a very serious impact on the type of defects created in GaN-based devices.

First, we will look at dislocation-related effects for deep traps. The most understood manifestation of dislocations in this respect is the effect of spatial correlation of electrons capture by deep traps decorating dislocations. Such correlated capture has been theoretically and experimentally treated in a number of papers (see e.g., [53, 54]). The most obvious effect of that sort is the logarithmic dependence of the DLTS peak amplitude of the trap on the injection pulse length t_p [53, 54]. Among the defects that we discussed above such behavior has been reported for the 1 eV N$_i$-related acceptors (Fig. 9.12 presents a typical example of this logarithmic dependence for 10 MeV electron-irradiated sample [55]). In [15] a similar behavior was reported for 0.78 eV and 0.95 eV electron traps introduced by 2 MeV proton irradiation. These traps probably are due to Ga$_i$ deep donors and N$_i$ deep acceptors.

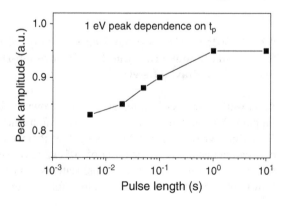

Fig. 9.12 The dependence of the DLTS peak height on the length of the filling pulse t_p for the 1 eV acceptor electron trap detected in 10 MeV electrons irradiated undoped n-GaN

This unambiguously shows that during irradiation at least part of the interstitial defects can travel to dislocation boundaries and decorate them. It has to be taken into account when constructing radiation defects models. It might also make us reconsider how we treat the carrier removal effects even in the cases when only point defects are produced because of the possible contribution of capture in the space charge regions surrounding dislocation cores.

Another factor that one has to bear in mind is the impact of dislocations on mobility of charge carriers. A credible theory describing the individual dislocations contribution to scattering of electrons in n-GaN was described in [9]. The theory predicts a strong increase of the electron mobility with decreasing the dislocation density and increasing the electron concentration (the latter due to enhanced screening of dislocations). However, the effect of dislocations on the carriers mobility (or rather, on resistivity of the sample) is not reduced to the sum of contributions from individual dislocations. For dislocation densities exceeding $\sim 10^8$ cm^{-2} in GaN dislocations form a cellular structure with dislocation boundaries formed mostly by the threading edge dislocations and the characteristic grain size determined by the dislocation density. The carriers travel over the grain boundaries involves in general overcoming a more or less high potential barrier. When the electron concentration is high, tunneling through the barriers is efficient and the material behaves as though electrical nonuniformities were not present. With increased doses of radiation, the electron concentration within the grains becomes lower and the potential barriers at the grains gain higher importance. For p-GaN tunneling of holes is difficult even at high acceptor densities and hole mobility seldom shows "theoretical" temperature or concentration dependence, see e.g., [56]. These effects have not been studied in any detail for GaN.

The cellular structure of GaN films with high dislocation density should also have a profound effect on diffusion length of nonequilibrium charge carriers. Chernyak et al. [57] even postulated that the diffusion length in n-GaN is determined by the distance between the low-angle dislocation boundaries and hence by the dislocation density. The rule worked very reasonably well for HVPE-grown n-GaN films with thickness varying from below 1 μm to 80 μm and the dislocation density decreasing, respectively, from about 10^{10} to about 10^8 cm^{-2} [57]. This is not always the case (we have observed relatively high diffusion lengths on the order of 1 μm for material with dislocation density of about 5×10^8 cm^{-2} [21]; clearly, in the model one has to account for the difference in recombination efficiency of dislocation boundaries in different materials [56]). And the model certainly is no longer operable for low-dislocation density material where the dislocations are distributed more or less randomly. In such material the diffusion length and the lifetime of charge carriers are determined mostly by other defects than dislocations.

Comparison of electrical characteristics, deep traps spectra, diffusion length, and MCL intensity of the low dislocation density ($\sim 5 \times 10^6$ cm^{-2}) ELOG wing regions and high dislocation density ($\sim 10^8$ cm^{-2}) ELOG window regions was carried out by combined DLTS, ODLTS, MCL, EBIC measurements in a series of papers [51, 58–60]. It was shown that, for undoped n-GaN ELOG films with the ELOG layer thickness of 12 μm, the residual donor concentration was much higher in the ELOG window region than in the ELOG wing region ($\sim 10^{15}\,\mathrm{cm}^{-3}$ versus $\sim 3 \times 10^{14}\,\mathrm{cm}^{-3}$), that the deep traps density in the ELOG wing region was about an order of magnitude lower than for standard MOCVD material, that the diffusion length in the low-dislocation-density ELOG wing was close to $0.3\,\mu\mathrm{m}$ versus $0.17\,\mu\mathrm{m}$ in the high-dislocation-density window (the lifetime difference of 0.4 ns versus 0.1 ns) and that the bandedge intensity was about five times higher in the ELOG windows [58]. Doping of ELOG GaN with Si up to about $10^{18}\,\mathrm{cm}^{-3}$ resulted in the donor concentration being about three times higher in the window region [59]. Decreasing the ELOG layer thickness to $6\,\mu\mathrm{m}$ instead of $12\,\mu\mathrm{m}$ resulted in the appearance of relatively high density of dislocation loops and greatly increased density of deep traps with activation energies of 0.25 eV, 0.6 eV, and 0.85 eV. Doping with Si suppressed the formation of the 0.6 eV traps, but enhanced the formation of the 0.85 eV traps with the result of the diffusion length L_d being relatively constant. It was slightly lower for thinner ELOG layers compared to thicker ELOG layers, $0.25\,\mu\mathrm{m}$ versus $0.3\,\mu\mathrm{m}$ [58, 59]. From these results it was inferred that the centers most likely to be the lifetime killers were the 0.6 eV and the 0.85 eV electron traps. It was observed that the ELOG wing region was also the region of strong band bending which resulted in anomalously long propagation of nonequilibrium charge carriers along the ELOG stripes even far outside the Schottky barrier [58, 59]. Only for relatively heavy donor doping this band bending effect was suppressed, most likely due to enhanced tunneling.

Neutron irradiation of undoped ELOG GaN films resulted in a much lower effective removal rate than for standard MOCVD material, $1\,\mathrm{cm}^{-1}$ versus $5\,\mathrm{cm}^{-1}$ [61]. Changes of ODLTS spectra upon neutron irradiation are illustrated by Fig. 9.1 and are very similar to those discussed earlier for standard MOCVD samples. As for them, the 0.5–0.6 eV pseudo-hole-trap band in ODLTS of irradiated samples comes from persistent photoconductivity due to DRs. The Fermi level position in heavily irradiated material is also the same suggesting that no strong changes in the structure of the core regions of DRs occur with changing the dislocation density.

Fig. 9.13 Changes in electron traps and hole traps concentrations in neutron-irradiated undoped ELOG n-GaN

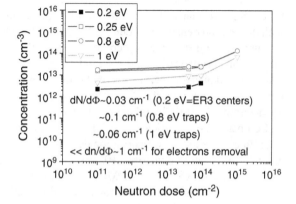

Fig. 9.14 EBIC signal I_c profiles taken across the n-GaN ELOG sample before irradiation (*solid line*) and after reactor neutrons irradiation with $10^{15}\,cm^{-2}$ (*dashed line*) and $2 \times 10^{15}\,cm^{-2}$ (*dash-dotted line*); the accelerating beam voltage 22 kV; variations in the EBIC signal from wing to window are mainly due to change in the local space charge region width and hence the local donor density, see [58,61]

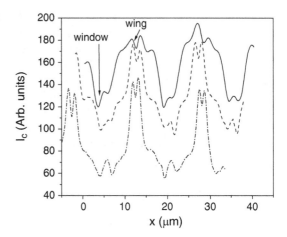

DLTS spectra of neutron-irradiated ELOG samples differed from their MOCVD opposite numbers in that the 1 eV N_i-related acceptor state could be clearly seen. However, as with MOCVD, the introduction rates of all traps were several times lower than the electron removal rate (see Fig. 9.13) again suggesting the dominant role of DRs in carrier removal. Mind also the much lower introduction rates of all traps compared to the MOCVD case.

EBIC profiling of the irradiated samples (see Fig. 9.14) showed that the strongest carrier removal occurred in the high-dislocation-density ELOG window regions and the effect slowly propagated into the ELOG wing region with lower donor concentration and low dislocation density. Estimates of the lifetime degradation constant from changes in bandedge MCL intensity give the values of $3 \times 10^{-16}\,cm^2$ for undoped ELOG material and $1.2 \times 10^{-16}\,cm^2$ for lightly doped ($2 \times 10^{16}\,cm^{-3}$) material. All in all the impact of the difference in dislocation density on the lifetime degradation does not seem to be strong for neutron irradiation.

It is interesting, of course, to check whether the difference in behavior between MOCVD and ELOG and between the wing and window regions of ELOG is related to the difference in dislocation density? In Fig. 9.11 we compare the electron concentration changes for neutron-irradiated ELOG samples with the donor concentration $2 \times 10^{16}\,cm^{-3}$ and $4 \times 10^{14}\,cm^{-3}$ and for the MOCVD samples with similar doping levels. It can be seen that the dependences are virtually identical which suggests that the main differences in the carrier removal rates of ELOG wing and ELOG window regions in Fig. 9.14 stem from the difference in doping level. However, gradual penetration of the damage from the boundary with high-dislocation-density window into the wing indicates that primary defects interaction with dislocations in the window region is also a factor. We would also like to note that the deep traps introduction rate increased for the doped ELOG samples.

Finally, we would like to say a few words on direct production of dislocations by fast neutrons. Clearly, for standard MOCVD material with a very high dislocation density it is hard to expect strong changes in that respect. However, for ELOG material we observed considerable broadening of triple-crystal X-ray rocking curves for the (0006) symmetric reflection and (11–20) asymmetric reflection from 165″ to 205″ (0006) and from 230″ to 270″ (11–20), respectively, as a result of neutron fluence irradiation of 1E18 cm^{-2} [55]. This broadening is a manifestation of increased density of extended defects which is confirmed by direct etch pits density (EPD) measurements that show EPD to increase by about 5×10^7 cm^{-2} in the high dislocation density regions and the appearance of new irradiation-induced inclined dislocation bands in the low dislocation density wings [55].

9.2.2.5   Thermal Stability of Radiation Defects in GaN

Annealing of defects introduced by light irradiating particles (2 MeV protons, 0.2–2.4 MeV electrons) was performed in [15]. It was found that the shallow radiation defects ER1, ER2, ER3 start annealing at 540 K and the annealing is complete after 620 K. Deeper electron traps ER5 associated with N_i were reported to start annealing also at 540 K, but a higher annealing temperature of 660 K was needed for complete removal of these defects [15]. For Ga_i deep donors it was found in [12] that they start moving at room temperature although the fate of these centers was not uncovered in the paper. The V_{Ga} centers responsible for the 0.95 eV PL band were found to be stable up to 500°C.

At the same time, it was found that in GaN samples with a high density of radiation defects (high doses of $\sim$100 keV-implanted hydrogen or heavier ions, neutron-irradiated material), the thermal stability of radiation damage was much higher. After heavy proton implantation the bandedge luminescence intensity could not be restored to the pre-irradiation value even after furnace annealing of 800°C [21]. For activation of ion implanted donors (Si) or acceptors (C) annealing to temperatures exceeding 1,000°C was necessary and even so the degree of impurity activation was relatively low [62]. Figure 9.15 presents the evolution of the sheet

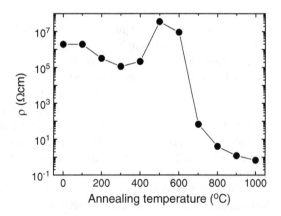

Fig. 9.15 Effect of annealing on resistivity of undoped n-GaN irradiated with 1.5×10^{17} cm^{-2} reactor neutrons (fast neutrons fluence equal to thermal neutrons fluence)

resistivity of undoped GaN sample irradiated with fast and thermal neutrons to fast neutron fluence of $1.5 \times 10^{17}\,\text{cm}^{-2}$ (the ratio of fast and thermal fluences 1:1) [23]. The as-irradiated resistivity was quite high, it measurably decreased at the $150 - 250°C$ stage, increased very strongly at the $250 - 450°C$ and then gradually decreased in a very broad stage 500–$1,000°C$ [23]. One could think that the first stage corresponds to reconstruction of the ER3 and ER5 acceptors as in [15] which perhaps explains the decrease in resistivity. The reverse annealing stage at 250–450°C is most likely due to movement of the N_i, Ga_i centers with forming new deep compensating centers. The onset of the third stage of recovery at 500°C correlates very reasonably with the V_{Ga} acceptors annealing stage in [12] which explains the decrease of the resistivity. Still, even after annealing at 800°C, the pre-irradiation resistivity was not reached, the Fermi level was pinned at relatively deep centers with activation energy 0.45 eV, the sample's series resistance was quite high which resulted in the appearance of DLTS peaks of the wrong sign. The most prominent electron traps were the 0.9 eV and the 1 eV traps that are very likely related to the Ga_i donors and the N_i acceptors, but with a very high binding energy, possibly trapped within DRs. The ODLTS spectra were dominated by the hole traps with activation energy close to the V_{Ga} 0.9 eV centers, but with a much lower capture cross-section. After $1,000°C$ annealing, the Fermi level was pinned near $E_c - 0.2\,\text{eV}$, DLTS spectra were dominated by the 0.6 eV and 0.9 eV traps in high concentration. It should be mentioned that the total concentrations of the 0.45 eV traps pinning the Fermi level after 800°C annealing and of the 0.2 eV traps dominant in the $1,000°C$ annealing are close to each other and equal to the number of donor Ge atoms converted from Ga by interaction with thermal neutrons ($2 \times 10^{16}\,\text{cm}^{-3}$). Hence, there is reason to believe that these relatively deep traps could be complexes of radiation defects with donor atoms [23]. These results show that, even after such not too high doses of neutron irradiation, it is very hard to break down the DRs and to restore the virgin conductivity.

For very high neutron fluences the resistivity of GaN passes through a maximum related to the onset of hopping conductivity (Fig. 9.16). The activation energy for

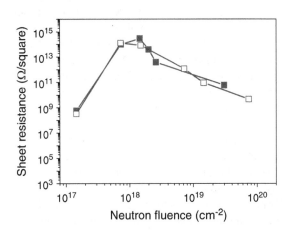

Fig. 9.16 Changes of the sheet resistivity of undoped n-GaN samples as a function of reactor neutrons fluence

the temperature dependence of resistivity for doses before the maximum showed the usual value of 0.9–1 eV. After the fluence corresponding to maximum, the temperature dependence was much weaker. Annealing of such heavily irradiated samples showed a strong reverse annealing stage up to 300°C where the density of radiation defects decreased and the activation energy returned to the 0.9 eV value. After that stage annealing proceeded in the same fashion for samples irradiated with fluences higher than the maximum in Fig. 9.16 and lower than the maximum. However, complete recovery could not be attained even after annealing at 1,000°C.

9.3 Radiation Effects in Other III-Nitrides

Radiation effects studies for III-nitrides other than GaN are very scarce. For InN Hall effect measurements on 150 keV protons and ^{60}Co γ-rays irradiated undoped n-InN films showed that, in contrast to n-GaN, irradiation increases electron concentration in InN. This difference seems to be closely linked to the difference in the position of the Fermi stabilization level in GaN and InN that in turn determines the band offsets in respective heterojunction and the difference in the Schottky barrier height between GaN and InN [63]. Based on calculations carried out in [63], with increasing the In composition in $In_xGa_{1-x}N$ solid solutions the Fermi stabilization level moves upwards to the conduction band edge. The cross-over point is close to $x = 0.34$ and this composition separates solid solutions in which the electron concentration decreases with irradiation from those in which irradiation increases the electron concentration. Blue LEDs are built on GaN/InGaN QW structures with In mole fraction in the QW close to 0.2 and their behavior should be reasonably close to GaN. For LEDs working in the longer wavelength the In composition is higher and one has to be careful when extrapolating GaN results to these green or red LEDs.

For AlGaN there are only a few papers describing the effects of proton and neutron irradiation. In [18] n-AlGaN films with Al mole fraction of $x = 0.12$ and free electron concentration of 10^{17} cm^{-3} were irradiated at room temperature and at 300°C with 2 MeV protons. It was found that the carrier removal rate in AlGaN was about twice as high as for GaN and decreased approximately by two times for high temperature irradiation. A broad feature near 200 K was observed in DLTS spectra after irradiation, but the authors failed to get the values of activation energy, capture cross-section and concentration for this center because of the strong concentration freeze-out effects.

For undoped n-AlGaN films with Al mole fraction of $x = 0.4$, the effects of 100 keV proton irradiation and fast reactor neutron irradiation were studied in [64, 65]. Before irradiation electrical properties of such samples were determined by the presence of a high (2.5×10^{18} cm^{-3}) relatively deep donors with ionization energy of 0.25 eV as follows from admittance spectra and low-frequency capacitance–voltage C–V measurements. The traps in question are believed to be due to defects that form resonance states within conduction band of GaN and emerge in the bandgap of AlGaN at compositions x $\sim$ 0.2–0.3 [6]. These traps were also detected by photoinduced current transients spectroscopy (PICTS). Even deeper states giving rise to defect bands peaked at 3.7 eV and 2.2–2.5 eV were detected in MCL spectra. Neutron irradiation led to compensation of the 0.25 eV traps and introduced instead deeper states with activation energy of 0.28 eV at neutron fluences of 10^{15} cm^{-2} to 2.5×10^{16} cm^{-2}. For higher neutron fluences deeper traps with activation energies 0.35 eV and 1 eV were formed according to PICTS. After irradiation with 1.7×10^{17} cm^{-2} neutrons the films became semi-insulating with the Fermi level pinned near 0.35 eV from the conduction band edge. Irradiation with higher neutron fluences increased the sheet resistivity of the layers so that it could not be measured (sheet resistivity in excess of $10^{14}\Omega$/square). If one goes by the value of the Schottky barrier height in such AlGaN films (2 eV [64,65]) one would expect the Fermi level to be pinned near E_c–2 eV in these samples after high doses of radiation which accounts for very high resistivity (see Sect. 9.2.2.2 devoted to Fermi level pinning in heavily disordered GaN).

Interestingly, the introduction rate for compensating defects for neutron irradiated n-AlGaN was very much higher than for undoped n-GaN (about 500 cm^{-1} versus 5 cm^{-1}) and even higher than in p-GaN ($\sim$100 cm^{-1} [21]). Measurable changes of electrical properties started at neutron fluence of 10^{15} cm^{-2}, i.e., similar to undoped n-GaN [20], even though the concentration of centers to be compensated was two orders of magnitude higher for n-AlGaN.

The effects of proton implantation were similar to the effects of neutron implantation, but the 100 keV protons started to change electrical properties of such AlGaN layers after the dose of 10^{12} cm^{-2}, again two orders of magnitude lower than for undoped n-GaN, even despite a much higher donor density in n-AlGaN [21]. As for neutron irradiation the resistivity of the samples rapidly rose with increasing the proton fluence and after irradiation with 10^{14} cm^{-2} of protons the sheet resistivity was $10^{13}\Omega$/square. We also observed that irradiation led to a decrease of the overall intensity of the two major defect bands and increase of the contribution of the 2.3 eV defect band at the expense of the 3.7 eV band [64].

For p-AlGaN with Al mole fraction of $x = 0.12$, the effect of 100 keV proton irradiation was studied in [66]. In the virgin state the electrical properties were determined by Mg acceptors with activation energy of 0.17 eV and concentration of 3×10^{18} cm^{-3} as follows from admittance spectra and C–V measurements. The concentration started to decrease after irradiation with a very very low proton fluence of 10^{12} cm^{-3}. After irradiation with 10^{13} cm^{-2}, the apparent activation energy of the dominant acceptors increased to 0.2 eV while the concentration further decreased. Irradiation with the fluence of 10^{14} cm^{-2} totally compensated the p-AlGaN film down to the depth corresponding to the range of 100 keV protons (see C–V concentration profiles in Fig. 9.17). Proton irradiation also led to a strong increase of the series resistance in the forward current–voltage I–V characteristics and decreased the reverse current of Au/p-AlGaN Schottky diodes. The Mg-related MCL band intensity decreased by about ten times after irradiation with 10^{14} cm^{-2} protons and after irradiation with higher proton doses we saw, alongside

Fig. 9.17 Acceptor
concentration profiles in
p-AlGaN before and after
irradiation with 100 keV
protons

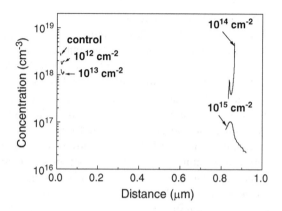

with decrease of the intensity of this band, the emergence of the yellow band. The observed changes, both qualitatively and quantitatively are similar to proton irradiation effects in p-GaN which is not altogether astonishing given the low Al composition of p-AlGaN.

9.4 Radiation Effects in GaN Schottky Diodes, in AlGaN/GaN and GaN/InGaN Heterojunctions and Quantum Wells

In this section we will look at radiation effects in the main building blocks of the III-nitrides devices. The simplest device is, of course, a Schottky diode. It can be a stand-alone device as in a Schottky diode photodetector or power rectifier or a part of a more complex device, such as a high electron mobility transistor (HEMT). The authors of [10] studied the effects occurring in Ni/n-GaN Schottky diodes upon ^{60}Co γ-rays irradiation and found a considerable, about 0.15 eV, increase in the Schottky barrier height after exposure to the dose of 20 MRad (Si). The effect was explained by the formation of a negatively charged defect interfacial layer near the Ni/GaN boundary. Similar increase by about 0.1 eV in the Ni/GaN Schottky barrier height was reported for the Schottky gates of AlGaN/GaN HEMTs irradiated with 1.8 MeV protons for protons doses in excess of 10^{14} cm^{-2} [67]. We observed similar slight Schottky barrier height increase for Au and Ni n-GaN films after irradiation with fast reactor neutrons and 10 MeV electrons. The increase of the specific ohmic contacts resistance reported in several papers (see e.g., [67,68]) for proton-irradiated AlGaN/GaN HEMTs is most likely also due to the increase of the Schottky barrier height upon irradiation: it is well known that good ohmic contacts are Schottky diodes with a low barrier height and strong tunneling of electrons provided by various means.

Fig. 9.18 *I–V* characteristics of Ni/GaN Schottky diodes after neutron irradiation

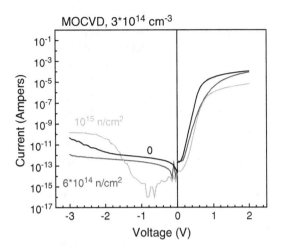

In most cases low and moderate doses of radiation decrease the reverse current of GaN Schottky diodes (see, e.g., the example for neutron-irradiated n-GaN in Fig. 9.18).

AlGaN/GaN heterojunction is the heart of any HEMT structure. Because of the strong electrical polarization field, the two-dimensional electron gas (2DEG) density in such structures is on the order of 10^{13} cm^{-2} even without selective doping of the AlGaN barrier which makes such devices preferable to their AlGaAs/GaAs opposite numbers for high-frequency/high-power applications (see e.g., [69]). Typically the structure consists of a substrate (sapphire, SiC, Si), a buffer layer (low temperature GaN or AlN for sapphire, AlN for SiC or Si), $\sim$1 μm of GaN that has to be highly resistive, $\sim$20–30-nm-thick AlGaN barrier with Al mole fraction typically in the $x = 0.2$–0.4 range and $\sim$2-nm-thick GaN cap. Multiple variations to this basic structure are possible. Selective donor doping of the AlGaN barrier to $\sim$10^{18} cm^{-3} Si donors is supposed to increase the 2DEG concentration. This selectively doped layer can be followed by undoped AlGaN spacer to suppress the scattering effect of donors in the barrier and to increase the 2DEG mobility. The barrier can be made of AlN to increase the 2DEG density or a composite AlGaN/AlN barrier can be used for the same purpose (see e.g., a recent review [70] and references therein). Effects of irradiation on characteristics of some of such structures was studied in a number of papers. Gaudreau et al. [71] irradiated AlGaN (undoped, 13 nm, $x = 0.3$)/GaN(200 nm)/SI-GaN(C) (2 μm)/AlN/sapphire with 2 MeV protons and reported that substantial changes of 2DEG concentration occurred after the proton fluence of 10^{15} cm^{-2}, while 2DEG mobility changes started after 10^{14} cm^{-2}. The results were explained by compensation of both the GaN channel layer and the AlGaN barrier layer and disordering of the AlGaN/GaN boundary. Perhaps the most revealing studies of the matter were performed in [67]. The authors irradiated the simplest HEMT structure consisting (from top to bottom) of 2 nm undoped GaN cap, 23 nm Si-doped AlGaN barrier ($x = 0.3$), and 2 microns undoped GaN (all films grown by MBE on sapphire) with 1.8 MeV protons at room temperature. It was found that the 2 DEG density started measurably decreasing after the proton fluence of 10^{14} cm^{-2}, while the 2DEG mobility started decreasing after the fluence of 10^{13} cm^{-2}. Secondary ions mass spectrometry (SIMS) measurements and MCL spectra measurements probing very thin layers of material near the AlGaN/GaN interface indicated that proton irradiation leads to mixing of AlGaN and GaN near the boundary which should decrease the strain in the structure and hence the polarization-field-induced charge in the 2DEG. The authors also postulated a strong decrease of donor doping in AlGaN that should contribute to the decrease of 2DEG density after the highest doses of irradiation. The mobility degradation was ascribed to the effects of AlGaN/GaN mixing at the boundary. Since the mobility is more sensitive to the irregularities of the boundary and to the interface roughness than the 2DEG concentration, the mobility changes occur earlier than concentration changes. Radiation defects introduced into the GaN channel also contribute to the mobility degradation although these effects should be strongly screened in the high-density 2DEG. Very similar changes of the 2DEG mobility and concentration were observed in [68] for more advanced HEMT structure with a combined barrier consisting of (top to bottom) a much thicker (250 nm) undoped GaN cap, 10 nm Si-doped graded AlGaN (x increases from 0 to 0.22), 40 nm undoped AlGaN ($x = 0.22$), and 7 nm AlN (the structure was grown by MOCVD on sapphire).

Effects of neutron irradiation on 2DEG concentration and mobility in a simple AlGaN/GaN HEMT structure were studied by us in [72]. As for proton irradiation, we observed that the decrease of the 2DEG mobility started from the neutron fluence of 10^{14} cm^{-2} for mobility measured at 77 K and after irradiation with 10^{15} cm^{-2} of fast neutrons for the 300 K 2DEG mobility. The degradation of the 2DEG concentration occurred only after the fluence of 2.5×10^{16} cm^{-2}. ODLTS and PICTS spectra measurements suggest that hole traps with activation energies 0.2 eV and 0.7 eV were produced by neutrons. The latter might be the pseudo-hole-trap peak due to DRs we discussed in Sect. 9.2.2.3.

The results of 10 MeV electrons irradiation of AlGaN/GaN HEMT structures grown by MBE were described by us in [73]. A simple structure with a combined SI-GaN(Fe)(0.7 mm)/undoped GaN (2.2 μm) buffer/undoped AlGaN ($x = 0.28$, 20 nm)/GaN cap(2 nm) grown by MBE was irradiated at room temperature. It was found that the 2DEG mobility started to decrease after irradiation with electron fluence of 5×10^{15} cm^{-2} (decrease of about 14%). After irradiation with the fluence of 3×10^{16} cm^{-2} the mobility decreased by six times. Since the 2DEG concentration was not affected by these doses of radiation the change in the sheet resistivity of the 2DEG channel was totally reflecting the changes in 2DEG mobility (see Fig. 9.19). An interesting observation also reported in [73] concerns the electron irradiation results for HEMT structures in which the barrier was 3-nm-thick AlN instead of 20 nm AlGaN. The 2DEG concentration in such AlN-barrier HEMT structures was 1.5 times higher than for their AlGaN-barrier opposite numbers (3×10^{13} cm^{-2} instead of 2×10^{13} cm^{-2}) with the room temperature mobility virtually the same (1,360 cm^2/Vs for AlN versus 1,450 for AlGaN). (Detailed description of electrical properties of these two types of heterojunctions can be found in [74].) As seen

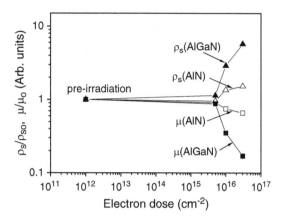

Fig. 9.19 Sheet resistivity R_s, and 2DEG mobility μ for electron irradiation of AlN HEMTs

from Fig. 9.19, for the HEMT structure with AlN barrier, the changes in mobility were much slower than in the structure with AlGaN barrier: even after the fluence of $3 \times 10^{16}\,\mathrm{cm}^{-2}$ the mobility was still 50% of its pre-irradiation value. This almost an order of magnitude higher radiation tolerance of AlGaN/GaN HEMTs is due to the fact that the AlN barrier is seven times thinner than its AlGaN counterpart with correspondingly lower energy deposited by 10 MeV electrons into the formation of defects in the barrier and near the barrier/channel interface. Similar considerations were invoked in [75] to explain an increased radiation tolerance of AlGaN/GaN HEMTs to proton irradiation as the energy of protons increased from 2 MeV to 108 MeV. In addition, one expects the displacement energy in AlN to be considerably higher than in AlGaN.

One more characteristic that is awfully important for AlGaN/GaN heterojunctions developed for HEMTs is the threshold voltage, i.e., the voltage you have to apply to your Schottky barrier gate to totally pinch off the 2DEG channel near the AlGaN/GaN interface. In simple theory this voltage V_T is determined by the magnitude of the electrical polarization field, the Schottky barrier height ϕ_B, and the donor doping level of the AlGaN channel N_d as [76]:

$$V_T = \phi_B - (qN_d t_{AlGaN}^{2/2\varepsilon}) - (\Delta E_c/q) - (\sigma_{pol} t_{AlGaN}/\varepsilon), \qquad (9.3)$$

where q is the electronic charge, ε is the dielectric permittivity, ΔE_c is the conduction band discontinuity, σ_{pol} is the polarization charge, t_{AlGaN} is the AlGaN barrier thickness. As discussed above in relation with proton-induced changes of 2DEG mobility and concentration there is reason to believe that the second and fourth terms in (9.3) could decrease after bombardment with high-energy particles which should lead to decreasing the threshold voltage. This indeed happens in many cases after proton irradiation (see [67, 68]). Typically the proton fluence after which the threshold voltage starts to decrease is close to $10^{14}\,\mathrm{cm}^{-2}$ for 2 MeV protons.

However, the situation could be more complicated. For example, in [73, 77] we reported the changes in C–V characteristics of 10-MeV irradiated AlGaN/GaN

Fig. 9.20 400 K $C-V$
characteristics as a function
of 10 MeV electrons dose for
Schottky diodes made on
AlGaN/GaN HJs

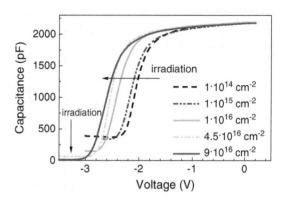

structures. The capacitance of Schottky diodes made on these heterojunctions was
high at low negative bias on the diode because of the very high charge accumulated
in the 2DEG region. As the negative bias increased the states in the AlGaN/GaN
quantum well (QW) were emptied which resulted in decreased capacitance. When
all electrons were driven out of the AlGaN/GaN QW the capacitance was the
capacitance of the GaN buffer and its change with voltage was determined by
the donor concentration in GaN. The voltage at which the capacitance becomes
the capacitance of the GaN is, of course, the same as the threshold voltage discussed
above for HEMT structures. Figure 9.20 shows such $C-V$ characteristics measured
at 400 K for MBE-grown AlGaN/GaN Schottky diode before and after irradiation
with increasing fluences of 10 MeV electrons. It can be seen that the threshold voltage in these experiments first increased after irradiation and only after that started
to decrease. In [77] we argue that the low-dose negative shift of threshold voltage
could be due to radiation-induced positive charge accumulation in the AlGaN
barrier that is then superseded by the above described processes of radiation-induced
interface disordering. We observed similar behavior for the initial stages of neutron
irradiation of AlGaN/GaN HEMT structures. The shift of the threshold voltage to
more negative values was also reported for γ-irradiated AlGaN/GaN HEMTs [78].
This positive trapped charge could also increase the 2DEG concentration and we
indeed reported some increase of the 2DEG concentration in neutron-irradiated
AlGaN/GaN heterojunctions for low neutron fluences [72].

For GaN LED structures perhaps the most important part for operation of the
device is the GaN/InGaN QW or multiple quantum well (MQW) region where
recombination of injected charge carriers should occur. Processes induced in such
MQW region by 10 MeV electron irradiation at room temperature were studied
by us in [79]. The MQWs were grown by MOCVD on basal plane sapphire
substrates. The samples consisted of a low-temperature GaN buffer, 2 μm undoped
n-GaN, 2 μm Si-doped n$^+$ GaN (n ∼ 2 × 10^{18} cm^{-3}), 5 QWs of InGaN (3 nm) with
undoped n-GaN barriers (15 nm) capped with undoped n-GaN (60 nm). Admittance
spectra measurements performed on the samples before irradiation showed a strong
influence of tunneling. $C-V$ profiling at 85 K revealed the presence of four QW
related peaks due to the four lowermost QWs (Fig. 9.21). At 400 K the profiles

Fig. 9.21 1 kHz $C–V$ profiles obtained for undoped MQW GaN/InGaN structure after various doses of 10 MeV electrons; measurements at 80 K

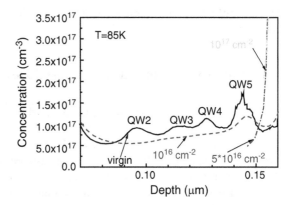

Fig. 9.22 The same as in Fig. 9.21, but measurements at 400 K

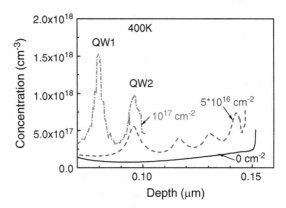

were strongly smeared and only a kimk due to the lowest QW could be seen (Fig. 9.22). It was argued that the observed profiles were due to interface traps decorating GaN/InGaN QW interfaces. Irradiation compensated the MQW region and increased the ionization energy of the traps decorating the QW interfaces from $20 − 60$ meV to 90 meV after irradiation with 5×10^{16} cm^{-2} electrons and to 190 meV after irradiation with 10^{17} cm^{-2} electrons. As a result at low temperature the whole MQW region behaved as dielectric layer (Fig. 9.21), while at 400 K the QW-related profile became more pronounced and showed a much enhanced concentration of interface traps (Fig. 9.22). The traps in question are quite similar to the electron traps introduced by electrons in n-GaN (see above) which probably points to the traps being located in the GaN barriers. DLTS spectra measured on irradiated samples indicated, in addition to these interface traps, a deeper trap with activation energy 1.1 eV reminiscent of the N_i-related acceptors in n-GaN. These traps were also located in the GaN barriers or at the GaN/InGaN interfaces on the GaN side of them [79].

Among other effects important for device applications we note a measurable increase of the series resistance of the MQW Schottky diodes and a strong decrease of the QW-related recombination. The latter indicates a lifetime degradation constant close to 6×10^{-16} cm^2 [79].

9.5   Radiation Effects in GaN-Based Devices

Studies for AlGaN/GaN HEMTs irradiated with 1.8 MeV protons showed that measurable decrease of transconductance, threshold voltage, and drain saturation current occurred after proton fluences of 10^{14} cm^{-2} [67, 68, 80]. This dose is about two orders of magnitude higher than for AlGaAs/GaAs HEMTs (see e.g. [81, 82]). Annealing at 800°C was shown to be efficient in partially restoring the electrical characteristics [80]. For higher proton energies of 40 MeV both DC characteristics (transconductance, threshold voltage, drain saturation current) and AC characteristics were little affected by proton fluences of up to 5×10^{10} cm^{-2} [83–85]. In fact, some improvement of DC characteristics (reverse leakage, breakdown voltage) was reported. Effects of proton irradiation of AlGaN/GaN HEMTs with protons of energies 1.8 MeV, 15 MeV, 40 MeV, 105 MeV and proton fluences up to 10^{13} cm^{-2} were compared in [75]. The strongest changes were observed for the lowest proton energy (but even those were only marginal changes) and the effect was explained by the decrease of the energy transferred to Al, Ga, and N atoms in elastic collisions occurring within the active region of devices as the range of protons increased with increasing energy. Gamma irradiation of AlGaN/GaN HEMTs up to the dose of 600 MRad did not substantially change the parameters [86]. All this is in line with the changes of the "bulk" properties of AlGaN/GaN heterojunctions with radiation discussed above and would suggest that the AlGaN/GaN HEMTs characteristics would not be affected by fast reactor neutrons irradiation up to fluences of 10^{15} cm^{-2}. However, it could be noted that the ohmic contacts resistance increase after proton irradiation starts at fluences an order of magnitude lower than for "bulk" parameters changes and for really low contact resistances needed for high-frequency HEMTs this could be a factor to reckon with when analyzing the radiation tolerance.

One more important thing to be noted in conjunction with AlGaN/GaN HEMTs performance as affected by radiation is the so-called current collapse phenomena in these devices. There is quite a literature on the subject due to its practical importance. We do not have time or space here to treat this matter in any detail. In brief, the phenomenon is a strong dependence of AC characteristics of transistor on frequency: a substantial loss of current that can be passed through a device at high frequency compared to DC characteristics, a substantial lag in the switching performance when rapidly changing the gate voltage, long-term drift of parameters after driving the device at high current (see e.g., [87–92]). It seems the most important among those is the loss of current transfer characteristics at high frequencies believed to occur because of the trapping of charge carriers in the AlGaN barrier and the formation of a virtual gate with a much increased area compared to the actual metallic gate (see e.g., [87–91]). It was shown that the effect can be mitigated by deposition of dielectric layers (Si_3N_4, Gd_2O_3, MgO) on top of the barrier (see e.g. [93–97]). The nature of this improvement is not thoroughly understood, but it seems that it has to do with the decrease of the surface traps density [98]. Hence, it is important that it has been demonstrated for Gd_2O_3 and MgO passivating layers that their beneficial effect is not affected by proton and γ-ray irradiation [84–86].

Another important class of GaN-based devices is, of course, LEDs. For double heterostructure blue GaN/InGaN LEDs measurable degradation of the light output started after irradiation with 10^{14} neutrons [99]. For proton-irradiated AlGaN/GaN QW LEDs it was found that the threshold dose for the start of degradation was two orders of magnitude higher than for AlGaAs/GaAs QW LEDs (10^{12} cm^{-2} versus 10^{10} cm^{-2} for 3 MeV protons). It was also observed that increasing the proton energy from 3 MeV to 5 MeV measurably increases the dose necessary for the onset of light output degradation, most likely due to a lower energy going into elastic collisions within the active region of devices. Even higher proton doses were found necessary for changing the characteristics of proton-irradiated blue GaN/InGaN LEDs in [100]. For green GaN/InGaN LEDs [101], it was reported that 2 MeV protons produce about 40% light output decrease after a fluence of 1.7×10^{12} cm^{-2}.

One of the most important characteristics of GaN-based solar-blind photodetectors is the spectral selectivity, i.e., the ratio of the signal in the UV region and the signal in the visible region (see e.g., [69]). Recently we measured the characteristics of Ni Schottky diodes photodetectors prepared on thick (12 μm) undoped ELOG GaN films and on thick (4 μm) undoped GaN films grown by MOCVD. The electron concentration in the active region was in low 10^{14} cm^{-3} in both cases. It was found that after irradiation with fast reactor neutrons the spectral sensitivity started to change after the neutron fluence of 10^{15} cm^{-2}. Figure 9.23 shows the actual changes in the photocurrent at −1 V before and after such irradiation for the ELOG sample (the neutron irradiation with lower fluences of 10^{14} cm^{-2} and 4×10^{14} cm^{-2} led to

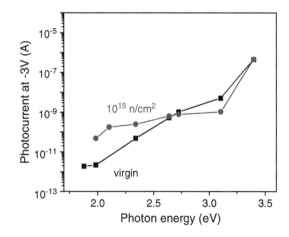

Fig. 9.23 Changes induced by neutron irradiation with the fluence of 10^{15} cm^{-2} in photocurrent at −3 V measured at different photon energies (excitation with 3 W GaN/InGaN LEDs, the data not corrected for the changing number of photons when varying the wavelength). The sample was 12-μm-thick undoped ELOG GaN Ni Schottky diode

no measurable changes in device characteristics; measurements in Fig. 9.23 were performed for a set of GaN/InGaN LEDs with photon energy varying from 1.9 eV to 3.4 eV operated at optical output power of 3 W). The observed variations in spectral characteristics are reminiscent of the persistent photoconductivity spectra in neutron-irradiated GaN (see above) and most likely come from the effect of charge carriers trapping in DRs. We also observed about 50% decrease of the short circuit current of these Schottky diodes produced by probing electron beam of 25 keV (Fig. 9.24). Again, these changes happen after at least an order of magnitude higher dose than for GaAs photodetectors and about two orders of magnitude higher dose than for typical Si photodetectors.

Finally, we would like to briefly discuss the irradiation-induced changes in performance of radiation detectors based on GaN. Compared to classical detector materials, such as Si, Ge and recently GaAs, GaN has, of course, serious disadvantages due to a much higher energy gap decreasing the number of electron–hole pairs produced in a single event, the lack of high-quality thick uniform bulk crystals or epitaxial layers, the much shorter, compared to Si and Ge, lifetimes. Therefore, one does not expect GaN to supplant these more established materials in standard radiation spectrometry tasks. However, the large bandgap and the much stronger crystal bonds of GaN suggest that it could have serious advantages in niche applications where one needs a very high radiation tolerance and/or high temperature of operation, whilst efficient charge collection for particles with long ranges is not required. Somewhat typical examples are α-particles detection in high-energy physics experiments, neutron spectrometry in reactor channels and some others. Initial attempts to build and test GaN radiation detectors were performed for undoped low-donor-density GaN films and Fe-compensated semi-insulating GaN films prepared by MOCVD. In [102, 103] it is reported that such detectors show charge collection efficiency between 50 and 90% for 241 Am 5.48 MeV α-particles and that irradiation with 24 GeV protons or 1 MeV neutrons decreases the charge collection efficiency to something on the order of 10–20% after the

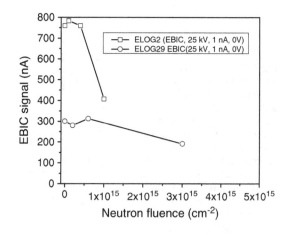

Fig. 9.24 Changes induced in EBIC signal of undoped ELOG GaN Ni Schottky diode by neutron irradiation; the probing beam accelerating voltage of 25 kV, the probing beam current of 1 nA (*open square*); also shown is the data for the lightly doped (2×10^{16} cm^{-3} ELOG sample) (*circle*); mind a higher radiation tolerance in the latter case

fluence of 10^{16} cm^{-2}. In our paper [104] it was demonstrated that Ni Schottky diodes α-particles detectors built on thick (3.6 μm) MOCVD undoped n-GaN films and thick (12 μm) undoped ELOG films have the charge collection efficiency close to 100%. Recently we have demonstrated that such detectors can operate at elevated temperature of 60°C and lose only about 50% of charge collection efficiency after irradiation with fast reactor neutron fluence of 10^{15} cm^{-2} which is two orders of magnitude higher than for standard Si detectors and one order of magnitude higher than for GaAs detectors.

9.6 Prospects of Radiation Technology for GaN

Traditionally, the areas in which radiation technology has been useful for semiconductor devices are the ion implantation for local doping and local isolation in planar device technology, neutron transmutation doping and improving the device characteristics with low-dose irradiation. Successful donor doping was achieved by Si implantation in GaN in [105–107]. The key to success here is to avoid amorphization during implantation with high doses (this is achieved by implanting Si at elevated temperature of around 350°C [108]) and to anneal the radiation damage at high enough temperature (rapid thermal annealing temperatures in excess of 1,000°C is needed for that). Co-implantation with heavy ions, such as P, to produce a high density of vacancies was also found very helpful, particularly for activation of Mg acceptors [105, 106]. Even under these conditions the donor activation ratio is commonly rather low, on the order of about 30% or lower, although more efficient activation can be achieved for higher annealing temperature [106]. Donor doping of AlGaN ($x = 0.2$) by Si ion implantation has also been demonstrated, but required annealing of radiation damage at even higher temperature close to 1, 100°C (for that masking the sample with AlN to prevent surface degradation is quite useful) [109]. Acceptor doping by ion implantation is much more difficult than donor doping. For Mg, this difficulty was attributed in a recent paper to effective formation of stacking faults [110]. For AlGaN with even moderate Al composition of $x = 0.2$ activation of ion-implanted Mg or C acceptors could not be achieved [109]. Because of these difficulties with annealing the radiation damage ion implantation doping has not yet found wide practical use in GaN device technology for ohmic contacts doping, p–n junctions fabrication, etc. Apparently, before that happens, reliable methods of annealing the samples at high temperatures without causing surface degradation have to be developed. The situation is somewhat similar to SiC prior to the introduction of high-pressure/high-temperature annealing (see e.g., a review in [111] and references therein).

On the other hand, device isolation by implanting high doses of various ion species does not meet with serious difficulties. Implantation of O, F, N, He at doses higher than $\sim 10^{15}$ cm^{-2} [4, 19, 112, 113] is effective in producing insulating layers with sheet resistivies higher than $10^{10}\Omega$/square. High activation energy of resistivity close to 0.8–1 eV in such layers could be a concern for high-temperature applications. This activation energy could be made much lower by increasing the implantation dose so as to get into the hopping conductivity regime (see our comments for neutron-irradiated GaN annealing experiments above), but at the expense of a lowered room temperature resistivity. Protons are, of course, preferred species for tasks in which the thickness of the insulation layer has to be relatively high. Proton implantation works very well for p-GaN and p-AlGaN [34, 66] producing films with very high sheet resistivity. For n-GaN proton implantation is only marginally acceptable because the sheet resistivity is on the order of 10^5–$10^6 \Omega$/square [21, 114]. Still, remarkable improvement of signal to noise ratio of AlGaN UV photodetectors was achieved by proton implantation at energies 2 or 3 MeV and dose higher than 10^{16} cm^2 due to a strong suppression of leakage current in GaN [115].

The main advantages of neutron transmutation doping NTD of classical semiconductors are related to a very high uniformity of doping. In GaN, as in GaAs, the main reaction to be considered is the conversion of Ga into Ge, a shallow donor in GaN. The concentration of Ge donors N_{Ge} introduced by such interaction can be estimated as $N_{Ge} = 0.125 \times F_{th}$, where F_{th} is the fluence of thermal neutrons [45]. For GaN, the main obstacle to a wide use of NTD is a very high temperature of annealing of radiation damage. Preliminary studies using Rutherford backscattering showed that heavily neutron-irradiated GaN films have a very high density of interstitial defects even after annealing at 800°C [45]. In our recent paper [23] we show that electrical activation of introduced donors can be achieved after annealing at 1, 000°C for irradiation with low neutron fluence of 1.5×10^{17} cm^{-3} of fast neutrons (the fluences of fast and thermal neutrons were equal). The concentration of NTD donors was close to the calculated value, but the donor activation energy was 0.2 eV, much higher than for Ge donors, which suggests that the centers we observed were complexes of radiation defects with Ge. If one would try to get a higher donor concentration by using higher neutron fluences annealing becomes really a problem.

As we have seen above, irradiation of GaN with low doses of electrons, protons, neutrons often decreases the leakage current of Schottky diodes which was proposed to use for improving the performance of GaN detectors in [116,117]. In AlGaN/GaN HEMTs one of the most challenging problems is to achieve the high resistivity of the GaN buffer without compromising the 2DEG concentration and mobility. In [73, 118] we show that, in MBE-grown AlN/GaN or AlGaN/GaN HEMTs, this can be achieved by carefully optimizing the concentration of Fe producing compensating deep centers near E_c–0.6 eV and changing the thickness of the undoped GaN subbuffer on top of GaN(Fe) layer. This is a rather technologically challenging process with not a 100% yield. As demonstrated above, electron irradiation with electron energy of 10 MeV can decrease the density of uncompensated shallow donors so that the Fermi level is shifted down to the E_c– 0.6 eV acceptors. In [72,79] we demonstrate that when the residual donors concentration in the buffers does not exceed some 10^{15} cm^{-2} there exists a useful technological window in such electron irradiation that allows to render the buffers semi-insulating without compromising the 2DEG concentration and mobility.

Also, for AlGaN/GaN HEMTs one would like to be able to decrease the threshold voltage so that the channel would be pinched-off at zero bias and would only pass current when positive voltage opens it (these so-called depletion mode devices are preferable in circuits with low power consumption in the off state). As we have seen above, proton irradiation can strongly decrease the threshold voltage of AlGaN/GaN HEMTs and could in principle be used to fabricate depletion-type HEMTs. Unfortunately, the doses necessary for that are so high that they also lead to strong degradation of the channel conductivity. However, implantation of F ions can be confined to the AlGaN barrier alone and can introduce a strong negative charge in the barrier shifting the threshold voltage to positive values (see e.g., [118]). This is one of the established ways of achieving depletion-type performance for AlGaN/GaN HEMTs.

9.7 Summary and Conclusions

We have seen that radiation effects in GaN can be reasonably well understood based on a rather simplistic picture in which the main radiation defects are due to shallow V_N and deep Ga_i donors and deep V_{Ga} and N_i acceptors. This picture places the V_N donors near E_c–0.06 eV, the Ga_i doubly charged donors near E_c–0.8 eV, the V_{Ga} acceptors near $E_v + 1$ eV, and the N_i acceptors near E_c–1 eV. Other prominent defects in n-GaN, relatively shallow ER1–ER3 traps, seem to be complexes of these primary defects, mostly of V_N, with unidentified species, possibly with donors in the case of ER3 if one considers the results of neutron doping and annealing experiments. In p-GaN there is evidence of formation of defects near E_c–0.5 eV that seem to be Mg acceptors complexes with native defects and also of defects of unidentified nature with a level near $E_v + 0.3$ eV. The carrier removal rate in GaN for light particles is well accounted for by the introduction of these simple defects. For particles, such as fast neutrons, bound to produce large recoil cascades the data suggests carrier removal by DRs in which the Fermi level in the core is pinned between the Ga_i donor level and the N_i acceptor level. In heavily irradiated samples these DRs pin the Fermi level near E_c–1 eV irrespective of initial type of conductivity and doping level. Charge trapping in DRs contributes to strong persistent photoconductivity effects in neutron-irradiated GaN. Both theoretical modeling and experiment show a much higher radiation tolerance of GaN compared to Si or GaAs. When comparing the results for different types of particles and different energies of particles one can see that where such comparison with modeling has been done it suggests that modeling provides a reasonable guide for predicting the relative effectiveness of different radiation types in changing the electrical properties of GaN. Thus the work carried out so far can serve as a firm basis for further studies and for developing predictive models of materials and device performance in III-nitrides. However, many issues still have to be addressed. Among them are (1) the strong asymmetry in carrier removal rates in n- and p-type materials and possible interaction of radiation defects with Mg acceptors; (2) the scarcity of our knowledge of radiation effects in III-nitrides other than GaN; (3) the poor state of understanding of radiation defects in nitrides with dislocations, particularly given the fact that the dislocation density can be very high; (4) the lack of proper understanding of radiation defects interaction with dopants and impurities; (5) poor understanding of the nature of recombination processes in irradiated nitrides; (6) very little effort devoted to studies of electrical and recombination effects in homojunctions, GaN/InGaN QW heterojunctions, effects at the AlGaN/GaN, GaN/InGaN interfaces for various compositions of HJs and QWs. Some results, such as apparently lower radiation stability of n-AlGaN films compared to GaN which does not fit theoretical predictions suggest checking possible effects of crystalline quality on performance. Much more has also to be done to properly understand defects transformation upon increasing irradiation temperature and upon annealing. For example, ODEPR experiments on low-temperature electron irradiation of GaN suggest that V_{Ga} acceptors are annealed at 500°C, yet these acceptors can be clearly seen in neutron-irradiated material even after annealing at 1,000°C.

If we now consider the state of affairs with radiation effects studies in GaN-based devices the main conclusion is that all these devices are at least an order of magnitude more radiation tolerant than their GaAs opposite numbers, to say nothing of Si devices. This means that in mixed systems GaN devices will not be a weak link; it also provides serious reasons for eventually substituting Si and GaAs devices in satellite and military systems for GaN devices once the issues of performance, reliability and manufacturability are solved. However, it should be mentioned that the devices studied so far have been in fact prototype devices that seriously differ from devices that will be used in real applications. For example, the majority of experiments on HEMTs were performed for structures grown on sapphire, whereas real high-power transistors will be grown on substrates with higher thermal conductivity, such as SiC. LED structures that have been studied in the early experiments are very different from high-power/high-brightness devices that dominate the field today. For military applications much more should be done in terms of neutron radiation damage assessment. Devices to be used in open space satellite modules will be subjected mostly to some hundred energy protons, low MeV energy protons, some MeV electrons of the van Allen radiation belts. This part of the irradiation particles spectrum has been somewhat covered in experimental studies. However, for systems operating in the atmosphere at high altitudes high-energy neutrons will be of comparable importance. This also is true for special devices such as radiation detectors operating in high-energy systems, in nuclear reactors, etc.

It will probably take the GaN radiation effects community some considerable time to solve all these problems and to get to the level of understanding comparable to that achieved in traditional semiconductor materials, but this work certainly has to be done and will be done and the author hopes his humble effort to survey the field as it is now will suggest ideas to researchers that are already involved in these studies and serve as an introduction to those who only start.

Acknowledgements I would like to thank my long time colleagues with whom I have been working on radiation effects in III-nitrides, Prof. S.J. Pearton of the University of Florida, Prof.

In-Hwan Lee of Chonbuk University in Korea, Dr. Amir Dabiran at SVT Technologies, Dr. Nikolai Smirnov and Dr. Anatoliy Govorkov at the Institute of Rare Metals, Dr. Nikolai Kolin at the Institute of Physical Chemistry in Obninsk for their help, support, and fruitful discussions. Also, useful discussions with Dr. D. C. Look and Dr. Z-Q. Fang at Wright State University are gratefully acknowledged. The work was supported in part by International Science and Technology Center ISTC grant #3870.

References

1. A. Ionascut-Nedelcsescu, C. Carlone, A. Houdayer, H.J. von Bardelleben, J.L. Cantin, S. Raymond, IEEE Trans. Nucl. Sci. **49**, 2733 (2002)
2. D.C. Look, D.C. Reynolds, J.W. Hemsky, J.R. Sizelove, R.L. Jones, R.J. Molnar, Defect donor and acceptor in GaN. Phys. Rev. Lett. **79**(12), 2273–2276 (1997)
3. J. Nord, K. Nordlund, J. Keininen, Molecular dynamics study of damage accumulation in GaN during ion beam irradiation. Phys. Rev. B. **68**, 184104 (2003)
4. S.O. Kucheyev, J.S. Williams, C. Jagadish, Effects of ion species on the accumulation of ion-beam damage in GaN. Phys. Rev. B. **64**, 035202 (2001)
5. D.W. Jenkins, J.D. Dow, M.-H. Tsai, J. Appl. Phys. **72**, 4130 (1992)
6. A.Y. Polyakov, M. Shin, J.A. Freitas, M. Skowronski, D.W. Greve, R.G. Wilson, On the origin of electrically active defects in AlGaN alloys grown by organometallic vapor phase epitaxy. J. Appl. Phys. **80**(11), 6349–6354 (1996)
7. J. Neugebauer, C.G. Van de Walle, Phys. Rev. B. **50**, 8067 (1994)
8. P. Boguslawski, E.I. Briggs, J. Bernholc, Phys. Rev. B. **51**, 17255 (1995)
9. D.C. Look, Z. Fang, L. Polenta, MRS Internet J. Nitride Semicond. Res. 5S1 (2000) paper W10.5
10. G.A. Umana-Membreno, J.M. Dell, G. Parish, B.N. Nener, L. Faraone, U.K. Mishra, 60Co g-irradiation on GaN Schottky diodes. IEEE Trans. Electron Dev. **50**(12), 2326 (2003)
11. A.Y. Polyakov, N.B. Smirnov, A.V. Govorkov, A.V. Markov, C.R. Lee, I-H. Lee, N.G. Kolin, D.I. Merkurisov, V.M. Boiko, J.S. Wright, S.J. Pearton, Defects in electron and neutron irradiated n-GaN: disordered regions versus point defects, Abstracts of the Fall MRS 2006 Meeting: Symposium I: Advances in III-Nitrides Semiconductor Materials and Devices (MRS, Warrington, 2006) paper I7.46
12. K.H. Chow, L.S. Vlasenko, P. Johannesen, C. Bozdog, G.D. Watkins, Intrinsic defects in GaN. I. Ga sublattice defects observed by optically detected electron paramagnetic resonance. Phys. Rev. B **69**, 045207 (2004)
13. L. Polenta, Z-Q. Fang, D.C. Look, On the main irradiation-induced defect in GaN. Appl. Phys. Lett. **76**(15) 2086–2088 (2000)
14. F.D. Auret, S.A. Goodman, F.K. Koschnick, J.-M. Spaeth, B. Beaumont, P. Gibart, Appl. Phys. Lett. **74**, 407 (1999)
15. S.A. Goodman, F.D. Auret, F.K. Koschnick, J.-M. Spaeth, B. Beaumont, P. Gibart Radiation induced defects in MOVPE grown n-GaN. Mater. Sci. Eng. B **71**, 100–103 (2000)
16. S.A. Goodman, F.D. Auret, F.K. Koschnick, J.-M. Spaeth, B. Beaumont, P. Gibart, Appl. Phys. Lett. **74**(6), 809 (1999)
17. F.D. Auret, S.A. Goodman, F.K. Koschnick, J.-M. Spaeth, B. Beaumont, P. Gibart, Appl. Phys. Lett. **73**(25), 3745 (1998)
18. M. Hayes, F.D. Auret, L. Wu, W.E. Meyer, J.M. Nel, M.J. Legodi, Electrical defects introduced during high-temperature irradiation of GaN and AlGaN. Phys. B, 340–342, 421–425 (2003)
19. P. Hacke, T. Detchprohm, K. Hiramatsu, N. Sawaki, Appl. Phys. Lett. **63**, 2676 (1993)
20. A.Y. Polyakov, N.B. Smirnov, A.V. Govorkov, A.V. Markov, S.J. Pearton, N.G. Kolin, D.I. Merkurisov, V.M. Boiko, Cheul-Ro Lee, In-Hwan Lee, Neutron irradiation effects in undoped n-GaN Films. J. Vac. Sci. Technol. **B25**(2), 436–442 (2007)

21. A.Y. Polyakov, A.S. Usikov, B. Theys, N.B. Smirnov, A.V. Govorkov, F. Jomard, N.M. Shmidt, W.V. Lundin, Effects of proton implantation on electrical and recombination properties of n-GaN. Solid-State Electron. **44**, 1971–1983 (2000)

22. J. Bourgoin, M. Lannoo, *Point Defects in Semiconductors II, Experimental Aspects* (Springer, Berlin, 1983) chapter 6

23. A.Y. Polyakov, N.B. Smirnov, A.V. Govorkov, N.G. Kolin, D.I. Merkurisov, V.M. Boiko, A.V. Korulin, S.J. Pearton, Neutron transmutation doping of GaN. J. Vac. Sci. Technol. B. **28**(3), 608–612 (2010)

24. A. Chantre, G. Vincent, D. Bois, Phys. Rev. B. **23**, 5335 (1981)

25. G.M. Martin, A. Mitonneau, D. Pons, A. Mircea, D.W. Woodard, J. Phys. C. **13**, 3855 (1980)

26. A.Y. Polyakov, N.B. Smirnov, A.V. Govorkov, Z.-Q. Fang, D.C. Look, R.J. Molnar, A.V. Osinsky, Deep hole traps in n-GaN films grown by hydride vapor phase epitaxy. J. Appl. Phys. **91**(10), 6580–6584 (2002)

27. M. Linde, S.J. Uftring, G.D. Watkins, Optical detection of magnetic resonance in electron irradiated GaN. Phys. Rev. B. **55**(16), R10177–10180 (1997)

28. I.A. Buyanova, M. Wagner, W.C. Chen, B. Monemar, J.I. Lindstrom, H. Amano, I. Akasaki, Photoluminescence of GaN: effect of electron irradiation. Appl. Phys. Lett. **73**(20), 2968–2970 (1998)

29. K.H. Chow, G.D. Watkins, A. Usui, M. Mizuta, Detection of interstitial Ga in GaN. Phys. Rev. Lett. **85**(13), 2761–2764 (2000)

30. F.J. Sanchez, D. Basak, M.A. Sanchez-Garcia, E. Calleja, E. Munoz, I. Izpura, F. Calle, J.M.G. Tijero, B. Beaumont, P. Lorenzini, P. Gibart, T.S. Cheng, C.T. Foxon, J.W. Orton, MRS Internet J. Nitride Semicond. Res. **1**(7), (1996)

31. A.Y. Polyakov, N.B. Smirnov, A.S. Usikov, A.V. Govorkov, B.V. Pushniy, Studies of the origin of the yellow luminescence band, the nature of nonradiative recombination and the origin of persistent photoconductivity in n-GaN films. Solid State Electron. **42**, 1959 (1998)

32. M.A. Reschikov, H. Morkoc, Luminescence properties of defects in GaN. J. Appl. Phys. **97**, 061301 (2005)

33. T. Ogino, M. Aoki, J. Appl. Phys. **19**, 2395 (1980)

34. A.Y. Polyakov, N.B. Smirnov, A.V. Govorkov, S.J. Pearton, J.M. Zavada, R.G. Wilson, Proton implantation effects on electrical and luminescent properties of p-GaN. J. Appl. Phys. **94**(5), 3069–3074 (2003)

35. A.Y. Polyakov, N.B. Smirnov, A.V. Govorkov, A.V. Markov, N.G. Kolin, D.I. Merkurisov, V.M. Boiko, K.D. Shcherbatchev, V.T. Bublik, M.I. Voronova, S.J. Pearton, A. Dabiran, A.V. Osinsky, Neutron irradiation effects in p-GaN. J. Vac. Sci. Technol. B. **24**(5), 2256–2261 (2006)

36. A.Y. Polyakov, N.B. Smirnov, A.V. Govorkov, J.M. Redwing, Deep traps in high resistivity AlGaN films. Solid State Electron. **42**, 831 (1998)

37. L.S. Berman, *Purity Control of Semiconductors by the Method of Capacitance Spectroscopy* (Electronic Integral Systems, St. Petersburg, 1995), p. 180.

38. P. Hacke, H. Nakayama, T. Detchprom, K. Hiramatsu, N. Sawaki, Deep levels in the upper band-gap region of lightly Mg-doped GaN. Appl. Phys. Lett. **68**, 1362 (1996)

39. G. Popovici, H. Morkoc, *Growth and doping of defects in III-nitrides, in GaN and Related Materials II*, ed. by S.J. Pearton (Gordon and Breach Science Publishers, the Netherlands, 1999) pp. 93–172 they produce blue band: my review on AlGaN

40. Z-Q. Fang, J.W. Hemsky, D.C. Look, M.P. Mack, Electron-irradiation-induced deep level in GaN. Appl. Phys. Lett. **72**(4), 448(1998)

41. B.R. Gossick, Disordered regions in semiconductors bombarded by fast neutrons. J. Appl. Phys. **30**(8), 1214 (1959)

42. A.Y. Polyakov, N.B. Smirnov, A.V. Govorkov, A.V. Markov, N.G. Kolin, D.I. Merkurisov, V.M. Boiko, K.D. Shcherbatchev, V.T. Bublik, M.I. Voronova, I.-H. Lee, C.R. Lee, Fermi level pinning in heavily neutron irradiated GaN. J. Appl. Phys. **100**(9), 093715–093719 (2006)

43. W. Walukiewicz, J. Vac. Sci. Technol. **B5**(4), 1062 (1987)

44. J. Tersoff, J. Vac. Sci. Technol. **B4**(4), 1066 (1986)

45. K. Kuriyama, T. Tokumasu, Jun Takahashi, H. Kondo, M. Okada, Lattice distortions and the transmuted-Ge related luminescence in neutron-transmutation-doped GaN. Appl. Phys. Lett. **80** (18), 3328–3330 (2002)
46. V.N. Brudnyi, A.V. Kosobutskiy, N.G. Kolin, Sov. Phys. Semicond. **43**(10), 1312 (2009) (in Russian)
47. V.V. Emtsev, V.Yu. Davydov, E.E. Haller, A.A. Klochikhin, V.V. Kozlovskii, G.A. Oganesyan, D.S. Poloskin, N.M. Shmidt, V.A. Vekshin, A.S. Usikov, Radiation-induced defects in n-type GaN and InN. Physica B 308–310, 58–61 (2001)
48. N.M. Shmidt, D.V. Davydov, V.V. Emtsev, I.L. Krestnikov, A.A. Lebedev, W.V. Lundin, D.S. Poloskin, A.V. Sakharov, A.S. Usikov, A.V. Osinsky, Effects of annealing on defects in as-grown and g-ray irradiated n-GaN layers. Phys. Stat. Sol. (b) **216**, 533–536 (1999)
49. B.H. Rose, C.E. Barnes, Proton damage effects on light emitting diodes. J. Appl. Phys. **53**, 1772–1780 (1982)
50. C.A. Usui, H. Sunakawa, A. Sakai, A.A. Yamaguchi, Jpn. J. Appl. Phys. **36**, L899 (1997).
51. In-Hwan Lee, A.Y. Polyakov, N.B. Smirnov, A.V. Govorkov, A.V. Markov, S.J. Pearton, Electrical and recombination properties and deep traps spectra in MOCVD ELOG GaN layers. Phys. Stat. Sol. (c), **3**(6), 2087–2090 (2006)
52. K. Fujito, K. Kiyomi, T. Mochizuki, H. Oota, H. Namita, S. Nagao, I. Fujimura, Phys. Stat. Sol. (a) **205**, 1056 (2008)
53. T. Wosinsky, J. Appl. Phys. **65**, 1566 (1988)
54. Y. Tokuda, Y. Matuoka, K. Yoshida, H. Ueda, O. Ishiguro, N. Soejima, T. Kachi, Evaluation of dislocation-related defects in GaN using deep-level transient spectroscopy. Phys. Stat. Sol. (c) **4**(7), 2568–2571 (2007)
55. A.Y. Polyakov, N.B. Smirnov, A.V. Govorkov, A.V. Markov, T.G. Yugova, C.R. Lee, I-H. Lee, N.G. Kolin, I.D. Merkurisov, V.M. Boiko, K.D. Scherbatchev, V.T. Bublik, M.I. Voronova, S.J. Pearton, Neutron irradiation effects on GaN films prepared by lateral overgrowth, Abstracts of the European Workshop on III-Nitride Semiconductor Materials, Heraklion, Crete, Greece, September 2006 (Heraklion University, Heraklion, 2006), pp. 91–92
56. A.Y. Polyakov, Structural and electronic properties of AlGaN, in *GaN and Related Materials II*, ed. by S. Pearton (Gordon and Breach, New York, 1999), pp. 173–233
57. L. Chernyak, A. Osinsky, G. Nootz, A. Schulye, J. Jasinski, M. Benamara, Z. Liliental-Weber, D.C. Look, R.J. Molnar, Appl. Phys. Lett. **77**, 2695 (2000)
58. E.B. Yakimov, P.S. Vergeles, A.Y. Polyakov, N.B. Smirnov, A.V. Govorkov, In-Hwan Lee, Cheul Ro Lee, S.J. Pearton, Spatial variations of doping and lifetime in epitaxial laterally overgrown GaN. Appl. Phys. Lett. **90**, 152114 (2007)
59. E.B. Yakimov, P.S. Vergeles, A.Y. Polyakov, N.B. Smirnov, A.V. Govorkov, In-Hwan Lee, Cheul Ro Lee, S.J. Pearton, Donor doping non-uniformity in undoped and Si doped n-GaN prepared by epitaxial lateral overgrowth. Appl. Phys. Lett. **94**(4), 042118-1–042118-3 (2008)
60. A.Y. Polyakov, N.B. Smirnov, A.V. Govorkov, A.V. Markov, E.B. Yakimov, P.S. Vergeles, n-Hwan Lee, Cheul Ro Lee, S.J. Pearton, Effects of changing the thickness of laterally overgrown n-GaN on deep traps spectra. J. Vac. Sci. Technol. (B). **26**(3), 990–994 (2008)
61. A.Y. Polyakov, N.B. Smirnov, A.V. Govorkov, A.V. Markov, E.B. Yakimov, P.S. Vergeles, N.G. Kolin, D.I. Merkurisov, V.M. Boiko, In-H. Lee, C.-R. Lee, S.J. Pearton, Neutron radiation effects in epitaxially laterally overgrown GaN films. J. Electron. Mater. **36**(10), 1320–1325 (2007)
62. S.J. Pearton, GaN device processing, in *GaN and Related Materials*, ed. by S.J. Pearton (Gordon and Breach, New York, 1999), pp. 475–540
63. J.W. Ager, R.E. Jones, D.M. Yamaguchi, K.M. Yu, W. Walukiewicz, S.W. Li, E.E. Haller, H. Lu, W.J. Schaff, Phys. Stat. Sol. (b). **244**, 1820 (2007)
64. A.Y. Polyakov, N.B. Smirnov, A.V. Govorkov, N.V. Pashkova, S.J. Pearton, J.M. Zavada, R.G. Wilson, Proton implantation effects on electrical and optical properties of undoped AlGaN with high Al mole fraction. J. Vac. Sci. Technol. **B21**(6), 2500–2505 (2004)

65. A.Y. Polyakov, N.B. Smirnov, A.V. Govorkov, A.V. Markov, N.G. Kolin, V.M. Boiko, D.I. Merkurisov, S.J. Pearton, Neutron irradiation effects in undoped n-AlGaN. J. Vac. Sci. Technol. **B24**(3), 1094–1097 (2006)

66. A.Y. Polyakov, N.B. Smirnov, A.V. Govorkov, K.H. Baik, S.J. Pearton, J.M. Zavada, Changes in electrical and optical properties of p-AlGaN due to proton implantation. J. Vac. Sci. Technol. **B22**(5), 2291–2294 (2004)

67. B.D. White, M. Bataiev, S.H. Gross, X. Hu, A. Karmarkar, D.M. Fleetwood, R.D. Schrimpf, W.J. Schaff, L.J. Brillson, Electrical, spectral and chemical properties of 1.8 MeV proton irradiated AlGaN/GaN HEMT structures. IEEE Trans. Nucl. Sci. **50**(6) 1934 (2003)

68. X. Hu, A. Karmarkar, B. Jun, D.M. Fleetwood, R.D. Schrimpf, R.D. Geil, R.A. Weller, B.D. White, M. Bataiev, L.J. Brillson, U.K. Mishra, Proton irradiation effects in AlGaN/AlN/GaN high electron mobility transistors, IEEE Trans. Nucl. Sci. **50**(6) 1791 (2003)

69. M.S. Shur, M.A. Khan, in *GaN and Related Materials II*, ed. by S.J. Pearton (Gordon and Breach Science, New York, 1999) pp.47–92

70. H. Miyamoto, Phys. Stat. Sol. (c). **3**, 2254 (2006)

71. F. Gaudreau, P. Fournier, C. Carlone, S.M. Khanna, H. Tang, J. Webb, A. Houdayer, Transport properties of proton irradiated gallium nitride-based two-dimensional electron-gas system. IEEE Trans. Nucl. Sci. **49**(6), 2702 (2002)

72. A.Y. Polyakov, N.B. Smirnov, A.V. Govorkov, A.V. Markov, S.J. Pearton, N.G. Kolin, D.I. Merkurisov, V.M. Boiko, Neutron irradiation effects on electrical properties and deep level spectra in undoped n-AlGaN/GaN heterostructures. J. Appl. Phys. **98**, 033529-1–033529-6 (2005)

73. A.Y. Polyakov, N.B. Smirnov, A.V. Govorkov, A.V. Markov, A.M. Dabiran, A.M. Wowchak, B. Cui, A.V. Osinsky, P.P. Chow, N.G. Kolin, V.M. Boiko, D.I. Merkurisov, Electron irradiation of AlGaN/GaN and AlN/GaN heterojunctions. Appl. Phys. Lett. **93**, 152101 (2008)

74. A.Y. Polyakov, N.B. Smirnov, A.V. Govorkov, A.V. Markov, A.M. Dabiran, A. M. Wowchak, B. Cui, A.V. Osinski, P.P. Chow, S.J. Pearton, Electrical properties and deep traps spectra in AlN/GaN and AlGaN/GaN heterojunctions, J. Appl. Phys. **104**, 053702 (2008)

75. X. Hu, B.X. Choi, H.J. Barnaby, D.M. Fleetwood, R.D. Schrimpf, S. Lee, S. Shojah-Ardalan, R. Wilkins, U. Mishra, R.W. Dettmer, The energy dependence of proton induced degradation in AlGaN/GaN HEMTs IEEE Trans. Nucl. Sci. **51**(2), 293 (2004)

76. A. Rashmi, S. Kranti, S. Haldar, P.S. Gupta, Impact of strain relaxation of AlGaN layer on 2-DEG sheet charge density and current-voltage characteristics of lattice mismatched AlGaN/GaN HEMTs. Microelectron. J. **33**(3), 205 (2002)

77. A.Y. Polyakov, N.B. Smirnov, A.V. Govorkov, N.G. Kolin, V.M. Boiko, D.I. Merkurisov, S.J. Pearton, Effects of electron irradiation on GaN-based transistor structures, in the extended abstracts of the 6th Russian Conference "Nitrides of gallium, indium and aluminum: structures and devices, S-Petersburg, Ioffe Physico-Technical Institute, 2008

78. B. Luo, J.W. Johnson, F. Ren, K.K. Alums, C.R. Abernathy, S.J. Pearton, A.M. Dabiran, A.M. Wowchak, C.J. Polley, P.P. Chow, D. Shoenfeld, A.G. Baca, Influence of ^{60}Co γ-rays on dc performance of AlGaN/GaN high electron mobility transistors. Appl. Phys. Lett. **80**(4), 604 (2002)

79. A.Y. Polyakov, N.B. Smirnov, A.V. Govorkov, In-H. Lee, J.H. Baek, N.G. Kolin, V.M. Boiko, D.I. Merkurisov, S.J. Pearton, Electron irradiation effects in GaN/InGaN multiple quantum well structures. J. Electrochem. Soc. **155**(1), H31–H35 (2008)

80. A.P. Karmarkar, B. Jun, D.M. Fleetwood, D.D. Schrimpf, R.A. Weller, B.D. White, L.S. Brillson, U.K. Mishra, Proton irradiation effects on GaN-based high electron mobility transistors with Si-doped AlGaN and thick GaN cap layers. IEEE Trans. Nucl. Sci. **51**(6), 3801 (2004)

81. S.J. Cai, J.S. Tang, R. Li, J. Wei, L. Wong, J.L. Chen, K.L. Wang, M. Chen, Y.F. Zhao, R.D. Schrimpf, J.C. Keay, K.F. Galloway, Annealing behavior of proton irradiated AlGaN/GaN high electron mobility transistors grown by MBE. IEEE Trans. Electron. Dev. **47**, 304 (2000)

82. Q. Wang, H.Q. Xu, P. Omling, C. Jang, G. Malmqvist, Transistor properties of proton-irradiated GaAs/AlGaAs two-dimensional electron gas structures. Nucl. Instrum. Methods Phys. Res. B **160**, 33 (2000)

83. B. Luo, J.W. Johnson, F. Ren, K.K. Allums, C.R. Abernathy, S.J. Pearton, R. Dwidevi, T.N. Fogarty, R. Wilkins, A.M. Dabiran, A.M. Wowchak, C.J. Polley, P.P. Chow, A.G. Baca, DC and RF performance of proton-irradiated AlGaN/GaN high electron mobility transistors, Appl. Phys. Lett. **79**(14), 2196 (2001)

84. B. Luo, J. Kim, F. Ren, J.K. Gillespie, R.C. Fitch, J. Sewell, R. Dettmer, G.D. Via, A. Crespo, T.J. Jenkins, B.P. Gila, A.H. Onstine, K.K. Allums, C.R. Abernathy, S.J. Pearton, R. Dwidevi, T.N. Fogarty, R. Wilkins, Electrical characteristics of proton-irradiated Sc2O3 passivated AlGaN/GaN high electron mobility transistors. Appl. Phys. Lett. **82**(9), 1428 (2003)

85. B. Luo, F. Ren, K.K. Allums, B.P. Gila, A.H. Onstine, C.R. Abernathy, S.J. Pearton, R. Dwivedi, T.N. Fogarty, R. Wilkins, R.C. Fitch, J.K. Gillespie, T.J. Jenkins, R. Dettmer, J. Sewell, G.D. Via, A. Crespo, A.G. Baca, R.J. Shul, Proton irradiation of MgO or Sc2O3 passivated high electron mobility transistors. Solid-State Electron. **47**, 1015 (2003)

86. B. Luo, J.M. Johnson, F. Ren, K.K. Allums, C.R. Abernathy, S.J. Pearton, A.M. Dabiran, A.M. Wowchak, C.J. Polley, P.P. Chow, D. Schoenfeld, A.G. Baca, Influence of 60Co g-rays on dc performance of AlGaN/GaN high electron mobility transistors. Appl. Phys. Lett. **80**(4), 604 (2002)

87. V. Tilak, B. Green, V. Kaper, H. Kim, T. Prunty, J. Smart, J. Shealy, L.F. Eastman, IEEE Electron Dev. Lett. **22**(11), 504 (2001)

88. I. Daumiller, D. Theron, C. Gaquiere, A. Vescan, R. Dietrich, A. Wieszt, H. Leier, R. Vetury, U. K. Mishra, I. P. Smorchkova, S. Keller, C. Nguyen, E. Kohn, IEEE Electron. Dev. Lett. **22**, 62 (2001)

89. U.K. Mishra, P. Parikh, Y.-F. Wu, Proc.IEEE **90**, 1022 (2002).

90. C. Lee, H. Wang, J. Yang, L. Witkowski, M. Muir, M. Asif Khan, P. Saunier, Electron. Lett. **38**, 924 (2002).

91. O. Mitrofanov, M. Manfra, N.G. Weimann, Appl.Phys.Lett. **82**, 4361 (2003).

92. S.C. Binari, K. Ikossi, J.A. Roussos, W. Kruppa, D. Park, H.B. Dietrich, D.D. Koleske, A.E. Wickenden, and R.L. Henry, IEEE Trans. Electron. Dev. **48**, 465 (2001)

93. B.M. Green, K.K. Chu, E.M. Chumbes, J.A. Smart, J.R. Shealy, L.F. Eastman, IEEE Electron. Dev. Lett. **21**, 268 (2000)

94. L.F. Eastman, V. Tilak, J. Smart, B.M. Green, E.M. Chumbes, R. Dimitrov, H. Kim; O.S. Ambacher, N. Weimann, T. Prunty, M. Murphy, W.J. Schaff, J.R. Shealy, IEEE Trans. Electron Dev. **48**, 479 (2001)

95. X. Hu, A. Koudymov, G. Simin, J. Yang, M. Asif Khan, A. Tarakji, M.S. Shur, R. Gaska, Appl. Phys. Lett. **79**, 2832 (2001).

96. B. Luo, J.W. Johnson, J. Kim, R.M. Mehandru, F. Ren, B.P. Gila, A.H. Onstine, C.R. Abernathy, S.J. Pearton, A.G. Baca, R.D. Briggs, R.J. Shul, C. Monier, J. Han, Appl. Phys. Lett. **80**, 1661 (2002)

97. A.Y. Polyakov, N.B. Smirnov, A.V. Govorkov, V.N. Danilin, T.A. Zhukova, B. Luo, F. Ren, B.P. Gila, A.H. Onstine, C.R. Abernathy, S.J. Pearton, Deep traps in unpassivated and Sc2O3 passivated AlGaN/GaN high electron mobility transistors. Appl. Phys. Lett. **83**(13), 2608–2610 (2003)

98. A.Y. Polyakov, N.B. Smirnov, B.P. Gila, M. Hlad, A.P. Gerger, C.R. Abernathy, S.J. Pearton, Studies of interface states in Sc2O3/GaN, MgO/GaN and MgScO/GaN structures. J. Electrochem. Soc. **154**(2), H115–H118 (2007)

99. S.M. Khanna, D. Estan, A. Houdayer, H. C. Liu, R. Dudek, Proton radiation damage at low temperature in GaAs and GaN light emitting diodes. IEEE Trans. Nucl. Sci. **51**(6), 3585–3594 (2004)

100. F Gaudraeau, C. Cardone, A. Noudayer, S.M. Khamna, Spectral properties of proton irradiated GaN blue diodes. IEEE Trans. Nucl. Sci. **48**, 1778–1784 (2001)

101. M. Osinsky, P. Perlin, H. Schone, A.H. Paxtone, E.W. Taylor, Effects of proton irradiation on AlGaN/InGaN/GaN green light emitting diodes. Electron. Lett. **33**, 1252–1254 (1997)

102. J. Grant, W. Cunningham, A. Blue, V. O'Shea, J. Vaitkus, E. Gaubas, M. Rahman, Nucl. Instr. Meth. Phys. Res. A. **546**, 213 (2005)

103. J. Grant, R. Bates, W. Cunningham, A. Blue, J. Melone, F. McEwan, J. Vaitkus, E. Gaubas, V. O'Shea, Nucl. Instr. Meth. Phys. Res. A **576**, 60 (2007)

104. A.Y. Polyakov, N.B. Smirnov, A.V. Govorkov, A.V. Markov, E.A. Kozhukhova, I.M. Gazizov, N.G. Kolin, D.I. Merkurisov, V.M. Boiko, A.V. Korulin, V.M. Zalyetin, S.J. Pearton, I.-H. Lee, A.M. Dabiran, P.P. Chow, Alpha particle detection with GaN Schottky diodes. J. Appl. Phys. **106**(10), 103708 (2009)

105. R.G. Wilson, S.J. Pearton, C.R. Abernathy, J.M. Zavada, Thermal stability of implanted dopants in GaN. Appl. Phys. Lett. **66**(17), 2238 (1995)

106. S.J. Pearton, C.B. Vartuli, J.C. Zolper, C. Yuan, R.A. Stall, Appl. Phys. Lett. **67**, 1435 (1995)

107. C.R. Abernathy, J.D. Mackenzie, S.J. Pearton, W.S. Hobson, Appl. Phys. Lett. **66**, 1969 (1995)

108. S.O. Kucheyev, M. Toth, M.R. Phillips, J.S. Williams, C. Jagadish, G. Li, Appl. Phys. Lett. **78**, 34 (2001)

109. A.Y. Polyakov, M. Shin, R.G. Wilson, M. Skowronski, D.W. Greve, S.J. Pearton, Ion implantation of Si, Mg and C into AlGaN. Solid-State Electron. **41**, 703 (1997)

110. K.A. Jones, C. Nguyen, M.A. Derenge, I. Batyrev, T.S. Zheleva, Electrical activation processes in Si and Mg implanted GaN"in the abstracts of the 8th International Conference on Nitride Semiconductors, October 18–23, 2009, Jeju Island, Korea, paper ThP37, p.1160

111. B.G. Svensson, A. Hallien, J. Wong-Leung, M.S. Janson, M.K. Linnarsson, A.Yu. Kuznetsov, G. Alfieri, U. Grossner, E.V. Monakhov, H.K. Nielsen, C. Jagadish, J. Grillenberger, Ion implantation processing and related effects in SiC. Mat. Sci. Forum. 527–529, 781–786 (2006)

112. S.J. Pearton, C.R. Abernathy, P.W. Wisk, W.S. Hobson, F. Ren, Appl. Phys. Lett. **63**, 2238 (1993)

113. J.C. Zolper, S.J. Pearton, C.R. Abernathy, C.B. Vartuli, Appl. Phys. Lett. **66**, 3042 (1995)

114. S.C. Binari, L.B. Rowland, W. Kruppa, H.B. Dietric, K. Doverspike, D.K. Gaskill, Electron. Lett. **30**, 1248 (1994)

115. S.M. Hearne, D.N. Jaimeson, C. Yang, S. Prawer, J. Salzman, O. Katz, Electrical characteristics of proton irradiated AlGaN devices. Nucl. Instrum. Methods. Phys. Res. B **190**, 873–877 (2002)

116. M. Lambsdorf, J. Kohl, J. Rosenzweig, A. Axmann, J. Schneider, Appl. Phys. Lett. **58**, 1881 (1991)

117. V.M. Rao, W.-P. Hong, C. Caneau, G.K. Chang, N. Papanicolau, H.B. Dietrich, J. Appl. Phys. **70**, 3943 (1991)

118. R. Wang, Y. Cai, W.C.W. Tang, K. M. Lau, K. J. Chen, Intergration of enhancement and depletion mode AlGaN/GaN MIS-HFET by fluoride-based plasma treatment. Phys. Stat. Sol. (a) **204**(6) 2024–2027 (2007)

Chapter 10
Recent Advances in GaN Nanowires: Surface-Controlled Conduction and Sensing Applications

Ruei-San Chen, Abhijit Ganguly, Li-Chyong Chen, and Kuei-Hsien Chen

Abstract Recent studies reveal an interesting surface-controlled conduction in the wide-bandgap single-crystalline GaN nanowires (NWs). The surface depletion and built-in electric field inherent to the NWs have led to the high-gain (long-lifetime) photoconduction and size-dependent transport properties. Efficient and selective sensing for ultraviolet light, gaseous and biological molecules based on the novel surface nature of nanostructure is presented.

10.1 Introduction

Gallium nitride (GaN), a wide direct bandgap semiconductor, has attracted intense attention on the fundamental researches and industrial applications due to its unique optoelectronic properties and excellent thermal and chemical stabilities [1, 2]. The capability to integrate with other binary III-nitrides as ternary compounds, such as InGaN and AlGaN, has substantially broadened its applicability for the electronic and optoelectronic devices. In the last decade, researchers have partially switched their focus to the one-dimensional (1D) nanostructures of GaN [3,4]. Semiconductor nanowires (NWs) have exhibited numerous novel and interesting properties [5–7].

R.-S. Chen
Graduate Institute of Applied Science and Technology, National Taiwan University of Science and Technology, Taipei 106, Taiwan

A. Ganguly · L.-C. Chen (✉)
Center for Condensed Matter Sciences, National Taiwan University, Taipei 106, Taiwan
e-mail: chenlc@ntu.edu.tw

K.-H. Chen (✉)
Center for Condensed Matter Sciences, National Taiwan University, Taipei 106, Taiwan

Institute of Atomic and Molecular Sciences, Academia Sinica, Taipei 106, Taiwan,
e-mail: chenkh@pub.iams.sinica.edu.tw

S. Pearton (ed.), *GaN and ZnO-based Materials and Devices*,
Springer Series in Materials Science 156, DOI 10.1007/978-3-642-23521-4_10,
© Springer-Verlag Berlin Heidelberg 2012

In particular, they could exhibit a semidiscrete density of states and at the same time maintain a continuous transport path for carriers. The development of imperative devices, such as rectifiers [8, 9], high electron mobility transistors [10], light-emitting diodes [11, 12], laser diodes [13], photodetectors [14–16], biosensors [17, 18], and solar cells [19], has benefited from the unique 1D structures of GaN.

Distinct from the traditional planer films, NWs with high surface-to-volume ratio and uniaxial charge transport path would be a better choice for advanced (nanoscale) applications. To exploit the full potential of GaN NWs, a thorough understanding of the transport properties and their correlations to the confined structure is essential. Different from the earlier understanding of bulk-dominant electrical properties in the materials, the GaN NW has been found to possess remarkable surface effects on its electronic transport in the dark [20–25] and in the light [24–29]. The surface property has been discriminated from the bulk measurement of the nanostructure. In this chapter, we have reviewed the advances in the knowledge of transport properties of GaN NWs. The latest understandings to the surface-controlled photoconductivity (PC), dark conductivity, and their underneath mechanisms are elucidated. The applications for the light, gaseous and biological molecules sensing upon the surface character are also presented.

10.2 Surface-Controlled Transport

10.2.1 Surface Photoconduction

10.2.1.1 High Responsivity of GaN Nanobridge Photodetectors

A design of GaN bridging nanowire (or nanobridge) devices, which avoid contact resistance and simultaneously integrate an ensemble of 1D nanostructures on-chip, has been demonstrated for ultraviolet (UV) light detection [16]. GaN NWs were laterally grown as bridges suspended over the trench between two n^+-doped GaN electrode posts. Contact resistance can be minimized via the homojunction between NW and heavily doped electrode post. As shown in Fig. 10.1a, via the photolithography, lift-off, and reactive ion etching (RIE) processing, the arrays of Ni/n^+-GaN electrode posts were patterned on the highly n^+-doped c-plane GaN on sapphire substrate. Subsequently, employing Au catalyst, NWs were grown by thermal chemical vapor deposition (CVD) technique.

Figure 10.1b shows a typical photocurrent (i_p) spectrum of the NW device in the range of photon energy (E) from 2.0 to 4.0 eV. The i_p revealed a steep rise at the energy position close to the bandgap of GaN at 3.4 eV. Below the threshold energy, no detectable photoresponse was observed, indicating the performance of visible–blind UV detection.

It is noted that under the low-intensity illumination ($I < 10\,\mathrm{W\,m^{-2}}$), the GaN nanobridge device with 1,500 suspending NWs could generate extraordinarily

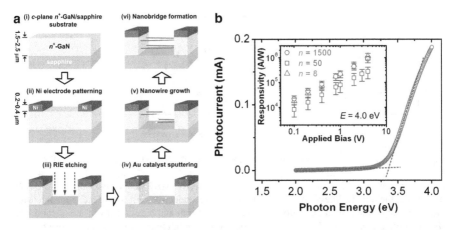

Fig. 10.1 (a) The schematic diagram of the fabrication process of the GaN nanobridge device. (b) A typical photocurrent spectrum of the bridging NWs device with wire number of 1,500 at a bias of 1.0 V. Inset: The responsivity versus applied bias for the bridging NWs devices with different wire numbers n under 4.0 eV excitation. [Reprinted with permission from [16], Copyright @ The Wiley-VCH Verlag GmbH & Co. KGaA 2008]

high i_p over 0.1 mA (for an effective detection area of about $700 \pm 500\,\mu m^2$) compared to the millimeter-sized photosensitive surface area in traditional thin film photodetectors [30–34]. Generation of such high i_p inside a nanosized transport path implies that the NW photoconductors have remarkably high photocurrent generation efficiency. To comprehend such distinctive phenomena, a physical quantity namely the responsivity (R), the i_p generated per unit power of light, P, has been employed [16, 35].

$$R = \frac{i_p}{P}. \tag{10.1}$$

As shown in the inset of Fig. 10.1b, the responsivity versus bias plots, overall the R values and $R–V$ slopes for the three devices with different wire numbers (n) were nearly the same irrespective of n. While changing the bias from 0.1 to 4.0 V, the R exhibited nearly two orders of magnitude increase from 1.6×10^4 to 7.1×10^5 A W^{-1}. It is worth noting that the maximal R, for the NWs detectors, could reach to the value above 10^6 A W^{-1}, which is nearly three to six orders of magnitude higher than the corresponding values of $1 - 3,000$ A W^{-1} reported so far for the photoconductive thin film detectors of GaN [30–34, 36–39]. To investigate the underneath mechanism of the superior photocurrent responsivity of NWs, there are a few PC parameters required of being defined and elaborated in the next section.

10.2.1.2   Photoconductive Gain

The photoconduction involves both optical absorption and carrier transport, depends on the combined efficiency of these two processes, hence R can be expressed as

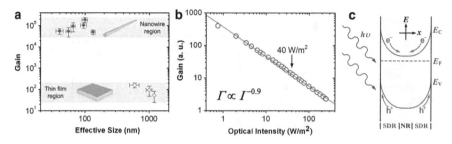

Fig. 10.2 (**a**) The photocurrent gain as a function of effective size for the GaN NWs (*blue solid circles*) and thin films (*open purple diamonds*) at $400\,\mathrm{V\,cm^{-1}}$ applied field and $4.0\,\mathrm{eV}$ excitation with $10 - 12\,\mathrm{W\,m^{-2}}$ power density. The "effective size" defines the values of the NW diameter or the film thickness. (**b**) The normalized photoconductive gain versus excitation intensity measurement of the NW at excitation wavelength of 325 nm and bias of $1.0\,\mathrm{V}$. The red solid line shows a fitting to the power law, $\Gamma \propto I^{-k}$, where $k = 0.9$, of the experimental data (*open circle*). [Reprinted with permission from [28], Copyright @ The American Institute of Physics 2007] (**c**) A schematic of the electron–hole (e–h) spatial separation mechanism induced by surface band bending (SBB) in a GaN NW under photoexcitation. *SDR* surface depletion region; *NR* neutral region

follows: [40]

$$R = \frac{e}{E}\eta\Gamma, \tag{10.2}$$

where e is the electron charge and E is the photon energy, η is the quantum efficiency of photon absorption [41, 42], and Γ is the photoconductive gain. The η values were estimated to be 26-59% for diameter $(d) = 40 - 135\,\mathrm{nm}$, at a photon energy of $4.0\,\mathrm{eV}$ [28]. From the values of η and R, the Γ, a factor determining the efficiency of electron transport and carrier collection during the photoconduction process, could been deduced from (10.2). Figure 10.2a shows the Γ values of the different-sized GaN NWs and of two thin film data points for comparison (Fig. 10.2a) [28]. From the Γ distribution, it is quite clear that the Γ values at $5.0 \times 10^4 - 1.9 \times 10^5$ of overall NWs with d spanning from 40 to 135 nm are three orders of magnitude higher than that at $5.2 \times 10^1 - 1.6 \times 10^2$ of the film photodetectors.

The current gain is related to the number of carriers circulating through the photoconductor per absorbed photon and per unit time, thus also written as:

$$\Gamma = \frac{\tau}{t} = \frac{F}{l}\tau\mu, \tag{10.3}$$

where τ is the excess carrier lifetime, t is the transit time of the carrier between electrodes, F is the applied electric field, and μ is the mobility [35, 40]. Higher Γ values for NWs imply longer lifetimes (i.e., higher τ) or higher μ than those of their thin film counterparts. As reported, GaN NWs possess moderate field-effect mobility values of $1 - 650\,\mathrm{cm^2\,V^{-1}\,s^{-1}}$ [20, 43–46]. Taking these values, the GaN NWs would have ultralong τ of up to 400 ms, which is three to seven orders of magnitude higher than the values derived from the thin film photoconductors ($\tau =$ 50 ns $-$ 360 μs) [31, 36, 38].

10.2.1.3 Spatially Separated Transport Mechanism

As GaN NWs without intentionally doping always reveals n-type conduction [44–46], the Γ value is predominantly determined by the lifetime of minority hole. Presence of hole traps, which prolong the lifetime of excess electron and contribute high gain, has been observed in GaN films [39, 47, 48]. However, according to Munoz's and Garrido's reports, strong surface band bending (SBB) in this wide-bandgap GaN would localize the transports of excess carriers and result in a similar long lifetime effect [38, 49].

Accordingly, the high surface-to-volume ratio of the 1D nanostructure could make the behavior of carrier spatial separation even more significant and dominate the transport in the NW. However, the influences of hole traps are difficult to be neglected from GaN and other material systems whatever the form is film or NW. A surface trapping effect making very high gain has also been reported for the ZnO NWs [50]. Besides, for the GaN bulks, Stevens et al. [47] and Binet et al. [39] have indicated that the contribution of trap can be identified by the intensity (I)-dependent behavior of R (and also Γ). Under low level excitation below a critical intensity ($I < 40\,\mathrm{W\,m^{-2}}$), the hole traps govern the Γ and make it independent of I, i.e., Γ and $\tau =$ const. Once exceeding the critical intensity, hole filling effect will change the recombination behavior and shorten the τ, making Γ follows an inverse power law, i.e., $\Gamma \propto I^{-k}$, where $k = 0.5$.

Based on this, Fig. 10.2b shows the Γ versus I relation of the bridging GaN NWs ($d = 20 - 60\,\mathrm{nm}$) to clarify the trap contribution, showing a clear power law ($\Gamma \propto I^{-0.9}$) in the intensity range over two decades from 0.75 to 250 W m^{-2} [28]. It is noted that the lowest intensity in this investigation was much weaker than the 40 W m^{-2} but no gain saturation was observed. This result strongly suggests a non-trap-dominant mechanism in the NWs. Moreover, according to Munoz's and Garrido's simulations, the change of Γ value induced by the modulation of surface space charge region ($k = 0.8 - 0.9$) [38, 49] would be even more sensitive to the photoexcitation power as compared to that induced by the intrinsic recombination mechanism ($k = 0.5$) [39, 47]. The higher k value at 0.9 for NWs further proved the high gain transport being dominated by the surface.

In addition to the power-dependent studies, the presence of internal electric field induced by Fermi level pinning at the GaN NW surface has also been reported by Cavallini et al. [27] Therein, the band tail absorption induced by the Franz–Keldysh effect was observed by the surface photovoltage spectroscopy and spectral PC measurements. Their results are in good agreement with the conclusion of surface-controlled photoconduction in the power-dependent studies. A schematic of the electron–hole (e–h) spatial separation under photoexcitation in a NW is depicted in Fig. 10.2c.

10.2.2   Size-Dependent Transport Properties

10.2.2.1   Photoconductivity and Dark Conductivity

Recently, size effects on the transport properties in the GaN NWs have been reported by a few groups [20–26, 28, 45]. Calarco et al. [26] have demonstrated the first observation on size dependence of i_p in the GaN NWs grown by molecular beam epitaxy (MBE). They observed a sharp decrease in i_p once the NW-diameter size below a critical value (d_{crt}) of 90 ± 10 nm (Fig. 10.3a). The model of e–h spatial separation was proposed. Since the τ of GaN NW is controlled by SBB, i.e., longer τ indicates stronger SBB, the size-dependent PC could be explained by the real-space band diagrams as shown in Fig. 10.3b. As the d is smaller than d_{crt}, the

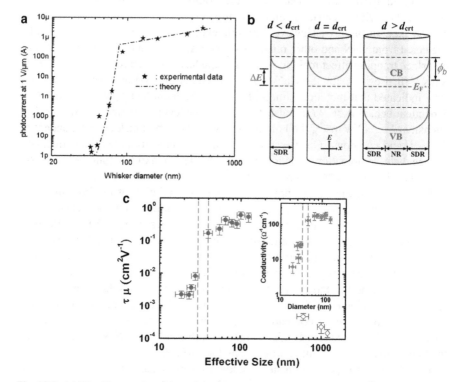

Fig. 10.3 (a) The diameter-dependent photocurrent measured under $15\,\mathrm{W\,cm^{-2}}$ UV light intensity of the GaN NWs grown by MBE. [Reprinted with permission from [26], Copyright @ The American Chemical Society 2005] (b) The schematic of real-space energy band diagrams for the NWs with different diameters d, where the ϕ_b is the maximal barrier height of SBB, ΔE is the energy difference between Fermi level and conduction band minimum. (c) The calculated product of carrier lifetime and mobility ($\tau\mu$) of individual NW (*solid blue circles*) and thin films (*open purple diamonds*). Inset: The dark conductivity as a function of diameter for NWs in the diameter range of 20–126 nm. [Reprinted with permission from [25], Copyright @ The American Institute of Physics 2009]

neutral region (NR) disappears and surface barrier height (ϕ_b) decreases, making the substantial increase of recombination rate and shortening of τ. This explains the orders of magnitude decrease of i_p at $d < d_{crt}$.

Subsequently, a similar size effect has also been observed by Chen et al. [25] from the CVD-grown GaN NWs, Fig. 10.3c. In their paper, the $\tau\mu$ product has been estimated of reflecting the exact change of lifetime due o the dimension variation. Two main regimes were observed from the $\tau\mu$ statistics. For the regime of bigger NWs ($d = 40 - 126$ nm), the $\tau\mu$ values showed only a slow decrease and overall kept at the level of $5 \pm 3 \times 10^{-1}$ cm^2 V^{-1}. While the d went below the d_{crt} of around 35 ± 5 nm, the $\tau\mu$ revealed a steep drop for nearly two orders of magnitude to $\sim 2 \times 10^{-3}$ cm^2 V^{-1}. As the μ value has been proved to be nearly constant compared to the change in τ [43,45]. This indicates the τ-dominant size effect, which further verifies the surface-controlled photoconduction in the GaN NWs. Additionally, a near three orders of magnitude difference in $\tau\mu$ values can be observed between the NWs (above d_{crt}) and planer films (data from [25–27]). This also implies the higher photocarrier collection efficiency of the NWs compared to their thin film counterparts.

In addition to PC, the size-dependent dark conductivity (σ) was also observed showing a rapid drop from a constant level of 150 ± 50 to 6 ± 2 Ω^{-1} cm^{-1} as the d decreases to the d_{crt} of 30-40 nm (inset of Fig. 10.3c) [25]. The position range of d_{crt} is rather coincident with the one observed from PC in Fig. 10.3c. The steep drop of σ is found to be mostly induced by the sharp decrease in the carrier concentration (n_e) for NWs with $d < d_{crt}$. As shown in Fig. 10.3b, above d_{crt}, the n-type NWs have big enough volume to supply sufficient electrons to surface states; the NR in the core exists. Once the d goes below d_{crt}, the smaller volume of NWs would not fill up the surface states, leading to less space charge and lower ϕ_b, hence the domination of surface depletion region (SDR) over NR. Since near the surface the n_e would decrease, due to the higher degree of depletion, NWs with size smaller than d_{crt} would become more insulating electrically. The lower ϕ_b would lead to the increase in ΔE (the energy difference between Fermi level and conduction band minimum) under thermal equilibrium, while the location of Fermi level remains unchanged. Such raise in ΔE with decreasing NW diameter could account for the sharp drop in n_e or conductivity.

Interestingly, there is a certain difference in the d_{crt} values for the GaN NWs grown by CVD (CVD-NW: $d_{crt} = 30 - 40$ nm) and MBE (MBE-NW: $d_{crt} = 80 - 100$ nm). The value of d_{crt} is actually determined by the maximal depletion width (w) at surface, i.e., $d_{crt} = 2w$, where the w is proportional to $n_e^{-1/2}$. According to the literatures [22, 25, 43, 45], the CVD-NWs possess higher n_e (5×10^{18}–5×10^{19} cm^{-3}) which is over one order of magnitude higher than the MBE-NWs ($n_e \sim 6 \times 10^{17}$ cm^{-3}) [26]. A rough estimation has exhibited narrower w (around 6-18 nm) of the CVD-NWs compared to the MBE-NWs ($w = 50 - 100$ nm), under similar assumption of ϕ_b (~ 0.6 eV) [26]. The theoretical values of d_{crt} for the CVD-NWs were found to be ranging from 12 to 36 nm. The upper bound at 36 nm agrees well with the mean value of the experimental observation of d_{crt} at 35 ± 5 nm. This result could properly clarify the smaller d_{crt} in CVD-NWs compared to MBE-NWs.

Furthermore, besides the n_e, the difference of long-axial orientations of GaN NWs could also be related to the d_{crt}. As the c-plane GaN is a polar surface which is different from the nonpolar m- and a-planes, the CVD-NWs with m-axial growth orientation will expose c-plane side walls, thus exhibit strong surface polarity. On the contrary, the c-axial NWs grown by MBE have no surface polarity on their side walls. The polar surface of GaN has been reported of revealing higher surface activity to the adsorption of ambient molecules, such as oxygen [51, 52]. Although so far, no evidence shows the direct correlation between the stronger polarity with higher SBB, the "polarity" of GaN NW is suspected of contributing partially to the difference of d_{crt} due to the different surface state density.

10.2.2.2   Electron Mobility

The electron mobility, an important parameter that determines the charge transport efficiency and the degree of scattering, has been investigated in the different-sized GaN NWs. A statistic result of μ value as a function of diameter, which is first done by Stern et al., has indicated no clear size dependence from plenty of NW samples [45]. The μ values are in a moderate level of $1 - 10\,\mathrm{cm^2V^{-1}\,s^{-1}}$ and the n_e are in the range of $2.5 \times 10^{19} - 2.3 \times 10^{20}\,\mathrm{cm^{-3}}$ for the wires sized between 20 and 200 nm. After that, Motayed et al. reported an observable trend of d dependence of μ from the NWs with relatively lower background concentration of $2 \times 10^{18}\,\mathrm{cm^{-3}}$ [20]. A near one order of magnitude decrease of μ from 319 to $40\,\mathrm{cm^2V^{-1}\,s^{-1}}$ is observed while the NW shrinks about half the diameter from 200 to 95 nm. The significant size effect on μ is mainly attributed to the increased scattering by surface depletion and surface defects in the thinner NWs. The presence of grain boundary that might enhance the scattering is also proposed in the intermediate diameter ($d = 120 - 180\,\mathrm{nm}$) NWs.

10.2.2.3   Thermal Activation Energy

Dark conductivity of materials comprises two intrinsic transport parameters including mobility and concentration of carriers. Earlier studies have revealed that the NW surface plays a crucial role in electron scattering that decides the μ value [20]. The surface effects on the n_e and its thermal activation mechanism, which determines the effective number of free carriers for electrical conduction in the material per unit volume, are also important issues. Recently, the size-dependent thermal activation energy (E_a) in the intrinsically n-type GaN NWs has been reported by Yoon et al. [23] and Chen et al. [25] The former group [23] has observed a continuous increase of E_a from 110 to 150 meV while the d decreases from 20 to 10 nm in the bare NWs. A similar trend but less dependence on size has also been observed in the NWs coated with the dielectric of SiO_2. The results agreed with their numerical calculation, suggesting the dielectric confinement effect. The surrounding dielectric could change the induced surface charge density and thus modify the thermal E_a in the GaN NWs [23].

In addition to the Coulomb interaction in the size-confined nanostructures, the trapping mechanism could also result in the E_a increase. Different from the Yoon and coworkers' study on smaller NWs (d mostly below 20 nn), Chen et al. [25] have observed a similar result from the NWs with bigger sizes (d ranging between 27 and 65 nm). The thinner wire ($d = 27$ nm) exhibited a stronger thermal quenching of conductivity compared to the thicker ones with d at 55 and 65 nm. The E_a value significantly increased from the level below 1–56 meV. The result has been attributed to the surface-controlled activation mechanism in the fully depleted NWs. While the sizes above the d_{crt}, the volume offers sufficient electrons that could fill up surface states and the NR in the wire core still remains. In this case, most of the thermal carriers transport along the NR core instead of the SDR. Thus, the activation process is decided by the shallow donor levels/bands with small E_a in the bulk. Once the d goes below the d_{crt}, as shown in Fig. 10.3b, the insufficient electrons from the very limited NW volume no more fill up the surface states, leaving electrons all captured on the surface and NW fully depleted. As now the surface electrons have to overcome the trapping energy to become conduction electrons, the thermal activation mechanism is no longer decided by the shallow donors but dominated by the deep-level surface states with E_a at 56 meV.

10.2.3   Persistent Photoconductivity

In a typical photocurrent response curve of the GaN NWs upon a single-shock UV excitation, as observed in Fig. 10.4a, the transient PC generally shows much shorter rise time (τ_r) in comparison with the very long decay time (τ_d). Theoretically, carrier lifetime can be obtained by measuring τ_r or τ_d, where for the intrinsic semiconductors $\tau_r = \tau_d$ [48]. For GaN system, the long τ_d, which outnumbers τ_r, was frequently observed and known as the persistent photoconductivity (PPC) due to hole trapping by certain metastable defects [53, 54]. Therefore, the PPC phenomenon has been attributed to the traditional bulk property in the past. However, a recent observation of the surface-controlled PPC from the GaN NWs has revealed a significant dependence of PPC on the ambience [29]. A much longer decay time was observed in vacuum ($\tau_d = 46$ ks) than those in air ($\tau_d = 4.8$ ks) and pure oxygen ($\tau_d = 1.1$ ks), as shown in the inset of Fig. 10.4a. The highly environment-sensitive behavior clearly implies that the PPC is governed by surface condition rather than trapping mechanism, generally observed in the bulk.

Based on the picture of the surface-dominant photoconduction, τ_r and τ_d can be understood as two different carrier lifetimes following the same $e–h$ separation mechanism but with different SBB phenomena. Since the SDR is quite sensitive to the photoexcitation intensity, photocurrent rise measurement under UV illumination would flatten the SBB leading to a shorter "steady-state lifetime" as τ_r [28]. When the excitation is off, SBB would recover to the original level at thermal equilibrium. In this dark condition, a much stronger SBB would pronouncedly slowdown the recombination rate and prolong the "equilibrium lifetime" as τ_d. Accordingly, the

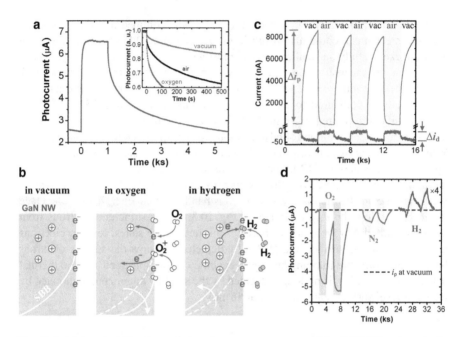

Fig. 10.4 (**a**) A typical line shape of photocurrent response curve of the GaN NW under a single shock UV illumination at 325 nm in air. Inset: The normalized photocurrent decay curve of the NW after removal of the UV light illumination in the vacuum, air, and pure oxygen. (**b**) The schematics of oxygen and hydrogen molecular effects on the surface of GaN NW. A pristine surface in vacuum with intrinsic SBB induced by the space charges, including trapped surface electrons and uncompensated donor ions, is also drawn for comparison. The oxygen and hydrogen molecules play, respectively, the donor- and acceptor-like surface states on the NW. (**c**) The in situ measurements of photocurrent (i_p) and dark current (i_d) of the NW during the repeated alternating exposure in vacuum and air. The i_p and i_d values of air have been subtracted to present the current change of Δi_p and Δi_d. The shadow region represents the duration of air exposure. (**d**) The in situ measurements of i_p of the NW during the repeated alternating exposure in the respective oxygen (O_2), nitrogen (N_2), and hydrogen (H_2) from the vacuum. The i_p value of vacuum is subtracted to be the zero baseline (the dash line at $i_p = 0$). [Reprinted with permission from [29], Copyright @ The American Institute of Physics 2009]

PPC phenomenon with huge τ_d to τ_r ratio ($\tau_d/\tau_r = 2,190$ in vacuum; $\tau_d/\tau_r = 370$ in air; $\tau_d/\tau_r = 100$ in oxygen) actually reflects the strong carrier separation effect in GaN NWs and τ_d can serve as an index for the extent of SBB without photo-flattening [29]. The decreasing τ_d/τ_r ratio (or SBB) from vacuum via air to pure oxygen can be associated with a molecular modulation effect on the SBB, which will be elucidated in the next section.

The surface-induced mechanism could give rise to the difference of photocurrent persistence in the NWs with different diameters. The slower decay of i_p in the thicker NWs has been observed by Calarco et al. [26] and Polenta et al. [24]. Such size effect was attributed to a hindered surface recombination due to the spatial separation of carriers induced by SBB. The bulk-located defects and their interplays with surface on the recombination mechanism are not correlated to the size-related conductivity and PPC. The opinions actually coincide with the aforementioned surface-controlled PC and PPC in the GaN NWs.

10.3   Molecular Sensing

GaN NWs, beside the benefits inheriting from its bulk predecessor, possess the nanomaterial-specific surface-dominated identities along with their own structural advantage with direct transport path. As discussed before, strong surface-induced spatial separation of charge carriers between core and surface of the NW leads to SBB-directed higher surface sensitivity compared to its planar film counterparts. Such spatial separation of carriers can easily be influenced by the ambient, especially in presence of charged bio/chemical molecules, leading to an alteration of SBB in the NWs hence its electronic behavior. Successful deployment of such molecular modulation effect on the GaN NWs has been recognized in recent advances in highly sensitive 1D GaN nanostructure-based sensing applications, where the detection technique is not only restricted to PC measurements [29], but reportedly extended to electrochemical (EC) [17, 18, 55], photoluminescence (PL) [17, 56], or field-effect transistor (FET) [57] based sensing methods.

10.3.1   Gain Amplified and Selective Gas Sensing

The SBB that governs the conduction properties of GaN NWs has been found of being rather sensitive to the exposure surroundings with oxygen. The smaller steady-state photocurrent and shorter PPC decay time detected in the pure oxygen actually indicates a lower SBB in NWs [29]. In vacuum, the pristine surface retains an intrinsic SBB induced by the space charges, including trapped surface electrons and uncompensated donor ions. Upwards of SBB in the intrinsic n-type GaN is often attributed to negatively charged surface states due to the native defects, dangling bonds, etc. While exposure to oxygen environment, the adsorbed oxygen could play as a donor-type molecules might release electrons and reduce space charges by simply adsorbing onto surface and/or filling the defect sites, resulting in the shortening of SDR width and lowering of ϕ_b. A schematic of band flattening by molecular adsorption is depicted in Fig. 10.4b [29].

The oxygen effect can be further manifested via the in situ photocurrent (i_p) and dark current (i_d) measurements of the GaN NWs during the repeated alternating exposures in vacuum and air (Fig. 10.4c) [29]. Interestingly, the i_d exhibited a positive response to the air exposure from vacuum, opposite to the i_p behavior, supporting the charge transfer phenomena between the oxygen molecules and NW. As proposed, the thermal electrons released by the adsorbed oxygen would lower the SBB, and hence decrease the i_p dramatically. The rest into the NR would contribute carriers for the i_d. This suggests the interesting inverse responses to air between i_p and i_d and further confirms the role of oxygen as a foreign donor-like surface state.

Since Δi_p is governed and magnified by the change of τ due to SBB variation and Δi_d is only contributed from partially released electrons without any gain enhancement (long τ mechanism), the NWs exhibited much higher value of Δi_p at $\sim 8,000$ nA than the Δi_d (~ 40 nA). This result demonstrated a gain-amplified molecule sensing mechanism at room temperature. The electrical signal, which might not be detectable in the dark, can be enhanced over two orders of magnitude by a simple photoexcitation to the semiconducting GaN NW sensors.

In addition to the oxidative gas like oxygen, two more gases with different nature including reductive hydrogen and neutral nitrogen are also tested, as shown in Fig. 10.4d [29]. The PC measurements with repeated exposures from vacuum to the nitrogen showed negative responses of i_p, similar to the oxygen although the i_p drop is much less. The more interesting one is that hydrogen exposure has led to positive response of i_p which is contrary to oxygen and nitrogen. The positive response with even higher i_p points to the stronger SBB of NWs in hydrogen than that in vacuum. The result implies that hydrogen could capture free electron on surface and further enhance SBB as an "acceptor-like" surface state, Fig. 10.4b, which is distinct from the "donor-like" oxygen. Moreover, the result has also demonstrated a high selectivity of photo-assisted gas sensing to distinguish the oxidative and reductive gases.

10.3.2   Biomolecular Sensing

Biosensing technology is a progressing area emerging by the interaction of biology, physics, chemistry, and material sciences. Basic feature of a biosensor is built upon the interaction of the probe/target biomolecule with its transducer support. Such bio-transducer coupling relies on the charge-transfer efficiency, and hence the alignments of the relevant electronic levels at their hybrid interface. Inevitably, the large bandgap semiconductors like GaN, which can cover almost the entire range of HOMO–LUMO positions of general bioorganic systems, are ideally suited compared to other "classical" semiconductors like Si and GaAs [58]. Additionally possessing the inherent chemical robustness, nontoxicity, and biocompatibility [59, 60], GaN have shown a promising possibility for biotechnology applications, though mainly constrained to its thin film configuration [61, 62]. Only recently, few approaches have utilized the 1D GaN nanostructure as sensing transducer, showing evidences of its great potential for the detection of DNA [17, 18, 56, 57, 63] or other biomolecules like cysteine [55].

10.3.2.1   Surface Functionalization of GaN Nanowires

To achieve an efficient, stable, and reliable biorecognition interface, the strategy of immobilization of probe biomolecules onto the transducer surface with an optimum

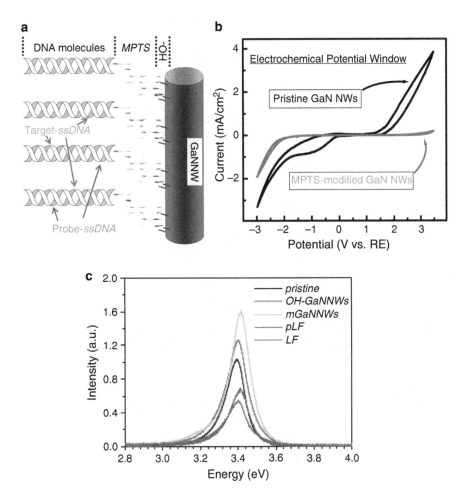

Fig. 10.5 (**a**) Schematic diagram of DNA-modified GaN NWs represents the functionalization stages: hydroxylation (–OH), MPTS modification (MPTS), DNA immobilization (Probe-ssDNA), and DNA hybridization (Target-ssDNA). (**b**) Typical CV spectra of the pristine and MPTS-modified NWs electrodes, in 5 ml phosphate buffer solution (pH=7.42), at a scan rate of 100 mV s^{-1}. [Reprinted with permission from [18], Copyright @ The Royal Society of Chemistry 2009] (**c**) Representative PL spectra of GaN NWs with different surface conditions. [Reprinted with permission from [56], Copyright @ American Institute of Physics 2011]

density is a key step. Chen et al. [17, 18] have adapted a simple silanization method to immobilize the probe biomolecules on GaN NWs through an intermediate organic linker like 3-mercaptopropyl trimethoxysilane (MPTMS or MPTS) (Fig. 10.5a). Firstly, the pristine NWs were hydroxylated by oxidizing acids in order to create –OH reactive sites for further anchoring of MPTS molecules. The resulting stable self-assembled MPTS monolayers would provide the linkage between the probe DNA and the NWs surface via covalent binding. In recent approach [64], the atomic layer deposition (ALD) coating of oxides (like SiO_2) on the NWs surfaces has been demonstrated as an improved strategy capable of creating –OH functional groups with much higher density compared to usual wet chemical methods. Moreover, employing a biotin–streptavidin model, it is shown that the surface with highest – OH content yielded the best protein attachment. An interesting posthydroxylation "passivation effect," as evident from the blue-shift of PL peak energy [56], is proposed; consequently, the negatively charged –OH functional groups could be bound to the depleted NWs surface to reduce surface barrier.

The effectiveness of the surface immobilized analytes specified electronic modulation of GaN NWs became profound while the NWs surfaces were functionalized by MPTS. Interestingly, the MPTS-modified CVD-NWs (MPTS-NWs) have significantly improved the EC characteristics of pristine GaN NWs [18], exhibiting wider potential window of 4.5 V with very low background current (Fig. 10.5b). Such characteristics of a transducer would provide an advantage for the sensitive EC detection of analytes like nucleic acids having high oxidation/reduction potential [65] Very recently, Hsu et al. [56] have also demonstrated that the MPTS could passivate the NWs surface effectively, as evidenced from the PL, Raman, and time-resolved PL investigations. The observed blue-shift of the PL peak with reduced bandwidth and enhanced intensity, and longer lifetime of the excited carriers could specify a prosperous passivation of MPTS-NWs (Fig. 10.5c). Decrease in the relative intensity of DARS (the disorder-activated Raman scattering) and the FWHM of E_2 (high) Raman modes have further justified the reduction of the (active surface) defect-related disorder and consequent improvement in the NWs quality. Although the role of MPTS still needs to be elucidated, a hypothesis associated with the modulation of SBB and defect levels near the band edges of NWs has been proposed to rationalize the suggested passivation effect. MPTS, being a soft donor, would tend to reduce the surface defect density and consequently results in significant decrease in the SBB of GaN NWs.

10.3.2.2 Label-Free In Situ Electrochemical Detection of DNA

The dual functionalities of MPTS, the biocompatibility linker and the surface passivator, open the possibility to develop a sensing platform, without any requirement of labeling of probes or target or the sensing template in advance [17, 18, 56]. Ganguly et al. [18] have successfully demonstrated a promising performance of the MPTS-NWs for label-free detection of DNA immobilization or hybridization phenomena.

The NWs sensor could achieve a limit of detection (LoD) below picomolar (pM) level concentration, via simply monitoring the in situ electrochemical EC oxidation of adenine and guanine, employing cyclic voltammetric (CV) technique. The study also demonstrated a clear distinction between NWs and thin film [18]. The application of the latter as an EC electrode was restricted by its inherent high resistivity and low surface sensitivity, even though the surface modifications from MPTS functionalization to DNA immobilization were successful. Concurrently, analogous experiments on carbon nanotubes electrodes (CNTs, as another potential nanomaterial candidate) failed to achieve any distinct voltammetric signal from DNA, due to inappropriately small potential window with high background current.

The "direct" detection, via CV technique, suffers from the irreversible redox activity of DNA bases, leading to the "consumption" of analytes hindering the sensor performance. EC impedance spectroscopy (EIS) can provide an alternate safer approach for in situ biosensing, being very sensitive to the changes in interfacial impedance of the electrodes, irrespective of its material types (conductor or semiconductor) [66,67]. Chen et al. [17] have demonstrated in situ label-free EIS detection of DNA (anthrax lethal factor sequence, LF), employing very low potential at the GaN NWs electrode avoiding the unnecessary oxidation of DNA bases.

Interestingly, the DNA molecules possess its own individual Faradaic characteristics alongside the pristine NWs (Fig. 10.6a), drawing a clear distinction from the conventional metallic electrodes, which cannot exhibit such Faradaic nature unless redox markers are present. DNA immobilization would create an additional impedimetric element (P2 in Fig. 10.6a), over the typical Faradaic spectrum of pristine NWs (P1 in Fig. 10.6a). Consequently, the electron would face two interfaces in series, GaN/DNA (GD) and DNA/electrolyte (DS), while transferring from electrolyte to electrode (Fig. 10.6c–e). Figure 10.6c, d show the equivalent circuit models of the electrode/electrolyte interfaces used for the best fit to the EIS results. Details of modeling and the parameters involved were elucidated in [17, 66]. The negatively charged DNA molecules would lead to an obvious flattening of SBB of GaN NWs, a consequence of trap state passivation and charge redistribution, and suppressing the resistance or barrier (R_{GD}) at the GD interface. With the capture of target LF, as more negative charges accumulate on NWs surface, R_{GD} decreased rapidly, as shown in Fig. 10.6b. Employing R_{GD} as a fingerprint, the sensor demonstrated the excellent selectivity to target DNA (human $p53$ gene), only which could trigger the sensor activity, down to pM concentration (LoD), in presence of noncomplementary and mismatched sequences.

10.3.2.3   Label-Free Photoluminescence Sensing of DNA

In parallel, room-temperature photoluminescence (PL) technique has been employed for ex situ DNA detection [17]. The excitonic emission of pristine GaN NWs (∼365 nm) showed a prominent quenching behavior with subsequent immobilization and hybridization of DNA. Here, the luminescence quenching was treated as a sensing parameter, since the DNA can absorb UV acting as a static quencher, and the phenomenon became more pronounced with increasing concentration of DNA. Likewise, Hsu et al. [56] have also explored the MPTS-NWs as a DNA sensor by tailoring its PL response (Fig. 10.5c), reporting the red-shift and enhanced FWHM of GaN emission, besides its quenching, with DNA modification. The features ascertained the ex situ interaction of DNA with GaN NWs (analogous to MPTS, but in an opposite way).

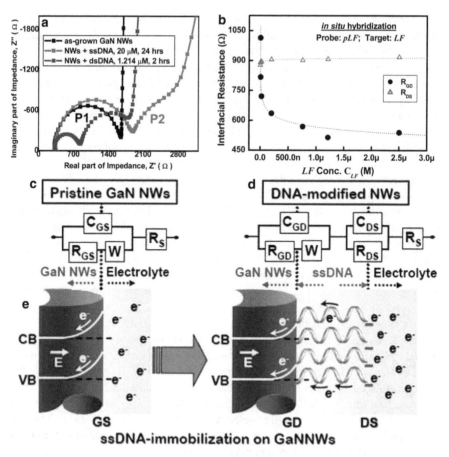

Fig. 10.6 (a) Nyquist plots of as-grown, probe ssDNA-modified, and dsDNA-modified GaN NWs for the EIS based in situ DNA detection. (b) Variation of the electron transfer resistances R_{GD} and R_{DS}, of the interfaces GD and DS, respectively, with the target concentration (C_{LF}). (c–d) The equivalent circuit models used for the (c) as-grown and (d) DNA-modified NWs systems, supporting the existence of two interfaces, the GaN/DNA interface (GD) and the DNA/electrolyte interface (DS), in series. (e) Proposed schematic illustration of flattening of SBB and enhanced charge transport at the GD interface with DNA modification. Symbols are explained in detail in [17]. [Reprinted with permission from [17], Copyright @ American Chemical Society 2009]

10.3.2.4 Ultra-Sensitive Field-Effect Transistor DNA Sensor

In a following study [57], a pronounced improvement has been achieved in the performance of GaN NW-based DNA sensor, simply by direct and easy assembling of as-synthesized NWs with a commercial FET device. The simple architecture of extended-gate FET (EGFET) biosensor (Fig. 10.7a), by completely isolating the FET device from sensing environment, could facilitate the design and maintenance of sensor, as well as the real-time sensing operation. Even GaN thin film could

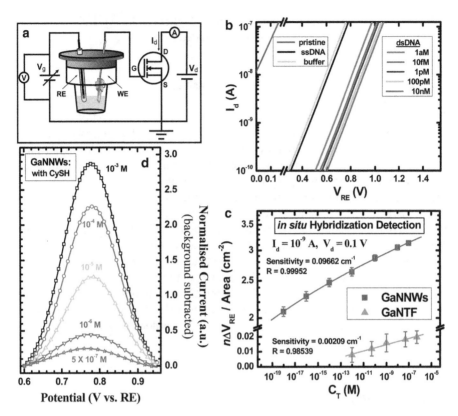

Fig. 10.7 (**a**) Schematic of GaN-based EGFET biosensor, using GaN NWs (or thin film) as an extended gate. V_{RE}: reference electrode voltage and I_d: drain current. Symbols are explicated in detail in [57]. (**b**) In situ DNA-hybridization sensing: I_d–V_{RE} characteristics of NWs-EGFET on the selected V_{RE}-span. (**c**) Comparison between NWs and thin film sensors: dependence of FET response ($n\Delta V_{RE}$) on target concentration (C_T), $n\Delta V_{RE} = [V_{RE}(\mathrm{dsDNA}) - V_{RE}(\mathrm{ssDNA})]/V_{RE}(\mathrm{ssDNA})$. [Reprinted with permission from [57], Copyright @ American Chemical Society 2011] (**d**) L-cysteine (CySH) oxidation peaks at the NWs electrode at different concentration of CySH. [Reprinted with permission from [55], Copyright @ Elsevier 2010]

achieve an excellent performance for label-free in situ detection of human $p53$ gene (LoD=pM). Detection of DNA has been visualized by a continuous shift of I_d–V_{RE} curve towards the positive V_{RE} direction caused by the probe immobilization or hybridization (Fig. 10.7b). Such positive shift in V_{RE} was attributed to the development of negative charges (of DNA molecules) on the gate surface (GaN). GaN NWs have clearly outperformed its thin film counterpart, with about two orders higher sensitivity, over a wide detection range, 10^{-19}–10^{-6} M, reaching about six orders lower LoD (Fig. 10.7c), revealing the enormous potential of NW structure for sensor applications. An excellent integration of a wide-bandgap semiconducting nature of GaN, surface sensitivity of NW structure, and high transducing performance of

EGFET design could achieve about three orders higher resolution (LoD=attomolar or aM) than other FET-based DNA sensors reported [57].

10.3.2.5 Direct Voltammetric l-Cysteine Sensing

Besides DNA, GaN NWs have also been utilized for direct sensing of L-cysteine. Lai et al. [55] have shown that the pristine CVD-grown NWs can response to the electrocatalytic EC oxidation of L-cysteine without any need of surface modification (Fig. 10.7d): a unique advantage over the conventional chemical functionalization (on electrode surface) strategies, eliminating the problems like involvement of complicated procedures or poor stability of chemical modifier, enhancing the sensitivity and reproducibility of sensor. Voltammetric studies on the effects of pH and potential scan rate made certain the catalytic behavior of NWs electrode: the oxidation mechanism was controlled by electroactive L-CyS$^-$ species, rather than by desorption of reaction products. GaN NWs sensor could achieve an optimum sensitivity of $42 \, nA \, \mu M^{-1}$ with an experimental LoD of 0.5 micromolar (μM), under physiological condition (pH=7.4), using CV techniques. Interestingly, GaN NWs have displayed impressive performances, by exhibiting improved sensitivity and superior LoD, in comparison to CNTs or boron-doped diamond, and other potential biosensing transducers [55].

10.4 Summary

Recent reports on GaN NWs have revealed their unprecedented surface properties and potential to be a transducer material for the future sensing applications. These studies, though so far at elementary stage, would definitely stimulate further researches on GaN NWs sensors in much wider scale for advanced and pragmatic applications.

References

1. S. Nakamura, S. Pearton, G. Fasol, *The Blue Laser Diode: The Complete Story*. (Springer, Berlin 2000)
2. H. Morkoc, *Nitride Semiconductors and Devices*. (Springer, Berlin 1999)
3. L.C. Chen, K.H. Chen, C.C. Chen, *Group III- and Group IV Nitride Nanorods and Nanowires, Ch. 9*. In: Z.L. Wang (ed.) Nanowires and Nanobelts: Materials, Properties and Devices, Vol. 1. (Kluwer Academic Publisher, Boston 2003)
4. S. Chattopadhyay, A. Ganguly, K.H. Chen, L.C. Chen, One-dimensional group III-nitrides: growth, properties, and applications in nanosensing and nano-optoelectronics. Crit. Rev. Solid State Mater. Sci. **34**, 224–279 (2009)

5. Y. Xia, P. Yang, Y. Sun, Y. Wu, B. Mayers, B. Gates, Y. Yin, F. Kim, H. Yan, One-dimensional nanostructures: synthesis, characterization, and applications. Adv. Mater. **15**, 353–389 (2003)
6. L. Samuelson, Self-forming nanoscale devices. Mater. Today **6**, 22–31 (2003)
7. Z.L. Wang, Nanostructures of zinc oxide. Mater. Today **7**, 26–33 (2004)
8. G. Cheng, A. Kolmakov, Y. Zhang, M. Moskovits, R. Munden, M.A. Reed, G. Wang, D. Moses, J. Zhang, Current rectification in a single GaN nanowire with a well-defined p–n junction. Appl. Phys. Lett. **83**, 1578 (2003)
9. S.Y. Lee, T.H. Kim, D.I. Suh, E.K. Suh, N.K. Cho, W.K. Seong, S.K. Lee, Dielectrophoretically aligned GaN nanowire rectifiers. Appl. Phys. A **87**, 739–742 (2007)
10. Y. Li, J. Xiang, F. Qian, S. Gradecak, Y. Wu, H. Yan, D.A. Blom, C.M. Lieber, Dopant-free GaN/AlN/AlGaN radial nanowire heterostructures as high electron mobility transistors. Nano Lett. **6**, 1468–1473 (2006)
11. H.M. Kim, Y.H. Cho, H. Lee, S.I. Kim, S.R. Ryu, D.Y. Kim, T.W. Kang, K.S. Chung, High-brightness light emitting diodes using dislocation-free indium gallium nitride/gallium nitride multiquantum-well nanorod arrays. Nano Lett. **4**, 1059–1062 (2004)
12. F. Qian, S. Gradecak, Y. Li, C.Y. Wen, C.M. Lieber, Core/multishell nanowire heterostructures as multicolor, high-efficiency light-emitting diodes. Nano Lett. **5**, 2287–2291 (2005)
13. F. Qian, Y. Li, S. Gradecak, H.G. Park, Y. Dong, Y. Ding, Z.L. Wang, C.M. Lieber, Multi-quantum-well nanowire heterostructures for wavelength-controlled lasers. Nat. Mater. **7**, 701–706 (2008)
14. S. Han, W. Jin, D. Zhang, T. Tang, C. Li, X. Liu, Z. Liu, B. Lei, C. Zhou, Photoconduction studies on GaN nanowire transistors under UV and polarized UV illumination. Chem. Phys. Lett. **389**, 176–180 (2004)
15. M.S. Son, S.I. Im, Y.S. Park, C.M. Park, T.W. Kang, K.H. Yoo, Ultraviolet photodetector based on single GaN nanorod p–n junctions. Mater. Sci. Eng. C **26**, 886–888 (2006)
16. R.S. Chen, S.W. Wang, Z.H. Lan, J.T.H. Tsai, C.T. Wu, L.C. Chen, K.H. Chen, Y.S. Huang, C.C. Chen, On-chip fabrication of well-aligned and contact-barrier-free GaN nanobridge devices with ultrahigh photocurrent responsivity. Small **4**, 925–929 (2008)
17. C.P. Chen, A. Ganguly, C.H. Wang, C.W. Hsu, S. Chattopadhyay, Y.K. Hsu, Y.C. Chang, K.H. Chen, L.C. Chen, Label-free dual sensing of DNA molecules using GaN nanowires. Anal. Chem. **81**, 36–42 (2009)
18. A. Ganguly, C.P. Chen, Y.T. Lai, C.C. Kuo, C.W. Hsu, K.H. Chen, L.C. Chen, Functionalized GaN nanowire-based electrode for direct label-free voltammetric detection of DNA hybridization. J. Mater. Chem. **19**, 928–933 (2009)
19. Y. Dong, B. Tian, T.J. Kempa, C.M. Lieber, Coaxial group III-nitride nanowire photovoltaics. Nano Lett. **9**, 2183–2187 (2009)
20. A. Motayed, M. Vaudin, A.V. Davydov, J. Melngailis, M. He, S.N. Mohammad, Diameter dependent transport properties of gallium nitride nanowire field effect transistors. Appl. Phys. Lett. **90**, 043104 (2007)
21. M. Niebelschütz, V. Cimalla, O. Ambacher, T. Machleidt, J. Ristic, E. Calleja, Electrical performance of gallium nitride nanocolumns. Physica E **37**, 200–203 (2007)
22. B.S. Simpkins, M.A. Mastro, C.R. Eddy, P.E. Pehrsson, Surface depletion effects in semiconducting nanowires. J. Appl. Phys. **103**, 104313 (2008)
23. J. Yoon, A.M. Girgis, I. Shalish, L.R. Ram-Mohan, V. Narayanamurti, Size-dependent impurity activation energy in GaN nanowires. Appl. Phys. Lett. **94**, 142102 (2009)
24. L. Polenta, M. Rossi, A. Cavallini, R. Calarco, M. Marso, R. Meijers, T. Richter, T. Stoica, H. Lüth, Investigation on localized states in GaN nanowires. ACS Nano **2**, 287–292 (2008)
25. H.Y. Chen, R.S. Chen, F.C. Chang, L.C. Chen, K.H. Chen, Y.J. Yang, Size-dependent photoconductivity and dark conductivity of m-axial GaN nanowires with small critical diameter. Appl. Phys. Lett. **95**, 143123 (2009)
26. R. Calarco, M. Marso, T. Richter, A.I. Aykanat, R. Meijers, A.v.d. Hart, T. Stoica, H. Lüth, Size-dependent Photoconductivity in MBE-Grown GaN-Nanowires. Nano Lett. **5**, 981–984 (2005)

27. A. Cavallini, L. Polenta, M. Rossi, T. Stoica, R. Calarco, R.J. Meijers, T. Richter, H. Lüth: Franz-Keldysh Effect in GaN Nanowires. Nano Lett. **7**, 2166–2170 (2007)

28. R.S. Chen, H.Y. Chen, C.Y. Lu, K.H. Chen, C.P. Chen, L.C. Chen, Y.J. Yang, Ultrahigh photocurrent gain in m-axial GaN nanowires. Appl. Phys. Lett. **91**, 223106 (2007)

29. R.S. Chen, C.Y. Lu, K.H. Chen, L.C. Chen, Molecule-modulated photoconductivity and gain-amplified selective gas sensing in polar GaN nanowires. Appl. Phys. Lett. **95**, 233119 (2009)

30. M. Asif Khan, J.N. Kuznia, D.T. Olson, J.M. Van Hove, M. Blasingame, L.F. Reitz, High-responsivity photoconductive ultraviolet sensors based on insulating single-crystal GaN epilayers. Appl. Phys. Lett. **60**, 2917 (1992)

31. D. Walker, X. Zhang, P. Kung, A. Saxler, S. Javadpour, J. Xu,, M. Razeghi, AlGaN ultraviolet photoconductors grown on sapphire. Appl. Phys. Lett. **68**, 2100 (1996)

32. B.W. Lim, Q.C. Chen, J.Y. Yang, M. Asif Khan, High responsivity intrinsic photoconductors based on $Al_xGa_{1-x}N$. Appl. Phys. Lett. **68**, 3761 (1996)

33. B. Shen, K. Yang, L. Zang, Z.Z. Chen, Y.G. Zhou, P. Chen, R. Zhang, Z.C. Huang, H.S. Zhou, Y.D. Zheng, Study of photocurrent properties of GaN ultraviolet photoconductor grown on 6H-SiC substrate. Jpn. J. Appl. Phys. **38**, 767–769 (1999)

34. C. Pernot, A. Hirano, M. Iwaya, T. Detchprohm, H. Amano, I. Akasaki, Low-intensity ultraviolet photodetectors based on AlGaN. Jpn. J. Appl. Phys. **38**, L487–L489 (1999)

35. P. Bhattacharya, Semiconductor optoelectronic devices, Ch. 8, pp. 346–347. (Prentice-Hall Inc., New Jersey 1997)

36. Z.M. Zhao, R.L. Jiang, P. Chen, D.J. Xi, Z.Y. Luo, R. Zhang, B. Shen, Z.Z. Chen, Y.D. Zheng, Metal–semiconductor–metal GaN ultraviolet photodetectors on Si(111). Appl. Phys. Lett. **77**, 444 (2000)

37. Z.C. Huang, D.B. Mott, P.K. Shu, R. Zhang, J.C. Chen, D.K. Wickenden, Optical quenching of photoconductivity in GaN photoconductors. J. Appl. Phys. **82**, 2707 (1997)

38. E. Munoz, E. Monroy, J.A. Garrido, I. Izpura, F.J. Sánchez, M.A. Sánchez-Garcia, E. Calleja, B. Beaumont, P. Gibart, Photoconductor gain mechanisms in GaN ultraviolet detectors. Appl. Phys. Lett. **71**, 870 (1997)

39. F. Binet, J.Y. Duboz, E. Rosencher, F. Scholz, V. Härle, Mechanisms of recombination in GaN photodetectors. Appl. Phys. Lett. **69**, 1202 (1996)

40. M. Razeghi, A. Rogalski, Semiconductor ultraviolet detectors. J. Appl. Phys. **79**, 7433 (1996)

41. T. Kawashima, H. Yoshikawa, S. Adachi, S. Fuke, K. Ohtsuka, Optical properties of hexagonal GaN. J. Appl. Phys. **82**, 3528 (1997)

42. J.F. Muth, J.H. Lee, I.K. Shmagin, R.M. Kolbas, H.C. Casey, B.P. Keller, U.K. Mishra, S.P. DenBaars, Absorption coefficient, energy gap, exciton binding energy, and recombination lifetime of GaN obtained from transmission measurements. Appl. Phys. Lett. **71**, 2572 (1997)

43. C.Y. Chang, G.C. Chi, W.M. Wang, L.C. Chen, K.H. Chen, F. Ren, S.J. Pearton, Electrical transport properties of single GaN and InN nanowires. J. Electro. Mater. **35**, 738–743 (2006)

44. T. Kuykendall, P. Pauzauskie, S. Lee, Y. Zhang, J. Goldberger, P. Yang: Metalorganic chemical vapor deposition route to GaN nanowires with triangular cross sections. Nano Lett. **3**, 1063–1066 (2003)

45. E. Stern, G. Cheng, E. Cimpoiasu, R. Klie, S. Guthrie, J. Klemic, I. Kretzschmar, E. Steinlauf, D. Turner-Evans, E. Broomfield, J. Hyland, R. Koudelka, T. Boone, M. Young, A. Sanders, R. Munden, T. Lee, D. Routenberg, M.A. Reed, Electrical characterization of single GaN nanowires. Nanotechnology **16**, 2941–2953 (2005)

46. Y. Huang, X. Duan, Y. Cui, C.M. Lieber, Gallium nitride nanowire nanodevices. Nano Lett. **2**, 101–104 (2002)

47. K.S. Stevens, M. Kinniburgh, R. Beresford, Photoconductive ultraviolet sensor using Mg-doped GaN on Si(111). Appl. Phys. Lett. **66**, 3518 (1995)

48. R.H. Bube, *Photoelectronic Properties of Semiconductors*. (Cambridge, New York 1992)

49. J.A. Garrido, E. Monroy, I. Izpura, E. Munoz, Photoconductive gain modelling of GaN photodetectors. Semicond. Sci. Technol. **13**, 563–568 (1998)

50. C. Soci, A. Zhang, B. Xiang, S.A. Dayeh, D.P.R. Aplin, J. Park, X.Y. Bao, Y.H. Lo, D. Wang, ZnO nanowire UV photodetectors with high internal gain. Nano Lett. **7**, 1003–1009 (2007)

51. T.K. Zywietz, J. Neugebauer, M. Scheffler, The adsorption of oxygen at GaN surfaces. Appl. Phys. Lett. **74**, 1695 (1999)
52. M. Sumiya, K. Yoshimura, K. Ohtsuka, S. Fuke, Dependence of impurity incorporation on the polar direction of GaN film growth. Appl. Phys. Lett. **76**, 2098 (2000)
53. T.Y. Lin, H.C. Yang, Y.F. Chen, Optical quenching of the photoconductivity in n-type GaN. J. Appl. Phys. **87**, 3404 (2000)
54. C.H. Qiu, J.I. Pankove, Deep levels and persistent photoconductivity in GaN thin films. Appl. Phys. Lett. **70**, 1983 (1997)
55. Y.T. Lai, A. Ganguly, K.H. Chen, L.C. Chen, Direct voltammetric sensing of L-cysteine at pristine GaN nanowires electrode. Biosens. Bioelectron. **26**, 1688–1691 (2010)
56. C.W. Hsu, A. Ganguly, C.P. Chen, C.C. Kuo, P.P. Paskov, P.O. Holtz, L.C. Chen, K.H. Chen, Optical properties of functionalized GaN nanowires. J. Appl. Phys. **109**, 053523 (2011)
57. C.P. Chen, A. Ganguly, C.Y. Lu, T.Y. Chen, C.C. Kuo, W.H. Tu, R.S. Chen, W.B. Fischer, K.H. Chen, L.C. Chen, Ultrasensitive in situ label-free DNA detection using a GaN nanowire-based extended-gate field-effect-transistor sensor. Anal. Chem. **83**, 1938–1943 (2011)
58. M. Stutzmann, J.A. Garrido, M. Eickhoff, M.S. Brandt, Direct biofunctionalization of semiconductors: A survey. Phys. Stat. Sol. A **203**, 3424–3437 (2006)
59. G. Steinhoff, O. Purrucker, M. Tanaka, M. Stutzmann, M. Eickhoff, $Al_xGa_{1-x}N$–a new material system for biosensors. Adv. Funct. Mater. **13**, 841–846 (2003)
60. T.H. Young, C.R. Chen, Assessment of GaN chips for culturing cerebellar granule neurons. Biomaterials **27**, 3361–3367 (2006)
61. N. Chaniotakis, N. Sofikiti, Novel semiconductor materials for the development of chemical sensors and biosensors: A review. Anal. Chim. Acta **615**, 1–9 (2008)
62. S.J. Pearton, D.P. Norton, F. Ren, The promise and perils of wide-bandgap semiconductor nanowires for sensing, electronic, and photonic applications. Small **3**, 1144–1150 (2007)
63. B.S. Simpkins, K.M. McCoy, L.J. Whitman, P.E. Pehrsson, Fabrication and characterization of DNA-functionalized GaN nanowires. Nanotechnology **18**, 355301 (2007)
64. D.J. Guo, A.I. Abdulagatov, D.M. Rourke, K.A. Bertness, S.M. George, Y.C. Lee, W. Tan, GaN nanowire functionalized with atomic layer deposition techniques for enhanced immobilization of biomolecules. Langmuir **26**, 18382–18391 (2010)
65. W.C. Poh, K.P. Loh, W.D. Zhang, S. Triparthy, J.S. Ye, F.S. Sheu, Biosensing properties of diamond and carbon nanotubes. Langmuir **20**, 5484–5492 (2004)
66. E. Katz, I. Willner, Probing biomolecular interactions at conductive and semiconductive surfaces by impedance spectroscopy: routes to impedimetric immunosensors, DNA-sensors, and enzyme biosensors. Electroanalysis **15**, 913–947 (2003)
67. J.Y. Park, S.M. Park, DNA hybridization sensors based on electrochemical impedance spectroscopy as a detection tool. Sensors **9**, 9513–9532 (2009)

Chapter 11
Minority Carrier Transport in ZnO and Related Materials

Elena Flitsyian, Zinovy Dashevsky, and Leonid Chernyak

Abstract Transport properties of minority carriers in ZnO and related compounds are of critical importance for the functionality of bipolar devices. This review summarizes the available information on the subject of minority carrier transport in ZnO-based semiconductors, focusing on its temperature dependence and the dynamics of nonequilibrium carrier recombination. The influence of deep traps on minority carrier diffusion length and lifetime is discussed. The experimental results, showing the impact of minority carrier transport on the performance of bipolar devices, as well as a discussion of techniques, used for measurements of the minority carrier diffusion length and lifetime, are provided.

11.1 Introduction

Recent developments in doping and growth of ZnO stimulated a renewal of interest in this material from the point of view of its applications in optoelectronic devices. As a direct wide bandgap semiconductor ($E_g \approx 3.35\,\text{eV}$ at room temperature) with high exciton binding energy ($60\,\text{meV}$, compared to $25\,\text{meV}$ in GaN), ZnO is a superior candidate for minority-carrier-based devices, such as light emitting diodes, laser diodes, and transparent p–n junctions. Moreover, it offers significant advantages over III-Nitride materials, which include availability of large area lattice-matched substrates, lower materials costs, and use of wet chemical etching, as opposed to reactive ion etching employed in GaN technology.

E. Flitsyian · L. Chernyak (✉)
Department of Physics, University of Central Florida, Orlando, FL 32816, USA
e-mail: chernyak@physics.ucf.edu; flitsyian@physics.ucf.edu

Z. Dashevsky
Department of Materials Engineering, Ben-Gurion University of the Negev, Beer-Sheva 84105, Israel

S. Pearton (ed.), *GaN and ZnO-based Materials and Devices*,
Springer Series in Materials Science 156, DOI 10.1007/978-3-642-23521-4_11,
© Springer-Verlag Berlin Heidelberg 2012

In the present state of the art, the development of the full potential of ZnO in bipolar device applications hinges in part on the availability of quality, highly conductive materials of both n- and p-type. Similar to GaN, achieving n-type conductivity in ZnO does not present a problem, since even nominally undoped material is generally n-type, due to the electrical activity of native defects, such as zinc interstitials, zinc antisites, and oxygen vacancies [1, 2], as well as hydrogen impurity [3]. Heterojunction devices employing n-ZnO have been demonstrated with p-AlGaN [4], p-SiC [5], p-diamond [6], and p-AlGaN/p-GaN heterostructure [7, 8] as hole emitters.

On the other hand, p-type conductivity with sufficiently high carrier concentrations appears to be much more elusive, mainly due to high ionization energies of potential acceptors, such as nitrogen, phosphorus, and arsenic. This problem is compounded by high concentrations of native and unintentional donors, which act as compensating centers, thus further reducing the free carrier concentration. It has been shown that the background donor concentration can be lowered if Mg is incorporated into the ZnO lattice, since each percent of Mg increases the bandgap of ZnO by 0.02 eV, suppressing the ionization of shallow donors [9]. The conversion to p-type can then be obtained by heavily doping the resultant material with phosphorus followed by annealing in O_2 atmosphere [10].

Recently, successful p-type doping of ZnO also has been attained using phosphorus [11, 12], nitrogen [13], arsenic [14], or antimony [15] yielding net hole concentrations up to 10^{18} cm^{-3}. It should be noted that while ZnO homojunction devices have been reported as early as the year 2000 [16–20], their quality is generally inferior to the heterojunctions employing n-ZnO.

Even with the advent of new technology enabling the production of viable p-type materials, the performance of bipolar devices is fundamentally limited by the transport properties of minority carriers. In direct bandgap semiconductors, including ZnO, minority carrier diffusion length is generally several orders of magnitude lower than in indirect gap materials such as silicon or germanium. In order to noticeably increase minority carrier diffusion length by reducing scattering by the dislocation walls, the edge threading dislocation density must be reduced by at least two orders of magnitude from a typical value of about 10^9 cm^{-2} in epitaxial ZnO layers [21–24].

Investigation of minority carrier diffusion lengths and lifetimes in both n- and p-type ZnO is an issue of practical importance, since it has direct implications on the performance of bipolar devices. Moreover, considering possible applications of these devices in high-temperature electronics, the insight into the temperature dependence of minority carrier properties is also of great value. Nonetheless, to the best of our knowledge, the reports on this subject are rather scarce. It is therefore the goal of this work to summarize the available information on the subject of minority carrier transport in ZnO and related compounds, focusing on its temperature dependence and the dynamics of nonequilibrium carrier recombination. This discussion will be preceded by a brief summary of the role of minority carrier transport in the performance of bipolar devices, as well as by the review of techniques of choice for measurement of the minority carrier diffusion length.

11.2   Role of Minority Carrier Diffusion Length in Bipolar Device Performance

In general, when nonequilibrium carriers are generated in a material due to external excitation in the absence of electric field, they diffuse over a certain distance before undergoing recombination. This parameter, namely the average distance traveled in a particular direction between generation and recombination, is characterized by the diffusion length, L. The diffusion length is related to the carrier lifetime, τ (i.e., the time between generation and recombination of nonequilibrium carriers) through carrier diffusivity, D:

$$L = \sqrt{D\tau}. \tag{11.1}$$

Diffusivity, or diffusion coefficient, is determined in turn by the mobility of the carriers, μ, according to the Einstein relation:

$$D = \frac{kT}{q}\mu, \tag{11.2}$$

where k is the Boltzmann's constant, T is absolute temperature, and q is fundamental charge.

The diffusion process is driven by concentration gradients; since external excitation has a much larger impact on the concentration of *minority* carriers than that of majority ones (because generation density is usually much lower than the majority carrier density), it is the minority carriers that are more susceptible to diffusion.

Diffusion of minority carriers is a process that is fundamental to the operation of bipolar photovoltaic devices, with minority carrier diffusion length being the central parameter defining the device performance. In the presence of a p–n junction or a Schottky barrier, the nonequilibrium minority carriers generated by external excitation (e.g., light incident on a photodiode) within a few diffusion lengths of the space-charge region can be collected by the built-in field and thus contribute to the current flow across the device. The greater the diffusion length of the carriers, the more current can be collected, leading to the higher efficiency of the device. In photodiodes, it is usually only one side of the p–n junction that contributes to photocurrent. If the light is absorbed in the p-region of the junction, the quantum efficiency, η, can be represented as follows:

$$\eta = (1 - r)\left(1 - \frac{e^{-aW}}{1 + aL_n}\right), \tag{11.3}$$

where r and a are the reflection and absorption coefficients, respectively, W is the width of the space-charge region, and L_n is the diffusion length of minority electrons.

Quantum efficiency is directly related to the spectral responsivity, R, of a photodiode:

$$R(E) = \frac{I_{\text{ph}}}{P_{\text{op}}} = \frac{q\eta}{E},\tag{11.4}$$

where I_{ph} is total photocurrent, P_{op} is optical power incident on the device, q is the fundamental charge, and E is the energy of the incident photons. The relationship between minority carrier diffusion length and responsivity of Schottky photodiodes has been examined in great detail in [25]. Schottky photodiodes are among the simplest photovoltaic devices, where the nonequilibrium minority carriers, generated in the bulk of the semiconductor due to light absorption, are collected by the built-in field of the Schottky barrier deposited on the surface of the semiconductor. In order for the carriers to contribute to device current, they have to be generated within a few diffusion lengths of the collector.

For incident energies greater than the bandgap of the absorber material, nonequilibrium electron–hole pairs are generated only in the thin layer next to the surface of incidence, with the maximum depth of $1/a$ This value is on the order of 100 nm in ZnO [26,27] and is generally much smaller than the thickness of the absorber layer. Considering a front-illuminated configuration (in which the incident light passes through the semitransparent Schottky contact), if L is greater than the generation depth, most of the nonequilibrium minority carriers are collected by the built-in field of the space-charge region. In this case, the internal quantum efficiency of the device approaches 100% and the responsivity is independent of the diffusion length value. Below this threshold, the responsivity decreases with L, provided that the width of the space-charge region is smaller than the generation depth.

If the energy of incident light is below the bandgap, light penetration depth is large (several micrometers), and a fair portion of the nonequilibrium carriers is generated in the neutral region of the semiconductor due to the ionization of the mid-gap levels. Since only the carriers within a few diffusion lengths of the space-charge region contribute to photocurrent, the responsivity at below-bandgap energies is limited by the diffusion length (unless the diffusion length exceeds the thickness of the absorber layer, in which case the latter is the limiting factor) [25].

11.3 Methods for Determination of Minority Carrier Lifetime and Diffusion Length

Although as of the date of this writing minority carrier transport in ZnO remains, with a few exceptions, essentially unexplored, the measurement of minority carrier diffusion length is a well-established subject. This section reviews three of the most widely used techniques, namely, Electron Beam-Induced Current (EBIC), Surface Photovoltage (SPV), and Time-Resolved Photoluminescence (TRPL). While, to the best of the authors' knowledge, only the EBIC method has been applied to the studies of ZnO, SPV and TRPL techniques are also reviewed in an effort to provide a more complete discussion.

Fig. 11.1 Experimental setup used for EBIC measurements in planar-collector configuration. The *dashed ellipse* represents the generation volume; d is the variable distance between electron beam and the Schottky contact, represented by a *solid rectangle*

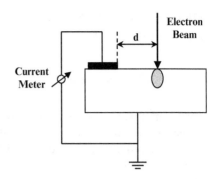

11.3.1   EBIC Technique

Due to a unique combination of convenience and reliability, EBIC method is among the most popular techniques for minority carrier diffusion length measurements. It requires comparatively simple sample preparation and is used in situ in a scanning electron microscope (SEM). Figure 11.1 shows a typical measurement configuration known as planar-collector geometry. The use of this configuration was pioneered by Ioannou, Davidson, and Dimitriadis [28, 29]. As a charge collection technique, EBIC method employs a Schottky barrier or a p–n junction to collect the current resulting from the nonequilibrium minority carriers generated by the beam of the SEM. As the beam is moved away from the barrier/junction in a line-scan mode, the current decays as fewer and fewer minority carriers are able to diffuse to the space-charge region.

The mathematical model for EBIC was further developed by Boersma et al. [30], who showed that the decay of current, I_{EBIC}, can be described by the following expression:

$$I_{EBIC} = Ad^{\alpha} \exp\left(-\frac{d}{L}\right), \tag{11.5}$$

where A is a scaling constant, d is beam-to-junction distance, and α is an exponent related to the surface recombination velocity, v_s.

The diffusion length is usually extracted by rearranging the terms of (11.5):

$$\ln(Id^{-\alpha}) = -\frac{d}{L} + \ln(A), \tag{11.6}$$

which yields a linear relationship between $\ln(Id^{-\alpha}$ and d with a slope equal to $-1/L$. This approach is taken to be accurate for $d > 2L$. It should be noted that in [30], the authors analyzed only the two asymptotic cases, namely $v_s = 0$ and $v_s = \infty$, and found that $\alpha = -1/2$ for the former and $-3/2$ for the latter. Later, Chan et al. [31] demonstrated that this approach can be applied to materials with arbitrary surface recombination velocity by selecting α such that a linear relationship between $\ln(Id^{-\alpha})$ and d is obtained. However, even if *any* value of α is used (such that $-3/2 \le \alpha \le -1/2$), the result for the diffusion length changes by less than 20%

[32] and this is due to the fact that the exponential term dominating the value of I_{EBIC} is independent of α.

11.3.2 SPV Technique

SPV is an attractive technique because it is nondestructive and contactless. The sample preparation does not require deposition of contacts, junctions, or high temperature processing. The use of SPV technique for minority carrier diffusion length measurements was first suggested by Moss [33] and is reviewed in great detail in [34].

SPV method generally requires a special treatment to create an accumulation of charge on the surface of a semiconductor. The options available for surface charging include chemical treatment and deposition of corona charge. The nature of the chemical treatment depends on the composition of the semiconductor of interest, but usually involves immersion in an oxidizing medium for n-type and a reducing medium for p-type substrates. Corona discharge method consists of the deposition of ions on the surface due to an electric field applied to a source of ions.

The surface charge induces equal and opposite charge density just below the surface, thus forming a space-charge region. If the surface of the semiconductor is then illuminated with monochromatic, above-bandgap light of a wavelength λ, the minority carriers generated in the space-charge region as well as within a few diffusion lengths of the latter diffuse and/or drift toward the surface, establishing a surface potential which is proportional to the nonequilibrium minority carrier concentration at the edge of the space-charge region. Changing the wavelength of excitation allows monitoring the diffusion of the carriers generated at different distances from the surface, since the maximum light penetration depth is determined by the absorption coefficient $a(\lambda)$. Throughout this process, the intensity of the incident light, I, is adjusted in order to keep the surface potential constant. Under certain assumptions [34], the relationship between I and λ has the following form:

$$I = C \left(\frac{1}{a(\lambda)} + L \right), \tag{11.7}$$

where C is an arbitrary constant. The diffusion length can be determined by extrapolating the data to obtain the $1/a(\lambda)$ intercept. Although SPV is a relatively straightforward and reliable method [35–38], it is clear from (11.7) that accurate analysis of the SPV data requires independent knowledge of $a(\lambda)$ in the measured specimen.

11.3.3 TRPL Technique

TRPL technique is an indirect method that relies upon measuring the minority carrier lifetime in order to determine the diffusion length according to (11.1) and

(11.2). It provides a time-efficient way for characterizing the transport of nonequilibrium minority carriers and is most useful for materials with good luminescence properties. A significant advantage of this method is that it also provides valuable insight into the nature of the recombination mechanisms governing minority carrier lifetime. The detailed description of the TRPL measurements and analysis is provided in [39].

TRPL measurements are based on recording the transient decay of near-bandgap-edge (NBE) photoluminescence (PL) excited by a short pulse of light, such as from a pulsed laser source. If the concentration of nonequilibrium minority carriers decreases exponentially from its initial value ρ_0, the intensity of the PL, I_{PL}, also follows an exponential decay:

$$I_{PL} = \frac{\rho_0}{\tau_R} \exp\left(-\frac{t}{\tau}\right), \tag{11.8}$$

where τ_R is the radiative recombination lifetime. In most cases, however, the PL intensity is affected not only by lifetime of minority carriers in the band, but also by the diffusion rate out of the absorber region, since the escape of the carriers from the active region is in this case equivalent to the reduction in their lifetime. Therefore, the lifetime obtained from the TRPL measurements is often less than the true minority carrier lifetime. In order to obtain an accurate measurement, it is often necessary to introduce a confinement structure to reduce the influence of diffusion. In such a confinement structure, the semiconductor of interest is "sandwiched" between two layers with wider bandgap and like doping type, so that the wide bandgap layers act as minority carrier mirrors.

11.4 Temperature Dependence of Minority Carrier Diffusion Length and Lifetime

11.4.1 Studies in n-Type ZnO

Because of its intrinsic thermal stability, ZnO is a good candidate for high-temperature optoelectronic devices. However, the subject of the temperature dependence of minority carrier transport properties has not yet been adequately addressed in the literature. This section reviews the results obtained from variable-temperature diffusion length measurements performed on n-type ZnO [40].

The samples under investigation were weakly n-type ZnO substrates with electron concentration of $\sim 10^{14}$ cm^{-3} and mobility of ~ 150 cm^2/V s. Secondary ion mass spectroscopy (SIMS) measurements revealed the Li concentration of about 3×10^{16} cm^{-3} [41]. The Schottky contacts for EBIC measurements were deposited on the nonpolar a-plane of ZnO crystal by electron beam evaporation of 80-nm-thick Au layer and subsequent liftoff. ZnO:N samples were grown using molecular beam epitaxy (MBE) by SVT Associates. Hall effect measurements revealed hole concentration of 4.5×10^{17} cm^{-3} and mobility of ~ 1 cm^2/V s.

The studies of L as a function of temperature were carried out using EBIC method in a planar-collector configuration with a Schottky barrier (Sect. 11.3.1). At each temperature, several measurements were taken by scanning the beam of the SEM along a line perpendicular to the edge of the Schottky contact and recording the exponential decay of current. The recorded data were fitted with (11.6) using $\alpha = -1/2$. This value corresponds to zero surface recombination velocity which, given the excellent luminescence properties of this sample and a good fit to the experimental results, is a reasonably good approximation. It should be noted that each EBIC line scan was recorded on a previously unexposed area in order to avoid the influence of electron irradiation on the value of diffusion length (cf. Sect. 11.5 below).

Table 11.1 summarizes the results of EBIC measurements performed on one of the bulk ZnO samples and shows that the diffusion length of minority holes in n-ZnO increases with increasing temperature, T. The increase of L with T is not unique to this semiconductor. Similar trends were previously observed in GaAs [42] and later in GaN epitaxial layers [43]. In all cases, this increase was exponential with temperature and was modeled with the following expression:

$$L = L_0 \exp\left(-\frac{\Delta E_{A,T}}{2kT}\right), \tag{11.9}$$

where L_0 is a scaling factor, $\Delta E_{A,T}$ is a thermal activation energy, and k is Boltzmann constant.

Figure 11.2 shows the experimental results for n-ZnO and the fit using (11.9), which yields activation energy of 45 ± 2 meV. This energy represents carrier delocalization energy, since it determines the increase of the diffusion length due to reduction of recombination efficiency [42]. The smaller is the activation energy, the more efficient is the de-trapping of captured carriers at a fixed temperature (see discussion below).

The role of increasing carrier lifetime is also supported by the results of cathodoluminescence (CL) measurements, carried out in situ in SEM, which are shown in Fig. 11.3. The inset of Fig. 11.3 shows a CL spectrum in the vicinity of the

Table 11.1 Temperature dependence of minority carrier diffusion length and cathodoluminescence intensity of the near-band-edge peak in n-ZnO (after [40])

Temperature (°C)	Diffusion length (μm)	CL intensity (10^3 counts)
25	0.438 ± 0.022	72.1 ± 3.7
50	0.472 ± 0.060	54.4 ± 3.8
75	0.493 ± 0.028	49.2 ± 2.4
100	0.520 ± 0.074	44.6 ± 4.7
125	0.547 ± 0.086	38.5 ± 6.8
E_A (eV)	0.045 ± 0.002	0.058 ± 0.007

Fig. 11.2 Experimental
dependence of minority
carrier diffusion length on
temperature (*open circles*).
The line shows the fit (11.9)
with activation energy of
45 ± 2 meV (after [40])

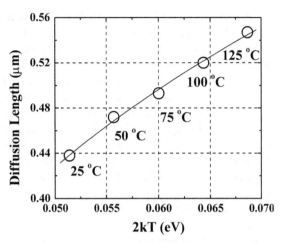

Fig. 11.3 Experimentally
obtained values for the peak
NBE CL intensity in ZnO:Li
as a function of temperature
(*open circles*) and the fit
[*solid line*; (11.10)], yielding
activation energy of
58 ± 7 meV. *Inset*: CL
spectrum on *n*-ZnO showing
the NBE transition at 25°C
(after [40])

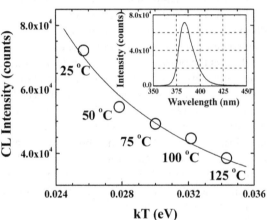

NBE transition at 383 nm (3.24 eV). This feature in bulk ZnO has been attributed to
the transition from the conduction band to a deep acceptor level [44]. It was observed
that the peak intensity, I_{CL}, of NBE luminescence decays systematically with
increasing temperature, providing direct evidence that the number of recombination
events decreases. Because the intensity of the NBE luminescence is inversely
proportional to the lifetime of carriers in the band, the decay of I_{CL} indicates the
increase of τ with temperature. The decay proceeds exponentially according to the
equation below [45]:

$$I_{CL} = \frac{A}{[1 + B \, \exp(-\Delta E_{A,T}/kT)]},\qquad (11.10)$$

where A and B are scaling factors, and $\Delta E_{A,T}$ is the activation energy, similar in
nature to that in (11.9). Based on the fit shown in Fig. 11.3, the activation energy

was determined to be 58 ± 7 meV. This value is in excellent agreement with that obtained by PL measurements in Li-doped ZnO films [46]. It is also consistent with the results of the variable-temperature EBIC measurements, which suggests that the same underlying process is responsible for both the increase in the diffusion length and the CL intensity decay. This process is outlined below.

The increase in minority hole lifetime in the valence band is likely associated with a smaller recombination capture cross-section for this carrier at elevated temperatures. In GaAs, for example, detailed analysis for temperature dependence of capture cross-section indicates an order of magnitude decrease of recombination efficiency, measured in terms of an "effective capture radius," in the temperature range from 100 to 300 K [42]. Nonequilibrium electron–hole pairs are generated by the beam of the SEM and subsequently annihilate by recombining with each other. Since the hole capture cross-section is inversely proportional to temperature [42, 47], the frequency of the recombination events (and, hence, the CL intensity) decreases as the temperature is raised. This means that nonequilibrium holes exist in the valence band for longer periods of time and, consequently, diffuse longer distances before undergoing recombination. Note that carrier diffusivity, D, is also a temperature-dependent quantity and, therefore, can affect the diffusion length [cf. (11.1)]. On the other hand, it has been demonstrated for n-ZnO that the mobility, μ, of majority carriers decreases in the temperature range of our experiments by about a factor of 2 [48]. Assuming that the mobility of the minority carriers exhibits the same behavior [43] and combining (11.1) and (11.2), it is clear that the value of the diffusion length is dominated by the growing lifetime of minority holes. From the Einstein relation (11.2), the above-referenced difference in mobility translates to about a 30% decrease in diffusivity at 125°C as compared to 25°C. Based on a 30% difference in diffusivity and using experimentally obtained values of diffusion length, we conclude that the lifetime of minority holes at 125°C is nearly 2.5 times greater than at room temperature.

Preliminary results indicate that temperature-induced increase in carrier lifetime also occurs in epitaxial ZnO. Nitrogen-doped ZnO samples grown using MBE were provided by SVT Associates. Hall effect measurements revealed hole concentration of 4.5×10^{17} cm^{-3} and mobility of ~ 1 cm^2/V s. Figure 11.4 shows the decay of NBE CL intensity fitted with (11.10). The measurements yielded activation energy of 118 ± 12 meV. This value is comparable to the activation energy of the nitrogen acceptor in ZnO [50, 51], which indicates possible nonequilibrium carrier trapping on nitrogen-related deep levels.

11.4.2   Studies in p-Type ZnO Doped with Antimony

The possibility of p-type doping with larger radii atoms, such as antimony, has been explored in [15, 52]. The studies demonstrated that despite the large size mismatch, which in principle should inhibit the substitution of this impurity on the oxygen site, effective p-type doping with hole concentrations up to 10^{20} cm^{-3}

Fig. 11.4 Maximum CL intensity of the NBE transition in *p*-ZnO:N as a function of temperature (*open circles*). *Solid line* shows the fit with (11.10), resulting in activation energy of $118 \pm 12\,\text{meV}$ (after [49])

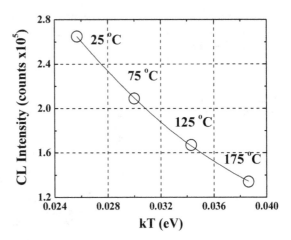

Table 11.2 Room-temperature electronic properties of Sb-doped p-type ZnO films (after [55])

Sample number	Hole concentration (cm^{-3})	Carrier mobility ($\text{cm}^2/\text{V s}$)
1	1.3×10^{17}	28.0
2	6.0×10^{17}	25.9
3	8.2×10^{17}	23.3
4	1.3×10^{18}	20.0

can be achieved [52]. These findings prompted the first-principles investigation by Limpijumnong et al., who suggested that the role of acceptors in size-mismatched impurity-doped ZnO is performed by a complex of the impurity with two zinc vacancies $\text{Sb}_{\text{Zn}}-2\text{V}_{\text{Zn}}$, the ionization energy of which is several-fold lower than that of a substitutional configuration and is consistent with the independent experimental observations [53]. Despite the encouraging predictions, however, very few attempts at achieving p-type conductivity in antimony-doped ZnO have been effective. Aoki et al. reported surprisingly high hole concentrations of up to $5 \times 10^{20}\,\text{cm}^{-3}$ in ZnO:Sb films prepared by excimer laser doping [52]. Some of the authors also obtained p-type ZnO:Sb by MBE [15].

The characteristics of an acceptor level in Sb-doped, p-type ZnO were studied using CL spectroscopy as a function of hole concentration. Variable-temperature CL measurements allowed to estimate the activation energy of a Sb-related acceptor from temperature-induced decay of CL intensity. The experiments were performed on ZnO:Sb layers grown on Si (100) substrates by an electron cyclotron resonance (ECR)-assisted MBE. The detailed growth procedures are available in [15]. Hall effect measurements revealed strong p-type conductivity, with hole concentrations up to $1.3 \times 10^{18}\,\text{cm}^{-3}$ and mobility up to $28.0\,\text{cm}^2/\text{V s}$ at room temperature (Table 11.2).

CL measurements were conducted in situ in the Philips XL30 SEM integrated with Gatan MonoCL CL system. The SEM is also fitted with a hot stage and an external temperature controller (Gatan) allowing for temperature-dependent

Fig. 11.5 NBE
cathodoluminescence spectra
of samples 1–4 taken at room
temperature. The peaks are at
382, 384, 385, and 387 nm,
respectively. *Inset*:
broad-range CL spectra of the
same samples (after [55])

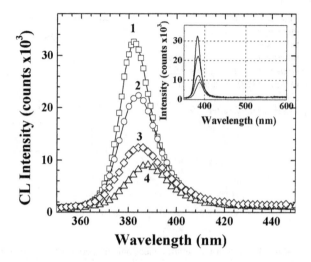

experiments. The decay of near-band-edge (NBE) luminescence intensity was monitored as a function of temperature in the range from 25 to 175°C. Accelerating voltage of 10 kV was used. Note that each measurement was taken in a previously unexposed area to avoid the potential influence of electron irradiation [40, 54].

The investigation of the luminescence properties of Sb-doped ZnO was started with the acquisition of room-temperature CL spectra shown in Fig. 11.5.

The inset of Fig. 11.5 reveals that the CL spectra of all three samples are dominated by the NBE band, which generally contains the band-to-band transition as well as the transition from the conduction band to a deep, neutral acceptor level (e, A^0) [44, 54]. Since acceptor levels form a band in the forbidden gap, the red shift of the NBE peak with increasing carrier concentration (i.e., higher doping levels) is consistent with the (e, A^0) emission and may indicate the broadening of the Sb-related acceptor band [56, 57]. Another observation that can be made from Fig. 11.5 is the systematic decay in intensity of the NBE luminescence with increasing doping level. This decrease may be attributed to the reduction in radiative recombination rates as more disorder is introduced into the ZnO lattice by large-radius Sb atoms. The increasing trend in the values of the full width at half maximum (FWHM) of the NBE spectra provide further evidence for the impact of the size-mismatched dopant – FWHM values were determined to be about 16.1, 19.4, 23.5, and 21.7 nm (corresponding to 136, 163, 196, and 178 meV) for samples 1, 2, 3, and 4, respectively. Note that while FWHM of NBE transitions in CL spectra tends to be greater than the width of PLpeaks, the above values are comparable to those obtained for (e, A^0) transitions in CL spectra of other ZnO and GaN materials [40, 54, 58].

The intensity of NBE luminescence was also monitored as a function of temperature. It was observed that the intensity decays with sample temperature, T, in agreement with expression (10) [45]. From (11.10), it can be deduced that the inverse intensity, $1/I$, should exhibit an exponential dependence on $1/kT$. This is

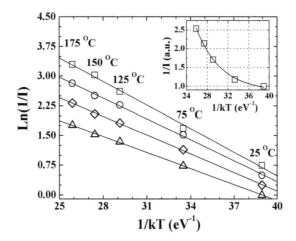

Fig. 11.6 Arrhenius plot showing the decay of normalized NBE luminescence intensity with increasing temperature for sample 1 (*open squares*), sample 2 (*open circles*), sample 3 (*open diamonds*), and sample 4 (*open triangles*). The linear fits (*solid lines*) yielded activation energies of 212 ± 28, 175 ± 20, 158 ± 22, and 135 ± 15 meV for samples 1, 2, 3, and 4, respectively. The data were vertically offset for clarity. *Inset*: exponential decrease of CL intensity for sample 1 (*open squares*) and the fit (*solid line*) (after [55])

shown in the inset of Fig. 11.6 on the example of sample 1. Note that the intensity in this and subsequent figures was normalized with respect to its room-temperature value for each of the samples. The activation energies, E_A, were obtained from the slopes of Arrhenius plot shown in Fig. 11.6. In case of a (e, A^0) transition, E_A is related to the ionization energy of acceptors: the lower the value of the activation energy, the more likely is the ionization of the acceptor by a valence band electron ($A^0 + e \rightarrow A^-$); since an ionized level does not participate in recombination via the (e, A^0) route, the rate of these transitions (i.e., the intensity of the luminescence) decreases with E_A at any given temperature. Conversely, for a constant E_A, the intensity decays with increasing temperature as more and more acceptors are ionized.

It is apparent from Fig. 11.6 that the activation energy shows a systematic dependence on the carrier concentration. The values of E_A are 212 ± 28, 175 ± 20, 158 ± 22, and 135 ± 15 meV for samples 1, 2, 3, and 4, respectively. These values are in reasonable agreement with the ionization energy of a Sb_{Zn}–$2V_{Zn}$ complex predicted by Limpijumnong et al. to have a value of about 160 meV [53]. Furthermore, the decay of activation energy with carrier density, p, follows a common pattern observed previously in other semiconductors [57–59] and is described by an equation of the type

$$E_A(N_A^-) = E_A(0) - \alpha(N_A^-)^{1/3}, \tag{11.11}$$

Fig. 11.7 Decrease of activation energy as a function of ionized acceptor concentration (after [55])

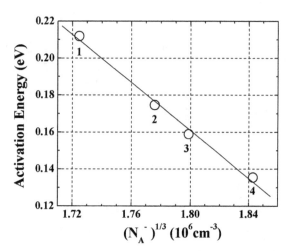

where N_A^- is the concentration of ionized acceptors, $E_A(0)$ is the ionization energy at very low doping levels, and α is a constant accounting for geometrical factors as well as for the properties of the material. Figure 11.7 demonstrates that (11.11) provides a reasonable fit to the experimentally obtained activation energies under the approximation that $N_A^- - N_D^+ = p$, where N_D^+ is the density of ionized shallow donors (due to compensation, the p-type conductivity is determined by the difference between the concentrations of ionized donors and acceptors).

The value of α was found to be equal to 6.4×10^{-7}, which is comparable to that in p-GaN and p-Si [57,59]. N_D^+ can be estimated from the electron concentration in undoped, n-type ZnO samples grown by the same method and is about 5×10^{18} cm^{-3} [15]. We note that this is a rough estimate and does not account for the earlier observation that background donor concentration in Sb-doped samples may be different from that in undoped ZnO films due to the creation of Zn vacancies induced by Sb doping. The concentration of Zn vacancies was shown to depend on Sb doping level [15], which would in turn lead to the variations in shallow donor density among the samples under investigation.

It should be noted that earlier PL measurements performed on sample 4 showed consistent activation energy of 140 meV [15]. Furthermore, temperature-dependent measurements of hole concentration in sample 4 shown in Fig. 11.8 suggest that the temperature dependence of luminescence intensity is associated with acceptors. The dependence of p on temperature can be modeled with a charge-balance equation of the following form:

$$p + N_D^+ = \frac{N_A^-}{1 + (p/\varphi)}, \qquad (11.12)$$

where N_D^+ and N_A^- were estimated as described earlier and $\varphi = AT^{3/2} \exp(-E_A/kT)$, with A being a factor accounting for the degeneracy of acceptor states and the density of states in the valence band [48]. The fit of the data shown in Fig. 11.8 revealed activation energy of about 100 meV [15], which is in reasonable

Fig. 11.8 Temperature dependence of hole concentration in sample 4. The fit yielded activation energy of about 100 meV (after [55])

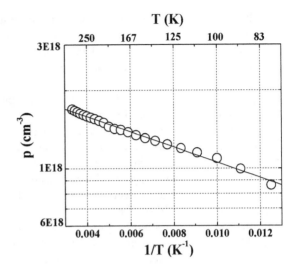

agreement with activation energy obtained by CL and is most likely related to the $Sb_{Zn}-2V_{Zn}$ complex.

Although the existence of other Sb-related acceptors cannot be categorically excluded, their involvement in the temperature-induced CL intensity decay is highly unlikely. The substitutional defect (Sb_O) as well as the single-vacancy complex ($Sb_{Zn}-V_{Zn}$) is predicted to have ionization energies about an order of magnitude greater than those obtained experimentally, while other defects can also be ruled out based on their electrical behavior and/or high formation energies [53].

The phenomenon of variation of the dopant activation energy with carrier concentration in semiconductors has been attributed to a number of causes. Among these are the formation of the band-tail states that extend into the forbidden gap, the broadening of the acceptor band in the gap, and the reduction of binding energy due to Coulomb interaction between the holes in the valence band and the ionized acceptor states [56–58]. The variable-temperature CL studies of Sb-doped p-type ZnO allowed estimating the activation energy of the Sb-related acceptor in the range of 135–212 meV. The activation energy was found to be strongly dependent upon the hole concentration. While the nature of the acceptor cannot be determined conclusively, evidence suggests that it is a $Sb_{Zn}-2V_{Zn}$ complex proposed by Limpijumnong et al. [53].

11.5   Studies of Minority Carrier Recombination

Deep carrier traps have pronounced implications on minority carrier transport and often give rise to such undesirable phenomena as radiation-induced optical metastability, persistent photoconductivity, and optical quenching of photocurrent

[60–62]. On the other hand, it has been demonstrated that capture of minority carriers by deep metastable traps is associated with the increase of minority carrier diffusion length and lifetime [40, 54, 63]. Furthermore, in GaN this increase was shown to result in a significant (several-fold) improvement of photoresponse in agreement with (11.3) [32, 63].

Extensive studies aimed at achieving p-type conductivity in ZnO reveal that most potential acceptors, such as nitrogen, phosphorus, and arsenic, tend to form acceptor levels far from the valence band maximum [64]. Since the ionization fraction of such acceptors is low (due to their high activation energy), there is a large concentration of neutral states that may act as traps for nonequilibrium electrons. The effects of electron trapping on the diffusion length and lifetime of minority carriers can be probed by subjecting the material of interest to the excitation by the electron beam of the SEM. The remainder of this review deals with the influence of electron irradiation on minority carrier diffusion length (Sect. 11.5) and lifetime (Sect. 11.5.2). The mechanism for the irradiation-induced effects and the role of deep carrier traps are discussed in detail in Sect. 11.5.3 [65].

11.5.1 Influence of Electron Trapping on Minority Carrier Diffusion Length

The measurements of diffusion length as a function of beam irradiation duration were carried out on bulk n-ZnO doped with lithium [40], MBE p-ZnO doped with nitrogen [65], phosphorus-doped ZnMgO grown by pulsed laser deposition (PLD) [54], and on Sb-doped epitaxial ZnO layers MBE grown on a Si substrate [55].

As was already mentioned, bulk ZnO samples (Tokyo Denpa Co.) were weakly n-type, showing electron concentrations of $\sim 10^{14}$ cm^{-3} and mobility of ~ 150 cm^2/V s at room temperature. SIMS measurements revealed the Li concentration of about 3×10^{16} cm^{-3} [41] (Li is often added to ZnO to increase the resistivity of initially n-type samples).

Phosphorus-doped $Zn_{0.9}Mg_{0.1}O$ layers were fabricated using PLD. Capacitance–voltage profiling of similar films, grown using the same procedure, resulted in net acceptor concentration of $\sim 2 \times 10^{18}$ cm^{-3} after annealing. Pt/Au (200/800 Å) layers were deposited on phosphorus-doped $Zn_{0.9}Mg_{0.1}O$ films by electron beam evaporation and patterned by liftoff with contact diameters ranging from 50 to 375μm. Circular electrode pairs with significantly different surface areas were employed for the EBIC measurements. The electrodes create an asymmetric rectifying junction, based on back-to-back Schottky diodes, with the larger area electrode being pseudo-ohmic.

EBIC experiments were conducted at room temperature in situ in a Philips XL30 SEM using a planar-collector configuration with a Schottky barrier (Sect. 11.3.1) to monitor the changes in minority carrier diffusion length as a function of time. The results of EBIC experiments are shown in Fig. 11.9. Note that the results shown

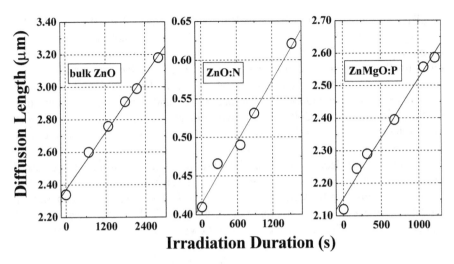

Fig. 11.9 Experimental dependence of minority carrier diffusion length on duration of electron beam irradiation at room temperature (*open circles*) and the linear fit (*solid line*) (after [40,54,65])

in the left panel of Fig. 11.9 and those discussed in Sect. 11.4 were obtained from different bulk ZnO samples, which may offer an explanation for the significant difference in the initial, room temperature values of diffusion length. Additionally, while the large diffusion length value of bulk ZnO can be attributed to the higher quality of the bulk material (compared to the epitaxial layers), the diffusion length of $\sim 2\,\mu m$ in PLD-grown ZnMgO is rather surprising because of polycrystalline nature of the layers. It appears that in the latter sample, the transport of minority carriers is not significantly influenced by scattering from the grain boundaries.

The experiments revealed that diffusion length follows a general trend in all samples studied. Figure 11.9 shows that irradiation by the electron beam clearly results in a significant increase of the carrier diffusion length, and that this increase is linear with respect to the duration of electron irradiation (t). Up to 50% increase of diffusion length was achieved [cf. Fig. 11.9 (center)], with the rates of diffusion length increase ranging from ~ 13 to $\sim 30\%$ per 1,000 s. This appears to be a common occurrence in wide bandgap semiconductors doped with species that create deep acceptor levels, as similar observations were made in (Al)GaN doped with Mg, Mn, Fe, and C [55,66,67]. It is also noteworthy that similar experiments conducted on bulk ZnO *without* any intentional dopants did not show any significant changes in minority carrier diffusion length [54].

The observed increase of L is ascribed to charging of the deep, neutral acceptor states by the electrons generated by the SEM beam, since trapping of nonequilibrium electrons prevents these levels from participating in recombination [54,55] (see also discussion in Sect. 11.5.3). Therefore, the difference in the rates of diffusion length increase is likely explained, at least in part, by the difference in the concentrations of these deep centers [68].

EBIC measurements in Sb-doped ZnO were performed on the samples by moving the electron beam of the SEM from the edge of the Schottky barrier (created on the top surface of ZnO:Sb samples by Ti/Au evaporation followed by liftoff) outwards (line scan) and recording an exponential decay of induced current.

After a single EBIC line scan was completed (12 s), the excitation of the sample was continued by moving the electron beam back and forth along the same line for the total time of $\sim 2,200$ s. EBIC measurements were periodically repeated to extract the minority carrier diffusion length [68], L, as a function of the duration of electron beam irradiation, t.

The effects of electron injection for Sb-doped $0.2 - \mu$m-thick p-type ZnO epitaxial layers ($p = 1.3 \times 10^{17}\,\mathrm{cm}^{-3}$; $\mu = 28\,\mathrm{cm}^2/\mathrm{V\,s}$ at room temperature) grown on Si substrate by MBE are shown in Fig. 11.10. The activation energy for the e-beam injection-induced increase of L, $\Delta E_{A,I} = 219 \pm 8\,\mathrm{meV}$, was obtained from the graphs in Fig. 11.10b, c using the following equation:

$$R = R_0 \exp\left(\frac{\Delta E_{A,I}}{kT}\right) \exp\left(-\frac{\Delta E_{A,T}}{2kT}\right), \tag{11.13}$$

where R_0 is a scaling constant, T is temperature, k is the Boltzmann's constant, $\Delta E_{A,I}$ is the activation energy of electron irradiation effect, and $\Delta E_{A,T}$ is the activation energy of thermally induced increase of L shown in Fig. 11.10a ($\Delta E_{A,T} = 184 \pm 10\,\mathrm{meV}$). The value of $\Delta E_{A,I}$ is in reasonable agreement with that for a $\mathrm{Sb_{Zn}}$–$2\mathrm{V_{Zn}}$ acceptor complex, predicted by Limpijumnong et al. [53] (see [69] for details).

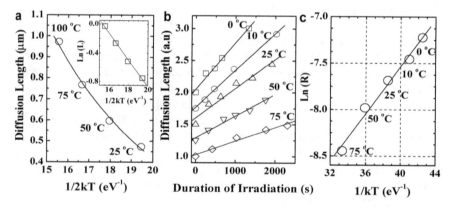

Fig. 11.10 (a) Diffusion length of minority electrons as a function of temperature (*open circles*) and the fit (*solid line*). *Inset*: Arrhenius plot of the same data yielding activation energy of $184 \pm 10\,\mathrm{meV}$. (b) Electron beam irradiation-induced increase of minority electron diffusion length at different temperatures. The values of the diffusion length were vertically offset for clarity and are not intended to illustrate the temperature dependence. (c) Rate of irradiation-induced increase of diffusion length as a function of temperature (*open circles*). The fit with (11.2) (*solid line*) gives activation energy of $219 \pm 8\,\mathrm{meV}$ (after [69])

Fig. 11.11 Saturation and relaxation dynamics of minority carrier diffusion length in *p*-ZnO:Sb at room temperature. The *arrow* marks the time at which the electron irradiation was discontinued. Note: a much steeper increase of *L* in this figure as compared to Fig. 11.6b is explained by higher electron beam accelerating voltage used in the first case (10 vs. 5 keV) (after [69])

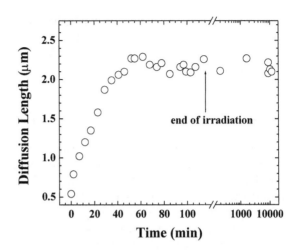

It can also be seen from Fig. 11.10 that the rate, R, of the diffusion length increase is reduced with increasing temperature. The increase of the diffusion length due to trapping is counteracted by the release of the trapped electrons that occurs if the carriers gain sufficient energy to escape the trap. As the temperature is raised, the likelihood of de-trapping increases, which dampens the irradiation-induced growth of the diffusion length.

The saturation and relaxation of irradiation-induced change of diffusion length was studied at room temperature. Figure 11.11 shows that L reaches its maximum value after about 50 min of continuous exposure to the electron beam. Further monitoring revealed that irradiation-induced increase persists for at least 1 week. Annealing the sample at 175°C for about 30 min resulted in a decrease of the diffusion length to about 1 μm. This behavior further supports the involvement of deep electron traps in the phenomenon of interest, since temperature-induced de-trapping of carriers (cf. Fig. 11.10c) re-activates the original recombination route (cf. Fig. 11.10a), thus reducing carrier lifetime and diffusion length.

11.5.2 *Optical Studies of the Effects of Electron Trapping on Minority Carrier Lifetime*

According to (11.1), electron irradiation-induced increase in minority carrier diffusion length discussed earlier is associated with the increase of lifetime of nonequilibrium carriers. Experimental evidence for this dependence was obtained from the CL measurements performed on the same samples. Time-dependent CL measurements were conducted in situ in SEM using setup described in Sect. 11.4.2. This setup allows combining periodic acquisition of CL spectra with continuous excitation of the sample by scanning the beam over the same location. For temperature-dependent CL measurements, the sample temperature was varied in situ using specially designed hot stage and an external temperature controller (Gatan). At each temperature, the electron beam irradiation and CL measurements were conducted at a different location.

11.5.2.1 Optical Studies of $Zn_{0.9}Mg_{0.1}O$ Doped with Phosphorus

Figure 11.12 shows a series of NBE transitions in p-type $Zn_{0.9}Mg_{0.1}O : P$ recorded under continuous electron irradiation and numbered in order of increasing irradiation duration. The edge of this spectrum at $\sim$355 nm (see also inset of Fig. 11.12) is in agreement with 10% Mg content in ZnO lattice, since each atomic percent of Mg is known to increase the ZnO bandgap by 0.02 eV [9]. The observed CL spectrum is attributed to the band-to-band as well as band-to-impurity (P-acceptor) optical transitions. The inset of Fig. 11.12 shows a wider range spectrum featuring a broad band, which is likely defect related [70].

While no changes were observed in the broad band CL (cf. inset of Fig. 11.12), the near band-edge luminescence in Fig. 11.12 exhibits a continuous decay with increasing duration of electron beam irradiation. This demonstrates that exposure to the electron beam results in the increase of carrier lifetime (τ), since I_{CL} is proportional to $1/\tau$. Similar phenomena were also observed in GaN, where the

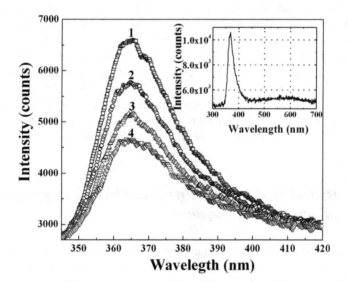

Fig. 11.12 Room-temperature CL spectra of $Zn_{0.9}Mg_{0.1}O : P$ measured in the same location at different times of electron injection. 1 is a preinjection spectrum; 2, 3, and 4 correspond to duration of electron injection of 359, 793, and 1,163 s, respectively. *Inset:* preinjection broad-range CL spectrum taken in a different location than measurements in Fig. 11.12 (after [54])

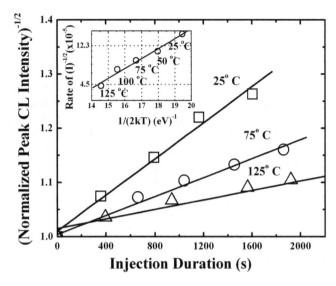

Fig. 11.13 Variable-temperature dependence for the square root of inverse normalized intensity on duration of electron injection in $Zn_{0.9}Mg_{0.1}O : P$. The rate at every temperature is obtained from the slope of a linear fit. *Inset*: temperature dependence for the rate of the square root of inverse normalized intensity (*open circles*) and the fit. The slope of the graph yields $\Delta E_A = 256 \pm 20$ meV (after [54])

decay of NBE CL intensity occurred concomitantly with increasing diffusion length [54, 66, 67].

To characterize the intensity decay, we relate it to the diffusion length, L, which is known to vary linearly with duration of irradiation (cf. Fig. 11.13). Since L is proportional to $\tau^{1/2}$ (11.1), the inverse square root of normalized (with respect to the initial maximum value) intensity must also be proportional to L, and consequently, would be expected to change linearly with duration of electron irradiation. Figure 11.13 shows that this is indeed the case, indicating that the observed increase of the diffusion length is attributable to the growing lifetime of nonequilibrium carriers.

In CL, the temperature of the samples was varied from 25 to 125°C. At each temperature, injection was performed on a site previously not exposed to electron beam. As is apparent from the inset of Fig. 11.13, the rate of the decrease in CL intensity (described by the slope, R, of the linear dependence of $I_{CL}^{-1/2}$ on t) diminishes with growing temperature. This suggests that while electron irradiation results in an increase of carrier lifetime, there exists another, thermally activated process that contributes to its decay.

Taking into account both injection-induced effect on $R[\propto \exp(\Delta E_A/kT)]$ and its temperature dependence [$\propto \exp(\Delta E_A/ - 2kT)$] [42], and assuming that activation energies are similar in both cases, the temperature dependence of R can be described as follows:

$$R = R_0 \exp\left(\frac{\Delta E_A}{2kT}\right), \tag{11.14}$$

where R_0 is a scaling constant and ΔE_A is the activation energy for the overall process. Fitting the experimental results with this expression (inset of Fig. 11.13) yielded activation energy of 256 ± 20 meV. This activation energy is in good agreement with that for the phosphorus acceptor obtained based on the simple hydrogenic model. The model assumes phosphorus substitution on the oxygen site and predicts the activation energy of about 250–300 meV [71]. The experimentally obtained value of the activation energy, combined with the fact that no electron irradiation effects were observed in undoped ZnO [54], suggests that carrier trapping on phosphorus acceptor levels plays a crucial role in this phenomenon (see Sect. 11.5.3 for detailed mechanism).

11.5.2.2 Optical Studies of Bulk ZnO

Similarly to ZnMgO, the electron irradiation-induced increase of minority carrier diffusion length in ZnO:Li (Sect. 11.5.1) was found to correlate with the increase of minority carrier lifetime. Figure 11.14 shows a series of room-temperature NBE spectra collected under continuous electron beam excitation, in which the intensity

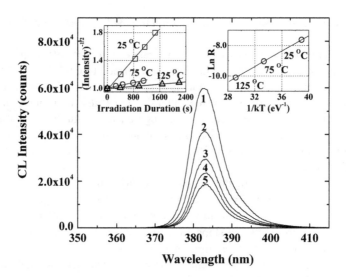

Fig. 11.14 Room-temperature cathodoluminescence spectra of ZnO:Li taken under continuous excitation by the electron beam. 1 is the preirradiation spectrum and 5 is the spectrum after 1,450 s of electron irradiation. *Left inset*: variable-temperature dependence of inverse square root of normalized intensity on duration of electron irradiation and the linear fit with the rate R. *Right inset*: arrhenius plot of R as a function of temperature yielding an activation energy $\Delta E_{A,l}$ of 283 ± 9 meV (after [40])

of the NBE transition can be seen to fall steadily with increasing t. The peak of this emission occurs at 383 nm (3.24 eV) and has been assigned to the transition from the conduction band to a deep acceptor level (e, A^0) [44]. The inverse square root of maximum CL intensity ($I_{CL}^{-1/2}$), which is proportional to $\sqrt{\tau}$ and, therefore, to L, changes linearly with duration of irradiation (cf. left inset of Fig. 11.14), which is consistent with the results of EBIC measurements and indicates that the increase in L occurs due to the irradiation-induced growth of carrier lifetime.

CL measurements conducted at elevated temperatures confirmed the same trend for the irradiation-induced change of luminescence intensity. It can be seen from the left inset of Fig. 11.14 that the inverse square root of intensity increases linearly for all temperatures. The temperature dependence of rate, R, can be used to determine the activation energy of the irradiation-induced processes according to (11.14). On the other hand, our earlier studies of the *temperature*-induced CL intensity decay (Sect. 11.4) yielded the activation energy of about 60 meV [54], thus allowing us to separate the two components as followed from (11.13): the activation energy of electron irradiation effect $\Delta E_{A,I}$; and, the previously determined, activation energy of thermally induced intensity decay $\Delta E_{A,T}$. This treatment yielded the value for $\Delta E_{A,I}$ of 283 ± 9 meV. Incidentally, [41] also reports high concentration of ~ 0.3 eV electron traps found in the same material by deep level transient spectroscopy (DLTS). The significant difference between $\Delta E_{A,I}$ and $\Delta E_{A,T}$ observed in bulk ZnO suggests that, unlike in other materials, temperature- and irradiation-induced changes of the minority carrier transport characteristics are two distinctly different processes.

Although several theoretical works have predicted a very shallow Li_{Zn} level [72, 73], these predictions have not been substantiated experimentally, as most studies find a rather deep Li acceptor with activation energies of several hundreds meV [50, 74]. In fact, recent first-principles calculations by Wardle et al. also suggest that the Li_{Zn} acceptor state lies at about 0.2 eV above the valence band maximum [74], which is in reasonable agreement with the $\Delta E_{A,I}$ of 283 ± 9 meV obtained in this work.

It should be clarified that the weak n-type character of the sample is not necessarily in contradiction with the dominant behavior of acceptor states observed in electron trapping phenomena. As was mentioned, the n-type conductivity in nominally undoped ZnO is due to the *shallow* donor states, whereas in presence of deep electron traps the Fermi level may lie far below these states. Although shallow donors may capture nonequilibrium electrons under excitation, those are quickly released if the temperature is sufficiently high. Therefore, if the difference in the energetic position between the donor and trap states is large, the latter dominate the kinetics of electron trapping [75].

11.5.2.3   Optical Studies of ZnO Doped with Nitrogen

CL measurements performed on MBE-grown, nitrogen-doped p-ZnO revealed behavior similar to that of bulk ZnO [40] and PLD-grown ZnMgO [54].

Fig. 11.15
Room-temperature CL
spectra of ZnO:N taken under
continuous excitation. Trace 1
corresponds to the
preirradiation spectrum and
trace 5 to the spectrum after
1,940 s of electron irradiation.
Inset: linear dependence of
the inverse square root of
normalized peak intensity at
room temperature (after [65])

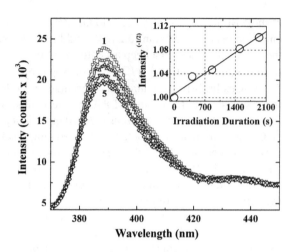

Room-temperature ZnO:N spectra are shown in Fig. 11.15 and feature a NBE luminescence band with a maximum at about 388 nm ($\sim$3.20 eV). This band includes the (e, A^0) transition as well as the donor–acceptor pair (DAP) recombination, with nitrogen identified as the acceptor in both processes [76]. Additionally, a violet band centered on 435 nm has been attributed to the radiative recombination of the electrons trapped at grain boundaries with the holes in the valence band [77].

As expected, irradiation with electron beam resulted in decay of the intensity of NBE luminescence, indicating increasing lifetime. One can observe from the inset of Fig. 11.15 that, in agreement with the diffusion length measurements (cf. Fig. 11.9 in Sect. 11.5.1), the inverse square root of the peak normalized intensity of the NBE transition changes linearly with irradiation time, yielding the rate R. Note that the intensity of the violet band is not affected by electron irradiation, which suggests that electron trapping at the grain boundaries does not play a significant role in the irradiation-induced increase of carrier lifetime.

CL measurements at elevated temperatures (Fig. 11.16) confirmed that R decreases with temperature, indicating a thermally activated process that counteracts the effects of electron injection, similar to what occurs in bulk ZnO and ZnMnO:P. Note that while the values of R were obtained based on the intensity normalized with respect to its initial value, the data displayed in Fig. 11.16 are offset by shifting the normalized results along the y-axis to avoid the overlap of the data points. The activation energy ($\Delta E_{A,I}$) of about 134 ± 10 meV was determined from the Arrhenius plot shown in the inset of Fig. 11.16, based on (11.13) and using $\Delta E_{A,T} = 118$ meV obtained from the temperature-dependent CL measurements (Sect. 11.4). This value is in reasonable agreement with the ionization energy of the nitrogen acceptor in ZnO [50, 51, 76, 78] and indicates that electron trapping by these levels plays an important role in the recombination dynamics of minority carriers.

Fig. 11.16
Variable-temperature
dependence of the inverse
square root of normalized
intensity in ZnO:N on
duration of electron
irradiation and the linear fit
with a rate R. The data are
offset for clarity. *Inset*:
arrhenius plot of R as a
function of temperature with
a fit yielding
$\Delta E_{A,I} = 134 \pm 10\,\text{meV}$
(after [65])

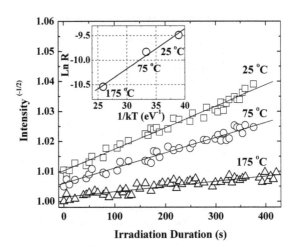

11.5.3 Mechanism of Electron Injection Effect

We have extensively applied the above-explained approach to investigate the influence of deep levels on minority carrier transport in acceptor-doped GaN [66,67] as well as ZnO compounds [40, 54, 55, 65]. In all cases, the activation energy determined from the temperature dependence of irradiation-induced intensity decay was in agreement with the ionization energy of the dominant deep acceptor species.

When exciting radiation, such as that from the beam of the SEM, generates nonequilibrium electron–hole pairs in acceptor-doped wide bandgap semiconductors, the dominant radiative recombination pathway generally involves the localized states of the acceptor atoms located far from the valence band edge (Fig. 11.17(I)). The deep nature of these acceptors prevents their complete ionization at room temperature, and the large concentration of neutral (nonionized) acceptor atoms act as traps for nonequilibrium electrons generated by the SEM beam. As a consequence of electron trapping, these states are removed from participating in carrier recombination (Fig. 11.17(II)). Since the transitions involving deep acceptor levels constitute a major part of radiative recombination events (cf. Sect. 11.5.2), electron trapping results in greater lifetime of nonequilibrium minority carriers in the band (due to a lower rate of recombination with nonequilibrium majority carriers) and, consequently, greater diffusion length.

The fact that the rate of the irradiation-induced lifetime increase, R, diminishes with increasing temperature indicates that there exists a competing, thermally activated process that counteracts the capture of nonequilibrium electrons by neutral acceptor levels. This process is the release of the trapped electrons, shown schematically in Fig. 11.17(III). As the temperature increases, the balance between capture and release shifts toward the latter, reducing the impact of electron irradiation on the carrier lifetime and diffusion length and resulting in slower increase of these parameters at elevated temperatures.

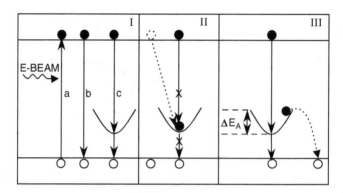

Fig. 11.17 Electron beam generates nonequilibrium electron–hole pairs (Ia). Nonequilibrium carriers recombine either via the band-to-band transition (Ib) or through unoccupied (neutral) acceptor states (Ic). If a nonequilibrium electron is trapped by the acceptor level, recombination cannot proceed (II), leading to increased lifetime of nonequilibrium minority carriers in the band. Release of the trapped electron with activation energy $\Delta E_{A,I}$ restores the original recombination pathway (III), resulting in a slower rate of lifetime increase at elevated temperatures

It is interesting to note that in p-type GaN the increase of minority electron diffusion length due to electron trapping can also be induced by application of forward bias to a p–n junction [63, 79]. This increase was shown to lead to the multifold enhancement of the peak responsivity in agreement with (11.3) and (11.4). Moreover, the increased responsivity persisted for at least several days at room temperature. Although conclusive evidence has not been obtained to date, the similarities in irradiation-induced phenomena, combined with the results of preliminary experiments, suggest that solid-state electron injection may produce the same results in ZnO.

11.5.4 Device Applications

If a p–n junction diode is biased in forward direction, the junction potential barrier decreases and the electrons from the n-type region are injected into p-type region. This should result in increase of minority carrier diffusion length in p-type ZnO and, therefore, enhancement of p–n junction's photoresponse.

Figure 11.18 shows a secondary electron image and superimposed EBIC line scan of ZnO p–n epitaxial structure grown on highly resistive (111)-oriented p-Si substrate and cleaved perpendicular to the growth plane.

The top inset of Fig. 11.18 presents a schematic of this structure, which was grown and fabricated at UCR. EBIC measurements on ZnO p–n junction (as well as spectral and temporal photoresponse measurements) were carried out at room temperature on the cleaved structures before and after forward bias electron

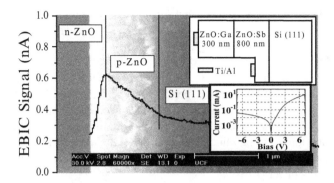

Fig. 11.18 Cross-sectional image of ZnO p–n junction in secondary electrons and superimposed EBIC line scan. *Inset (top)*: schematic of ZnO epitaxial structure (not to scale). *Inset (bottom)*: I–V curve of ZnO p–n junction (after [80])

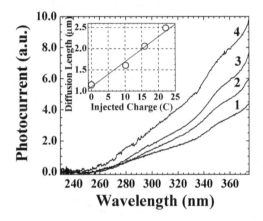

Fig. 11.19 Spectral response of ZnO p–n junction photodiode as a function of forward bias electron injection. Edge-illuminated configuration (as in Fig. 11.18) was employed for measurements. A shoulder observed on the spectra at ~350 nm corresponds to the effective bandgap of ZnO (~ 3.54 eV). An increase of the photoresponse beyond this wavelength into visible region (ideally the response from device should end there) is due to the collection of photogenerated carriers in the Si substrate. Spectrum 1 corresponds to the preinjected state; spectrum 2 – 2 C injected; spectrum 3 – 12.6 C injected; spectrum 4 – 25.5 C injected. *Inset*: L dependence on charge in *p*-ZnO:Sb layer due to forward current injection (after [80])

injection (see I–V curve in the bottom inset of Fig. 11.18). The values of L in the p-type ZnO layer were extracted from the line scans.

The inset of Fig. 11.19 shows diffusion length dependence on injected charge (forward bias, resulting in currents from 7 to 48 mA, was applied for the duration of about 1,500 s, in 300–600 s increment; L saturation was observed with increasing duration of injection, but it is not shown on the plot). *Note that the time (12 s) of p–n junction diode exposure to an electron beam during EBIC measurements is negligible as compared to that of forward bias application (up to ~1,500 s).*

Therefore, the impact of electron beam injection on increase of minority carrier diffusion length, observed in the inset of Fig. 11.19 for p-ZnO layer, is minimal. We also note that the diffusion length of minority holes in n-type ZnO (which was determined to be around several hundred nm) was not affected by forward bias.

A significant increase of L with forward bias electron injection is consistent with a pronounced and *long-lasting* (persisted at the same level for at least several days!) enhancement of ZnO p–n junction diode spectral photoresponse as shown in Fig. 11.19. For lateral collection devices (side-illuminated, as in Fig. 11.18), the photocurrent is known to vary linearly with L [81], which results in improved carrier collection and elimination of detector's "dead space" [82], where carriers recombine before they are collected.

To cut-off a p–n junction response in the visible light spectral region (cf. Fig. 11.19) a Si substrate was substituted by a wider bandgap bulk ZnO. While an increase in L (due to an increase in lifetime) leads to a significant increase of quantum efficiency, the temporal photoresponse becomes slower as is seen in Fig. 11.20 for ZnO p–n junction diode excited by pulsed fs laser tuned at 355 nm. Elongation of decay constant (Fig. 11.20, lower inset) is likely related to an increase of minority carrier lifetime due to electron injection. The result in Fig. 11.20 also supports the general rule that more sensitive detectors are slower [84]. While the decay of the photoresponse is limited by the RC constant of the measuring circuitry, the fundamental decay constant of the device *after electron injection* is about 15 μs.

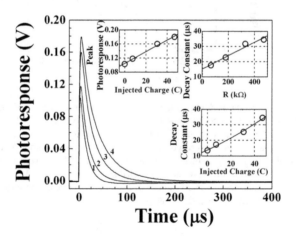

Fig. 11.20 Temporal response of ZnO p–n junction photodiode (cf. Fig. 11.18) as a function of forward bias electron injection. *Upper left inset*: dependence of peak photoresponse on injected charge. *Upper right inset*: decay constant of the improved photodiode (trace 4) as a function of external resistance (used for the measurements), indicating that the intrinsic decay time due to diode's series resistance is about 15 μs. *Lower inset*: dependence of decay constant on injected charge (after [83])

11.6   Summary

Issues affecting minority carrier transport in ZnO have been discussed, with special attention given to the temperature dependence of minority carrier diffusion length and lifetime, as well as to the recombination dynamics of nonequilibrium minority carriers. The mechanisms governing temperature- and irradiation-induced effects have been presented. Additionally, device implications of minority carrier diffusion length control were demonstrated.

Acknowledgements This research is partially supported by the National Science Foundation (ECCS #0900971), US-Israel Binational Science Foundation (award #2008328) and NATO (SfP #981939).

References

1. F. Oba, S.R. Nishitani, S. Isotani, H. Adachi, I. Tanaka, J. Appl. Phys. **90**, 824 (2001)
2. G.W. Tomlins, J.L. Routbort, T.O. Mason, J. Appl. Phys. **87**, 117 (2000)
3. C.G. Van de Walle, Phys. Rev. Lett. **85**, 1012 (2000)
4. Y.I. Alivov, E.V. Kalinina, A.E. Cherenkov, D.C. Look, B.M. Ataev, A.K. Omaev, M.V. Chukichev, D.M. Bagnall, Appl. Phys. Lett. **83**, 4719–4721 (2003)
5. Y.I. Alivov, U. Ozgur, S. Dogan, D. Johnstone, V. Avrutin, N. Onogima, C. Liu, J. Xie, Q. Fan, H. Markos, Appl. Phys. Lett. **86**, 241108 (2005)
6. C.X. Wang, G.W. Yang, C.X. Gao, H.W. Liu, Y.H. Han, J.F. Luo, G.T. Zou, Carbon **42**, 317 (2004)
7. A. Osinsky, J.W. Dong, M.Z. Kauser, B. Hertog, A.M. Dabiran, P.P. Chow, S.J. Pearton, O. Lopatiuk, L. Chernyak, Appl. Phys. Lett. **85**, 4272–4274 (2004)
8. J.W. Dong, A. Osinsky, B. Hertog, A.M. Dabiran, P.P. Chow, Y.W. Heo, D.P. Norton, S.J. Pearton, J. Electron. Mater. **34**, 416–423 (2005)
9. T. Gruber, C. Kirchner, R. Kling, F. Reuss, A. Waag, Appl. Phys. Lett. **84**, 5359 (2004)
10. K. Ip, Y.W. Heo, D.P. Norton, S.J. Pearton, J.R. LaRoche, F. Ren, Appl. Phys. Lett. **85**, 1169 (2004)
11. K.G. Chen, Z.Z. Ye, W.Z. Xu, B.H. Zhao, L.P. Zhu, J.G. Lv, J. Cryst. Growth **281**, 458 (2005)
12. V. Vaithianathan, B.T. Lee, S.S. Kim, J. Appl. Phys. **98**, 043519 (2005)
13. E.J. Egerton, A.K. Sood, R. Singh, Y.R. Puri, R.F. Davis, J. Pierce, D.C. Look, T. Steiner, J. Electron. Mater. **34**, 949 (2005)
14. V. Vaithianathan, B.T. Lee, S.S. Kim, Appl. Phys. Lett. **86**, 062101 (2005)
15. F.X. Xiu, Z. Yang, L.J. Mandalapu, D.T. Zhao, J.L. Liu, W.P. Beyermann, Appl. Phys. Lett. **87**, 152101 (2005)
16. Y.R. Ryu, W.J. Kim, H.W. White, J. Cryst. Growth **219**, 419 (2000)
17. J.M. Bian, X.M. Li, C.Y. Zhang, W.D. Yu, X.D. Gao, Appl. Phys. Lett. **85**, 4070 (2004)
18. F. Zhuge, L.P. Zhu, Z.Z. Ye, D.W. Ma, J.G. Lu, J.Y. Huang, F.Z. Wang, Z.G. Ji, S.B. Zhang, Appl. Phys. Lett. **87**, 092103 (2005)
19. S.K. Hazra, S. Basu, Solid State Electron. **49**, 1158 (2005)
20. Y.W. Heo, Y.W. Kwon, Y. Li, S.J. Pearton, D.P. Norton, J. Electron. Mater. **34**, 409 (2005)
21. S.H. Lim, D. Shindo, H.B. Kang, K. Nakamura, J. Vac. Sci. Tech. B **19**, 506 (2001)
22. K. Miyamoto, M. Sano, H. Kato, T. Yao, J. Cryst. Growth**265**, 34 (2004)
23. Y. Wang, X.L. Du, Z.X. Mei, Z.Q. Zeng, Q.Y. Xu, Q.K. Xue, Z. Zhang, J. Cryst. Growth **273**, 100 (2004)

24. M.W. Cho, A. Setiawan, H.J. Ko, S.K. Hong, T. Yao, Semicond. Sci. Tech. **20**, S13 (2005)
25. E. Monroy, F. Calle, J.L. Pau, F.J. Sanchez, E. Munoz, F. Omnes, B. Beaumont, P. Gibart, J. Appl. Phys. **88**, 2081 (2000)
26. J.F. Muth, R.M. Kolbas, A.K. Sharma, S. Oktyabrsky, J. Narayan, J. Appl. Phys. **85**, 7884 (1999)
27. H. Yoshikawa, S. Adachi, Jpn. J. Appl. Phys. Part 1 – Regul. Pap. Short Notes Rev. Pap. **36**, 6237 (1997)
28. D.E. Ioannou, S.M. Davidson, J. Phys. D – Appl. Phys. **12**, 1339 (1979)
29. D.E. Ioannou, C.A. Dimitriadis, IEEE Trans. Electron Dev. **29**, 445 (1982)
30. J. Boersma, J.J.E. Iindenkleef, H.K. Kuiken, J. Eng. Math.**18**, 315 (1984)
31. D.S.H. Chan, V.K.S. Ong, J.C.H. Phang, IEEE Trans. Electron Dev. **42**, 963 (1995)
32. L. Chernyak, A. Osinsky, A. Schulte, Solid State Electron. **45**, 1687 (2001)
33. T.S. Moss, J. Electron. Control **1**, 126 (1955)
34. D.K. Schroder, Meas. Sci. Tech. **12**, R16 (2001)
35. R. Chakrabarti, J. Dutta, S. Bandyopadhyay, D. Bhattacharyya, S. Chaudhuri, A.K. Pal, Sol. Energ. Mater. Sol. Cell. **61**, 113 (2000)
36. J. Lagowski, A. Aleynikov, A. Savtchouk, P. Edelman, Eur. Phys. J. – Appl. Phys. **27**, 503 (2004)
37. H.O. Olafsson, J.T. Gudmundsson, H.G. Svavarsson, H.P. Gislason, Physica B **274**, 689 (1999)
38. R. Shikler, N. Fried, T. Meoded, Y. Rosenwaks, Phys. Rev. B **61**, 11041 (2000)
39. R.K. Ahrenkiel, Solid State Electron. **35**, 239 (1992)
40. O. Lopatiuk, L. Chernyak, A. Osinsky, J.Q. Xie, P.P. Chow, Appl. Phys. Lett. **87**, 214110 (2005)
41. A.Y. Polyakov, N.B. Smirnov, A.V. Govorkov, E.A. Kozhukhova, S.J. Pearton, D.P. Norton, A. Osinsky, A. Dabiran, J. Electron. Mater. **35**, 663 (2006)
42. M. Eckstein, H.U. Habermeier, J. Phys. IV **1**, 23 (1991)
43. L. Chernyak, A. Osinsky, H. Temkin, J.W. Yang, Q. Chen, M.A. Khan, Appl. Phys. Lett. **69**, 2531 (1996)
44. K. Thonke, T. Gruber, N. Teofilov, R. Schonfelder, A. Waag, R. Sauer, Physica B **308**, 945 (2001)
45. D.S. Jiang, H. Jung, K. Ploog, J. Appl. Phys. **64**, 1371 (1988)
46. A. Ortiz, C. Falcony, J. Hernandez, M. Garcia, J.C. Alonso, Thin Solid Films **293**, 103 (1997)
47. J.I. Pankove, *Optical Processes in Semiconductors*. (Prentice-Hall, Englewood Cliffs, New Jersey, 1971)
48. D.C. Look, D.C. Reynolds, J.R. Sizelove, R.L. Jones, C.W. Litton, G. Cantwell, W.C. Harsch, Solid State Commun.**105**, 399 (1998)
49. O. Lopatiuk-Tirpak, *Influence of Electron Trapping on Minority Carrier Transport Properties of Wide Band Gap Semiconductors*, Ph.D. Dissertation, University of Central Florida, Orlando, 2007
50. B.K. Meyer, H. Alves, D.M. Hofmann, W. Kriegseis, D. Forster, F. Bertram, J. Christen, A. Hoffmann, M. Strassburg, M. Dworzak, U. Haboeck, A.V. Rodina, Phys. Status Solidi B – Basic Res. **241**, 231 (2004)
51. S. Yamauchi, Y. Goto, T. Hariu, J. Cryst. Growth **260**, 1 (2004)
52. T. Aoki, Y. Shimizu, A. Miyake, A. Nakamura, Y. Nakanishi, Y. Hatanaka, Phys. Status Solidi B – Basic Res. **229**, 911 (2002)
53. S. Limpijumnong, S.B. Zhang, S.H. Wei, C.H. Park, Phys. Rev. Lett. **92**, 155504 (2004)
54. O. Lopatiuk, W. Burdett, L. Chernyak, K.P. Ip, Y.W. Heo, D.P. Norton, S.J. Pearton, B. Hertog, P.P. Chow, A. Osinsky, Appl. Phys. Lett. **86**, 012105 (2005)
55. O. Lopatiuk-Tirpak, W.V. Schoenfeld, L. Chernyak, F.X. Xiu, J.L. Liu, S. Jang, F. Ren, S.J. Pearton, A. Osinsky, P. Chow, Appl. Phys. Lett. **88**, 202110 (2006)
56. M.G. Cheong, K.S. Kim, C.S. Kim, R.J. Choi, H.S. Yoon, N.W. Namgung, E.K. Suh, H.J. Lee, Appl. Phys. Lett. **80**, 1001 (2002)
57. P. Kozodoy, H.L. Xing, S.P. DenBaars, U.K. Mishra, A. Saxler, R. Perrin, S. Elhamri, W.C. Mitchel, J. Appl. Phys. **87**, 1832 (2000)

58. W. Gotz, R.S. Kern, C.H. Chen, H. Liu, D.A. Steigerwald, R.M. Fletcher, Mater. Sci. Eng. B – Solid State Mater. Adv. Technol. **59**, 211 (1999)
59. P.P. Debye, E.M. Conwell, Phys. Rev. **93**, 693 (1954)
60. S. Dhar, S. Ghosh, Appl. Phys. Lett. **80**, 4519 (2002)
61. V. Ursaki, I.M. Tiginyanu, P.C. Ricci, A. Anedda, S. Hubbard, D. Pavlidis, J. Appl. Phys. **94**, 3875 (2003)
62. J. Ryan, D.P. Lowney, M.O. Henry, P.J. McNally, E. McGlynn, K. Jacobs, L. Considine, Thin Solid Films **473**, 308 (2005)
63. L. Chernyak, G. Nootz, A. Osinsky, Electron. Lett. **37**, 922 (2001)
64. C. Look, B. Claftin, Phys. Status Solidi B – Basic Res. **241**, 624 (2004)
65. O. Lopatiuk, A. Osinsky, L. Chernyak, in *Zinc Oxide Bulk, Thin Films and Nanostructures*, ed. by C. Jagadish, S. Pearton (Elsevier Ltd., Oxford, 2006), pp. 241–265
66. L. Chernyak, W. Burdett, M. Klimov, A. Osinsky, Appl. Phys. Lett. **82**, 3680 (2003)
67. W. Burdett, O. Lopatiuk, L. Chernyak, M. Hermann, M. Stutzmann, M. Eickhoff, J. Appl. Phys. **96**, 3556 (2004)
68. L. Chernyak, A. Osinsky, V. Fuflyigin, E.F. Schubert, Appl. Phys. Lett. **77**, 875 (2000)
69. O. Lopatiuk-Tirpak, L. Chernyak, F.X. Xiu, J.L. Liu, S. Jang, F. Ren, S.J. Pearton, K. Gartsman, Y. Feldman, A. Osinsky, P. Chow, J. Appl. Phys. **100**, 086101 (2006)
70. Y.W. Heo, K. Ip, S.J. Pearton, D.P. Norton, Phys. Status Solidi A – Appl. Res. **201**, 1500 (2004)
71. S.J. Pearton, D.P. Norton, K. Ip, Y.W. Heo, T. Steiner, Prog. Mater. Sci **50**, 293 (2005)
72. C.H. Park, S.B. Zhang, S.H. Wei, Phys. Rev. B **66**, 073202 (2002)
73. E.C. Lee, K.J. Chang, Phys. Rev. B **70**, 115210 (2004)
74. M.G. Wardle, J.P. Goss, P.R. Briddon, Phys. Rev. B **71**, 155205 (2005)
75. M. Salis, A. Anedda, F. Quarati, A.J. Blue, W. Cunningham, J. Appl. Phys. **97**, 033709 (2005)
76. F. Reuss, C. Kirchner, T. Gruber, R. Kling, S. Maschek, W. Limmer, A. Waag, P. Ziemann, J. Appl. Phys. **95**, 3385 (2004)
77. R. Ghosh, B. Mallik, S. Fujihara, D. Basak, Chem. Phys. Lett. **403**, 415 (2005)
78. G. Xiong, K.B. Ucer, R.T. Williams, J. Lee, D. Bhattacharyya, J. Metson, P. Evans, J. Appl. Phys. **97**, 043528 (2005)
79. L. Chernyak, A. Schulte, A. Osinsky, J. Graff, E.F. Schubert, Appl. Phys. Lett. **80**, 926 (2002)
80. O. Lopatiuk-Tirpak, L. Chernyak, L.J. Mandalapu, Z. Yang, J.L. Liu, K. Gartzman, Y. Feldman, Z. Dashevsky, Appl. Phys. Lett. **89**, 142114 (2006)
81. H. Holloway, J. Appl. Phys. **49**, 4264 (1978)
82. L. Chernyak, A. Schulte, A. Osinsky, in *Encyclopedia of Sensors*, vol. 10, ed. by C.A. Grimes, E.C. Dickey (American Scientific Publishers, 2006), pp. 1–14
83. O. Lopatiuk-Tirpak, G. Nootz, E. Flitsyian, L. Chernyak, L.J. Mandalapu, Z. Yang, J.L. Liu, K. Gartsman, A. Osinsky, Appl. Phys. Lett. **91**, 042115 (2007)
84. M. Razeghi, A. Rogalski, J. Appl. Phys. **79**, 7433–7473 (1996)

Chapter 12
Conduction in Degenerately Doped $Zn_{1-x}Al_xO$ Thin Films

Michael Snure, David Toledo, Paul Slusser and Ashutosh Tiwari

Abstract This chapter reports the electrical and optical properties of degenerately doped $Zn_{1-x}Al_xO$ films. Thin films of $Zn_{1-x}Al_xO$ ($x = 0$–0.1) were grown on single crystal c-plane sapphire substrates by pulsed laser deposition (PLD) technique. From electrical and optical characterizations, all film compositions were found to be highly transparent over the visible spectrum and exhibit a metal-like temperature dependence of resistivity. The carrier concentration, conductivity, and optical band gap of these films were not observed to increase continuously with increased Al concentrations. Detailed characterization showed that these properties initially increased reaching a maximum at 4 at.% Al. Beyond this Al concentration a decrease in these properties was observed. Through temperature-dependent electrical transport measurements it was found that at higher Al concentrations, scattering due to electron–electron interactions increased. Electron–electron interactions arise due to the introduction of disorder in the crystal potential caused by high concentrations of Al dopants and oxygen vacancies. High levels of disorder can significantly disrupt the band structure of ZnO reducing the density of states at the Fermi level ($N(E_F)$). The decrease in $N(E_F)$ directly corresponds to the decrease in carrier concentration affecting the optical band gap and conductivity.

12.1 Introduction

Transparent conducting oxides (TCOs) are used in a wide number of applications as transparent metals for heaters, electromagnetic shielding, and interconnects for photovoltaics (PVs), flat panel displays, and light emitting diodes (LEDs) [1–

M. Snure · D. Toledo · P. Slusser · A. Tiwari (✉)
Nanostructured Materials Research Laboratory, Department of Materials Science and
Engineering, University of Utah, SLC, UT 84112
e-mail: tiwari@eng.utah.edu

S. Pearton (ed.), *GaN and ZnO-based Materials and Devices*,
Springer Series in Materials Science 156, DOI 10.1007/978-3-642-23521-4_12,
© Springer-Verlag Berlin Heidelberg 2012

3]. Indium tin oxide (ITO), F-doped SnO_2 (FTO), and Al-doped ZnO are the major three TCOs commercially used, where ITO has been considered the industry work horse [4, 5]. However, concerns about the high cost, poor chemical stability, and toxicity of ITO have motivated researchers to find alternative TCOs. Of the alternative TCOs, Al-doped ZnO is the most suitable replacement for ITO due to its large band gap (3.3 eV), high conductivity, low cost, good refractive index, low toxicity and good thermal and chemical stability [6, 7].

Al-doped ZnO thin films have been widely studied for their excellent optical and electrical properties. Highly conductive and transparent Al-doped ZnO films have been grown using chemical vapor deposition (CVD), sputtering, and PLD techniques, where films grown by PLD have the lowest resistivity and highest transparency [8–11]. Much of this work has been performed on an empirical basis due to the lack of a clear understanding of the doping mechanism in ZnO and conduction in heavily doped semiconductors [12]. For many TCO applications, it is important to understand the doping limit of resistivity in Al-doped ZnO. Much of the work on heavily doped Al-doped ZnO does not deal with the physical mechanisms limiting resistivity [13–15]. Singh et al. [16] reported a maximum in carrier concentration and conductivity in 2 wt.% Al_2O_3-doped ZnO films deposited by PLD. While in films deposited by magnetron sputtering 4 at.% Al gave a minimum room temperature resistivity ($4 \times 10^{-4}\Omega$ − cm) and carrier concentration (7.5×10^{20} cm^{-3}) [17]. In these studies, it was found that increased scattering due to ionized impurities limited the conductivity.

In the present study, we have grown heavily doped $Zn_{1-x}Al_xO$ thin films by PLD on sapphire substrates to explore the effect of disorder on the electrical and optical properties. $Zn_{1-x}Al_xO(x = 0–0.1)$ thin films were deposited under vacuum to achieve a high defect density and carrier concentration. Temperature-dependent resistivity and thermoelectric properties were measured to study the nature of the charge carriers.

12.2 Experimental Procedure

PLD was used to deposit $Zn_{1-x}Al_xO(x = 0–0.1)$ thin films on sapphire (0001). Targets used for ablation were prepared using a solid state powder processing technique. The targets were sintered at $1,200°C$ for 14 h. The targets and substrates were then loaded into a stainless steel vacuum chamber and evacuated to 10^{-6} Torr. A KrF excimer laser (248 nm wavelength and 25 ns pulse width) delivered 6,000 pulses to a 1 mm^2 spot on the target at a frequency of 10 Hz and energy of 350 mJ. The substrates were held at a temperature of 600°C during deposition. The resulting films thicknesses were confirmed to be about 300 nm by cross sectional scanning electron microscopy.

All samples were characterized using X-ray diffraction (XRD), optical transmission spectroscopy, Hall effect measurements, temperature-dependent electrical resistivity measurements, and thermoelectric power measurements. $\theta − 2\theta$ scans were performed using CuKα radiation. The optical transmission measurements were performed using a DU 730 UV/visible scanning spectrophotometer in wavelength scanning mode over the range of 190–1,100 nm. Electrical resistivity of the films was measured using a four-probe technique in the temperature range of 77–300 K. Thermoelectric power was measured over the temperature range of 85–300 K. For these measurements, a temperature difference (ΔT) of 3 K was applied across the samples and the voltage difference (ΔV), thus developed across the sample was measured using copper thermocouple wires. The ratio $\Delta V/\Delta T$ gives the thermoelectric power of the sample with respect to copper. Absolute thermoelectric power was obtained by adding to this the absolute thermoelectric power of copper. Hall effect measurements were performed at room temperature to obtain the carrier concentration and mobility of charge carriers in the samples using a four-probe Van der Pauw configuration.

12.3 Results

Figure 12.1 shows $\theta - 2\theta$ XRD patterns of $Zn_{1-x}Al_xO$ films with $x = 0-0.1$. Over the scan range of $20-100°2\theta$, only two peaks (((0002) and (0004)) corresponding to the (0001) family of planes for wurtzite $Zn_{1-x}Al_xO$ were detected. This indicates a strong c-axis alignment of the films. The absence of any additional peaks also indicates that the films do not contain secondary phases such as Al_2O_3.

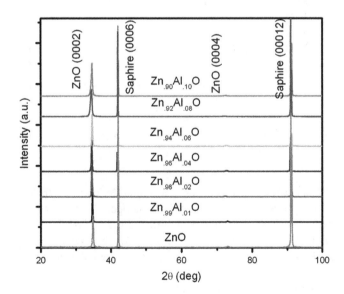

Fig. 12.1 X-ray $\theta - 2\theta$ diffraction pattern of $Zn_{1-x}Al_xO$ thin films with $x = 0-0.1$

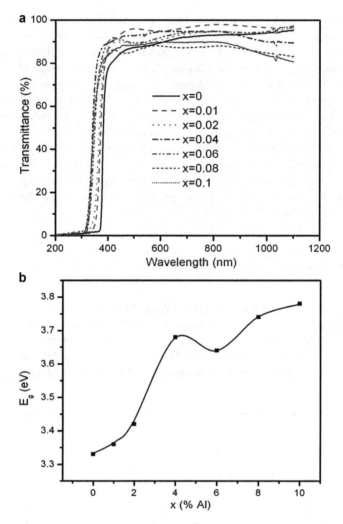

Fig. 12.2 Carrier concentration and Hall mobility of $Zn_{1-x}Al_xO$ ($x = 0–0.1$) thin films. *Solid lines* are intended as a guide to eye

Figure 12.2 shows electron carrier concentration and electron mobility of $Zn_{1-x}Al_xO$ films at room temperature. The undoped ZnO sample has a carrier concentration of 5×10^{19} cm^{-3} and a mobility of 4.2 cm^2/(Vs). The high carrier concentration is due to the high concentration of oxygen vacancies formed under vacuum deposition. As Al was introduced into the system the carrier concentration increased reaching a maximum of 7.5×10^{22} cm^{-3} at $x = 0.04$ and decreased on further Al doping. The mobility steadily decreased with Al concentration from 4.2 cm^2/(Vs) in the $x = 0$ films to 0.25 cm^2/(Vs) in the $x = 0.1$ films. This is expected because of the combined effect of an increased number of scattering

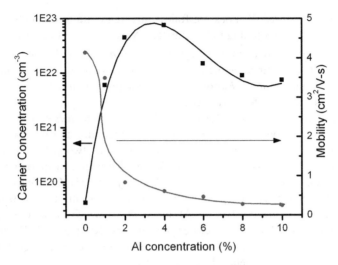

Fig. 12.3 Optical characterization of $Zn_{1-x}Al_xO (x = 0–0.1)$ (**a**) shows optical transmission data and (**b**) shows the variation of the measured band gap as a function of Al concentration

centers (Al) and an increase in free charge carrier scattering. The observed decrease in carrier concentration at $x > 0.04$ is quite interesting and can be explained by treating these materials as highly disordered metals. Further explanation will be given in the discussion section.

Figure 12.3a shows optical transmission spectra for $Zn_{1-x}Al_xO$ ($x = 0–0.1$) films over the wavelength range of 190–1,100 nm. The average optical transparency of the undoped ZnO film is 90% and increases to 95% for films with $x = 0.01$ and 0.02. This increase is understood to be caused by a decrease in deep level defects, which can absorb in the visible portion of the spectrum. Group-IIIb dopants have been shown to reduce the formation of deep level defects, which arise from interstitial atoms (Zn or O) and oxygen vacancies [18, 19]. As the doping level was increased to $x = 0.1$ the average transmittance decreases to 85%. At wavelengths between 380 and 400 nm a strong absorption band is observed, which corresponds to the direct band gap (E_g) of ZnO. The absorption band shifts to lower wavelengths as the Al concentration is increased, which is attributed to an increase in E_g. From the transmission data at the absorption edge, E_g can be estimated from $\alpha \sim (hv - E_g)^{1/2}$ [20]. By plotting α^2 vs. hv, E_g is obtained by extrapolating the straight line portion of the absorption band to zero. Figure 12.3b shows the variation of E_g with Al concentration. As seen in the figure, E_g increases with Al concentration. At 4% Al doping a local E_g maximum is observed and as the Al concentration increases to 6% E_g decreases by 0.04 eV. At still higher Al concentrations E_g continues to increase. The increase in E_g is driven by a combination of effects: band filling (Burstein–Moss effect) [21, 22] and alloying [23]. At higher Al concentrations alloying will play a significant role and according to Vegard's rule an increase in E_g with increasing Al should be seen due to the large E_g of Al_2O_3 (9.9 eV) [23].

However, the measured E_g does not follow this trend due to the additional band filling contribution. According to the Burstein–Moss effect (BM effect) E_g should increase with carrier concentration (n_c) at a rate of $n_c^{2/3}$ [21]. The local maximum in E_g at 4% Al is explained by the combination of band filling and alloying effects, where the local maximum in E_g coincides with the maximum carrier concentration.

Figure 12.4 shows temperature-dependent electrical resistivity data for $Zn_{1-x}Al_xO$ films. The undoped ZnO film has a room temperature resistivity of 36 mΩ-cm and exhibits a semiconductor-like temperature dependence. With the addition of Al the room temperature resistivity drops by two orders of magnitude and follows a trend consistent with the measured carrier concentrations. Where the $x = 0.04$ film showed the lowest resistivity (0.205 mΩ − cm) as x was increased further the resistivity increases to 5.02 mΩ − cm at $x = 0.1$. Metal-like conductivity is observed in all Al-doped films, which arises due to carrier concentrations exceeding Mott's critical density ($> 10^{19}$ cm^{-3}) [24]. In heavily doped semiconductors, this results in a shifting of the Fermi level into the conduction band causing metal-like conductivity.

Temperature-dependent thermoelectric power measurements were performed to further characterize the charge transport. Figure 12.5 shows the Seebeck coefficient (S) as a function of temperature for $Zn_{1-x}Al_xO$ films in the range 85–300 K. The thermoelectric power of all films is negative confirming that electrons are the dominate charge carrier. The magnitude of S at room temperature sharply decreases with the addition of Al until a minimum is reached around 2%. This overall trend is consistent with the observed variation in the carrier concentration. It should also be noted that in all samples S decreases with temperature. In metals S decreases linearly with temperature, while in semiconductors and insulators S increases following a hyperbolic temperature dependence [25]. Therefore, it is expected that the Al-doped films follow a metal-like temperature dependence. Since the undoped ZnO sample follows a semiconductor-like temperature dependence of resistivity it is unexpected that S would decrease with temperature. However, in the pure ZnO sample the carrier concentration was found to be 4×10^{19} cm^{-3}, which is greater then Mott's critical density needed to push a material to metallic-like conductivity. In addition to metals, Anderson-localized insulators have also been shown to display a negative temperature dependence of S [26]. In Anderson-localized insulators, materials with carrier concentrations above Mott's critical density can show a semiconductor-like conductivity caused by localization of charge carriers in the conduction band. This phenomenon explains the observed semiconductor-like electrical resistivity and metal-like thermoelectric power observed in the pure ZnO film.

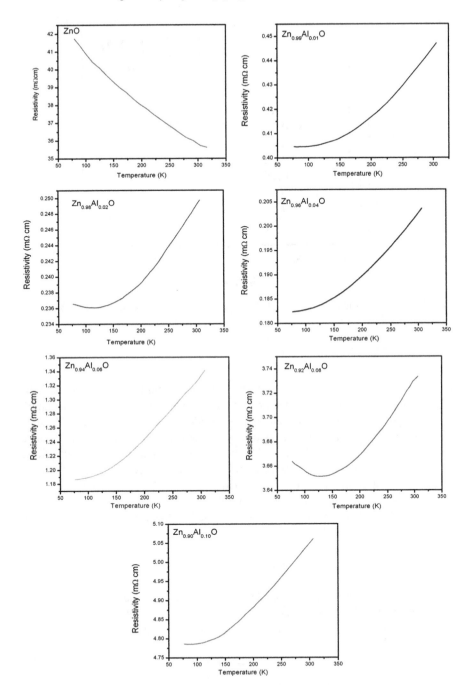

Fig. 12.4 Temperature-dependent electrical resistivity of $Zn_{1-x}Al_xO$ thin films over the temperature range of 77–300 K

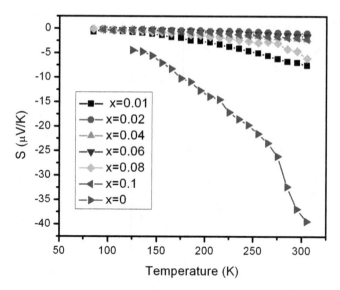

Fig. 12.5 Seebeck coefficient of $Zn_{1-x}Al_xO(x = 0\text{--}0.1)$ over the temperature range of 85–300 K

12.4   Discussion

In these $Zn_{1-x}Al_xO$ films, high concentrations of impurities were purposefully added to increase the carrier concentration and conductivity. However, these impurities also produced a significant amount of disorder, which can cause considerable electron scattering. Therefore, standard conduction mechanisms for metals and band gap insulators cannot be used. Materials with significant disorder, like those examined here, can see considerable elastic scattering from defects, which can lead to diffusive motion of charge carriers reducing mobility. This can lead to electron–electron (e–e) interactions that result from the inability of the electron screening cloud to easily follow scattered electrons. Conduction due to e–e interactions at low temperature follows as [27, 28]:

$$\sigma = \sigma_0 + m_\sigma \sqrt{T}, \tag{12.1}$$

where σ_0 and m_σ are constants related to diffusivity (D) and the density of states at the Fermi level $(N(E_F))$. At higher temperatures additional inelastic scattering also becomes significant, which gives rise to the modified expression for conduction [27, 28]:

$$\sigma = \sigma_0 + m_\sigma \sqrt{T} - aT^p, \tag{12.2}$$

where a is a constant and p is between 1.5 and 3. Figure 12.6 shows the $x = 0.01\text{--}0.1$ samples fit to (12.2). Table 12.1 summarizes the fitting parameters. For all samples, χ^2 is small indicating a good fit. It is indeed quite interesting that conductivity in these samples follow (12.2) in this temperature range (77–

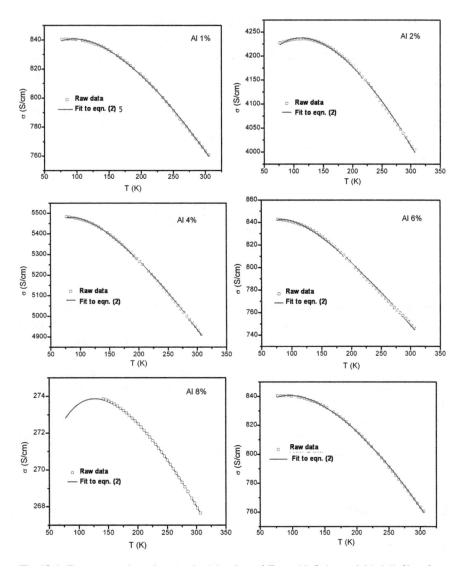

Fig. 12.6 Temperature-dependent conductivity data of $Zn_{1-x}Al_xO$ ($x = 0.01$–0.1) films fit to (12.2)

300 K) since e–e interaction effects are usually observed at lower temperature; at higher temperature these effects are generally obliterated due to other faster dephasing processes. However, similar observations have been reported in some other correlated oxide systems [29, 30]. From the fitting parameters in Table 12.1, the $N(E_F)$ can be obtained from the following [28]:

Table 12.1 Values of fitting parameters and χ^2 of temperature-dependent conductivity data fit to (12.2)

x	$\sigma_0(S/cm)$	$m_\sigma(S/(cm - K^{1/2}))$	$a(S/(cm - K^{1/p}))$	p	χ^2
0.01	761.9	10.4	0.579	1.20	1.8×10^{-5}
0.02	3852.4	58	0.683	1.3	4.5×10^{-5}
0.04	4488.2	299	34.6	0.82	4.4×10^{-5}
0.06	357.1	264.3	24.9	0.611	6.5×10^{-4}
0.08	249.7	5.1	0.477	0.87	3.5×10^{-5}
0.1	184.2	7.9	0.384	0.89	6.7×10^{-4}

$$D = \left(\frac{0.915e^2}{2\pi^2 hm_\sigma}\right)^2 \left(\frac{2}{3} - \frac{3}{4}\tilde{F}_\sigma\right)^2 \frac{k_B}{\hbar}, \tag{12.3}$$

$$N(E_F) = \frac{\sigma_0}{e^2 D}, \tag{12.4}$$

where is $\tilde{F}_\sigma \approx 0.808$ [31]. Figure 12.7 plots $N(E_F)$ as a function of the Al concentration as obtained from (12.3) and (12.4). Initially, an increase in $N(E_F)$ is observed until a maximum is reached at $x = 0.04$. The increase is caused by the increased number of electrons in the conduction band due to doping, which is consistent with Fig. 12.3. However, as x is increased further $N(E_F)$ decreases. According to Altshuler and Aronov this is due to e–e interactions, which is a sign of increased disorder [27]. It is because of this decrease in $N(E_F)$ that we explain the measured decrease in the free carrier concentration in samples with $x > 0.04$. The decrease in $N(E_F)$ can be related to the decrease in E_g, as seen in Fig. 12.2b, due to reduced band filling according to the BM effect.

12.5 Summary

We have observed an upper doping limit of 4 at.% in Al-doped ZnO thin films grown by PLD on sapphire under vacuum at a substrate temperature of 600°C. At Al concentration less than 4 at.% a continuous increase in conductivity and carrier concentration was observed. As the Al concentration was increased further the conductivity and carrier concentration decreased. From temperature-dependent electrical characterization we found that the incorporation of Al at higher concentrations increased scattering due to e–e interactions. This decreased $N(E_F)$ as the Al concentration was increased beyond 4% causing the decrease in the carrier concentration, resistivity, and E_g.

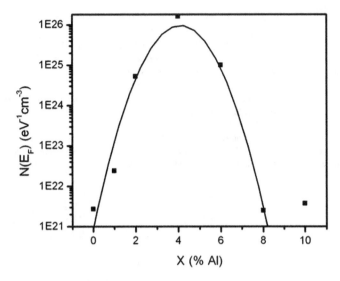

Fig. 12.7 $N(E_F)$ as a function of x (Al concentration) calculated from (12.3) and (12.4) using values given in Table 12.1. *The solid line* is intended as a guide to eye

References

1. K.L. Chopra, S. Major , D.K. Pandya, Thin Solid Films **102** 1 (1983)
2. C.G. Granqvist, A. Azens, A. Hjelm, L. Kullman, G.A. Niklasson, D. Ronnow, M. Stromme Mattsson, M. Veszelei , G. Vaivars , Sol. Energy **63** 199 (1998)
3. Y. Nakato, K. Kai , Kawabe, Sol. Energy Mater. Sol. Cells **37**323 (1995)
4. B.G. Lewis , D.C. Paine , MRS Bull. **24** 22 (2000)
5. T. Minami, T. Miyata, T. Yamamoto, J. Vac. Sci. Technol. A **17** 1822 (1999)
6. U. Özgür, I. Alivov, C. Liu, A. Tekeb, M.A. Reshchikov, S. Dogan, V. Avrutin, S.J. Cho , M. Morkocd , J. Appl. Phys. **98** 041310 (2005)
7. W. Beyer, J. Hupkes , H. Stiebig , Thin Solid Films **516** 147 (2007)
8. W. Water , S.Y. Chu, Mater. Lett. **55** 67 (2002)
9. T. Ohgaki, N. Ohashi, H. Kakemoto, S. Wada, Y. Adachi, H. Haneda, T. Tsurumi, J. Appl. Phys. **93** 1961 (2003)
10. M. Snure , A. Tiwari, J. Appl. Phys. **101** 124912 (2007)
11. V. Craciun, J. Elders, J.G.E. Gardeniers, I.W. Boyd, Appl. Phys. Lett. **65** 2963 (1994)
12. K. Ellmer, J. Phys. D: Appl. Phys. **34** 3097 (2001)
13. H.L. Hartnagel, A.L. Dawar, A.K. Jain , C. Jagadish, *Semiconducting Transparent Thin Films* (Institute of Physics Publishing, Bristol, 1995)
14. A.J. Freeman, K.R. Poeppelmeier, T.O.Mason, R.P.H.Chang , T.J. Marks, MRS Bull. **45** (2000)
15. K.L. Chopra, S. Major , D.K. Pandya, Thin Solid Films **102**1 (1983)
16. A.V. Singh, R.M. Mehra, N. Buthrath, A. Wakahara, A. Yoshida, J. Appl. Phys. **90** 5661 (2001)
17. K.H. Kim, K.C. Park , D.Y. Ma, J. Appl. Phys. **81** 7764 (1997)
18. Y. Liu , J. Lian , Appl Surf. Sci. **253** 3727 (2007)
19. P.K. Nayak, J. Yang, J. Kim, Chung, J. Jeong, C. Lee , Y. Hong J. Phys. D: Appl. Phys. **42** 035102 (2009)
20. A. Tiburcio-Silver, J. Joubert , M. Labeau, J. Appl. Phys. **76** 1992 (1994)
21. T. Moss, Proc. Soc. London Sect. B **67** 775 (1954)

22. E. Burstein, Phys. Rev. **93** 632 (1954)
23. R. Ghosh , D. Basak, J. Appl. Phys. **101**023507 (2007)
24. N.F. Mott, *Metal-Insulator Transition* (Taylor& Francis, London, 1974)
25. J.S. Dugdale, *The Electrical Properties of Metals and Alloys* (Adward Arnold, London, 1977)
26. A. Tiwari, C. Jin, J. Narayan, M. Park, J. Appl. Phys. **96** 3827 (2004)
27. P.A. Lee , T.V. Ramakrishnan, Rev. Mod. Phys. **57** 287 (1985)
28. B.L. Altshuler , A.G. Aronov, *Electron-Electron Interactions in Disordered Systems ed AL Efros and M Polak* (North Holand, Amsterdam, 1985)
29. A. Tiwari, K.P. Rajeev, J. Phys. Condensed Mater. **11**, 3291 (1999)
30. A. Tiwari, K.P. Rajeev, Solid State Commun. **111** 33 (1999)
31. H. Fukuyama A.L.Efros, M. Polak, *Electron-Electron Interactions in Disordered Systems ed AL Efros and M Polak* (Amsterdam: North Holland, 1985)

Chapter 13
Multifunctional ZnO Nanostructure-Based Devices

Yicheng Lu, Pavel I. Reyes, Jian Zhong, and Hannhong Chen

Abstract Functional ZnO-based nanostructures are reviewed. Particular emphasis is placed on applications in biosensing and conducting electrodes which require control of morphology and conductivity.

13.1 Introduction

ZnO is a wide bandgap ($E_g \sim 3.3\,\text{eV}$ at room temperature) semiconductor that has received increasing attention. Through proper doping and alloying, ZnO can be made piezoelectric, ferromagnetic, transparent and conducting, or resistive switching, leading to multifunctional applications. Piezoelectric ZnO possesses a large electromechanical coupling coefficient and is widely used in bulk and surface acoustic wave (BAW and SAW) devices for wireless communication and signal processing, micro- and nano-electromechanical system (MEMS and NEMS) technology. Transition metal (TM)-doped ZnO is an oxide-diluted magnetic semiconductor which is an excellent candidate material for room temperature spintronics. Recently, ZnO and its nanostructure-based memristors have also attracted significant interests as it possesses advantages such as well-controlled resistive switching properties through in situ doping and alloying, simple fabrication process, and radiation hardness. These properties have stimulated the research interests and opened up new opportunities for novel device applications.

Nanotips are of strong interest for applications such as field emission devices for flat panel display and sensors. In traditional micro-tip field-emission devices, tip fatigue due to radiation damage and oxidation causes a major reliability issue. Therefore, it is preferred to use wide bandgap semiconductor material

Y. Lu (✉) · P.I. Reyes · J. Zhong · H. Chen
Department of Electrical and Computer Engineering, Rutgers University, 94 Brett Road, Piscataway, NJ 08854-9058, USA
e-mail: ylu@rci.rutgers.edu

S. Pearton (ed.), *GaN and ZnO-based Materials and Devices*, 361
Springer Series in Materials Science 156, DOI 10.1007/978-3-642-23521-4_13,
© Springer-Verlag Berlin Heidelberg 2012

nanostructures for field emission, particularly those with low work function. There have been several reports on SiC [1] and GaN [2, 3] based nanowires. For device applications, it is desirable to have highly oriented nanotips grown on a selectively patterned area. Vertically aligned ZnO nanowire arrays can be grown on various substrates at relatively low deposition temperatures, giving it a unique advantage over other wide bandgap semiconductors. Catalyst-free growth of ZnO nanotips can be achieved by metalorganic chemical vapor deposition (MOCVD), leading to superior optical and electrical properties. MOCVD growth of these nanotips also provides atomically-sharp interface control and possible end-facet morphology modulation by altering the growth conditions, which is particularly useful for biosensing applications. This chapter presents the examples of several novel devices using multifunctional ZnO and its nanostructures.

13.2   Multifunctional ZnO Nanostructures for Biosensing

Multifunctional ZnO and its nanostructures possess unique properties that are particularly suitable for biosensor technology. ZnO-based sensors have demonstrated high sensitivity to various gases and liquids such as CH_4, CO, H_2O, H_2, NH_3, trimethylamine, ethanol, and NO_2 [4,5]. ZnO and its ternary alloy, $Mg_xZn_{1-x}O$, are known to be the biocompatible oxides, in which Zn and Mg are important elements for neurotransmitter production and enzyme function [6]. ZnO nanostructures with different morphologies (nanometer-scale rods, belts, tips, etc.) can be grown on various substrates including Si, quartz, glass, and metals. ZnO nanostructures are also considered as a coating material for medical implants [7]. ZnO nanostructures are also compatible with intracellular material and are highly sensitive to pH changes inside cellular environments [8] with $Mg_xZn_{1-x}O$ having a large range of pH compatibility. ZnO nanorods are also used for detection of enzymatic reactions [9]. The biocompatibility of ZnO and its feasibility for biosensing applications are further demonstrated in detections of proteins, antibodies, and DNA through the proper surface functionalizations [10–13]. The control of the surface wettability of ZnO nanotips between the superhydrophilic and superhydrophobic states is used to dramatically enhance the sensitivity of the biosensors [14].

ZnO and its nanostructures can also serve as the basic sensing platform for detection of biophysical activity of larger biospecies such as cells, bacteria, and viruses. The variation in morphology of the ZnO nanostructured surfaces can impact the attachment of certain biological cell lines (i.e., NIH 3T3 fibroblasts, umbilical vein endothelial cells, and capillary endothelial cells) and control the extent of cellular adhesion [15]. ZnO nanostructures were also used to bind with bacterial and viral cultures for reaction with enzymes and antibodies for applications in immunosensing [16, 17]. ZnO nanostructures have also been employed in intracellular measurements where ZnO nanotips were used as coatings for microprobes or serve as the probes themselves. These intracellular probes can sense conductivity changes due to the ZnO coating's reaction to various target biochemicals within the cellular environment. This setup was used widely for pH determination of intracellular environments which was first reported by Al-Hilli et al. [8]. More recently, microtubes coated with ionophore-functionalized ZnO nanorods were used as an intracellular probe to selectively sense the Ca^{2+} ions within the cell membrane [18]. This selective binding with ZnO nanorods causes the probe tip to measure the voltage gradient between the membranes, and consequently giving a highly accurate measurement of intracellular pH levels. A similar method was also reported to determine the membrane potential across a human fat cell [19]. The use of ZnO nanorods as fluorescence enhancing substrates has been reported where biomolecular detection sensitivity of subpicomolar and attomolar levels was obtained by using a conventional fluorescence microscope [20]. This ultrasensitive detection was due to presence of ZnO nanomaterials which contributed greatly to the increased signal-to-noise ratio of biomolecular fluorescence. Moreover, *Escherichia coli* cells were monitored optically through surface enhanced Raman spectroscopy by introducing ZnO nanoparticles into the cells to increase the optical signal [21].

The design of a smart multifunctional ZnO-based biosensing platform is due to three important components of ZnO's properties: (1) reversible wettability control, (2) bi-functional biochemical binding (serves as both active biological attachment and optical platform, i.e., fluorescence emission), (3) nanostructure morphology control. These controllable properties can be tailored to simultaneously enhance and optimize liquid sample intake, biochemical sensitivity and selectivity, and biospecies binding and distribution.

13.2.1 Wettability Control on ZnO Nanostructures

ZnO nanostructures have shown reversible and controllable surface wettability. The ability to switch from superhydrophilic to superhydrophobic reversibly is a highly desirable property of sensor surfaces and will greatly aid in optimizing liquid sample intake, transport, and storage within a single chip. In this section, the wettability property of ZnO will be discussed. The levels of the surface wettability of these various ZnO surfaces are most conveniently determined from the measurement of contact angles (CAs) between the ZnO surfaces and the water droplet. To demonstrate the different extents of the surface wettability in ZnO, the following ZnO surface morphologies are examined: ZnO nanotips grown on c-sapphire substrate (sharp nanotips), ZnO nanotip grown on (100) Si substrate (rounded nanotips), and ZnO epitaxial film grown on r-sapphire (flat surface) as shown in Fig. 13.1a–c.

A fresh untreated as-grown sample of ZnO nanotips surface is naturally hydrophilic with a low CA of about 10°. Once the same sample is stored in the dark in a nitrogen box at room temperature the CA rises slowly until it reaches a maximum value of 120°. This transition to the hydrophobic state takes 1–2 months to reach the maximum CA value if no special surface treatment is applied.

Fig. 13.1 Surface
morphologies of ZnO (**a**)
nanotips on c-sapphire, (**b**)
nanotips on Si, and (**c**) film
on r-sapphire, taken by
field-emission scanning
electron microscope
(SEM) [14]

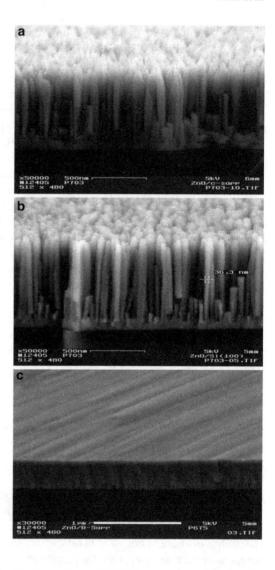

Special surface treatment of the ZnO samples will facilitate more rapid, controllable, and reversible wettability transitions. UV illumination of the ZnO surface induces transitions from hydrophobic to hydrophilic states. On the other hand, low temperature O_2 annealing of the ZnO samples induces the transition from hydrophilic to hydrophobic. Successive application of UV illumination and O_2 annealing shows reversible wettability state transitions as shown in Fig. 13.2a. The hydrophobic status of the ZnO nanotips, where CA > 120°, and the superhydrophilic status, where CA is close to 0°, are highly repeatable after the five cycles, implying a stable change of the surface status.

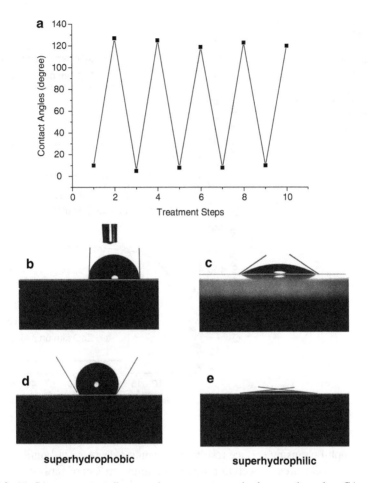

Fig. 13.2 (**a**) CAs at corresponding sample treatment steps in the growth → low CA → dark storage → high CA → UV irradiation → low CA → dark storage → high CA... (**b**) Hydrophobic surface of ZnO film with contact angle (CA) ~95°. (**c**) Hydrophilic surface of ZnO film with CA ~37°. (**d**) Hydrophobic surface of ZnO nanotips with CA ~130°. (**e**) Superhydrophilic surface of ZnO nanotips with CA ~0° [14]

The ZnO surface morphology also changes the extent of hydrophobic and hydrophilic levels. The flat ZnO samples only exhibited 95° CA for the hydrophobic state after O_2 annealing and 37° CA for the hydrophilic state after UV illumination (Fig. 13.2b–c). In comparison, using the same O_2 annealing and UV treatment levels on the sharp ZnO nanotips causes a more pronounced change. The nanotips surface attained a superhydrophobic state with CA of 130° (Fig. 13.2d) after oxygen anneal and attained a superhydrophilic state with CA of 0° (Fig. 13.2e). The high hydrophilicity and high hydrophobicity of ZnO nanostructures are important for the chemical and biochemical sensor areas. For example, it is found that

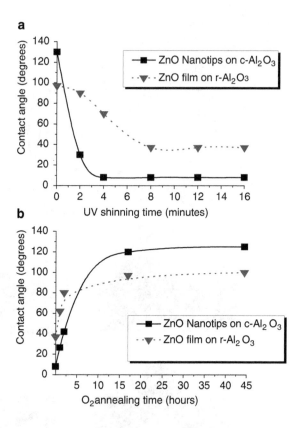

Fig. 13.3 (a) Contact angle change under UV light as a function of time. The substrates are ZnO nanotips and film, respectively. (b) Contact angle change under O_2 annealing at 50°C as functions of time. The substrates are ZnO nanotips and film, respectively

superhydrophilic sensing surface results in a dramatic decrease of liquid consumption on the sensing area, as well as a significant improvement of sensitivity.

The special surface treatments to control the ZnO surface wettability also improves the wettability transition rate, which is crucial for real device applications, particularly, for the effective control of the micro- or nano-fluid motion. As shown in Fig. 13.3a, the transition time for CA to drop from 130° to 8° on ZnO nanotips (Fig. 13.1a) only takes 4 min. Similarly, the UV shinning takes 6 min to decrease CA for the nanotips with a rounded top end shown in Fig. 13.1b. The transition rates measured here is ~30 times faster than untreated results, where it took ~2 h for the sample to change from hydrophobic to hydrophilic state. The CA recovering rates are also plotted in Fig. 13.3b, as measured on the same sample with O_2 annealing at 50°C. The time taken for full CA recovery is around 17 h on ZnO nanotips, ten times shorter than the reported CA recovery time [22, 23].

The wettability transition process can be described by (13.1–13.3). UV irradiation generates electron–hole pairs in the ZnO surface, and some of the holes can react with lattice oxygen to form surface oxygen vacancies. Meanwhile, water and oxygen may compete to dissociatively adsorb on them. The defective sites are kinetically more favorable for hydroxyl adsorption than oxygen adsorption. Similar as TiO_2 [24, 25], the process can be described in (13.1–13.3):

$$ZnO + 2hv \rightarrow 2h^+ + 2e^- \tag{13.1}$$

$$O^{2-} + h^+ \rightarrow O_1^- \tag{13.2}$$

$$O_1^- + h^+ \rightarrow \frac{1}{2}O_2 + [O]_v \, (\text{Oxygen Vacancy}) \tag{13.3}$$

It has been demonstrated that the surface becomes energetically unstable after the hydroxyl adsorption. Oxygen adsorption is thermodynamically favored, and it is more strongly bonded on the defect sites than the hydroxyl group does [24]. Therefore, the hydroxyl groups adsorbed on the defective sites can be replaced gradually by oxygen atoms when the UV-irradiated samples were placed in the oxygen ambient. Subsequently, the surface evolves back to its state before UV illumination, and the wettability is reconverted from superhydrophilic status to hydrophobic status. The surface chemical composition provides a photosensitive surface, which can be switched between hydrophilicity and hydrophobicity, and the nanostructure further enhances these properties due to the roughness effect. Furthermore, the reversible transitions of the surface wettability of ZnO are proceeded only by the adsorption and desorption of surface hydroxyl groups at the outmost layer of oxide films [26]. The structure below the outmost layer remains stable, free from changes in chemical conditions. It would take a month to change the ZnO superhydrophilic surface to the hydrophobic surface if the samples were stored in the nitrogen box. In order to reduce the transition time from low CA to high CA, annealing is required for the superhydrophilic samples in the controlled environmental chamber. The samples were annealed at 50°C in three ambients: O_2, N_2, and vacuum. For the sample annealed in O_2, the CA was fully recovered to 130° in less than 17 h. Increasing the temperature does not further shorten the recovery time. The samples annealed in N_2 and vacuum did not have the CA changes at all after 24 h annealing. This result implies that oxygen adsorption is thermodynamically favored, and it is more strongly bonded on the defect sites than the hydroxyl group. The hydroxyl groups adsorbed on the defective sites can be replaced gradually by oxygen atoms when the hydrophilic samples are annealed in oxygen ambient at low temperature. Subsequently, the surface evolves back to its original state (before UV irradiation), and the wettability is reconverted from superhydrophilicity to hydrophobicity.

13.2.2 Biofunctionalization of ZnO Nanostructures

Biofunctionalization is a critical issue for ZnO nanostructured surface to obtain the high sensitivity and selectivity to various types of biospecies. The binding properties of a number of small molecules with functional groups are reported to form covalent bonds with metal oxides nanoparticles, including ZnO [27], COOH, SH, SiOMe$_3$,

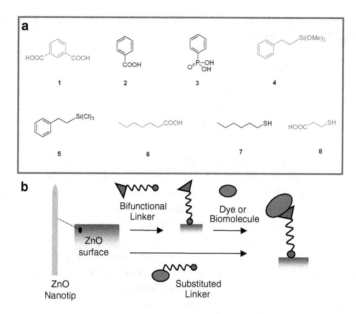

Fig. 13.4 (**a**) Compounds employed in the binding study. The compounds that did bind to the ZnO nanotips are shown in red. (**b**) The schematic of the functionalization of ZnO nanotip using bifunctional linker [28]

SiCl$_3$, and PO(OH)$_2$. The structures of the compounds are shown in Fig. 13.4a [28]. To achieve the selective biochemical attachment to the ZnO nanostructures, a family of linkers called *bifunctional linkers* are employed, having one functional group that attaches to the ZnO nanotip surface while leaving other end available for further biological functionalization. The surface functionalization could be done in one step by using a functionalized linker, as shown in Fig. 13.4b [28]. The ability to functionalize the nanotips would lead to numerous applications, for instance, new ZnO-based integrated, multimode and multifunctional sensor technology to achieve higher accuracy and selectivity than existing sensors. For comparison, the binding study was also performed on the epitaxial ZnO films.

Table 13.1 summarizes the results for the binding of linkers **1–8**. To study the binding we have taken into consideration (1) the anchoring groups, (2) the ZnO material, specifically ZnO nanotips and ZnO films, and (3) pH effects.

The binding experiments indicate that the carboxylic acid group (COOH) is the most stable anchoring group for binding to the ZnO-N but that the number and position of the COOH groups, the acidity of the anchoring group, and solution pH also play a role in the binding. Another potentially useful, but flexible, bifunctional linker is 3-mercaptopropionic acid since it did bind to ZnO with its COOH end, leaving the SH group available for selective binding with various biomolecule targets. Increased resistance to acids can also be obtained by alloying ZnO with MgO (5–10%). Mg$_x$Zn$_{1-x}$O are a novel material that exhibits increased stability

Table 13.1 Compounds bound (Compounds can be detected by FT-IR-ATR and cannot be rinsed off) (b), physisorbed (Compound can be easily rinsed off during the wash) (p), or nonbound (nb) to ZnO [28]

Compound	Anchoring group	Concentration	ZnO material	Result	Detection methods
1	COOH	18 mM	Nanotip	B	FT-IR-ATR
2[a]	COOH	2–100 mM	Nanotip	nb[e]	IR
3[c]	P(O)(OH)$_2$	10–20 mM	Nanotip	p[e]	IR
		10–20 mM	Film	nb[e]	
4	Si(OMe)$_3$	2 mM	Nanotip	B	FT-IR-ATR
5[d]	SiCl$_3$	2–25 mM	Nanotip	nb[e]	IR
6	COOH	25 mM	Nanotip	B	FT-IR-ATR
7[b,d]	SH	2–25 mM	Nanotip	Nb	IR
		2–25 mM	Film	Nb	
8[b]	COOH and SH	2 mM	Nanotip	B	FT-IR-ATR
		2–25 mM	Film	Nb	

[a] Solution prepared from EtOH
[b] Solution prepared from CH$_3$CN
[c] Solution prepared from H$_2$O
[d] Solution prepared from CH$_2$Cl$_2$
[e] Sample etched by the acidic solution

toward acidic linkers, while staying identical to the ZnO in other respects. Alloyed films could therefore be employed with a larger variety of binding groups and at wider pH ranges. In conclusion, we have demonstrated that ZnO-N can be functionalized with bifunctional linkers. This opens the possibility to develop ZnO nanostructures functionalized with a variety of dyes, redox-active molecules, or biomolecules, eventually leading to novel types of anodes for solar cells or to a new ZnO-based integrated, multimode and multifunctional sensor technology, which combines different sensing mechanisms (acoustic, electrical, optical, biological, etc.) to achieve higher accuracy and sensitivity and has a broad impact in biological and environmental applications.

13.2.3 Morphology Effects of ZnO Nanostructures on Adhesion of Biospecies

It is important to determine the most suitable ZnO nanostructure morphology to facilitate the optimal adhesion to ZnO of certain biospecies like, for instance, living biological cells.

The biochemical/biospecies binding resulting from the optimal surface morphology of ZnO facilitates high sensitivity in the sensor. It also provides the samples being sensed (especially living samples like cells) a biocompatible platform for them to live and grow. In order to determine the optimum morphology for the

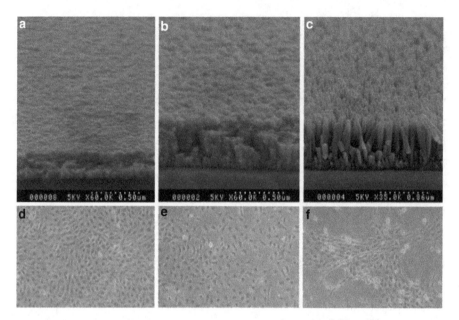

Fig. 13.5 The FESEM images of the different surface morphologies of ZnO nanostrcutures: (**a**) flat film grown on r-sapphire, (**b**) rough surface formed by dense nanocolumns grown on glass, (**c**) sharp nanotips grown on glass. These surfaces were treated with fibronectin and seeded with BAEC cells. Transmission type optical microscope images of the growing BAEC cells on (**d**) flat ZnO film, (**e**) rough surface of dense ZnO nanocolumns, and (**f**) sharp ZnO nanotips[29]

cell adhesion, ZnO nanostructures with three different surface morphologies were examined. Figure 13.5a–c shows the three different surface morphologies of ZnO films, taken by the field emission scanning electron microscope (FESEM). The relatively flat surface, shown in Fig. 13.5a, was grown at a low temperature of ∼250°C, while the relatively rough one (Fig. 13.5b) was grown at ∼330°C. For the sample with morphology of nanotips (Fig. 13.5c), which is referred to as the "sharp" surface later, a relatively high temperature (>400°C) was applied for its growth. The surface roughness (root mean square) of ZnO films was characterized by the atomic force microscopy (AFM), which is 1.39 nm for flat ZnO (Fig. 13.5a), 7.48 nm for rough one (Fig. 13.5b), and 11.4 nm for sharp one (Fig. 13.5c). The three different samples were treated with fibronectin which serves as the biochemical layer that facilitates initial cell anchorage to ZnO. The standard transmission type optical microscope images of the three ZnO samples were taken after 60 h of incubation. For the flat ZnO surface, the cell culture is close to 100% confluency and uniformly spreading and proliferating on the ZnO substrate (Fig. 13.5d). However, it is found that after the entire duration of the monitoring cycle the cells have crowded with each other, competing for nutrients and space. This condition will eventually induce the cells to die and detach from the ZnO surface. On the other hand, the cells on the rough ZnO surface have reached about 75% confluency, uniform proliferation, and the individual cells have considerable amount of spreading as shown in Fig. 13.5e. The cells adhered to the sharp ZnO surface, but the cells did not establish good focal adhesion to facilitate uniform proliferation nor proper individual cell spreading. The cell culture on the sharp ZnO surface only attained 40% confluency and a clumped cell distribution as shown in Fig. 13.5f. This clumped distribution might cause localized areas on the ZnO surface where cell death is induced. The results of poor cell adhesion on sharp nanorods are consistent with that of Lee et al. [15] wherein the same cell line (BAEC cells) lacked the ability to establish strong initial adhesion to the sharp nanostructures thus prohibited them to produce lamellipodia (cell-anchoring mechanism). To obtain a good cell culture for monitoring purposes, three components of cell growth need to be satisfied (1) good initial cell adhesion, (2) uniform cell proliferation, and (3) considerably large individual cell spreading. The flat and rough ZnO surfaces displayed all three while the sharp ZnO surface failed to fulfill the last two requirements. It would be a natural choice to use the flat ZnO surface since it gives us a large yield in cell growth while fulfilling all three cell culture requirements. However, there is an inherent tradeoff in choosing the optimal ZnO morphology for the sensing surface. In terms of cell attachment, the cells favor the adhesion to flatter surfaces, but in terms of device performance the sharper surface provides the highest sensitivity due to the large effective sensing area provided by the nanostructures [10, 14]. It is determined that the rough ZnO surface is the most suitable morphology for adhesion and viability for cell growth without sacrificing the device sensitivity.

13.2.4 ZnO Nanostructure-Based Acoustic Biosensors

We designed and fabricated the novel ZnO nanotip-LiNbO$_3$ SAW (N-SAW) biosensor and ZnO nanotip-Quartz QCM (N-QCM) biosensor, which integrate ZnO nanostructures with LiNbO$_3$ SAW delay lines and quartz crystal microbalance, respectively. The ZnO nanotip array is used for surface modification and immobilizing the bioreceptor, while the SAW device fabricated on 128° Y-cut LiNbO$_3$ or the quartz crystal microbalance is used as the transducer to detect the mass loading effect resulting from the analyte–receptor reaction. This chapter is focused on the design, modeling, fabrication, and test of ZnO N-SAW devices, while the next chapter is mainly discussing the ZnO N-QCM device and the effects of surface modifications for biological sensing.

13.2.4.1 ZnO Nano-SAW for Oligonucleotide Immobilization Detection

The ZnO Nano-SAW biosensor integrates ZnO nanostructures with LiNbO$_3$-based SAW delay lines. The multifunctional properties of the ZnO nanotip array, namely, the controllable wettability and biofunctionalization are used for immobilizing the bioreceptor and increasing the sensitivity of the bare SAW device. The bare SAW

Fig. 13.6 Schematic of the
ZnO nanotip-based SAW
biosensor structures: (**a**)
cross-section, (**b**) top
view [12]

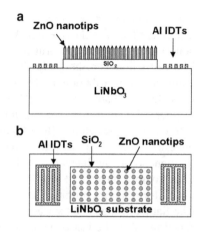

device is fabricated on 128° Y-cut LiNbO$_3$ and is used as the transducer to detect the
mass loading effect resulting from the analyte–receptor reaction. The LiNbO$_3$ SAW
device with the ZnO nanotip array is schematically shown in Fig. 13.6. The device
consists of two SAW channels, a reference and a sensing channel. The test devices
are regular SAW delay lines. Two inter-digital transducers (IDTs) are fabricated
on a 128° Y-cut LiNbO$_3$ substrate with a ZnO nanotip array/SiO$_2$ layer structure
deposited on the SAW propagation path. A 150-nm SiO$_2$ film was first deposited on
the LiNbO$_3$ followed by a layer of ZnO nanotips serving as the sensing platform. For
128° Y-cut LiNbO$_3$ substrates, the IDT period of 1,200 μm gives a center frequency
of 305.6 MHz and a 3 dB Bandwidth (BW$_{3dB}$) of 1.8%. Due to its high coupling
coefficient, the device on 128° Y-cut LiNbO$_3$ has about 50 dB less insertion loss than
the same device structure on ST cut quartz. It makes 128° Y-cut LiNbO$_3$ substrate to
be a good option for the sensor templates, since the ZnO nanotip and extra binding
layer would introduce some more insertion loss on the bare device due to the mass
loading effect and charge changes on the delay path. The biosensing device consists
of two SAW channels, a reference and a sensing channel. The reference channel
had no biological component bonding, whereas the sensor channel was bonded with
DNA oligonucleotide or protein on the ZnO nanotip array.

The sensing surfaces, the ZnO nanotip array, are first functionalized by immobilizing DNA oligonucleotide then a second complementary DNA strand is introduced
into the device to accomplish hybridization. A schematic representation of the
DNA immobilization and hybridization process is shown in Fig. 13.7a–d. Different
surfaces, such as LiNbO$_3$ and ZnO flat film, were treated with the same method for
comparison of the immobilization efficiency. A comparative study of immobilization of DNA oligonucleotide on different surfaces was carried out by radioactivity
testing. Followed by incubation and wash steps mentioned earlier, 1 μg/μl DNA
oligonucleotide solution was applied over 0.1 cm^2 substrates of LiNbO$_3$, (0001)
ZnO film, and ZnO nanotip array. The LiNbO$_3$ substrate and the ZnO film have
little immobilization of oligonucleotide (27 ng/cm^2 and 80 ng/cm^2, respectively).
In comparison, the oligonucleotide immobilization on the ZnO nanotip arrays

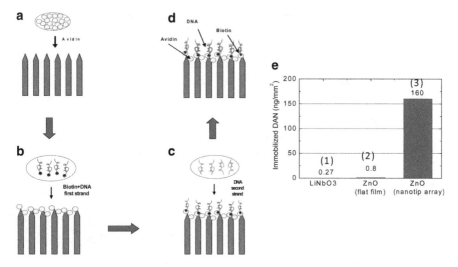

Fig. 13.7 Schematic representation of the DNA immobilization and hybridization process: (**a**) Applying Avidin (protein immobilization); (**b**) Applying Biotin–DNA primary strand; (**c**) Applying DNA complementary secondary strand (or non-complementary secondary strand APE, the control solution); (**d**) After DNA hybridization. (**e**) Immobilization of DNA oligonucleotide on different device surfaces. (1) bare device; (2) bare device with ZnO flat film; (3) bare device ZnO nanotip array. The oligonucleotide immobilization on the device with ZnO nanotip arrays is approximately 200 times larger than the flat ZnO film surface and approximately 600 times larger than the LiNbO$_3$ substrate [12]

(1.6×10^4 ng/cm^2) is approximately 200 times larger than the flat ZnO film surface and approximately 600 times larger than the LiNbO$_3$ substrate, shown in Fig. 13.7e. The ZnO nanostructures possess a giant surface area, which provide more binding sites and larger effective sensing area than the planar surface.

In the DNA oligonucleotide immobilization test, when a 1 μg/μl DNA oligonucleotide solution was applied on the sensor channel, a 191° phase shift (Fig. 13.8) and 6.5 dB of additional insertion loss at the center frequency of 327.94 MHz was observed compared with the reference channel. Mass sensitivity of a SAW sensor in the feedback loop of an oscillator circuit can be calculated by determining the mass on the sensor surface and the total frequency change caused by this mass accumulation:

$$\text{Mass Sensitivity} = \frac{\Delta f / f_c}{\Delta m / A}, \tag{13.4}$$

where m is the total mass on the active surface with area of A, f_c is the center frequency of the SAW device, and Δf is the frequency shift of the oscillator circuit. As the SAW sensors were tested using a network analyzer, we used the phase shift, $\Delta\Phi$s across the device, instead of frequency shift. For $\Delta\Phi < 360°$, Δf is related to $\Delta\Phi$ by the relation:

$$Df = \frac{\Delta\Phi \times V_{\text{SAW}}}{360° \times L}, \tag{13.5}$$

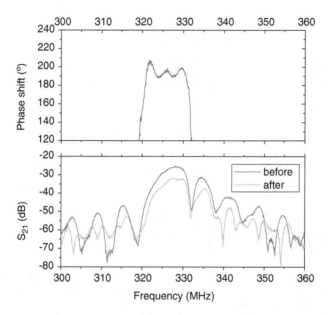

Fig. 13.8 S21 (transmission) spectra for the ZnO nanotip SAW sensor, before and after DNA oligonucleotide immobilization. The phase shift at the center frequency of 327.94 MHz is 191° [12]

where L is the center-to-center distance of SAW delay line and V_{SAW} is the acoustic velocity. Thus, in terms of phase shift, mass sensitivity is derived as:

$$\text{Mass Sensitivity} = \frac{A}{\Delta m} \frac{\Delta \Phi \times V_{SAW}}{360° \times L \times f_c}. \tag{13.6}$$

The sensitivity to immobilization is thus $357.1 \, \text{cm}^2/\text{g}$, while the sensitivity to the target second strand DNA oligonucleotide is $526.3 \, \text{cm}^2/\text{g}$.

13.2.4.2 ZnO Nano-TFBAR for DNA Immobilization and Hybridization Detection

Piezoelectric ZnO thin-film transducers are used in a variety of microwave acoustic device applications for the generation and detection of SAW and BAW due to their high electromechanical coupling coefficients. Various RF devices utilizing piezoelectric ZnO-based TFBARs have been reported, ranging from high-frequency filters to compact, low-power Global System for Mobile Communication (GSM) receivers [30–33]. In addition to their applications for communications, BAW devices have also been used for sensor applications, such as temperature, pressure, and ambient gaseous atmosphere detection [29–34]. Since TFBARs allow the implementation of BAW devices from several hundred megahertz up to 20 GHz, a higher sensitivity manifested in a large frequency shift per unit mass loading is expected. In addition, TFBAR sensors are much smaller and can be readily integrated into arrays. Solidly mounted TFBAR sensors can be fabricated on silicon, glass, or other esubstrates using planar technology. The sensors can be integrated with other Si-based electronic components on the same Si substrate. Another distinct advantage of BAW sensors is that they can be used for wireless distance probing in ecological applications. In this work, $Mg_xZn_{1-x}O$-based single-mode thin film bulk acoustic wave resonators (TFBARs) are built on Si substrates. In order to achieve the single-mode TFBAR, an acoustic mirror reflector structure is used to prevent the acoustic wave from transmitting into the Si substrate. The piezoelectric properties are tailored by controlling the Mg composition in the $Mg_xZn_{1-x}O$ film. The TFBARs are adapted for mass-sensitive sensors. ZnO nanotips are grown on TFBAR surface using MOCVD to make the ZnO nano-TFBAR with high mass sensitivity. The ZnO nanostructured surfaces are functionalized through selective DNA immobilization and hybridization for biosensing.

A complete ZnO-based nano-TFBAR biosensor device structure consisting of $Al/Mg_xZn_{1-x}O/Au/$acoustic mirror$/Si$ is shown in Fig. 13.9a. Here, Al and Au serve as the top and bottom electrode, respectively. Au is chosen as the bottom electrode as high quality ZnO and $Mg_xZn_{1-x}O$ thin films can be deposited by RF sputtering technique. Al is chosen as the top electrode to minimize electrode mass loading. The acoustic mirror, composed of two periods of alternating quarter-wavelength Bragg reflector, is used to isolate the resonator from the substrate. RF sputter-deposited $Mg_xZn_{1-x}O$ thin films have a preferred c-axis orientation. When a signal is applied between the top and the bottom electrodes, a longitudinal acoustic wave mode is excited. Figure 13.9b shows an FESEM image of a $Mg_{0.2}Zn_{0.8}O$ thin film deposited on the mirror/Si structure. The image is taken from the active region of the device where the film is directly on the acoustic mirror, rather than on the Au bottom electrode. The $Mg_{0.2}Zn_{0.8}O$ film has a dense columnar structure, with smooth surface. The two-period acoustic mirror structure (750 nm SiO_2 and 720 nm W) is clearly visible as alternating light and dark layers. Single crystalline ZnO nanostructures of 200 nm height and 70 nm diameter were directly grown on a Au electrode by MOCVD. An FESEM image of the vertically aligned ZnO nanostructures is shown in Fig. 13.9c

A stepwise biofunctionalization scheme was developed to demonstrate the biosensing ability of the above-mentioned ZnO nano-TFBAR. The multistep bio-chemical process is outlined in Fig. 13.10. The sensor's biosensing performance is demonstrated in Fig. 13.11a which shows the frequency shift of ZnO TFBAR using the developed functionalization scheme. The TFBAR sensing area is 0.25 mm × 0.25 mm. The solid black line represents the frequency response of TFBAR before step 1. The frequency peak at minimum insertion loss is 1,562.8 MHz. The dashed purple line is the resonance characterization of TFBAR after step 1 (SH linker coating). The frequency peak at minimum insertion loss is 1,553.7 MHz. The dotted green line is the frequency response measured after step 2 (SH-ssDNA coating). The frequency peak at minimum insertion loss is 1,540.9 MHz. The blue line represents

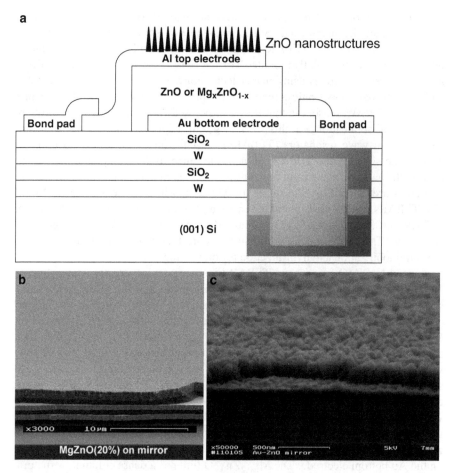

Fig. 13.9 (**a**) A schematic diagram of a $Mg_xZn_{1-x}O$ nano-TFBAR structure inset: top view optical microscope image of the device; (**b**) $Mg_{0.2}Zn_{0.8}O$ film deposited on the mirror/Si structure; (**c**) SEM image of MOCVD grown ZnO layer on Au/piezo-ZnO/Au/mirror/Si [13]

the frequency response measured after step 3 (SH-ssDNA-DNAFl). The frequency peak at minimum insertion loss is 1,535.9 MHz. The preliminary results showed that the resonance frequency decreased with increasing mass loaded on top electrode of TFBAR. The total frequency shift was about 26 MHz. As the device sensitivity ($S_f A$) is about 1 KHz cm^2/ng, the loaded mass after step 3 is about 16.25 ng. The nano-TFBAR does not only output acoustic signals to indicate the biosensing, but it also operates optically through the selective fluorescence emission from the immobilized and hybridized DNA. The fluorescence image of the nano-TFBAR in Fig. 13.11b shows that only the ZnO nanostructured sensing electrode area (bright gray squares) is positively fluorescing. This confirms that the DNA molecules only attached themselves to the ZnO nanostructured sensing area of the nano-TFBAR

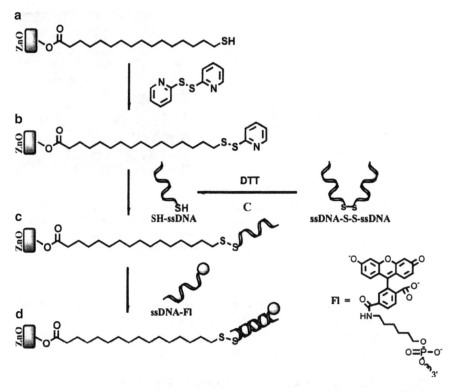

Fig. 13.10 Schematic presentation of the ZnO nanotips surface modification using SH-based linker (**a**) MHDA modified surface (the formation of sulfhydryl surface); (**b**) an activated sulfhydryl surface with pyridyl groups; (**c**) a reduction of ssDNA-S-S-ssDNA; (**d**) an exchange reaction between disulfide surface and sulfhydryl modified DNA (SH-ssDNA); and (**e**) hybridization of SH-ssDNA modified surface with ssDNA-Fl [11]

device. These results demonstrate that the ZnO nano-TFBAR is a promising device for biosensing applications.

13.2.4.3 ZnO Nano-QCM for Real-Time Viscoelastic Cell Monitoring

A dynamic and noninvasive cell monitoring technology can be realized by combining the unique sensing ability and biocompatibility of ZnO nanostructures with a QCM device to form the ZnO nanostructure-modified QCM sensor. In contrast with the reported methods, the nano-QCM has ZnO nanostructures that serve as both the biomolecular interface by the surface functionalization and as the sensitivity-enhancing layer resulted from the controlled morphology and giant surface area. This combination is expected to greatly enhance the device sensitivity, allow for simultaneous measurements of multiple parameters in a single test, and enable noninvasive and dynamic monitoring. The system has portable feature and is

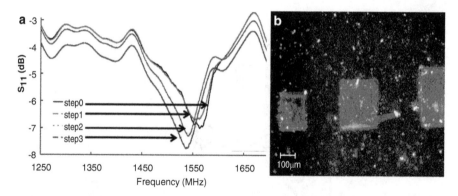

Fig. 13.11 (a) Frequency shift due to mass loading on the nano-TFBAR, step 0: nano TFBAR only, step 1: adding bifunctional linker, step 2: DNA immobilization, and step 3: DNA hybridization. (b) The fluorescence image of three nano-TFBAR sensors on the same substrate where the gray squares correspond to the sensing areas of the nano-TFBAR devices containing the fluorescing hybridized DNA [13]

the cost-effective design. The biophysical properties, mainly viscoelastic transitions, mass accumulation, and cell monolayer adhesion and proliferation relating to cell activity are monitored in real time through the specific time-evolving sensor spectral signatures, including spectral shape, Nyquist characteristics, and peak frequency shift.

The nano-QCM device schematic is shown in Fig. 13.12a and its multilayer structure in Fig. 13.12b. The piezoelectric AT-cut quartz layer is sandwiched between two 100-nm gold electrodes. The quartz substrates have a diameter of 1.37 cm and the sensing area is 0.2047 cm². The ZnO nano-QCM device consists of ZnO nanostructured arrays that are integrated on a standard AT-cut QCM. The ZnO nanostructures are directly grown on the sensing area of the QCM through a shadow mask using MOCVD. The optimal ZnO nanostructure morphology was determined through the control of the MOCVD growth conditions such as chamber pressure and substrate temperature. ZnO nanostructure layer has thickness of ∼500 nm. The ZnO-covered sensing area was exposed to UV light to make it superhydrophilic [14]. The combined effect of giant surface area of the nanostructures and superhydrophilic state makes the ZnO nano-QCM a very sensitive mass-measuring device as well as a monitoring device for viscoelastic transitions in the sample. The ZnO nano-QCM was then deployed inside a Teflon cell-growth well to serve as the test device (Fig. 13.12c), while a standard QCM was inserted in a similar Teflon cell-growth well to serve as the control device.

The time varying admittance parameters $Y(\omega, t)$ will provide dynamic information corresponding to viscoelatic transitions due to proliferation, adhesion, and growth of the cells on the nano-QCM biosensor.

Figure 13.13a–e shows the time evolution of the acoustic transmission spectra of both devices. Figure 13.13a shows the time-evolving acoustic spectrum of the control device (standard QCM). The inset of Fig. 13.13a shows the fluorescence imaging of the cells reaching full confluency on the control device; however, its

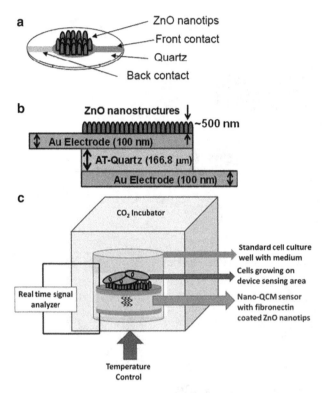

Fig. 13.12 (**a**) The ZnO nanostructure-based nano-QCM biosensor schematic, (**b**) its multilayer (**c**) time-evolving acoustic spectrum of the ZnO nano-QCM shows both amplitude reduction (**c**) setup for deploying the ZnO nano-QCM biosensor for noninvasive and dynamic cell growth monitoring [35]

acoustic spectrum does not exhibit any changes. We applied the same growth conditions to the ZnO nano-QCM, which exhibited more information than the reference device. The time-evolving $Y(\omega)$ spectrum of the ZnO nano-QCM shows both periodic amplitude reduction and increase (Fig. 13.13c) and upward frequency shifting (Fig. 13.13e) within the 60-h incubation. The amplitude reduction indicates that the sample being monitored exhibits elastic properties and is dampening the acoustic waves resonating within the device. An increase in amplitude on the other hand indicates the sample's stiffening, making it act "solid-like" which contributes less to the acoustic wave dampening. An increase in the peak frequency also indicates the increase in stiffness of the samples [36]. In Fig. 13.13c we see an initial increase in the amplitude of the $Y(\omega)$ spectrum for the first 10 h of the cell incubation then starts decreasing up to around 20 h. This indicates that upon cell seeding the cells have not attached to the ZnO nanostructures yet but are just settled on the surface, resulting in a lower signal amplitude. However, as the cells start to adhere and proliferate on the nanostructures, the cells form a more solidly adhering layer that reduces the dampening of the acoustic waves in the sensor, causing the

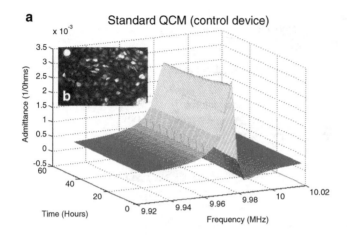

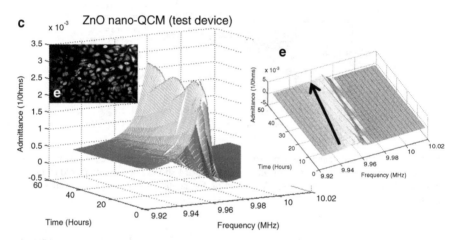

Fig. 13.13 (**a**) The time-evolving admittance spectra of the standard QCM showing very little changes to amplitude modulation during the 60-h cell monitoring cycle, (**b**) 20× fluorescence image of the BAEC cells in full confluency on the sensing area of the standard QCM; (**c**) time-evolving admittance spectra of the ZnO nano-QCM with rough ZnO columns which show both amplitude modulation and (**e**) upward frequency-shifting which indicates variation in stiffness of the cells due to adhesion and proliferation. The cells also reached full confluency on the ZnO nano-QCM as shown on the 20× fluorescence image of its sensing area (**d**) [35]

signal to increase amplitude. The signal amplitude reduction and increase happen again at 25–40 h and then increase again and stay at that amplitude until the end of the monitoring cycle at 60 h. The stabilized amplitude in the last 20 h indicates that the cells have fully proliferated and reach full confluency on the entire sensing area. The cells have indeed reached full confluency on the ZnO nano-QCM as shown on the fluorescence image of the sensing area (Fig. 13.13d). It is difficult to determine the reasons of the amplitude increase and reduction between 25 and 40 h because the amplitude variation may or may not be due to the cell proliferation and adhesion alone; it might reflect viscoelastic transitions due to cell cycle transitions. At this stage, the seeded cells have not been synchronized in growth during seeding so each individual cell will be transitioning in different cell growth phases while they are adhering and proliferating on the ZnO nanostructures. A study of signal variation with a synchronized cell culture will be conducted in the future.

13.3 The 3D Electrodes Consisting of ZnO TCO Films and Nanostructures for Optoelectronic Devices

Integration of a two-dimensional (2D) ZnO transparent conductive oxide (TCO) film with an one-dimensional (1D) ZnO nanotip array forms the three-dimensional (3D) nanostructured electrode or photoelectrode. Such integration can be realized through the sequential MOCVD growth with the in situ doping process. The ZnO 3D electrodes enable performance enhancement of optoelectronic devices, such as dye-sensitized solar cells (DSSCs) and GaN light emitting diodes (LEDs).

13.3.1 Integration of ZnO TCO Films and ZnO Nanotip Arrays

13.3.1.1 Ga-Doped ZnO as Transparent Conducting Films

Transparent conducting oxide (TCO) films play a critical role in many optoelectronic devices, such as flat panel displays, LEDs, photovoltaics, and energy efficient windows [37, 38]. Tin-doped indium oxide (ITO) films have been most widely used as the TCO films because of their high transmittance in the visible wavelength region and low resistivity comparable to metals [39]. For example, ITO was used in liquid crystal display (LCD). However, the indium raw material is expensive and has led to increasing manufacture cost. Nowadays, the real concern is whether there is going to be enough indium material to meet increasing industry demand. These concerns have stimulated extensive research for low cost alternatives. ZnO, doped with group III elements (B, Al, Ga, In), is a promising relatively inexpensive alternative TCO material with electrical and optical properties comparable to those of ITO.

Unintentionally doped semiconductor ZnO shows n-type conductivity due to the presence of point defects and impurities, many of which act as shallow donors in ZnO. Doping ZnO with group III elements (B, Al, In, Ga) leads to stable n-type conductivity in ZnO films [40–42]. The deposition temperature of ZnO-based TCO ranges from room temperature to several $100°$. The electron carrier concentration can be tuned by adjusting the concentration of group III dopants. Aluminum-doped ZnO (AZO) films with a visible transmittance of 85% and a resistivity of $\sim 3 \times 10^{-4}\,\Omega$ cm, comparable to that of ITO films, have been achieved by magnetron sputtering [43]. AZO films are stable in Ar gas ambience at temperatures up to 600°C, and in air up to 400°C [41]. The optimized AZO films, i.e., low stress, low resistivity, high Hall mobility, and high visible wavelength transmittance, are obtained at substrate temperatures ~250°C [40]. Ga-doped ZnO (GZO) films grown on c-plane sapphire substrates by electron beam evaporator are highly transparent (>80%) in the visible-near infrared ranges, and the optical bandgap exhibits a blue shift for the as-deposited films. Ga-doped ZnO films with 2 at% Ga doping have achieved resistivities approaching those of ITO [44, 45]. The transition from an insulator to a metallic conductor can be explained by the increase of carrier concentration and the formation of a degenerate conduction band due to the shallow ionization energy of the donors, as suggested by Mott [46]. Ga-doped ZnO TCO films exhibit a metallic-semiconductor transition (MST) behavior at low temperature. The $Zn_{0.95}Ga_{0.05}O$ films exhibited a resistivity close to $1.4 \times 10^{-4}\,\Omega \cdot cm$ and transmittance $> 80\%$ in the visible wavelength region [47]. It is interesting to find that heavily Ga-doped ZnO TCO films exhibited better reliability. The resistivity of 12.4 wt% Ga-doped ZnO films is $1.3 \times 10^{-3}\,\Omega$ cm and changes by less than 3% over a 2,000-h reliability test at a temperature of 85°C and a humidity of 85% [48].

ZnO TCO films have been widely used in thin film solar cells as the front windows and transparent electrodes [49]. In Cu(In, Ga)Se$_2$ devices, textured ZnO TCO can significantly improve the light absorption [50]. Recently, it was found that ZnO TCO film could be used as the transparent electrode for p-type GaN [51,52]. The efficiency of GaN light-emitting devices is significantly affected by the transparent Ohmic contacts to p-type GaN. Contacts with low specific resistance, high thermal stability, and high transparency are of great importance in improving the efficiency of the light output of GaN LEDs. Indium-oxide-doped ZnO (IZO) has also been reported for transparent ohmic contact to the p-GaN. The light output power of an LED with an IZO/p-GaN contact was improved by 34% compared to that of a GaN LED with a Ni/Au ohmic contact [53].

Undoped ZnO shows n-type conductivity due to the existence of native defects and hydrogen interstitials in the ZnO lattice [54–56]. Even though undoped ZnO films can achieve fairly high carrier concentration, the defects and hydrogen interstitials induced conductivity is thermally unstable and the conductivity is significantly reduced when annealed in oxygen ambience at a temperature of 400°C [57]. In order to obtain thermally stable conductivity, it is necessary to introduce intentional dopants in the ZnO lattice. According to semiconductor physics theory, the substitution of Zn^{2+} ions with group III ions (B^{3+}, Al^{3+}, Ga^{3+}, In^{3+}) generates extra electrons, resulting in n-type conductivity. It has been reported that the doping of group III elements can significantly increase the conductivity of ZnO material due to the relatively low ionization energy of the dopants [58–60]. Among group III elements, gallium has an ionic radius of 0.62 Å and a covalent radius of 1.26 Å, which is the closest to those of zinc (0.74 Å and 1.31 Å). The bond length of Ga-O (1.97 Å) is close to that of Zn-O (1.86 Å) and offers the advantage of minimizing the deformation of ZnO lattice at high substitutional gallium concentration [61]. Furthermore, gallium has a higher equilibrium oxidation potential and is less reactive with oxygen ambience, making it a better choice of doping candidate.

In MOCVD growth of GZO films, at low growth temperature, the Zn and O reacting species have low kinetic energy. As the Zn and O atoms are adsorbed onto the growth planes, it takes long time to diffuse to the lattice position and the crystallite barely has change to increase in size before surrounded by the newly formed nuclei. Therefore, the growth of GZO film is kinetically limited and the growth rate is independent of the precursor flow rate. As the temperature increases, the kinetic energy of the reacting species increases. The diffusion rates of Zn and O at the growth surface increase, resulting in shorter relaxation time and higher incorporation rate into ZnO lattice and the polycrystalline ZnO has a larger crystalline size. As growth temperature further increases, due to the lowest surface energy of ZnO (002) crystalline planes, the ZnO crystallites quickly elongate along the c-axis direction. The growth rate along c-axis orientation becomes dominant, and the GZO films form textured films with c-axis normal pointing to the direction of the substrate surface normal. As the substrate temperature further increases beyond 450°C, the growth rate no longer increases, which implies that the GZO growth is mass flow rate controlled. In order to obtain dense films, we keep the growth temperature below 400°C. The as-grown polycrystalline films are composed of small size crystallites. A typical morphology of GZO film grown at 380°C is shown in Fig. 13.14a. The GZO films form a dense film structure with relative

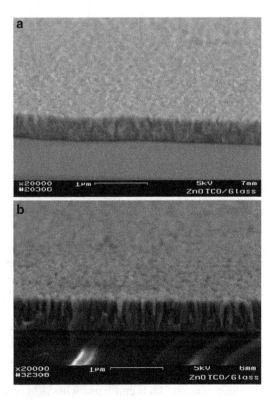

Fig. 13.14 Field emission scanning electron microscope (FE-SEM) images of GZO films grown on different nominal substrate temperatures: (**a**) 380°C, (**b**) 450°C [64]

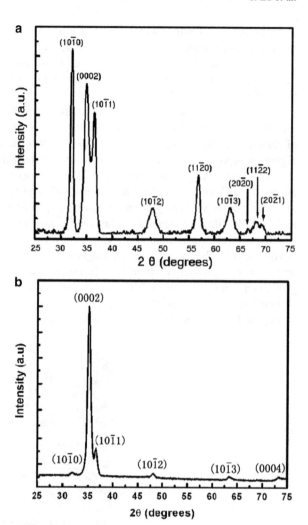

Fig. 13.15 X-ray diffraction (XRD) spectra of Ga-doped ZnO films grown on glass at substrate temperature of (**a**) 380°C, (**b**) 450°C [62]

smooth surface. At substrate temperature of ∼400°C, the GZO films start forming columnar structure as shown in Fig. 13.14b.

Figure 13.15 shows the X-ray diffraction spectra of GZO films grown at two different temperatures. At low growth temperature 380°C, besides the peaks from ZnO (0002) and ZnO (0004), the peaks from other low-indexed ZnO crystalline planes, such as ZnO(10$\bar{1}$0), ZnO(10$\bar{1}$1), and ZnO(10$\bar{1}$2), are also present with strong intensity, indicating that the GZO film has a polycrystalline structure. As the temperature is increased to 450°C, the dominant diffraction peak is from the ZnO (0002) crystalline plane, and the intensity of other low-indexed planes, such as ZnO(10$\bar{1}$0) and ZnO(10$\bar{1}$1)is significantly reduced, which indicates that the GZO films have a preferred *c*-axis orientation. The change from polycrystalline structure to *c*-axis textured film is consistent with SEM observations.

Fig. 13.16 Transmission spectra of undoped ZnO and Ga-doped ZnO films grown at the similar conditions. The thickness of undoped ZnO and GZO films is ∼400 nm and the GZO film has a sheet resistance of ∼25 Ω/sq [62]

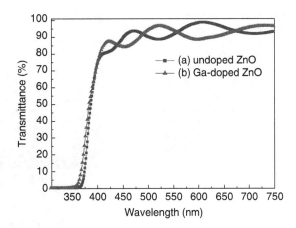

The optical transmittance and sheet resistance are two important parameters for TCO films. In solar cells, the TCO films act as both the optical window and the collecting electrode, which determine the amount of light entering the absorbers and influence the extraction of photo-induced charges. Reduction of the resistivity requires an increase in the carrier concentration and the increase of carrier concentration reduces the transmittance. High conductivity (or low sheet resistance) is balanced against high transmittance; therefore, a tradeoff between transmittance and sheet resistance has to be considered.

Figure 13.16 shows the transmittance of undoped ZnO and GZO films. The undoped ZnO (400 nm) has a transmittance in visible wavelength over 89% as shown in curve a. In the UV region, the optical transmittance of ZnO film falls sharply due to the onset of the fundamental absorption in this region. The energy gap (E_g) can be estimated by assuming a direct transition between the valence and conduction band. For the undoped ZnO, the absorption edge is around 370 nm. For the GZO films (400 nm, sheet resistance of ∼25 Ω/sq) grown at optimized conditions, a transmittance above 85% in the visible wavelength is obtained. The absorption edge of the GZO films shifts to a shorter wavelength of ∼359 nm. A blue shift of the absorption edge with increasing Ga doping level is observed, which is due to the Burstein–Moss effect associated with higher electron concentration in the conduction band of ZnO [63].

13.3.1.2 Growth of ZnO Nanotip Arrays on Ga-Doped ZnO Films

ZnO nanotips can be grown directly on GZO films, and the combination of ZnO nanotips and GZO films would provide a novel multifunctional photoelectrode for applications such as DSSCs, polymer-based solar cells, and transparent electrode for light emitting devices. The n-type semiconductor ZnO nanotips provide large surface area for dye anchoring in conjunction with direct conduction pathways for charge transport, while the GZO film acts as the transparent electrode. Figure 13.17a

Fig. 13.17 A field emission
scanning electron microscope
(FE-SEM) images of (**a**)
Ga-doped ZnO TCO films;
(**b**) sequentially grown ZnO
nanotips/Ga-doped ZnO film.
The Ga-doped ZnO film is
grown at low temperature of
∼400°C and the undoped
ZnO nanotip array is grown at
temperature of ∼470°C [64]

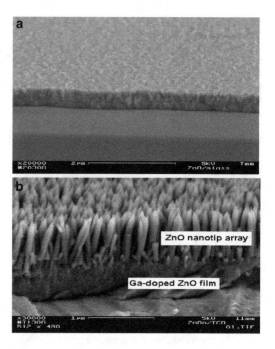

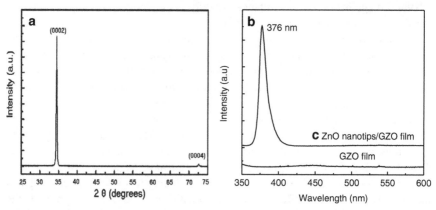

Fig. 13.18 (**a**) X-ray diffraction (XRD) spectrum of ZnO nanotips grown on glass substrate, which
is grown under the same growth conditions of the ZnO nanotips grown on Ga-doped ZnO film.
Room temperature photoluminescence (PL) spectra; (**b**) GZO film, (**c**) ZnO nanotips grown on the
GZO film [64]

shows the GZO film growth at low temperature of ∼400°C. The GZO film
exhibits a dense film structure with smooth surface. To grow ZnO nanotips on
GZO films, the growth temperature is increased to above ∼470°C. Figure 13.17b
shows a SEM image of ZnO nanotips/GZO structure sequentially grown on a
glass substrate. The ZnO nanotips are vertically aligned along the substrate normal
direction. Figure 13.18a shows the XRD spectrum of the ZnO nanotips grown

on accompanying glass substrate. Only peaks from ZnO (0002) and ZnO (0004) are observed, indicating that ZnO nanotips are c-axis preferably oriented. We speculate that the ZnO nanotips grown on the GZO films also have c-axis preferred orientation.

Shown in Fig. 13.18a–b are the room temperature PL spectra of ZnO nanotips/GZO film structure. The near band edge (NBE) emission of GZO films is significantly quenched due to the heavy Ga doping (curve a). The peak at ~376 nm (3.30 eV) from the ZnO nanotips/GZO film structure is predominantly from the free exciton emissions of ZnO nanotips (curve b). Due to the larger exciton binding energy of 60 meV, free excitons in high quality MOCVD-grown ZnO nanotips or nanorods can survive up to room temperature. The strong peak intensity in PL indicates good optical property of ZnO nanotips.

13.3.2 ZnO 3D Photoelectrodes for Dye-Sensitized Solar Cells

Photovoltaic technology has stimulated growing research interest due to the increasing demand for renewable and clean energy. In comparison to conventional solid-state *solar cells*, the DSSC has attractive features, such as simple process and low-cost fabrication on conductive glass substrate and even plastic substrates. Since Grätzel et al. demonstrated a DSSC with overall light-to-electric energy conversion efficiency of 10.4% in 1991, TiO$_2$ nanoparticles have been extensively investigated for DSSCs [65]. ZnO is another attractive, but less explored wideband gap semiconductor oxide. DSSCs built from ZnO nanoparticles have achieved the second highest efficiencies after TiO$_2$ [66]. Various ZnO nanotips of high crystalline quality can be synthesized at low growth temperature. As the MOCVD-grown ZnO is concerned, the combination of well-aligned ZnO nanotips and ZnO TCO film provides a promising photoelectrode for DSSCs and polymer-based organic solar cells. Therefore, it is interesting to explore ZnO nanotip/GZO films hybrid structure for solar cell applications.

13.3.2.1 Device Structure and Operating Principles

Figure 13.19a shows a DSSC structure using the 3D photoelectrode consisting of ZnO nanotips grown on the GZO film. The photoactive electrode consists of a layer of GZO film (~400 nm) and an array of ZnO nanotips (several μm), which are vertically aligned along the substrate surface normal. The GZO film and ZnO nanotips are sequentially grown on glass substrates using temperature-modulated MOCVD growth. To fabricate the solar cells, ZnO nanotips are sensitized by immersing in a 0.3 mM ruthenium polypyridine dye (RuN3, Dyesol) ethanolic solution. A 25 − μm microporous polyolefin battery separator (Celgard) soaked with 0.5 M LiI, 0.05 M I$_2$, 0.5 M *tert*-Butyl pyridine in methoxypropionitrile electrolyte is placed on the ZnO nanotip array. A platinum foil is clamped with a glass slide

Fig. 13.19 (**a**) A schematic
illustration of dye sensitized
solar cell using well-aligned
ZnO nanotip array and
Ga-doped ZnO TCO film as
the photoactive electrode. (**b**)
A schematic illustration of
the energy band diagram of
ZnO nanotip/GZO-based dye
sensitized solar cells

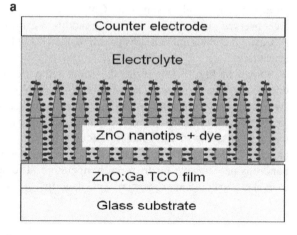

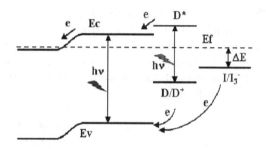

on the separator as the counter electrode. The solar cell area is calculated to
be 1.0 cm^2.

The energy band diagram of ZnO nanotip/GZO-based DSSC is shown in
Fig. 13.19b. With an electron affinity of 4.35 eV and a bandgap of 3.30 eV, the
valence band of ZnO nanotips is 7.65 eV below the vacuum energy level [67]. The
redox potential of an electrolyte based on 0.5 M LiI and 0.05 M I_2 is about +0.3 V
versus normal hydrogen electrode (NHE), corresponding to the 5.0 eV below the
vacuum energy level. Since the light is incident from the backside of TCO electrode,
UV photons with energy higher than the bandgap energy of ZnO (3.30 eV) are
absorbed by the ZnO photoelectrode and raise electrons from the valence band to
conduction band of ZnO. The photogenerated electrons diffuse through the ZnO
nanotip array to the TCO electrode, leaving the holes in the ZnO valence band.
As the valence band energy level of ZnO is lower than that of the electrolyte and
the electrolyte is in close contact with the ZnO nanotips, the holes in the ZnO
valence band are replenished with electrons donated from ions in the electrolyte. The
photons with energy lower than the bandgap energy of ZnO are harvested by the dye
molecules, which are anchored at the surface of ZnO nanotips. The photogenerated

charges are separated at the dye/ZnO interface. The oxidized dye molecules are reduced by the electrons donated from the ions in the liquid electrolyte.

For good photovoltaic performance, it is important to maximize the absorption of incident photons and ensure the efficient electron injection into the conduction band of semiconductor oxide. The basic photovoltaic processes that occur in the photoactive dye/oxide interface and electron transfer between RuN3 dye molecules and ZnO semiconductor oxide in redox electrolyte are described as following:

$$RuN3ZnO \rightarrow RuN3^{**}ZnO \rightarrow RuN3^{++}ZnO + 2e^- \tag{13.7}$$

$$RuN3^{**}ZnO \rightarrow RuN3^*ZnO \rightarrow RuN3^+ZnO + e^- \tag{13.8}$$

$$RuN3^+ZnO + e^- \rightarrow RuN3ZnO \tag{13.9}$$

The RuN3 dye captures an incident photon and generates a long-lived exciton [(13.7) and (13.8)]. The excitons are separated at the dye/oxide interface and the electron is injected into the conduction band of ZnO nanotip and holes diffused to the opposite side of the device by the means of redox species in the liquid electrolyte. The two possible electron transfer pathways are electron injection from the fully relaxed excited state (RuN3*), and from nonthermalized, higher lying excited states (RuN3**). When an electron is injected, the dye cation (RuN3$^+$) is formed together with an electron in the conduction band of ZnO (ZnO(e$^-$)). Backward electron transfer from the conduction band of ZnO semiconductor to the dye cation is the unwanted reverse reaction, which restores the original ground state of dye molecules (13.9), and does not contribute to the photocurrent. The ratio of the forward injection and backward electron transfer depends on the relaxation time. It is reported that the forward electron injection process occurs in femtosecond to picosecond time scale, while the backward electron transfer occurs in the range of microsecond to millisecond time scale [68]. The difference of several orders of magnitude between the time constants of electron injection and backward electron transfer is one of the important properties, which makes the RuN3–ZnO interface an efficient light-to-electricity converter. A key issue in the photo-induced processes of dye-sensitized semiconductor films has been the mechanism of ultrafast, subhundred femtosecond time scale electron injection, whether it occurs from nonthermalized states prior to energy equilibration or over all electronic, vibrational, and rotational degrees of freedom. The absorption of a photon by a dye molecule happens via an excitation between the electronic states of the molecule. The excitation of the Ru complexes via photon absorption is of metal to ligand charge transfer (MLCT) type. This means that the highest occupied molecular orbital (HOMO) of the dye is localized near the metal Ru ion, whereas the lowest unoccupied molecular orbital (LUMO) is localized at the ligand species, in this case, at the bipyridyl rings. At the photo-excitation, an electron is lifted from the HOMO level to the LUMO level. The LUMO energy level, extending even to the –COOH anchoring groups, is spatially close to the ZnO surface, which means that there is significant overlap between the electron wave functions of the LUMO energy level of dye molecules and the conduction band of ZnO. This directionality of the excitation is one of the reasons for the fast electron transfer process in the dye–ZnO system.

In the DSSCs, the electron transport in the semiconductor oxide is dominated by diffusion since there is no significant electrical field present. In nanoparticle films, the nature of electron transport in oxide nanoparticle films is well understood using time-resolved photocurrent and photovoltage measurements and the corresponding modeling study [69–71]. It is found that the electron transport in nanoparticle films is through a trapping–detrapping diffusion process, in which the photogenerated electrons interact with charge traps as they undertake a random walk through the nanoparticle network. In comparison to nanoparticle films, the ZnO nanotips are vertically aligned along the substrate surface normal. They provide direct conduction pathways from the point of charge injection to the TCO electrode. Furthermore, ZnO nanotips are of single crystalline quality with a low defect density. The diffusion of electrons in the ZnO nanotips does not suffer from the scattering at the crystalline boundaries and avoids being trapped in the charge traps, therefore, electron transport speed in ZnO nanotips is expected to be much faster.

Most of the reported DSSCs are built from TiO_2 particle films deposited on FTO substrates. At the FTO/TiO_2 interface, an electrostatic field creates an energy barrier for electron transfer from the TiO_2 into FTO [72]. Unlike the FTO/TiO_2, ZnO nanotips and GZO electrode belong to the same material system. They have the same values of electron affinity and the energy band bending of the ZnO/GZO homojunction is aligned without the presence of a significant energy barrier. The Fermi level difference of ZnO and GZO results in a built-in electric field that favors the electron transfer from the ZnO nanotips into the GZO electrode as shown in Fig. 13.19b.

13.3.2.2 DSSCs Using Liquid Electrolyte

ZnO Nanotips/Liquid Electrolyte Interface

In the conventional solid-state solar cells, two materials with different types of conductivity form a pn junction and Fermi levels are aligned in equilibrium conditions, which is associated with the displacement of majority carriers within semiconductor near the pn interface and the creation of a depletion region. In the DSSC, when the redox liquid electrolyte is in contact with ZnO surface, there is an analogy with the Fermi level in solid-state pn junction interface. The ions in the redox electrolyte have two different states, i.e., an oxidized state (-ox) I^- and a reduced state (-red) I_3^-. The reduced state I_3^- ion is converted to an oxidized state I^- ion by acquisition of a negative charge at the anodic electrode, and the process is reversible at the cathode. The potential, at which no current passes at a reversible electrode, is the redox potential, at which both ionic components are at equilibrium.

Figure 13.20a shows the Incident Photon to Current Efficiency (IPCE) spectrum of ZnO nanotip/GZO photoelectrode immersed in I^-/I_3^- redox electrolyte without dye sensitization. The cell built from $0.8 - \mu$m-long ZnO nanotips exhibits an absorption peak at $\sim$378 nm with an IPCE value of 16.5%. The UV light harvesting

Fig. 13.20 (**a**) An incident
photon-to-current conversion
efficiency (IPCE) spectrum of
the ZnO nanotips in contact
with the liquid I-/I3-redox
electrolyte. (**b**) Incident
photon-to-current conversion
efficiency (IPCE) spectra of
the ZnO nanotip-based dye
sensitized solar cells with
three different ZnO nanotip
lengths of 0.8 μm, 1.6 μm,
and 3.2 μm [64]

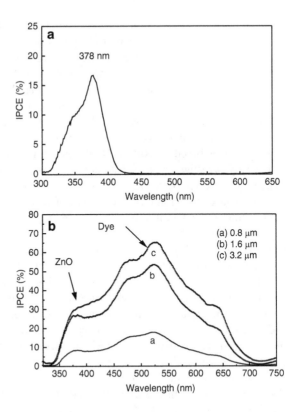

is realized through direct charge exchange between the ZnO photoelectrode and
redox electrolyte without the involvement of dye molecules. The UV photon is
absorbed by the ZnO photoelectrode and the photo-induced electrons in ZnO
conduction band diffuse through ZnO nanotips to the TCO electrode, leaving the
holes in the ZnO valence band, which are replenished with electrons donated from
the red-I_3^- ions. The high IPCE peak centered at 378 nm indicated that the back
transfer of conduction band electron of ZnO nanotips to the electrolyte is much
slower in comparison to the forward electron transfer to the TCO electrode. The UV
light harvesting could be exploited to improve the device stability of DSSCs under
high illumination by preventing dye molecules from UV exposure.

Photoresponse of ZnO 3D Photoelectrode-Based DSSCs

Figure 13.7b shows the IPCE spectra of DSSCs with three different ZnO nanotip
lengths. The maximum absorption peaks are observed at the incident wavelength
of ∼530 nm. It was found that IPCE increases with the ZnO nanotips length, with
values of 18%, 53%, 65% for 0.8 μm, 1.6 μm, and 3.2 μm ZnO nanotip length,
respectively. In addition, another photocurrent peak at 378 nm is clearly observed

in all the curves. The values of IPCE at 378 nm are 8%, 26%, and 31% for 0.8 μm, 1.6 μm, and 3.2 μm nanotip length, respectively. This photocurrent peak is attributed to the UV light harvesting by the ZnO nanotips and GZO films. ZnO has a bandgap of ∼3.3 eV at room temperature and exhibits a strong UV absorption. This charge transfer pathway is confirmed by the observation of the photocurrent peak at ∼378 nm in the absence of dye sensitization. The oxidized dye molecules are restored by electrons donated and transferred from the I_3^-/I^- electrolyte. Many dyes, including organic dyes and TM complexes, have been employed in DSSCs. TM complexes based on Ru(II), Os(II), Pt(II), Fe(II), Re(I), and Cu(I) have been extensively investigated for DSSC applications. Up to date, Ru(II)-based charge-transfer (CT) polypyridyl complexes show the highest energy conversion efficiency. This is because of their intense CT absorption in the whole visible range, moderately intense emission with fairly long lifetime in fluid solutions at ambient temperatures, high quantum yield for the formation of lowest CT excited state, and redox reactivity and ease of tunability of redox properties. Photoexcitation of the metal-to-ligand charge-transfer (MLCT)-excited states of adsorbed dye leads to an efficient injection of electrons into the conduction band of oxide. From Fig. 13.20b, it is observed that at the wavelength above ∼630 nm, the IPCE is dramatically decreased, which indicates a low light harvesting percentage at longer visible wavelength. The low IPCE value at longer visible wavelength is a major drawback of Ru-based dye molecules. In order to improve the light harvesting efficiency, there is a need to develop dyes whose absorption spectra have a large overlap with the solar spectrum.

Figure 13.21a shows the I–V characteristics of a DSSC built from 3.2-μm-long ZnO nanotips. In the dark, the cell shows rectifying characteristics. Under the illumination of AM 1.5 G (100 mA/cm²) simulated sunlight, the I–V curve is shifted downwards to the fourth quadrant, indicating power generation. The short circuit current density J_{sc} and power efficiency η as a function of ZnO nanotip length are shown in Fig. 13.21b.

For the cell with 0.8 μm ZnO nanotips, the short-circuit current density J_{sc} and the open circuit voltage V_{oc} are 1.0 mA/cm² and 0.367 V, respectively. The low value of V_{oc} might be caused by the insufficient dye coverage and the direct contact of liquid electrolyte with the GZO electrode. The low value of J_{sc} is attributed to less available ZnO surface area for dye anchoring and inefficient light harvesting, as the surface area of ZnO nanotips is approximately proportional to the ZnO nanotip length. As the length of ZnO nanotip increases, the open circuit voltage increases. The typical open circuit voltage for 3.2 μm ZnO nanotip is 610–630 mV. The short circuit current density of the cells built from 0.8 μm, 1.6 μm, 3.2 μm, and 4.8-μm-long ZnO nanotips are 1.0, 1.7, 2.8, and 4.1 mA/cm², respectively. The corresponding power efficiency is calculated to be 0.14%, 0.24%, 0.5%, and 0.77% for 0.8, 1.6, 3.2, and 4.8 μm ZnO nanotips. The monotonous increase of power efficiency with nanotips length indicates that dye loading is the main limiting factor, and further improvement in power efficiency can be achieved by increasing the ZnO nanotip length. The fill factor is calculated to be ∼0.40 for cells built from $3.2 - \mu m$ ZnO nanotips and is relatively constant at light intensity from 40 to 260 mW/cm². This value of fill factor is lower than typical TiO₂ DSSC solar cells (0.6–0.7).

Fig. 13.21 (a) $I–V$ characteristics of 3.2 μm ZnO nanotip-based dye sensitized cells in the dark and under the illumination of Xe AM 1.5 G (100 mA/cm²) simulated sunlight. (b) Power conversion efficiency (η) and short-circuit current density (J_{sc}) of DSSCs as a function of ZnO nanotip length. The photoresponse is measured under the illumination of Xe AM 1.5 G simulated sunlight (100 mW/cm²) with a 1.0 cm² cell area [64]

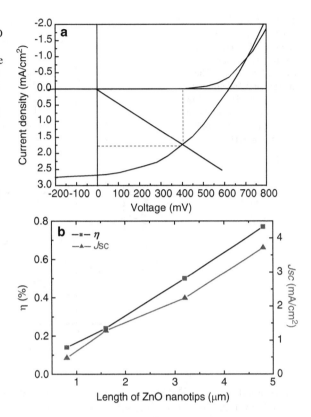

This might be attributed to the recombination of photo-excited carriers between the ZnO nanotips and triiodide ions in the electrolyte. It may also be attributed to the relatively large area of the solar cell (1.0 cm × 1.0 cm). As the device area is reduced to 0.28 × 0.28 cm, the fill factor is increased to 0.55 and the power efficiency is increased from 0.77 to 1.11% for the 4.8 − μm ZnO nanotips. Lowering the sheet resistance of the GZO electrode and using grid electrode are expected to achieve better fill factors.

Equivalent Circuit Model of DSSCs

The $I–V$ characteristics of the DSSC can be analyzed using a simple equivalent circuit as shown in Fig. 13.22a. In this equivalent circuit, the current source I_{ph} results from the excitation of excess carriers by incident solar radiation. A diode in parallel with a shunt resistor (R_{sh}) represents the oxide/oxide interface, and a series resistor (R_s) represents the resistance around the circuit loop. The shunt resistance is due to leakage current across the pn junction around the cell edge and the presence of crystal defects and/or impurities in the bulk and at the surface of

Fig. 13.22 (a) An electrical equivalent circuit model of solar cells. Rs and Rsh are resistors that represent electrical losses. (b) The experimental and simulated I–V characteristics of ZnO cells built from ZnO nanotips with length of 0.8, 1.6, 3.2, and 4.8 μm [64]

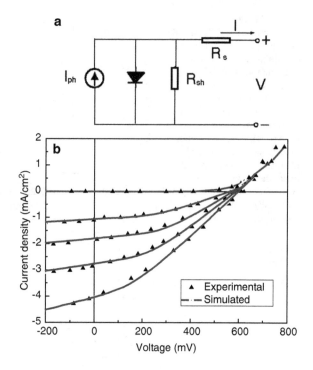

semiconductor oxide. The series resistance is mainly caused by the bulk resistance of semiconductor materials, metallic contacts, and interconnections, and contact resistance between the metallic contacts and ZnO.

The light harvesting efficiency of DSSCs depends on (1) the properties of dye molecules, such as the bandgap, absorbance, and extinction coefficient; (2) dye loading efficiency in the oxide surface; (3) the optical path within the electrode. The I–V curves of DSSCs are described with a one-diode equivalent circuit

$$I = I_{ph} - I_0 \cdot \left(\exp \left(\frac{V + I R_s}{\eta V_T} \right) - 1 \right) - \left(\frac{V + I R_s}{R_{sh}} \right), \qquad (13.10)$$

where $I_{ph}(mA/cm^2)$ is the photocurrent, $I_0(mA/cm^2)$ is the diode saturation current, η is the diode ideality factor, R_s is the series resistance, and R_{sh} is the shunt resistance. The short circuit current is given by

$$I_{sc} = I_{ph} - I_0 \cdot \left[\exp(\frac{I_{sc} \cdot R_s}{nV_T}) - 1 \right] - \frac{I_{sc} \cdot R_s}{R_{sh}}. \qquad (13.11)$$

Shown in Fig. 13.22b are the experimental and simulated I–V characteristics of the cells built from four different ZnO nanotips lengths, i.e., 0.8, 1.6, 3.2, and 4.8 μm. The curves represented by the triangle symbol are from the experimental results and the curve in solid line is the simulated results using the one-diode equivalent model. It is found that the two sets of $I-V$ curves agree well at the biasing voltage from -200 to 650 mV. The ideality factor of the diode is around 1.05, indicating good rectifying properties of the dye/oxide interface. The shunt resistance is of the order of $\sim 10^3 \ \Omega \ cm^2$. As the photocurrent density increases, the shunt resistance decreases, photoconductive current is observed. The photoconductive current may be attributed to the backward electron transfer from the ZnO conduction band to the ZnO electrolyte. The series resistance of the cells is $\sim 100 \ \Omega \ cm^2$ and is relatively high in comparison to solar cells built from TiO_2/FTO, which results in a low fill factor. The series resistance of the cells is $\sim 100 \ \Omega \ cm^2$ and is relative high, which results in a low fill factor of 0.31. The series resistance of the solar cells is mainly attributed to the relatively high sheet resistance of the GZO films. As the device area is reduced from 1.0 cm × 1.0 cm to 0.28 cm × 0.28 cm, the fill factor is increased to 0.55 and the power efficiency is increased from 0.77 to 1.11% for the $4.8-\mu m$ ZnO nanotips. The recombination rate at the dye/ZnO does not increase significantly with the ZnO nanotip length as attested by the constant value of fill factor (resp. 0.28, 0.33, and 0.32 for 1.6, 3.2, 4.8 − μm ZnO nanotips). We attribute the low recombination rate to the nontrap limited electron transport in the ZnO nanotips. The improvement of short circuit current could be achieved by increasing the length of ZnO nanotips, which improves the dye/oxide interface for better light absorption. By increasing the length of ZnO to 6.4 μm and reducing the shunt resistance to $10 \ \Omega.cm^2$, power conversion efficiency of 3.94% is predicted from the simulation using the equivalent circuit model.

13.3.2.3   DSSCs Using Gel Electrolyte

Grätzel et al. [65] demonstrated a DSSC with overall light-to-electric energy conversion efficiency of 10.4% in 1991. Since then, little progress has been made in absolute power conversion efficiency. This efficiency value is competitive with amorphous silicon thin film solar cells. However, besides power conversion efficiency, two other requirements must be met for ensuring a successful commercial future of this technology, i.e. (1) very low-cost fabrication and (2) high device stability. In recent years, DSSC has stimulated growing research interest due to the increasing demand for renewable and clean energy. Although DSSCs based on liquid redox electrolytes have reached high efficiency; however, the liquid electrolyte has low thermal and chemical stability. Furthermore, the liquid may evaporate when the cell is imperfectly sealed. The permeation of water or oxygen molecules and their reaction with electrolytes may also degrade cell performance. Liquid electrolytes also reduce the manufacturability of multicell modules because cells must be series connected electrically but separated chemically to achieve desired output voltage, preferably on a single panel [73]. Recently, many attempts have been made to solve the above problems by replacing liquid electrolyte with solid or quasi solid-state hole conductors. Compared with those using liquid-phase redox electrolytes, the efficiency of solid-state DSSCs utilizing p-type semiconductors has yet to be improved.

An alternative choice to improve the device stability is to use semisolid gel electrolyte, which has the properties of relatively high ionic conductivity and easy solidification. The use of conducting polymers as electrolytes or electrode materials for the development of plastic-like electrochemical devices is an appealing concept that dates back to the late 1980s. Ionically conducting polymers, such as poly (ethylene oxide) (POE) could be used in solid-state lithium-ion batteries in conjunction with the common inorganic oxide intercalation cathodes. From then on, polymer electrolytes have been widely used in lithium batteries, redox type laminated supercapacitors, and multichromatic optical windows.

The feasibility of replacement of liquid electrolyte with semisolid gel electrolyte in the DSSCs has been studied. A polymer based on N-methyl pyrrolidinone (NMP) gelled with poly(vinyl-difluoroethylene-hexafluororpropylene) copolymer (PVDF–HFP) is used as the semisolid gel electrolyte. NMP is an ideal solvent for PVDF–HFP copolymers and PVDF homopolymers as stable gels can be formed by their mixtures. Gel electrolytes containing liquid electrolyte and gelator are first injected into photoelectrode. The impregnation of the solar cells is achieved by heating the cells at temperature of $\sim 80°C$.

DSSCs are built with $3.2-\mu m$ ZnO nanotip arrays grown on GZO films. Gel electrolyte provides high device stability compared with liquid electrolyte. However, the fill-in process in nanoparticle-sensitized solar cells presents a challenge as the gel electrolyte usually has higher viscosity than the liquid electrolyte. Lower power conversion efficiency in TiO_2 nanoparticle-sensitized solar cells is observed when using the gel electrolyte. Compared with nanoparticle films, MOCVD-grown ZnO nanotips are well aligned along the substrate surface normal; therefore, form straight "trenches" from the top to the base of the ZnO nanotip array. The diameters of the "trenches" between the neighboring nanotips are in the range of 5–10 nm, which is larger than the porous size of typical nanoparticle film. Therefore, the gel electrolyte can easily penetrate into the ZnO nanotips array and efficiently contact with the dye-sensitized ZnO surface.

Figure 13.23a–b compares the incident photo-to-current conversion efficiency (IPCE) spectra of DSSC using gel electrolyte and liquid electrolyte built from the same ZnO nanotips under the same impregnation conditions. The peak IPCE values (at ~ 530 nm) for the gel and liquid electrolyte are 70 and 65%, respectively. The higher IPCE value in gel electrolyte indicates that the backward electron transfer from ZnO conduction band to the electrolyte is lower than that using liquid electrolyte. The higher internal quantum efficiency in semisolid gel electrolyte-based DSSC is achieved.

Figure 13.23c compares the photoresponse of DSSCs built from $3.2\,\mu m$ ZnO nanotips impregnated with MPN-based liquid electrolyte and NMP-based gel electrolyte. The short-circuit current density is 1.9 and $2.9\,mA/cm^2$ for liquid and gel electrolyte, respectively. The open circuit voltage, V_{oc} using gel electrolyte is ~ 726 mV in comparison to ~ 610 mV for the liquid electrolyte counterpart. The V_{oc} increases could be attributed to the negative shift of the conduction band edge

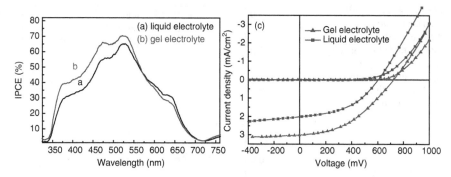

Fig. 13.23 Incident photo-to-electrical conversion efficiency (IPCE) spectra of $3.2 - \mu m$ ZnO nanotip-based dye sensitized solar cells using gel (**a**) and liquid (**b**) electrolyte. (**c**) $I-V$ characteristics of DSSCs built from $3.2 - \mu m$ ZnO nanotips impregnated with both liquid electrolyte and gel electrolyte. The photoresponse is measured under the illumination of Xe AM 1.5 G simulated sunlight ($100\,mW/cm^2$) with a $1.0\,cm^2$ cell area [62]

of ZnO with respect to the redox potential of I_3^-/I^- couple due to the basicity of NMP molecules that adsorb to ZnO surface. This is similar to the case of the observed increase in V_{oc} within TiO$_2$ nanoparticle-based DSSCs using NMP gel electrolyte [74]. An improvement in fill factor is also observed with the NMP gel electrolyte, which leads to higher overall power conversion efficiency. The increase in fill factor is also reflected in the higher parallel resistance of the cells using NMP gel electrolyte. The cells using liquid electrolyte exhibit a power efficiency of 0.77%, and higher power conversion efficiency of 0.89% is achieved when the NMP electrolyte is gelled with 20 wt.% PVDF-HFP (at $100\,mW/cm^2$). The adsorption of NMP molecules at the ZnO surface may also reduce back electron transfer.

The use of NMP gel electrolyte also improves the stability of DSSCs, by slowing down the solvent evaporation and minimizing the dye detachment from the ZnO surface. To evaluate the device stability, aging test is conducted with both liquid NMP and NMP-gelled electrolyte under thermal stress conditions. Both cells are illuminated with halogen light source producing the equivalent of one sun irradiation at the temperature of $\sim$100°C. Figure 13.24 shows power conversion efficiency changes as a function of thermal stress time. Under the same conditions, the power efficiency of the cells with NMP liquid electrolyte rapidly drops by 80% because of rapid solvent evaporation as shown in curve (a). In the case of unsealed cells with NMP-gelled electrolyte, the power efficiency of the cell drops rapidly first, then recovers and is stabilized at 60% of the initial power efficiency as shown in curve (b), which is attributed to better gel impregnation of the ZnO nanotips, and retention of a fraction of the initial solvent contents. The initial decrease of power efficiency may be attributed to the loss of liquid electrolyte contacting the ZnO nanotips; however, after further irradiation (>300 s), the melting of the gel onto the surface of the ZnO nanotips causes the efficiency recovery. For DSSC using liquid electrolyte, the sealing is the challenging issue and yet to be resolved. In comparison, the sealing

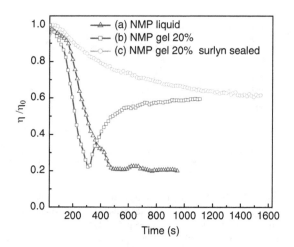

Fig. 13.24 Power efficiency changes as a function of thermal stress time under the illumination of halogen light source producing the equivalent of one sun irradiation at the temperature of ∼100°C [62]

of the DSSC using NMP-gelled electrolyte can be achieved using epoxy. Curve (c) shows the power efficiency changes of the cell sealed with epoxy as a function of thermal stress time. The decrease of power efficiency with stress time is slow and rapid efficiency drop is not observed. When epoxy-sealed cells are continuously exposed to one sun of AM 1.5 G light, the power conversion efficiency is stabilized at 60% over weeks.

13.3.3 ZnO 3D Electrodes for Enhanced Emission Efficiency in GaN LED

An interesting topic in solid-state electronics is the fabrication of low dimensional material systems to introduce quantum confinement effect. The electron–hole oscillator strength will have orders of magnitude enhancement in a nanostructure, due to the dramatically increased joint density of states near the van Hove singularity. Nanostructured optoelectronic and electronics devices have attracted much interest due to improved device performance in comparison with their counterparts without nanostructures. For instance, low lasing threshold, low temperature sensitivity, high differential gain, and high modulation bandwidth have been obtained in infrared semiconductor laser when quantum dots layer is used as the active layer [75]. A Si photoconductive photodetector with nanostructured grating in the active region has shown increased responsivity, higher internal quantum efficiency, and faster photoresponse [76]. In addition, miniaturized single nanowire devices are the building blocks for the integrated electronics and photonics of next generation.

Among various nanoscale materials, ZnO nanostructure has been placed among the most desirable candidates, which stimulated increased research interests. Bulk ZnO has a small exciton Bohr radius in a range of 1.8–2.3 nm [77–79]. Therefore, the quantum confinement effect in ZnO nanowires should be observable at the scale of an exciton Bohr radius. The well-width-dependent PL blue shift has been observed in ZnO/MgZnO MQW epitaxial nanorods [80, 81], with the ZnO well widths ranging from 1 to 5 nm. Quantum size effect also caused a blue shift in free excitonic emission in ZnO nanorods with diameters smaller than 10 nm [82]. The extremely giant oscillator strengths of free exciton and biexciton emission lead to a thermally stable low threshold lasing in ZnO nanowires. ZnO is a polar semiconductor. It shows strong exciton–polariton interactions and has been identified as the *most suitable* semiconductor for a room temperature polariton laser [83]. In a polariton laser, no population inversion is required to achieve optical amplifier. The strong light-matter coupling in a resonant microcavity produces coherent light amplification due to Bose condensation of exciton polaritons, forming the basis of a theoretically zero threshold and spin-polarization conserved stimulated emission.

Compared with bulk materials, another significant characteristic of nanostructures is the high surface-to-volume ratio. ZnO nanowires have the same lattice constants and crystal structure of bulk, confirmed by powder XRD and TEM data [84–86]. Therefore, many bulk properties are still preserved. Furthermore, doping and energy band engineering can be made in the ZnO nanotips, which are supported by the epitaxial MOCVD growth technique. The Zn–O bond is half ionic and half covalent. Doping in ZnO is much easier compared with other covalent-bond wide bandgap semiconductors, such as GaN. By appropriate doping, the electrical conductivity of ZnO can be tailored from semiconducting to semimetal, keeping high optical transparency to the visible and UV wavelength. An important consideration for space communication applications is that the material should be radiation hard, resistant against high energy particle bombardment. ZnO nanotips are attractive for field emission due to antioxidization, low emission barrier, high saturation velocity, and high aspect ratio. These make ZnO an ideal candidate among TCOs for field emission displays and transparent electronics.

UV light sources are of strong research interest because of their broad applications, including non-line-of-sight covert communication, bioagent detection, high-density optical storage, and UV photonics. ZnO is closely lattice matched with GaN. Therefore, when integrated with mainstream GaN LED and laser diode (LD) technology, ZnO nanostructure can be used for high efficient solid-state light source, such as a high light extraction efficient UV/blue nanoLED. We have reported that single crystalline and highly oriented ZnO nanotips can be grown on Si and GaN substrates by a catalyst-free growth process, providing a potential integration of ZnO nanotip arrays with Si and GaN devices [87, 88]. Because of the diversified choices of the substrate materials for ZnO nanotips, including glass, metals, Si, and GaN, novel vertical nanotip-based devices can be developed, which are significant in both scientific research and engineering application.

One of the major technical issues for nitride LEDs is low external quantum efficiency, even with much higher internal quantum efficiency. GaN has a high refractive index $n \sim 2.5$. Emission incident at the GaN–air interface at an angle greater than $23.6°$ is trapped by the total internal reflection. The light out coupling efficiency from a flat semiconductor surface is approximately given by

$$\frac{P_{escape}}{P_{source}} = \frac{1}{4}\left(\frac{1}{n}\right)^2\left[1 - \left(\frac{n-1}{n+1}\right)^2\right]. \tag{13.12}$$

From (13.12), only $\sim$3.3% of internal generated light can be extracted from a GaN LED. Due to the narrow escape cone, the parasitic nonradiative losses during photon recycling degrade the external quantum efficiency. Extensive research efforts have been made to improve the light extraction from GaN LEDs, such as surface roughening and forming photonic crystals on the top layer [89, 90]. The light extraction dynamics has been discussed in regard to modifying the spontaneous emission by the microcavity effect [91]. Other techniques include light output coupling through surface plasmons, corrugated Bragg gratings, or using random surface texturing [92–97], to convert wave-guided modes into free-space modes. Typical GaN LEDs have p-side-up structures, in which a thick p-GaN is undesirable due to the high resistivity. Manipulation such as dry etching is often difficult in a thin layer without degrading the electrical properties. Flip-chip bonding and laser lift-off technologies have been used in the LED fabrication [89]. These approaches significantly complicate the GaN LED fabrication.

13.3.3.1   Integration of ZnO 3D Electrodes with GaN LED

The integration of the ZnO 3D electrode which consists of an 2D GZO film and 1D nanotips, with GaN LED is realized by directly growing ZnO nanotips/GZO on top of a GaN LED using MOCVD. In this structure, a Ga-doped ZnO (GZO) layer is first deposited on top of a conventional GaN LED, which is used as a transparent conductive layer to p-GaN. ZnO nanotips are then grown on GZO-coated GaN as a light extraction layer for the LED. Unlike other technologies used to enhance light extraction, such as using rough surfaces and photonic crystals, this approach does not require e-beam lithography or etching, making it suitable for low-cost and large-scale fabrication. The results promise the integration of ZnO nanotips with GaN-based optoelectronic devices using epitaxial growth technology.

Figure 13.25 shows a schematic of a cross-sectional diagram of an integrated ZnO nanotips/GZO/GaN LED. InGaN/GaN multiple quantum well (MQW) LED templates were grown on c-plane sapphire by MOCVD. The GaN LEDs consist of undoped and Si-doped n-type GaN base layers, InGaN/GaN MQW active region, followed by Mg-doped p-type layers. A vertical flow rotating disk MOCVD reactor is used for growth of the GZO film and ZnO nanotips. The chamber pressure is kept at 50 torr. The growth temperature for ZnO ranges from 400 to 500°C with a growth rate of 1–2 μm/h. Diethylzinc (DEZn) and oxygen were used as the Zn metalorganic source and oxidizer, respectively. Triethylgallium (TEGa) was used as the n-type doping source for growth of GZO films. Exposure of samples to the oxygen gas before the growth was minimized in order to reduce the oxidation at the ZnO/GaN interface.

Fig. 13.25 Schematic cross-sectional diagram of an integrated ZnO nanotips/GZO/GaN LED [99]

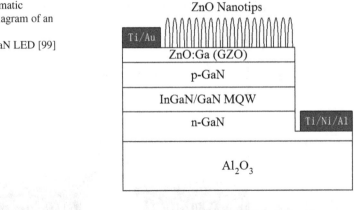

Fig. 13.26 FESEM image of ZnO nanotips/GZO/GaN (**a**) perspective view, and (**b**) top view [99]

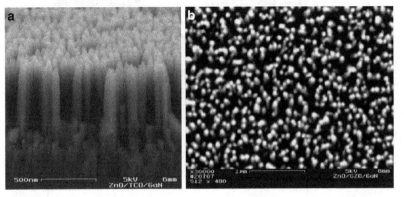

By suppressing the fast growth rate along ZnO c-axis direction, GZO film with thickness 150 nm is first grown on GaN. Four-point probe measurements show that the GZO film has a resistivity of $\sim 5 \times 10^{-3} \, \Omega$ cm. The subsequent growth of ZnO nanotips is governed by a self-assembling process. ZnO columnar structures are formed due to the highest growth rate along the c-axis from the difference in surface relaxation energies between (0001) and (000–1) planes. The columnar growth has an epitaxial relationship between ZnO and GaN: (0002) ZnO ‖ (0002) GaN and (10–10) ZnO ‖ (10–10) GaN. FESEM pictures of ZnO nanotips grown on a GZO-coated GaN LED are shown in Fig. 13.26a, b. Figure 13.26a is a perspective view, showing that ZnO nanotips are grown and distributed over the entire GZO/GaN template surface. They were found to be vertically aligned with the c-axis, normal to the basal plane of the GZO/GaN/c-sapphire template. The nanotips have an average diameter of 60 nm at the bottom and ~ 10 nm at the top, and a height of ~ 400 nm. Figure 13.26b is a top view of ZnO nanotips grown on GZO/GaN.

XRD $\theta - 2\theta$ measurements were used to determine the structural relationship between the ZnO nanotips, the GZO and the GaN. Figure 13.27a shows a $\theta - 2\theta$

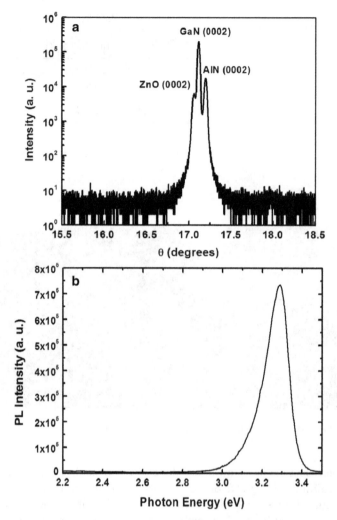

Fig. 13.27 (a) XRD $\theta - 2\theta$ scan of ZnO nanotips grown on GZO/GaN/c $-$ Al$_2$O$_3$. (b) Room temperature PL spectrum of ZnO nanotips/GZO/GaN [99]

scan of ZnO nanotips grown on GZO/GaN/c $-$ Al$_2$O$_3$. Only ZnO(0002) and GaN (0002) peaks are observed, indicating that these nanotips are preferentially oriented along the c-axis (denoted by the z direction), normal to the basal plane of the GZO/GaN/c-sapphire template (denoted by the $x-y$ plane). It was difficult to separate the possible peaks of (0002) oriented GZO film from (0002) ZnO nanotips. This was expected, as the crystal structure of GZO film is similar to ZnO nanotips due to the similar atomic radii of Ga and Zn. Since no other ZnO peaks are observed, the GZO film must be oriented in the (0002) orientation, maintaining an epitaxial relationship with the GaN template.

Optical properties of ZnO nanotips/GZO/GaN were investigated by PL. Figure 13.27b shows the room-temperature PL spectrum of ZnO nanotips grown on GZO/GaN. A strong NBE PL peak is seen at 3.29 eV with a full width half maximum of 135 meV, while deep level emission is negligible. We have noted that PL is quenched in the heavily doped Ga-doped GZO film due to impurity-assisted nonradiative recombination centers created by excess Ga atoms. The dominant free excitonic emission indicates high optical quality of ZnO nanotips.

To fabricate an integrated ZnO 3D electrode/GaN LED, the ZnO nanotips/GZO structure was selectively etched to expose the n-type GaN layer for its n-electrode contact. Ti/Ni/Al and Ti/Au contacts were deposited on the exposed n-type GaN and ZnO for n- and p-electrodes, respectively. A conventional p-metal layer of Ni/Au was also deposited on a GaN LED wafer, serving as a reference. For comparison, a GZO ($\sim$150 nm)/GaN LED without top nanotips layer was fabricated. Light output power of LED chips was measured by CAS 140B LED tester at various current injection. Figure 13.28 shows light output power of an integrated ZnO nanotips/GZO/GaN LED as a function of forward injection current. Emission from a Ni/Au p-contacting GaN LED is also plotted in Fig. 13.28. The EL intensity from a ZnO nanotips/GZO/GaN LED has increased by a factor of 1.7 at a forward current of 20 mA, in comparison to the Ni/Au contact GaN LED with the same size. The light output power of a ZnO nanotips/GZO/GaN LED is also compared with the reference GZO/GaN LED without ZnO nanotips, showing significantly increased light extraction efficiency due to ZnO nanotips. Shown in the inset of Fig. 13.28 are the EL spectra of a ZnO nanotips/GZO/GaN LED and of a conventional Ni/Au GaN LED at a current of 20 mA. The two EL spectra have essentially identical normalized spectra shape and line width.

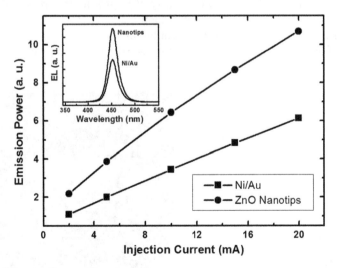

Fig. 13.28 Light output power versus forward injection current for Ni/Au and ZnO nanotips/GZO/GaN LED; inset is EL spectra of Ni/Au p-contact GaN LED and ZnO nanotips/GZO/GaN LED at a forward current of 20 mA [99]

The external quantum efficiency of a p–n junction LED is defined as the ratio of the rate of photons emitted into free space to the rate of electrons injection:

$$\eta_{\text{ext}} = \eta_{\text{int}}\eta_{\text{extraction}} = \frac{P/(h\nu)}{I/e}, \tag{13.13}$$

where η_{int} is the internal quantum efficiency, $\eta_{\text{extraction}}$ is the light extraction efficiency, P is the optical power emitted into free space, and I is the injection current. An optical power of 6 mW is measured at a current of 20 mA for a conventional Ni/Au p-contact GaN LED, corresponding to an external quantum efficiency of 7%. The external quantum efficiency is increased to 11.4% in a ZnO nanotips/GZO/GaN LED.

13.3.3.2 Analysis of Light Extraction Enhancement from ZnO Nanostructures

The enhancement in light extraction efficiency can be explained by the interaction between the spontaneous emission from the GaN LED and ZnO nanostructures. To analyze the light extraction properties of ZnO nanotips, we performed a two-dimensional Fourier transform on the top view shown in Fig. 13.29. No distinguishing nonzero spatial frequency has been observed, indicating that the nanotips are randomly oriented in the basal plane of the GZO/GaN template, i.e., the x–y plane. The absence of spatial periodicity suggests that no photonic crystal is formed that contributes to the light extraction enhancement.

At the dominant EL wavelength of 452 nm the refractive index of ZnO is ∼2.0, while that of GaN is ∼2.5. The area density of as-grown nanotips is estimated to be on the order of 10^{10} cm^{-2}. As the first approximation, we consider the nanotips layer as a disordered subwavelength assembly of ZnO and air. Effective medium theory [98] then gives the effective refractive index of the nanotip layer as

Fig. 13.29 Two-dimensional Fourier transformation of top view of ZnO nanotips grown on a GZO/GaN template

$$n_{\text{eff}} = \left[n_1^2 f + n_1^2(1 - f)\right]^{1/2}, \qquad\qquad (13.14)$$

where f is the filling factor, and n_1 and n_2 are the refractive indices of ZnO and air, respectively. Note that f here is a function of z. The effective refractive index of the nanotip layer along the z direction is gradually reduced from 1.57 at the bottom (corresponding to a filling factor of ~ 0.49) to 1.0 at the top (Fig. 13.30a). The numerical results show that an increase of $\sim 20\%$ in light extraction can be attributed to enhanced optical transmission, in comparison with a GaN layer without ZnO nanotips.

The ZnO nanotips also introduce surface roughness in the LED. As can be seen in Fig. 13.26, height variation exists among the nanotips. The nanotips surface has a correlation length around 100 nm. It has been shown that light scattering occurs from a random rough surface in the case of correlation length smaller than the wavelength, leading to an increase of transmitted evanescent waves beyond the critical angle [100]. In addition, our FESEM photographs indicate that there is a ~ 5 nm deviation in an average diameter of 60 nm for the ZnO nanotip bases. This diameter fluctuation gives rise to $\pm 7\%$ variation in the effective refractive index in the x–y plane of the nanotip region (Fig. 13.30b). The combination of surface roughening and nonuniform distribution of dielectric presumably disperses the angular distribution of spontaneous emission in the optical phase space, leading to a larger escape cone in the nanotips layer over the planar structure (Fig. 13.30c). Numerical simulation indicates that most of the spontaneous emission from the GaN

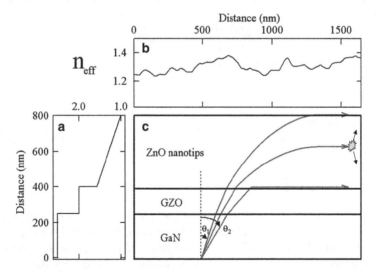

Fig. 13.30 (a) Effective refractive index for ZnO nanotips/GZO/GaN LED structure, assuming a uniformly distributed nanotip layer; (b) illustration of the variation of effective refractive index in a randomly distributed ZnO nanotips layer 50 nm away from the interface; (c) photon trajectories in a ZnO nanotips structure. Light extraction is enhanced by surface roughening (not shown) and scattering events caused by nonuniform dielectric [99]

layer within a cone of $\theta_1 = 23.6°$ will escape to the air. The emission outside a cone of $\theta_2 = 38.9°$ will not escape the ZnO layer at all. For emission within 23.6–38.9°, it gradually bends within the ZnO nanotip region until it propagates parallel to the surface.

A Monte Carlo simulation has been developed to calculate the variation in the effective refractive index of the nanotip layer [101]. We have assumed Gaussian distribution for the diameters of ZnO nanotips with an average diameter of 60 nm at the bottom. Average effective refractive index is thus numerically calculated. It is found that the diameter fluctuation gives rise to $\pm 7\%$ variation in the effective refractive index in the x–y plane of the nanotip region (Fig. 13.30b).

We have deduced the fraction of light scattering out in the ZnO nanotip layer based on a simplified light scattering model. The first assumption is for each scattering event in the nanotip layer, 50% of the photons go down to the lower hemisphere, which is almost lost. The other 50% go up to the upper hemisphere. Among this up part, the emission within the critical angle will escape to the air. The second assumption is that the rest part of light remains in the nanotip layer and is repeatedly scattered by the variation in the effective refractive index in the x–y plane.

The fraction of light within the critical angle without the scattering is given as

$$\frac{P_{escape(0)}}{P_{source}} = \frac{1}{2}(1 - \cos\theta_C) = \frac{A}{2}. \tag{13.15}$$

After the first scattering event, the fraction of remained light within the critical angle becomes

$$\frac{P_{escape(1)}}{P_{source}} = \frac{A}{2}\frac{(1-A)}{2}. \tag{13.16}$$

By adding up all the scattering events, the total fraction of light scattering out the nanotip layer is calculated by

$$\eta_{out} = \frac{\sum_{n=0}^{\infty} P_{escape(n)}}{P_{source}} = \frac{A}{2} + \frac{A}{2}\frac{(1-A)}{2} + \frac{A}{2}\left(\frac{1-A}{2}\right)^2 + \ldots + \frac{A}{2}\left(\frac{1-A}{2}\right)^n$$

$$+ \ldots = \frac{A}{1+A}. \tag{13.17}$$

Finally, we estimate that ~30% will escape upward averagely. Shown in Fig. 13.31, is the simulated angular dependent of transmitted light with ZnO nanotips grown on top of a GaN layer. An effective refractive of 1.57 is used in the calculation. For comparison, the optical transmission for standard GaN sample without top ZnO nanotips is also plot.

Combining all effects, an enhancement factor of 1.8 in light output power is deduced, compared to the standard uncoated GaN structure. In our case, a significantly increased light output from a ZnO nanotips/GZO/GaN LED, in comparison

Fig. 13.31 Simulated angular distribution of transmitted light with ZnO nanotips on top of GaN. For comparison, optical transmission of a standard GaN layer without top ZnO nanotips layer is also shown

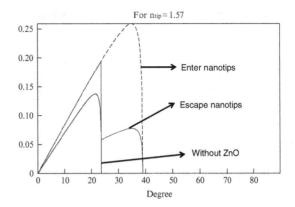

with a GZO/GaN LED reference, suggests that out coupling enhancement results from the nanostructure morphology.

Enhancement of light extraction from an integrated ZnO nanotips/GaN LED is demonstrated. The device is composed of a GaN LED with a Ga-doped ZnO (GZO) transparent conductive layer, and ZnO nanotips grown on GZO for light extraction. The light output power of a ZnO nanotips/GZO/GaN LED exhibits 1.7 times enhancement, in comparison with a conventional Ni/Au p-metal LED. The higher emission efficiency is attributed to the enhanced light transmission and scattering in the ZnO/GaN multilayer. With an optimized GZO layer on p-GaN, significantly higher light emission efficiency is expected to be achieved from the integrated ZnO nanotips/GZO/GaN LED structure. The integration of ZnO nanotips with GaN shows a promising application for UV/blue optoelectronics.

13.4 Conclusion

Multifunctional ZnO and its nanostructures are promising to make various novel electronic and optoelectronic devices. ZnO nanostructures are the ideal choice of biosensing material because its various functionalities can be obtained and changed through in situ doping during the crystal growth, morphology tuning, surface wettabilty control, and biochemical functionalization. ZnO is biocompatible, useful for neurotransmitter production and enzyme function; therefore, it is suitable for working inside the human body. This chapter presents the examples of the integration of ZnO nanotip arrays with various acoustic wave sensors to achieve high sensitivity and dual-mode operations. Currently, ZnO and its nanostructures have also been used to make other types of biosensors, including the conductive-resistive and optical sensors. The integration of 2D ZnO TCO films and 1D ZnO nanotip array forms 3D electrodes, which enable performance enhancement for various solar cells and GaN LEDs. More research efforts have been made on integration of multifunctional ZnO nanostructures with other materials including semiconductors, and with the organic materials, which will pave the way for broad device applications.

Acknowledgements The authors gratefully acknowledge support for this work by NSF (Grants No. ECS 1002178), AFOSR (Grant No. FA9550–08–0100452), and NJ Commission of Science and Technology Excellence Center Program. LY would like to show his high appreciation to his Ph.D students Zheng Zhang, Gaurav Saraf, Ying Chen, and Ziqing Duan for their valuable contributions.

References

1. K.W. Wong, X.T. Zhou, F.C.K. Au, H.L. Lai, C.S. Lee, S.T. Lee, Appl. Phys. Lett., **75**(19) p. 2918 (1999)
2. J.Y. Li, X.L. Chen, Z.Y. Qiao, Y.G. Cao, M. He, T. Xu, Appl. Phys. A. Mater. Sci. Process Springer-Verlag (2000)
3. G.S. Cheng, L.D. Zhang, Y. Zhu, G.T. Fei, L. Li, C.M. Mo, Y.Q. Mao, Appl. Phys. Lett. **75**(16): p. 2455 (1999)
4. V.I. Anisimkin, M. Penza, A. Valentini, F. Quaranta, L. Vasanelli, Sens. Actuators B: Chem., **23** (2–3), 197–201 (1995)
5. T.-J. Hsueh, S.-J. Chang, C.-L. Hsu, Y.-R. Lin, I.-C. Chen, Appl. Phys. Lett. **91**, 053111 (2007)
6. B.G. Miller, T.W. Trantim, J. Am. Chem. Soc. **120**, 2666 (1998)
7. J. Lee, B.S. Kang, B. Hicks, T.F. Chancellor, Jr., B.H. Chu, H.-T. Wang, , B.G. Keselowsky, F. Rena, T.P. Lele, Biomaterials **29** 3743–3749 (2008)
8. S.M. Al-Hilli, R.T. Al-Mofarji, M. Willander, Appl. Phys. Lett. **17**, 89 173119 (2006)
9. A. Wei, X.W. Sun, J.X. Wang, Y. Lei, X.P. Cai, C.M. Li, Z.L. Dong, W. Huang, Appl. Phys. Lett. **89**(12), 123902 (2006)
10. P.I. Reyes, Z. Zhang, H. Chen, Z. Duan, J. Zhong, G. Saraf, Y. Lu, O. Taratula, E. Galoppini, N.N. Boustany, IEEE J. Sens. **10**(10) 1302–1307 (2009)
11. O. Taratula, E. Galoppini, R. Mendelsohn, P.I. Reyes, Z. Zhang, Z. Duan, J. Zhong, Y. Lu, Langmuir **25**, 2107–2113 (2009)
12. Z. Zhang, N.W. Emanetoglu, G. Saraf, Y. Chen, P. Wu, J. Zhong, Y.Lu, J. Chen, O. Mirochnitchenko, M. Inouye, IEEE Trans. Ultrasonics, Ferroelec. Freq. Contr. **53**(4), 786–792 (2006)
13. Y. Chen, P.I. Reyes, Z. Duan, G. Saraf, R. Wittstruck, Y. Lu, O. Taratula, E. Galoppini, J. Electron. Mater., **38**(8), 1605–1611, (2009)
14. Z. Zhang, H. Chen, J. Zhong, G. Saraf, Y. Lu, TMS IEEE J. Electronic Mater. **36**(8), 895 (2007)
15. J. Lee, B.S. Kang, B. Hicks, T.F. Chancellor, Jr., B.H. Chu, H.-T. Wang, B.G. Keselowsky, F. Rena, T.P. Lele, Biomaterials **29** 3743–3749 (2008)
16. R. Yoshica, D. Kitamura, S. Maenosono, J. Toxicol. Sci., Vol. **34**, No. 1, pp. 119–122, (2009)
17. J. Liu, P.M. Goud, M. Raj, Z. Iyer, Lin, Wang, R.R. Tummala, Electronic Components and Technology Conference (2008)
18. M.H. Asif, A. Fulati, O. Nur, M. Willander, Cecilia Brännmark, Peter Strålfors, Sara I. Börjesson, Fredrik Elinder, Appl. Phys. Lett., **95**, 023703 (2009)
19. S. Al-Hilli and M. Willander, Nanotechnology **20**, 175103 (2009)
20. Adam Dorfman, Nitin Kumar, Jong-in Hahm, Langmuir, **22**, 4890–4895 (2006)
21. R. Dutta, P. Sharma, A. Pandey, "Surface enhanced Raman spectra of *E. Coli* cell using ZnO Nanoparticles" Digest J. Nanomat. Biostruct., Vol. **4**, No. 1, March 2009, p. 83–87 (2009)
22. X. Feng, L. Feng, M. Jin, J. Zhai, L. Jiang, D. Zhu, J. Am. Chem. Soc., vol. **126**, pp.62, (2004)

23. Y. Li, W. Cai, G. Duan, B. Cao, F. Sun, F. Lu, J. Colloid Interface Sci., vol. **287**, pp.634, (2005)
24. R.D. Sun, A. Nakajima, A. Fujishima, T. Watanabe, K. Hashimotoh, J. Phys. Chem. B, vol. **105**, pp.1984, (2001)
25. R. Wang, K. Hashimoto, A. Fujishima, M. Chikuni, E. Kojima, A. Kitamura, M. Shimohigoshi, T. Watanabe, Nature, vol. **388**, pp.431, (1997)
26. R. Wang, N. Sakai, A. Fujishima, T. Watanabe, K. Hashimoto, J. Phys. Chem. B, vol. **103**, pp.2188, (1999)
27. E. Galoppini, Coord. Chem. Rev., vol. **248**, pp.1283, (2004)
28. O. Taratula, E. Galoppini, D. Wang, D. Chu, Z. Zhang, H. Chen, G. Saraf, Y. Lu, J. Phys. Chem. B, vol. **110**, pp.6506, (2006)
29. G.D. Mansfeld, I.M. Kotelyansky, *Proc. 2002 IEEE Int. Ultrasonics Symposium*, pp.909–912, (2002)
30. L. Mang and F. Hickernell, *Proc.1996 IEEE Int. Frequency Control Symposium*, pp.363–5, 1996
31. C.W. Seabury, P.H. Kobrin, R. Addison, D.P. Havens, *IEEE MTT-S Digest*, pp.181–184, (1997)
32. J. Kaitila, M. Ylilammi, J. Molarius, J. Ella, T. Makkonen, *Proc. 2001 IEEE Int. Ultrasonics Symposium*, pp.803–806, (2001)
33. Y.S. Park, S. Pinkett, J.S. Kenney, W.D. Hunt,*Proc. 2001 IEEE Int. Ultrasonics Symposium*, pp.839–842, (2001)
34. P. Hauptmann, R. Lucklum and Jens Schröder, *Proc. 2003 IEEE Int. Ultrasonics Symposium*, pp.56–65, (2003)
35. P.I. Reyes, Z. Duan, Y. Lu, D. Khavulya, N.N. Boustany, J. Biosens. Bioelectron., June 2011 Submitted
36. A. Alessandrini, M.A. Croce, R. Tiozzo, P. Facci, Appl. Phys. Lett. **88**, 083905 (2006)
37. D.S. Ginley, C. Bright. MRS Bull. **25**, 15 (2000)
38. T. Minami, MRS Bull. **25**, 38 (2000)
39. T. Minami, Semicond. Sci. Technol. **20**, 35 (2005)
40. S. Bose, S. Ray, A.K. Barua, J. Phys. D: Appl. Phys. **29**, 1873 (1996)
41. T. Minami, T. Miyata, T. Yamamoto, J. Vac. Sci. Technol. A, **17**, 1822 (1999)
42. N. Malkomes, M. Vergohl, B. Szyszka, J. Vac. Sci. Technol. A **19**, 414 (2001)
43. B. Szyszka, Thin Solid Films **351**, 164 (1999)
44. N.R. Aghamalyan, E.A. Kafadaryan, R.K. Hovsepyan, S.I. Petrosyan, Semi. Sci. Technol. **20**, 80 (2005)
45. G.A. Hirata, J. McKittrick, J. Siqueiros, O.A. Lopez, T. Cheeks, O. Contreras, J.Y. Yi, J. Vac. Sci. Technol. A **14**, 791 (1996)
46. N.F. Mott, "Metal-Insulator Transition", (Tailor and Francis, London, 1974)
47. V. Bhosle, A. Tiwari, J. Narayan, Appl. Phys. Lett. **88**, 032106 (2006)
48. O. Nakagawara, Y. Kishimoto, H. Seto, Y. Koshido, Appl. Phys. Lett. **89**, 091904 (2006)
49. J. Hupkes, B. Rech, S. Calnan, O. Kluth, U. Zastrow, H. Siekmann, M. Wuttig, Thin Solid Films **502**, 286 (2006)
50. J. Springer, B. Rech, W. Reetz, J. Muller, M. Vanecek, Solar Energ. Mater. Solar Cells **85**, 1 (2005)
51. J. Song, K. Kim, S. Park, T. Seong, Appl. Phys. Lett. **83**, 479 (2003)
52. J. Lim, D. Hwang, H. Kim, J. Oh, J. Yang, R. Navamathavan, S. Park, Appl. Phys. Lett. **85**, 6191 (2004)
53. J. Lim, D. Hwang, H. Kim, J. Oh, J. Yang, R. Navamathavan, Seong-Ju Park, Appl. Phys. Lett. 85, 6191 (2004)
54. K. Vanheusden, C.H. Seager, W.L. Warren, D.R. Tallant, J.A. Voigt, Appl. Phys. Lett. **68**, 403 (1996)
55. T. Yamamoto, H.K. Yoshida, Jpn. J. Appl. Phys. **38**, L166 (1999)
56. D.G. Thomas, J.J. Lander, J. Chem. Phys. **25**, 1136 (1956)
57. G.A. Shi, M. Saboktakin, M. Stavola, S.J. Pearton, Appl. Phys. Lett. **85**, 5601 (2004)

58. K. Ellmer, J. Phys. D: Appl. Phys. **34**, 3097 (2001)
59. K. Kakinuma, K. Kanda, H. Yamamura, J. Mater. Sci. **38**, 7, (2003).
60. M. Hiramatsu, K. Imaeda, N. Horio, M. Nawata, J. Vac. Sci. Technol. A**16**, 669 (1998)
61. V. Assuncao, E. Fortunato, A. Marques, A. Goncalves, I. Ferreira, H. Aguas, R. Martins, Thin Solid Films **442**, 102 (2003)
62. H. Chen, Z. Duan, Y. Lu, A. Du Pasquier, J. Eletron. Mat. **38**(8), (2009)
63. N.R. Aghamalyan, E.A. Kafadaryan, R.K. Hovsepyan, S.I. Petrosyan, Semi. Sci. Technol. **20**, 80 (2005)
64. H. Chen, A. Du Pasquier, G. Saraf, J. Zhong, Y. Lu, Semicond. Sci. Technol. **23**, 045004 (2008)
65. B. O'Regan, M.A. Grätzel, Nature (London) **353**, 737 (1991)
66. M.K. Nazeeruddin, A. Kay, I. Rodicio, R. Humphrybaker, E. Muller, P. Liska, N. Vlachopoulos, M. Grätzel, J. Am. Chem. Soc. **115**, 6382 (1993)
67. J.A. Aranovich, D.G. Golmayo, A.L. Fahrenbruch, R.H. Bube, J. Appl. Phys. **51**, 4260 (1980)
68. Neil A. Anderson, T. Lian, Annu. Rev. Phys. Chem. **56**, 491, (2005)
69. J. Krüger, R. Plass, M. Grätzel, P.J. Cameron, L.M. Peter, J. Phys. Chem. B **107**, 7536 (2003)
70. J. Nelson, Phys. Rev. B **59**, 15374 (1999)
71. J. Van de Lagemaat, A.J. Frank, J. Phys. Chem. B **105**, 11194 (2001)
72. S. Rühle, D. Cahen, J. Phys. Chem. B **108**, 17946 (2004)
73. K. Tennakone, V.P.S. Perera, I.R.M. Kottegoda, G. Kumara, J. Phys. D-Appl. Phys. **32**, 374 (1999)
74. M.G. Kang, K.M. Kim, K.S. Ryu, S.H. Chang, N.G. Park, K.J. Kim, J. Electochem. Soc. **151**, E257 (2004)
75. H.C. Liu, M. Gao, J. McCaffrey, Z.R. Wasilewski, S. Fafard, Appl. Phys. Lett. **78**, 79 (2001)
76. A.K. Sharma, S.H. Zaidi, G. Liecthy, S.R. Brueck, *Proceedings 1st IEEE Conference on Nanotechnology (IEEE-NANO'2001)*, pp. 368–373, Piscataway, NJ, USA
77. B. Gil, A.V. Kavokin, Appl. Phys. Lett. **81**(4), 748 (2002)
78. E.M. Wong, P.C. Searson, Appl. Phys. Lett. **74**, 2939 (1999)
79. R.T. Senger, K.K. Bajaj, Phys. Rev. B **68**, 045313 (2003)
80. A. Ohtomo, M. Kawasaki, I. Ohkubo, H. Koinuma, T. Yasuda, Y. Segawa, Appl. Phys. Lett. **75** (7), 980 (1999)
81. W.I. Park, G. Yi, M. Kim, S.J. Pennycook, Adv. Mater. **15** (6), 526 (2003)
82. K. Maejima, M. Ueda, S. Fujita, S. Fujita, Jpn. J. Appl. Phys. Part 1, **42** (5A), p. 2600–2604
83. M. Zamfirescu, A. Kavokin, B. Gil, G. Malpuech, M. Kaliteevski, Phys. Rev. B **65**, 161205 (2002)
84. V.A.L. Roy, A.B. Djurisic, W.K. Chan, J. Gao, H.F. Lui, C. Surya, Appl. Phys. Lett. **83**(1), 141 (2003)
85. Y.B. Li, Y. Bando, T. Sato, K. Kurashima, Appl. Phys. Lett. **81**(1), 144 (2002)
86. Z.W. Pan, Z.R. Dai, Z.W. Wang, Science **291**, 1947 (2001)
87. S. Muthukumar, H. Sheng, J. Zhong, Z. Zhang, N.W. Emanaetoglu, Y. Lu, IEEE Trans. Nanotech 2 (1), 50(2003)
88. J. Zhong, G. Saraf, S. Muthukumar, H. Chen, Y. Chen, Y. Lu, TMS IEEE J. Electron. Mater. **33** (6), 654 (2004)
89. J.A. Edmond, K. Das, R.F. Davis, J. Appl. Phys **63**, 922 (1988)
90. Ya I. Alivov, U. Ozgur, S. Dogan, D. Johnstone, V. Avrutin, N. Onojima, C. Liu, J. Xie Q. Fan, H. Morkoc, Appl. Phys. Lett. **86**, 241108 (2005)
91. T. Fujii, Y. Gao, R. Sharma, E. L. Hu, S. P. DenBaars, S. Nakamura, Appl. Phys. Lett. **84**, 855 (2004)
92. T.N. Oder, K.H. Kim, J.Y. Lin, H.X. Jiang, Appl. Phys. Lett. **84**, 466 (2004)
93. E.F. Schubert, Y.H. Wang, A.Y. Cho, L.W. Tu, G.J. Zydzik, Appl. Phys. Lett. **60**, 921 (1992)
94. W.L. Barnes, J. Lightwave Technol. **17**, 2170 (1999)
95. B.J. Matterson, J.M. Lupton, A.F. Safonov, M.G. Salt, W.L. Barnes, I.D.W. Samul, Adv. Mater. **13**, 123 (2001)
96. H.J. Peng, Y.L. Ho, X.J. Xu, H.S. Kwok, J. Appl. Phys. **96** (3), 1649 (2004)

97. I. Schnitzer, E. Yablonovitch, C. Caneau, T.J. Gmitter, A. Scherer, Appl. Phys. Lett. **63**, 2174 (1993)
98. R. Windisch, C. Rooman, S. Meinlschmidt, P. Kiesel, D. Zipperer, G.H. Döhler, B. Dutta, M. Kuijk, G. Borghs, P. Heremans, Appl. Phys. Lett. **79**, 2315 (2001)
99. J. Zhong, H. Chen, G. Saraf, Y. Lu, C.K. Choi, J.J. Song, D.M. Mackie, H. Shen, Appl. Phys. Lett, **90**, 203515 (2007)
100. Y. Ono, Y. Kimura, Y. Ohta, N. Nishida, Appl. Optics **26** (6), 1142 (1987)
101. M. Nieto-Vesperinas, J.A. Sánchez-Gil, J. Opt. Soc. Am. A **9** (3), 424 (1992)
102. J.N. Mait, D.W. Prather, M.K. Mirotznik, Optics Lett. **23** (17), 1343 (1988)

Chapter 14
ZnO/MgZnO Quantum Wells

Jeffrey Davis and Chennupati Jagadish

Abstract In this review, we discuss first some of the recent works to reveal properties of conventional ZnO/ZnMgO QWs grown c-axis oriented. This will include the properties of the quantum confined Stark effect (QCSE) that results from the internal electric field in the unit cell. We will then discuss various unconventional QW growths, including non-polar ZnO QWs, graded barrier QWs and double QWs. We finish with a review of current progress towards light emitting devices based on ZnO/ZnMgO QWs.

ZnO has been a material of interest for over 50 years; however, the ability to grow high-quality epilayers of ZnO and ZnO-based ternary systems[1–4] has led to renewed interest over the past decade in ZnO for device applications. The demand for optoelectronic devices in the blue-UV region of the electromagnetic spectrum has been well established and ZnO possesses several properties that are superior to GaN for many applications [5,6].

The large exciton binding energy of $\sim 60\,\mathrm{meV}$ suggests excitonic emission that is very efficient above room temperature, leading to great potential for light emitting devices. The large piezoelectric and pyroelectric coefficients suggest potential for applications as piezoelectric sensors or actuators. Other advantages of ZnO are its comparatively low growth temperatures [5], low optical power threshold for lasing [7], radiation hardness [8,9] and biocompatibility [10]. ZnO may also find application as a transparent conductive oxide [11] to replace indium-tin-oxide (ITO) in photovoltaic applications because it remains transparent even when doped above the level of degeneracy [12].

J. Davis
Centre for Atom Optic and Ultrafast Spectroscopy, Swinburne University of Technology, Hawthorn, VIC 3122, Australia
e-mail: JDavis@swin.edu.au

C. Jagadish (✉)
College of Physical and Mathematical Sciences, Australian National University, Canberra, ACT 0200, Australia
e-mail: Chennupati.Jagadish@anu.edu.au

S. Pearton (ed.), *GaN and ZnO-based Materials and Devices*,
Springer Series in Materials Science 156, DOI 10.1007/978-3-642-23521-4_14,
© Springer-Verlag Berlin Heidelberg 2012

This ability to grow high-quality epilayers and thin films rapidly led to the development of ZnO-based quantum wells (QWs). The two main attractions of developing QWs are the tunability offered for the transition energy, and the increased exciton binding energy, which typically leads to increased oscillator strength and greater efficiency for light emitting devices. The two most common types of ZnO-based QWs are Zno/ZnMgo and ZnO/ZnCdO-based systems, where alloying with Mg leads to an increase in the band gap and with Cd leads to a reduction in the band gap. In this chapter, we will focus on the more common ZnO/ZnMgO QWs.

The band gap of $Zn_{x-1}Mg_xO$ is given by $3.37 + 2.51x$ eV [13], which together with a conduction to valence band offset ratio of approximately 70:30 leads to type I confinement in $ZnO/Zn_{x-1}Mg_xO$ QWs. The value of x is limited to less than 0.43 [13] in these systems as above this value phase separation tends to occur. Nevertheless, even with values significantly less than 0.43, strong confinement is obtained.

In such ZnO/ZnMgO QWs, exciton binding energies up to 120 meV have been reported [14]. The biexciton binding energy is also enhanced in QWs going from 15 meV in bulk ZnO to $\sim$ 25 meV in ZnO/ZnMgO QWs [15–17]. With biexciton binding energies greater than the thermal energy, it is possible that biexcitons may play a major role in the optical properties at room temperature.

Since the first ZnO epitaxial layers were grown, high-quality ZnO quantum wells have been grown by several different methods, including molecular beam epitaxy (MBE), metal-organic vapour phase epitaxy (MOVPE) and pulsed laser deposition (PLD). The growth in almost all cases is c-axis oriented despite growth on a range of single crystal substrates, including sapphire, ZnO and Si oriented along various crystal planes.

14.1 Properties of Conventional ZnO/ZnMgO Quantum Wells

In bulk ZnO, the valence band is split by the crystal field and spin–orbit coupling into three states, labelled A, B and C, as depicted in Fig. 14.1. The highest energy A band has Γ_9 symmetry whist the B and C bands, as well as the conduction band, have Γ_7 symmetry [18–20]. Excitons formed with holes in the A band can have Γ_5 or Γ_6 symmetry, which are optically allowed and forbidden, respectively. Excitons formed with holes in the B or C band can have Γ_5, Γ_1 or Γ_2 symmetry, the first two of which are optically allowed and the Γ_2 exciton is optically forbidden [18]. For electric fields polarised perpendicular to the c-axis only the three Γ_5 excitons (A, B and C) are optically active, whereas for electric fields polarised parallel to the c-axis only the two Γ_1 excitons (B and C) are optically active [18]. In ZnO quantum wells that are grown c-axis oriented, only the three Γ_5 excitons are observed for emission perpendicular to the plain of the well, with the A exciton typically dominating the optical spectrum [21].

Fig. 14.1 A schematic of the
band structure of ZnO
showing the valence band
split due to spin–orbit
interaction and crystal field
splitting, giving A, B and C
excitons

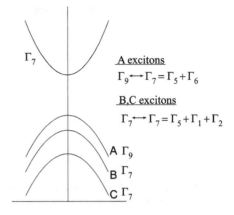

Perhaps the greatest impact of the c-axis growth of ZnO/ZnMgO QWs, however, results from the large polarisation across the unit cell in this direction. The spontaneous polarisation across the $Zn_{1-x}Mg_xO$ unit cell along the c-axis is $0.057 + 0.066x$ C/m^2. The difference between the values in the ZnMgO barrier and ZnO QW leads to a mismatch of $\Delta P_{SP} = 66x$ mC/m^2 [22–24] at the interfaces and a charge imbalance that causes an electric field across the well. Due to the lack of an inversion centre in wurtzite crystals and the strong electromechanical coupling there is also potential for a large piezoelectric field. In ZnO/ZnMgO QWs, substantial strain is present at the interfaces due to a significant lattice mismatch. The lattice constants for ZnO are 3.250 Å along the a-axis and 3.34 Å along the c-axis [13]. For $Zn_{x-1}Mg_xO$ these are given by $3.250 + 0.035x$ Å along the a-axis and $3.34–0.063x$ Å along the c-axis [13]. At the interfaces, the substantial strain and large ZnO piezoelectric constant lead to a polarisation at the interface of $\Delta P_{PZ} = -34x$ mC/m^2 [22–24]. This piezoelectric field is in the opposite direction to the spontaneous polarisation and so reduces the overall electric field strength. In such c-axis oriented QWs, the net internal electric field across the QWs is in the opposite direction to the piezoelectric field and in the direction of the growth. The effects of such an internal electric field across a QW are collectively known as the quantum confined Stark effect (QCSE).

The QCSE typically causes a red-shift of the transition energy and reduction of the electron–hole overlap, which in turn causes a reduction in oscillator strength and increase in radiative lifetime. This is depicted in Fig. 14.2, which shows the calculated electron and hole wavefunctions for a 1D potential profile (equivalent to the confinement in a QW) with and without an internal electric field. A band-offset ratio of 70:30 and effective masses for the electron and hole of 0.28 and $0.78m_e$, respectively, were used to calculate the electron and hole wavefunctions and energies [25]. The electron and hole wavefunctions are clearly shown to be separated by the electric field and the overlap integral significantly reduced. The electron and hole energies are also moved closer together by the electric field, leading to a red-shift of the transition energy.

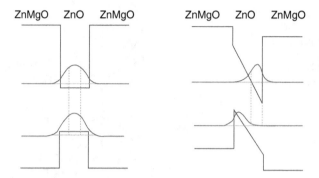

Fig. 14.2 The potential profile and electron and hole wavefunctions for a square well with and without an internal electric field

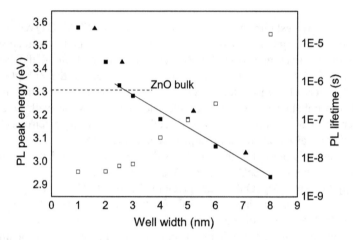

Fig. 14.3 The PL peak energy (*filled squares* [29] and *triangles* [25]) and lifetime (*open squares* [29]) are plotted as a function of well width, revealing the impact of the QCSE. [25, 29]

For a uniform electric field across a QW, it is expected that the extent of the QCSE increases as the well width is increased. This is the case in practice and several groups have used this property to determine the strength of the internal electric field by measuring the transition energy as a function of well width [25–31]. The transition energy as a function of well width is shown in Fig. 14.3 for ZnO/Zn$_{1-x}$Mg$_x$O QWs for $x = 0.3$ and 0.22 [25, 29]. For QW widths varying from 2.5 to 8 nm, the transition energy is red-shifted by more than 300 meV. For QWs wider than ~ 3.5 nm, the transition energy is pushed below the bulk ZnO transition energy as the QCSE begins to dominate over confinement effects. Based on these values, the internal electric field was determined to be 0.8–0.9 MeV/cm. The magnitude of the internal field does, however, vary as a function of Mg concentration, since the mismatch of spontaneous and piezoelectric polarisations reduces as the compositions of the well and barrier become more similar. The internal electric field in ZnO/Zn$_{1-x}$Mg$_x$O QWs grown by MBE or MOVPE is given by $4.1x$ MV/cm [24, 27], in agreement with the results presented above.

The decrease in oscillator strength and increase in radiative lifetime as a result of the electron–hole separation is also represented in Fig. 14.3 in the form of lifetimes from time-resolved photoluminescence measurements. Once again, in QWs less than ~ 3.5 nm wide confinement effects are seen to dominate and the measured lifetimes vary little. For QWs wider than 3.5 nm, however, the lifetimes increase rapidly as the QCSE begins to dominate. In going from 3 to 8 nm, the lifetime increases by over three orders of magnitude. This rapid increase in the lifetime is in good agreement with the expected trends associated with the increased electron–hole separation and corresponding reduction of the oscillator strength.

The lifetime values plotted in Fig. 14.3 are treated as the equilibrium values for the QWs, or in other words, the transition energy and lifetime of a single exciton in the QW. In practice, many carriers are typically excited in the QWs and these are able to screen the internal field to some extent. As a result, the expected red-shift is reduced and the expected lifetime shortened by amounts proportional to the excitation density. Furthermore, as the excitons relax and the extent of screening of the internal field changes, the lifetime and transition energy will change. This can be seen clearly in Fig. 14.4a which depicts the emission intensity as a function of time after excitation, with the decay rate decreasing at longer times as the carrier density reduces and the extent of screening is diminished. The corresponding effect on the emission energy can be seen in cw photoluminescence spectra taken as a function of laser intensity, as shown in Fig. 14.4b, where the transition was blue-shifted as the intensity was increased.

To determine the carrier density required for substantial screening of the internal field Hall et al. [32] performed calculations for a 6-nm wide QW, solving the Schrödinger and Poisson equations self-consistently. In these calculations,

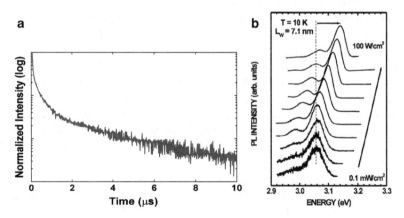

Fig. 14.4 (a) The time-resolved photoluminescence data plotted on a log scale shows the constantly changing lifetime [32]. (b) Increasing the intensity of the excitation and hence the density of carriers induces a blue-shift [25]

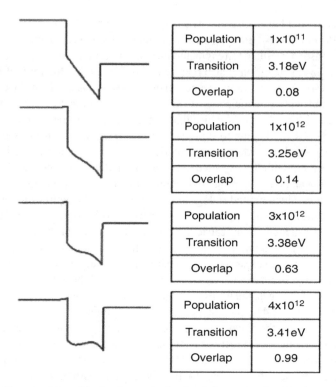

| Population | 1x10¹¹ |

Population	1×10^{11}
Transition	3.18eV
Overlap	0.08

Population	1×10^{12}
Transition	3.25eV
Overlap	0.14

Population	3×10^{12}
Transition	3.38eV
Overlap	0.63

Population	4×10^{12}
Transition	3.41eV
Overlap	0.99

Fig. 14.5 The potential profile for the conduction band for carrier densities between 10^{11} and 4×10^{12} cm^{-2}. The accompanying tables show the calculated transition energy and electron–hole wavefunction overlap

the electron and hole wavefunctions were calculated and the associated charge distribution subsequently used to determine the change to the total electric field and potential profile of the QW. This was then repeated for the new potential profile with incrementally increasing carrier density, until the internal field was completely shielded. The charge density incorporated at each step was $< 10^{10}$ cm^{-2}, which was sufficiently small to ensure there were no sharp changes in the profile as the extra carriers were added. It was found that for densities up to 10^{12} cm^{-2}, the potential profile changed very little. For densities of 4×10^{12} cm^{-2}, however, the internal electric field was nearly completely screened, with the transition energy blue-shifted and its lifetime reduced. The potential profile for the conduction band is shown in Fig. 14.5 for carrier densities of 1×10^{12}, 2×10^{12}, 3×10^{12} and 4×10^{12} cm^{-2}. The corresponding transition energy (ignoring the exciton binding energy) and electron–hole overlap integrals are shown for each density in the table in Fig. 14.5.

The evolution of the transition energy and electron–hole overlap integral is plotted in Fig. 14.6 as a function of excitation density. For densities above $\sim 4 \times 10^{12}$ cm^{-2}, both the transition energy and the overlap integral reach a plateau at values of 3.41 eV and 0.99, respectively. The overlap integral can be used to determine the lifetime by

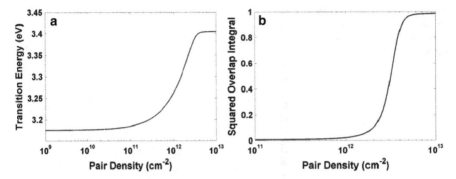

Fig. 14.6 (a) The transition energy ignoring exciton binding effects, as a function of carrier density behaves as expected. (b) The overlap integral as a function of pair density shows two plateau regions, one below 10^{12} cm^{-2} and one at $\sim 4 \times 10^{12}$ cm^{-2}

$$T_1(n) = T_1(0) \left(\int_{-\infty}^{\infty} \psi_e(n)\psi_h(n)\mathrm{d}z \right)^{-2}, \tag{14.1}$$

where T_1 is the density-dependent lifetime, $\psi_{e,h}$ represent the electron and hole wavefunction envelopes and $T_1(0)$ is the lifetime in the case of perfect electron–hole wavefunction overlap. From this density-dependent lifetime, the evolution of the exciton density, n, can then be determined using:

$$\frac{\mathrm{d}n}{\mathrm{d}t} = -\frac{n}{T_1(n)}. \tag{14.2}$$

If it is assumed that the dominant relaxation pathway is the radiative one, then this calculated decay of the pair density should represent the decay of the time-resolved photoluminescence signal.

In (14.1), however, there remains one undetermined constant – the lifetime for the case of perfect electron–hole overlap. In order to determine this value, comparison to the experimental data is required. This could be achieved by determining the carrier density in the QW immediately following excitation, and matching the overlap integral for that density to the initial lifetime measured. Given the steep slope of the curve in Fig. 14.6b, however, the significant uncertainty in the QW capture efficiency would be magnified when converted to electron–hole overlap and compared to the lifetime, which also varies rapidly in the time shortly after excitation. Furthermore, immediately following excitation it is possible that the carrier density in the QWs exceeds the regime in which this model is valid. Similarly, the longest time-constant measured cannot be assumed to be the asymptotic value of the lifetime because the measured signal is limited by detector sensitivity. Hall et al. [32] therefore chose to fit the simulation results to the experimental TRPL data using

Fig. 14.7 The time-resolved
PL shown is used in a fitting
procedure to determine the
complete carrier dynamics

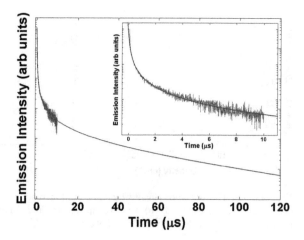

$$S_{FIT} = A \times S_{CALC}(t \times T + t_0), \qquad (14.3)$$

where $S_{CALC}(t)$ represents the calculated decay curve as a function of delay; A
the amplitude scaling factor, which corresponds to the detection efficiency; T is
the scaling factor for the time axis and effectively fits for $T_1(0)$, the unscreened
lifetime; and t_0 offsets the calculated decay curve along the t-axis to correct for
the initial pair density within the QW immediately following excitation. The fitting
error is determined from the relative difference between the calculated and measured
curves.

The resulting calculated recombination dynamics are shown in Fig. 14.7 and
fit remarkably well to the experimental data. This calculation may then be used
to determine the carrier density and transition energy at any time after excitation.
Based on such calculations the initial pair density was found to be $(2.5 \pm 0.2) \times 10^{12} cm^{-2}$, giving a combined absorption and QW capture efficiency of between
0.06 and 0.074% per well, following excitation high into the QW barrier.

In those experiments the initial lifetime was determined to be 2 ns, however
pump–probe measurements for similar excitation conditions reveals an additional
fast component that decays within the first 200 ps, suggesting that perhaps the
excitation led to the carrier density surpassing the Mott density, and putting a lower
bound on the Mott density in such ZnO/ZnMgO QWs. The asymptotic lifetime,
which corresponds to the lifetime in the completely unshielded QW, was calculated
to be $35 \pm 10 \mu s$, a factor of 6 greater than the longest time constant measured,
highlighting the limitations of the detection.

Previous works [33, 34] have reported lifetimes far shorter than those in [29] and
[35], whilst others have attributed multi-exponential decays to the transition from
free to bound exciton emission [35]. In these works, the potential role played by the
effects of screening the internal electric field is not taken into account and may be
an alternative explanation for the effects seen. This work by Hall et al. provides a
guide for determining the maximum excitation densities that should be used if the

effects of carrier-induced screening of the internal electric field are to be minimised, allowing them to be ignored in the analysis of experiments in similar ZnO/ZnMgO QWs.

In multiple QW samples, it is also possible that the internal electric field is screened to some extent by neighbouring QWs, with the extent of the screening dependent on barrier width as $L_B/(L_B + L_W)$. Many of the initial reports on the QCSE in ZnO-based QWs underestimated the electric field strength due to this screening [25].

For QWs with small barriers, that is, barriers with low Mg concentration, the QCSE remains small and confinement effects dominate for well widths less than 5 nm [1]. In this case, recombination times and oscillator strengths become much more sample-dependent as the role of defects and well width fluctuations in localising excitons becomes more important. This type of QW can therefore provide a sensitive measure of the quality of the epitaxial growth of ZnO and ZnMgO.

Another effect observed in narrow QWs is the mixing of A and B excitons induced by the enhanced electron–hole exchange interaction [21, 36, 37]. Based on calculations, a splitting of 12 meV is predicted but unobservable due to inhomogeneous broadening. A corollary of such a strong exchange interaction and significant mixing of the A and B exciton states is that there should be a redistribution of oscillator strengths, which should also be reflected in the radiative lifetime of the excitons. In time-resolved photoluminescence experiments, Gil et al. [21, 36] report bi-exponential decay curves which they attribute to the mixed A- and B-like excitons. In QWs wider than 3 nm, the internal electric field reduces the exchange interaction and the splitting is no longer seen.

All of these results looking at the recombination dynamics are based on experiments performed at low temperature. At room temperature, where any devices would operate, non-radiative recombination becomes much more important. A corollary of this is that the quality of the sample and the density of defects also become increasingly important, as these can greatly assist non-radiative recombination. The role and nature of defects in bulk ZnO has been extensively studied and is fairly well understood, see for example the review by Janotti and Van de Walle [9]. In QWs additional types of defects associated with the interfaces and lattice mismatch are possible, however these are particularly sample-dependent, as can be seen by the extensive literature reporting different defects in ZnO/ZnMgO QWs, for example [31, 33, 38–42]. The precise nature, cause, effect, and control of different types of defects remain, however, open questions of much interest and importance for future device applications.

As the temperature is increased, interactions with phonons also play an increasingly important role in the optical properties of ZnO/ZnMgO QWs. In low temperature photoluminescence measurements strong LO phonon replica peaks can be seen out to a fifth replica. These LO phonon replicas arise as a result of the strong LO phonon–exciton coupling, not only in ZnO QWs, but also in bulk ZnO. The energy of the LO phonon is 72 meV in bulk ZnO and approximately the same in ZnO/ZnMgO QWs. In addition to generating phonon replicas, phonons also contribute to the transition linewidth, an effect that is predictably temperature-dependent. Studies of photoluminescence spectra as a function of temperature are therefore used to determine the properties of the exciton–phonon interaction. At low temperatures, it is expected that acoustic phonon interactions will dominate the homogeneous linewidth as the LO phonons are frozen out. In some cases, however, the inhomogeneous contribution, which is strongly sample-dependent, may be the main contributor to the total linewidth up to 100 K. As the temperature increases, the effect of LO phonon scattering increases, leading to the broadening of the transition, and at high temperatures LO phonon Frölich scattering dominates. Combining these three main processes, the full width at half maximum (FWHM) of the emission/absorption spectrum is approximately described by:

$$\Gamma(T) = \Gamma_{inh} + \gamma_{ph}T + \Gamma_{LO}\left[\exp\left(\frac{\hbar\Omega_{LO}}{k_B T}\right) - 1\right]^{-1}, \qquad (14.4)$$

where Γ_{inh} is the temperature-independent inhomogeneous linewidth, $\gamma_{ph}T$ is due to acoustic phonon scattering, with γ_{ph} the acoustic phonon coupling strength. The final term is due to LO phonon scattering, where Γ_{LO} is the exciton–LO phonon coupling strength, and $\hbar\Omega_{LO}$ is the LO phonon energy, which has been shown to be 72 meV [43]. This final term could be modified to include the effect of all optical phonons, or, as done by Klingshirn et al. [44], be replaced by an effective optical phonon energy and average coupling strength. In quantum wells in particular, however, it is expected that LO phonons are the dominant optical phonon interaction, and so the effect of the other optical phonons can be ignored.

For a 4.7-nm wide QW with low barriers (i.e., low Mg concentration in the barriers), the terms representing the coupling strengths of the acoustic and LO phonons have been calculated to be $\gamma_{ph} = 31\,\mu eV/K$ and $\Gamma_{LO} = 341.5\,meV$, respectively [39]. As the well width is increased, the LO phonon coupling strength is enhanced [39, 45–47] due to the increased separation of the electron and hole. Similarly, if the height of the barrier is increased, the LO phonon–exciton coupling strength is enhanced due to the increased internal electric field and corresponding electron–hole separation. In bulk ZnO, the exciton binding energy is smaller than in even the widest ZnO QWs, and so the LO phonon–exciton coupling strength should be enhanced further, as has been reported by Makino et al. [39, 46]. These values differ greatly from values reported by Klinghshirn, where $\gamma_{ph} = 16\,\mu eV/K$ and $\Gamma = 47\,meV$ [44, 48]. The large difference here stems from the assumptions used in fitting the data. Klingshirn et al. use an effective optical phonon energy of 33 meV rather than the LO phonon energy of 72 meV, and take into account the longitudinal transverse splitting which dominates the linewidth at low temperature, both of which lead to lower coupling strengths. This highlights that different measurements of phonon coupling strengths can give significantly different values based on the assumptions made; thus, care needs to be taken in making comparisons between different measurements. Nonetheless, it is clear that the phonon coupling in ZnO and ZnO QWs is very strong and much stronger than, for example, in GaAs QWs [44].

14.2   Unconventional QW Structures

The reduced electron–hole wavefunction overlap and hence reduced oscillator strength caused by the internal electric field may lead to reduced efficiency in ZnO/ZnMgO-based optical devices. As a result, there have been many attempts to reduce the quantum confined Stark effects of the field by growing QW structures that are non-polar, or with potential profiles to minimise the effect of the internal filed.

14.2.1   Non-Polar ZnO

The large internal electric field that is present in most ZnO QWs arises because the growth of ZnO epitaxial layers is almost always is c-axis oriented. The wurtzite structure of ZnO is very polar in this direction, leading to strongly polarised ZnO unit cells in the growth direction. Along several of the other crystallographic directions the unit cell is non-polar and much work has been devoted to growing ZnO epitaxial layers in these directions. The growth of ZnO-based quantum wells in non-polar directions is desirable to reduce the effects of the internal electric field on the optical properties as discussed above [49–52].

Growths of ZnO epitaxial layers along the a-axis, m-axis and r-axis have all been reported; however, high quality and repeatable growth of heterostructures along these directions is still a major challenge. In both a- and m-plane ZnO, the c-axis is perpendicular to the growth direction and leads to anisotropy of both the strain and optical properties. As discussed above and shown in Fig. 14.1, there are typically five types of excitons in bulk ZnO that can contribute to the optical signal, three Γ_5 excitons (which couple to light polarised parallel to c-axis oriented QWs) and two Γ_1 excitons (which couple to light polarised perpendicular to c-axis oriented QWs). In a- and m-plane ZnO, the excitons that contribute to the total spectrum depends on the polarisation of the light, with the Γ_1 (B and C) excitons and Γ_5 (A, B and C) excitons dominating for polarisations parallel and perpendicular to the c-axis, respectively. For polarisations perpendicular to the c-axis, the lowest energy A exciton typically dominates the optical spectrum. The absence of this exciton in the spectrum for light polarised parallel to the c-axis leads to a significant drop in emission intensity and slight wavelength shift, giving anisotropic optical properties. This type of optical anisotropy is desirable for some applications, such as backlighting liquid crystal displays, adding further motivation to achieve the growth of high-quality non-polar ZnO QWs.

In most reports of non-polar ZnO growth, the layer surfaces typically show anisotropic morphology elongated along the c-axis [53, 54], and anisotropic strain. The surface roughness, which in most reports to date is typically $> 15\,\mathrm{nm}$ RMS provides a major barrier to high-quality ZnO QW growth. Fluctuations in the surface layer will lead to fluctuations in the QW width and localised states will dominate the optical properties, meaning that atomically flat interfaces are required. The extent of this strain anisotropy, which has recently been well characterised by Chauveau et al. [53], makes the growth of high-quality ZnO layers very difficult. When growing non-polar QWs, this strain causes even more problems, leading to high densities of defect states, and emission spectra where defect peaks are comparable in intensity to the QW peaks. In the case of ZnO grown on $Zn_{1-x}Mg_xO$ this strain has been shown to be compressive in one direction and tensile perpendicular to it due to the different changes in lattice constants, further increasing the level of difficulty in growing high-quality ZnO/ZnMgO QWs.

Non-polar ZnO/ZnMgO QWs have, however, been grown with a-plane ZnO on r-plane sapphire by plasma-assisted MBE [28, 55, 56], and with m-plane ZnO using hydrothermal m-plane ZnO substrates by PLD [54].

In the work of Chauveau et al. [53], they observe the expected anisotropic emission and plot the emission energy as a function of well width, revealing strong confinement effects, but no evidence of any QCSE. On this basis, they confirm the much reduced internal electric field strength in the a-plane ZnO QWs. The PL spectra, however, also reveal two main defect-related lines that are comparable in intensity to the QW exciton transitions. It is expected that these defects would cause significant problems with the use of such materials in device applications, and further work is needed to improve the quality of such non-polar QWs.

Recent work by Lautenschlaeger et al. [52] has shown the growth of non-polar a-plane ZnO epilayers on a range of substrates, with very high morphological and structural quality observed only for growth on a-plane ZnO substrates. These samples were grown using a home-built CVD system, and they suggest that homoepitaxial growth is required for high-quality non-polar ZnO growth. The RMS roughness of the surface of these a-axis ZnO layers was 6 Å, comparable to the best CVD grown c-axis ZnO. Furthermore, in these samples the A exciton peak was clearly observed in the emission spectrum with strength comparable to the best c-axis ZnO layers, something not previously seen for a-plane ZnO. This growth of high-quality non-polar ZnO layers with nearly atomically flat surfaces presents the best candidate yet for the growth of high-quality non-polar ZnO QWs.

14.2.2 Effects of Varying the Potential Profile

The traditional picture of a quantum well is a square potential with vertical barriers. In ZnO/ZnMgO QWs, this is not an accurate picture as the internal electric field imposes a slope on the bottom of the quantum well to the detriment of several optical properties. This poses the question of whether control of the optical properties can be achieved by carefully controlling the potential profile of QWs.

Coleman et al. [57] used low-energy oxygen ion implantation and a subsequent rapid thermal anneal to induce intermixing of Zn and Mg atoms in order to alter the potential profile. The intermixing is driven by the creation of defects, which encourages the diffusion of Mg from the barrier layers into the ZnO QWs, and

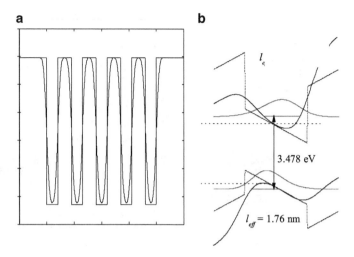

Fig. 14.8 (**a**) The effect of the intermixing can be easily seen in this multiple QW sample. (**b**) When the internal electric field is introduced the effects of the QCSE are mitigated

although the luminescence is initially quenched by these defects, good recovery was observed following rapid thermal annealing. The extent of the intermixing is strongly dependent on the ion dose, allowing a good level of control over the resultant potential profile. An example of the potential profiles obtained (ignoring internal electric field effects) is shown in Fig. 14.8a, where the previously square potentials are smoothed. The internal electric field in such wells is also no longer constant across the entire QW profile, but varies with Mg concentration. If the samples are assumed to be homogeneous in their composition and no additional strain is introduced by the implantation and rapid thermal anneal, then the internal field can be determined as a function of position in the well, and the resultant potential profile calculated, as shown in Fig. 14.8b.

As a result of the changes to the potential profiles, the effects of the QCSE are reduced. The transition is blue-shifted, as shown in the emission spectra in Fig. 14.9a, and the lifetime reduced, as indicated in Fig. 14.9b [29]. It is worth noting that whilst the transition is blue-shifted for each implantation dose, corresponding to increased intermixing, the lifetime is reduced significantly only in going from the reference square well to the sample with lowest implantation dose. This suggests that the effects of the QCSE are mediated by even the smallest amount of intermixing observed, and subsequent blue-shifts are due to the increased Mg concentration across the QW [29].

The biggest limitation of intermixing for altering the QW profile is that only limited control over the potential profile can be achieved. The shape is determined by the diffusion parameters, meaning that the profiles must always be symmetric. A further problem is that in multiple QW samples the ion dose received by the different QWs will differ across the sample, and the QW profiles will no longer be identical,

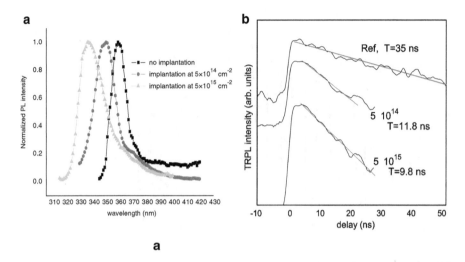

Fig. 14.9 (**a**) The photoluminescence spectra for a series of QWs that have undergone different amounts of ion implantation. (**b**) The time-resolved data show the lifetime decreasing for the smallest amount of intermixing but not changing much after that [29]

unless multiple energy implantations are carried out to obtain uniform defect profile across the multiple quantum wells.

An alternative approach to controlling the potential profile was recently reported by Hall et al. [58], where they have grown QW structures with graded barriers. This follows previous work on GaN QWs, where structured QWs have recently been demonstrated as a possible method for countering an intrinsic QCSE [59, 60]. These graded barrier samples were grown by MBE on a-plane Al_2O_3 substrates. During the graded barrier growth, a growth interruption of 2 min was introduced at each heterointerface to make the interfaces abrupt. The Mg composition of the samples, which was stepped every 1 nm of growth, was controlled by changing Mg flux under a Zn-excess condition with constant O_2 and Zn fluxes [5, 61]. Two of the structures grown are shown in Fig. 14.10. The only difference between the two is the growth direction relative to the gradient of the steps, and in the absence of any internal electric field they should have identical properties. In the presence of an internal electric field, however, the potential profiles differ quite substantially, as shown in Fig. 14.11. Calculations of the electron and hole wavefunctions are also shown, along with the relative eigen energies and transition energies (ignoring exciton binding energies). The transition for sample B is blue-shifted with respect to A and the electron–hole overlap increased. The photoluminescence spectra and time-resolved PL evolution of these two samples are shown in Fig. 14.12, along with those for square 2 and 4 nm wide ZnO/ZnMgO QWs for comparison.

The calculated transition energies and electron–hole wavefunction overlap integrals are shown for each sample in Table 14.1, alongside the measured peak luminescence energy and lifetimes. It is clear that each of the measured values

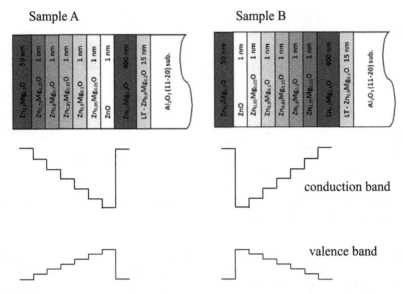

Fig. 14.10 The schematic of the layer structure for two graded barrier ZnO QWs. Potential profiles without electric field are also shown

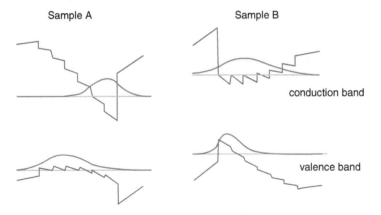

Fig. 14.11 The internal electric field causes the potential profiles and electron and hole wavefunctions to be significantly altered

follows the trends expected from the calculations, though the absolute values of the lifetimes are not fully accounted for. Comparisons between the two asymmetric QWs and the 4-nm square well suggest that the effects of the QCSE are reduced in sample B but increased in sample A. The reason for comparing to the 4-nm square well is that the volumes are essentially equivalent, however, for many applications, the more important parameter is the transition energy, and so comparison of the two graded barrier QWs to the 2-nm wide square QW may be more relevant. In this case, the lifetimes are both increased compared to the 2-nm wide QW with transition

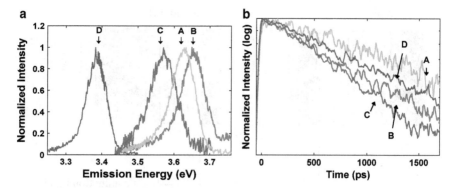

Fig. 14.12 (**a**) The photoluminescence spectra for each of the samples discussed in the text, and (**b**) the corresponding time-resolved PL profiles

Table 14.1 Calculated transition energies and wave function overlaps, compared to experimental values for the lifetimes, measured by TRPL, and transition energies. The calculated transition energies do not include exciton binding energy

Sample	Measured T_1 (ps)	Measured E (eV)	Calculated overlap	Calculated E (eV)
Sample A	1791 ($\pm$50)	3.56 ($\pm$0.01)	0.25	3.52
Sample B	847 ($\pm$12)	3.62 ($\pm$0.01)	0.71	3.57
Sample C (2 nm QW)	781 ($\pm$10)	3.65 ($\pm$0.01)	0.86	3.58
Sample D (4 nm QW)	1041 ($\pm$13)	3.39 ($\pm$0.01)	0.36	3.38

energies that are similar. The implication of this is that the oscillator strength, and hence radiative efficiency for any light emitting devices, is better in the square 2-nm QW than the graded barrier samples for the same transition energy. It is clear, however, that by using this technique of varying the potential profile an additional level of control over the QW properties can be engineered with predictable results. This may open the door to further growth designs that increase overlap and oscillator strength, or devices that make use of the enhanced charge separation as observed in sample A discussed above.

The optical properties and the potential profile are also dependent on the growth technique used. Until recently, there were no reports of a pronounced QCSE in

ZnO/ZnMgO QWs grown by PLD, despite reports of QWs of sufficient width to expect substantial QCSE [62–65]. In a recent report, Brandt et al. [62] attribute this to the high kinetic energy of impinging plasma particles in the PLD process. The PLD plasma can be as high as several hundred electron volts, with the topmost layer constantly subject to ion erosion, leading to intermixing between layers. This in turn will lead to smoothed barriers, as obtained with ion implantation and as discussed above. This smoothing of the barriers leads to a reduced QCSE [29].

Brandt et al. confirm these speculations, showing that the strength of the QCSE is very sensitive to the sharpness of the interfaces and that the PLD fluence has a major effect on this. They grew a series of $ZnO/Zn_{1-x}Mg_xO$ ($x > 0.2$) QWs by PLD with two different laser fluences. In the case of the samples grown with the greater laser fluence of $2.4\,J/cm^2$, only a negligible QCSE was observed. When the laser fluence was reduced to $1.8\,J/cm^2$, however, a strong QCSE was observed with the transition energy shifting by up to 230 meV below the bulk ZnO transition, and the lifetime increasing up to 4.2 μs. In these samples, the electric field was calculated to be between 0.31 and 0.52 MV/cm for samples with $x = 0.32$. This electric field is approximately half the value determined for MBE and MOCVD grown samples [27, 66] and suggests that even at the lower laser fluence the interfaces are still not perfectly abrupt.

14.2.3 Coupled Quantum Wells

Further attempts to control the properties of ZnO/ZnMgO QWs have involved the growth of double QWs [67, 68]. Double quantum wells become interesting when the barrier between them is sufficiently narrow to allow significant tunnelling and coupling between the transitions in the two QWs, as depicted in Fig. 14.13. When the barrier between the QWs is large, the transition energies of the two identical QWs are the same. When the barrier width is reduced and significant tunnelling allowed, the degenerate energy levels are forced to split due to the coupling between them. The lower energy state then becomes the dominant emitter, leading to a redshift of the peak emission [67, 68].

Further work by Zippel et al. [68] studied the time-resolved luminescence from a series of coupled QWs with different barrier widths, as depicted in Fig. 14.13. In these experiments they reveal a clear two-component decay for barrier widths 2.2,

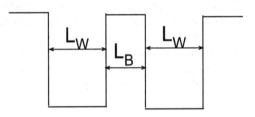

Fig. 14.13 Coupled quantum wells exhibit significant tunnelling when the barrier width, L_B, is sufficiently narrow

4, and 6 nm, but not as clearly for the 1 and 14 nm barriers. The authors attribute the additional decay component to a spatially indirect exciton, which occurs only when the wells are coupled, and support this by spectra at different delays which show two peaks at long time, the higher energy of which they again attribute to an indirect exciton. This, however, is not a clear description of the physics, as in these samples where the two coupled wells are identical, the electron and hole densities for each state should be equally distributed across the two QWs, and the only energy splitting due to symmetric and antisymmetric electron or hole wavefunctions. We would suggest, therefore, that the additional decay component and additional peak in the long delay spectra are due to recombination between antisymmetric electron and symmetric hole or vice versa. The same arguments regarding reduced electron–hole overlap integral and reduced oscillator strength still apply, but this does not invoke the picture of carrier density differences in the different QWs, which is unphysical assuming the wells are truly identical. Nonetheless, the ability to manipulate the optical properties on ZnO/ZnMgO QWs by controlling the coupling between a symmetric or antisymmetric pair of QWs has been clearly demonstrated and provides another means of control over the optical properties. Further control can be achieved with asymmetric coupled QWs where the two wells are of different widths and spatially indirect excitons are realised, with increased lifetime and reduced oscillator strength as expected [67–69].

14.3 Progress Towards ZnO/ZnMgO QW Devices

The ultimate aim of much of this research on ZnO QWs has been to develop devices based on an in-depth understanding of the relevant properties. In the pursuit of ZnO/ZnMgO QW-based devices there are two main hurdles that need to be overcome, the reliability of p-type doping and the complete understanding of the optical properties, including the role played by defects and the mechanism of stimulated emission. There have, however, been several successful attempts at making ZnO/ZnMgO QW-based devices [70–73], and the development and observation of optical gain and stimulated emission in such materials has been increasingly reported.

The large exciton binding energy in ZnO strongly suggests that the mechanism for stimulated emission is likely to be exciton–exciton scattering, and at low temperature this is clearly the case. At room temperature it is less clear and alternative mechanisms, such as from an electron–hole plasma, have been proposed for stimulated emission in bulk ZnO [44]. In ZnO/ZnMgO QWs the mechanism of lasing is more clearly excitonic up to room temperature; due largely to enhanced exciton binding energies up to 120 meV [14, 39, 43, 53, 74], though with sufficiently high optical excitation stimulated emission from electron–hole plasma recombination has also been observed [3].

Reports of stimulated emission and optical gain at room temperature from ZnO/ZnMgO QWs first appeared almost immediately after the initial reports of the growth of such materials [2, 3]. The threshold first reported for excitonic lasing in ZnO/ZnMgO QWs was $\sim 80\,kW/cm^2$ at room temperature [2]. Over the ensuing years, the threshold for optical gain has been reduced little [14, 73] with recent gain curves [14] showing the intensity increasing as the excitation density to the power of 5.5. More recently, Tsang et al. [70] reported optical gain in ZnO/ZnMgO QW structures grown at 200°C by filtered vacuum arc technique. The threshold was $500\,kW/cm^2$ at room temperature, however, they were able to lower this to $400\,kW/cm^2$ by etching a ridge structure in the sample. In addition to the reduced threshold, the slope efficiency was doubled, and the emission linewidth reduced as a result of the ridge waveguide structure. Whilst the efficiency and threshold are still not as good as those achieved with other growth methods, the lower growth temperature provides great opportunities for lower cost devices and the ridge waveguide may be a structure that can enhance the optical gain in other ZnO/ZnMgO QW light emitting devices.

One of the major goals, however, is to integrate these materials in electrically driven devices. Recent work by Polyakov et al. [42] has studied the nature of defects in ZnO/ZnMgO QWs embedded in a p–n diode structure where effective the p-type doping was not realised. In addition to examining the role of defects in preventing the realisation of effective p-type doping, they also reveal deep trap states in the QW region with activation energies of 0.32 and 0.57 eV following annealing of the sample; however, this is highly dependent on the growth parameters and the quality of the samples. The role of defects in both the QW and the p-type material clearly requires further systematic studies to allow the routine production of ZnO/ZnMgO QW-based light emitting devices.

Optical emission and gain has, however, recently been achieved with electrically pumped ZnO/ZnMgO QW-based samples [71, 72]. Shukla [71] reported light emitting diodes with ZnO/$Zn_{0.9}Mg_{0.1}$O QWs as the active medium with phosphorous-doped p-type and gallium-doped n-type layers. Efficient electroluminescence was observed at room temperature for bias voltages greater than 7 V. The efficiency was observed to decrease over time with exposure to ambient air. It was, however, recovered following thermal annealing; suggesting that surface passivation may prevent the degradation of the emission properties.

Another example of electrically pumped ZnO/ZnMgO QW-based devices was the ZnO diode lasers on silicon reported by Chu et al. [72]. They report lasing at room temperature with a very low threshold injection current of $10\,A/cm^2$. In this case the p-type dopant was antimony and the n-type dopant gallium. Due to the lattice mismatch with the Si, the ZnO layers actually formed closely packed nanopillars. In this structure it is suggested that closed loops are formed as a result of scattering from the grain boundaries, establishing gain pathways, and random lasing. This is further evidenced by the observed angle-independent emission, multiple narrow modes in the emission spectrum, and images identifying spatially localised emission. In this system recombination in the QW is essential for the gain, and whilst this is not a typical QW-based laser, it may provide a template for ZnO QW-based devices.

14.4 Summary

The highly desirable optical properties of ZnO/ZnMgO QWs has made them a material system of significant interest over the past decade and will continue to drive research towards efficient light emitting devices in the blue-UV spectral range. In this chapter, we have discussed some of the recent advances in the understanding of the optoelectronic properties of these systems and some of the work undertaken to improve the radiative recombination efficiency. We concluded with a brief look at current progress towards device structures and note that the greatest limitation for electrically injected light emitting devices remains the availability of reproducible high-quality p-doped ZnO. Advances in this area, do however, offer great promise for ZnO/ZnMgO-based optoelectronic devices in the near future.

References

1. T. Makino, C.H. Chia, N.T. Tuan, H.D. Sun, Y. Segawa, M. Kawasaki, A. Ohtomo, K. Tamura, H. Koinuma, Appl. Phys. Lett. **77**(7), 975–977 (2000)
2. A. Ohtomo, K. Tamura, M. Kawasaki, T. Makino, Y. Segawa, Z.K. Tang, G.K.L. Wong, Y. Matsumoto, H. Koinuma. Appl. Phys. Lett. **77**(14), 2204–2206 (2000)
3. H.D. Sun, T. Makino, N.T. Tuan, Y. Segawa, Z.K. Tang, G.K.L. Wong, M. Kawasaki, A. Ohtomo, K. Tamura, H. Koinuma, Appl. Phys. Lett. **77**(26), 4250–4252 (2000)
4. A. Ohtomo, M. Kawasaki, I. Ohkubo, H. Koinuma, T. Yasuda, Y. Segawa, Appl. Phys. Lett. **75**(7), 980–982 (1999)
5. C. Jagadish, S.J. Pearton, *Zinc Oxide Bulk, Thin Films and Nanostructures* (Elsevier, Oxford, 2006)
6. S.H. Park, D. Ahn, T.W. Kang, S.J. Lee, in *Optoelectronic applications of ZnO/ZnMgO quantum well lasers in the blue and the UV spectral regions*, ed. by W. Jantsch, F. Schaffler. Physics of Semiconductors, Pts A and B, vol. 893 (2007), pp. 261–262
7. U. Ozgur, H. Morkoc, in *Zinc Oxide Bulk, Thin Films and Nanostructures*, ed. by C. Jagadish, S.J. Pearton (Elsevier, Oxford, 2006), p. 175
8. S.O. Kucheyev, J.S. Williams, C. Jagadish, J. Zou, C. Evans, A.J. Nelson, A.V. Hamza, Phys. Rev. B **67**(9), 094115 (2003)
9. A. Janotti, C.G. Van de Walle, Rep. Prog. Phys. **72**(12), 126501 (2009)
10. Y.W. Heo, F. Ren, D.P. Norton, in *Zinc Oxide Bulk, Thin Films and Nanostructures*, ed. by C. Jagadish, S.J. Pearton (Elsevier, Oxford, 491) p. 491
11. K. Koike, I. Nakashima, K. Hashimoto, S. Sasa, M. Inoue, M. Yano, Appl. Phys. Lett. **87**(11), 112106 (2005)
12. A. Dev, A. Elshaer, T. Voss, IEEE Journal of Selected Topics in Quantum Electronics **17**(4), 896–906 (2010)
13. V.A. Coleman; C. Jagadish. Basic Properties and Applications of ZnO, ed. by C. Jagadish, S.J. Pearton. In *Zinc Oxide Bulk, Thin Films and Nanostructures* (Elsevier, Oxford, 2006)
14. J.W. Sun, Y.M. Lu, Y.C. Liu, D.Z. Shen, Z.Z. Zhang, B.H. Li, J.Y. Zhang, B. Yao, D.X. Zhao, X.W. Fan, J. Phys. D-Appl. Phys. **40**(21), 6541–6544 (2007)

15. C.H. Chia, T. Makino, K. Tamura, Y. Segawa, M. Kawasaki, A. Ohtomo, H. Koinuma, Appl. Phys. Lett. **82**(12), 1848–1850 (2003)
16. J. Davis, L. Van Dao, X. Wen, P. Hannaford, V. Coleman, H. Tan, C. Jagadish, K. Koike, S. Sasa, M. Inoue, M. Yano, Appl. Phys. Lett. **89**(18), 182109 (2006)
17. H.D. Sun, T. Makino, Y. Segawa, M. Kawasaki, A. Ohtomo, K. Tamura, H. Koinuma, Appl. Phys. Lett. **78**(22), 3385–3387 (2001)
18. D.C. Reynolds, D.C. Look, B. Jogai, C.W. Litton, G. Cantwell, W.C. Harsch, Phys. Rev. B **60**(4) 2340 (1999)
19. Y.S. Park, C.W. Litton, T.C. Collins, D.C. Reynolds, Phys. Rev. **143**(2), 512 (1966)
20. B.K. Meyer, H. Alves, D.M. Hofmann, W. Kriegseis, D. Forster, F. Bertram, J. Christen, A. Hoffmann, M. Strassburg, M. Dworzak, U. Haboeck, A.V. Rodina, Phys. Status Solidi B Basic Res. **241**(2), 231–260 (2004)
21. B. Gil, P. Lefebvre, T. Bretagnon, T. Guillet, J.A. Sans, T. Taliercio, C. Morhain, Phys. Rev. B **74**(15), 153302 (2006)
22. M. Yano, K. Hashimoto, K. Fujimoto, K. Koike, S. Sasa, M. Inoue, Y. Uetsuji, T. Ohnishi, K. Inaba, J. Cryst. Growth **301**, 353–357 (2007)
23. M.W. Allen, P. Miller, R.J. Reeves, S.M. Durbin, Appl. Phys. Lett. **90**(6), 062104 (2007)
24. J. Davis, C. Jagadish, Laser Photon. Rev. **3**(1–2), 85–96 (2009)
25. C. Morhain, T. Bretagnon, P. Lefebvre, X. Tang, P. Valvin, T. Guillet, B. Gil, T. Taliercio, M. Teisseire-Doninelli, B. Vinter, C. Deparis, Phys. Rev. B **72**(24), 241305 (2005)
26. T. Bretagnon, P. Lefebvre, P. Valvin, B. Gil, C. Morhain, X.D. Tang, J. Cryst. Growth **287**(1), 12–15 (2006)
27. T. Bretagnon, P. Lefebvre, T. Guillet, T. Taliercio, B. Gil, C. Morhain, Appl. Phys. Lett. **90**(20), 201912 (2007)
28. J.M. Chauveau, D.A. Buell, M. Laugt, P. Vennegues, M. Teisseire-Doninelli, S. Berard-Bergery, C. Deparis, B. Lo, B. Vinter, C. Morhain, J. Cryst. Growth **301**, 366–369 (2007)
29. J. Davis, L. Dao, X. Wen, C. Ticknor, P. Hannaford, V. Coleman, H. Tan, C. Jagadish, K. Koike, S. Sasa, M. Inoue, M. Yano, Nanotechnology **19**(5), 055205 (2008)
30. T. Makino, K. Tamura, C.H. Chia, Y. Segawa, M. Kawasaki, A. Ohtomo, H. Koinuma, Appl. Phys. Lett. **81**(13), 2355–2357 (2002)
31. T. Makino, A. Ohtomo, C.H. Chia, Y. Segawa, H. Koinuma, M. Kawasaki, Phys. E Low-Dimensional Syst. Nanostruct. **21**(2–4), 671–675 (2004)
32. C.R. Hall, L. Dao, K. Koike, S. Sasa, H.H. Tan, M. Inoue, M. Yano, P. Hannaford, C. Jagadish, J.A. Davis, Phys. Rev. B **80**(23), 235316 (2009)
33. C.H. Chia, T. Makino, Y. Segawa, M. Kawasaki, A. Ohtomo, K. Tamura, H. Koinuma, J. Appl. Phys. **90**(7), 3650–3652 (2001)
34. B.P. Zhang, N.T. Binh, K. Wakatsuki, C.Y. Liu, Y. Segawa, Appl. Phys. Lett. **86**(3), 032105 (2005)
35. Z.P. Wei, Y.M. Lu, D.Z. Shen, C.X. Wu, Z.Z. Zhang, D.X. Zhao, J.Y. Zhang, X.W. Fan, J. Lumin. **119**, 551–555 (2006)
36. T. Guillet, T. Bretagnon, T. Taliercio, P. Lefebvre, B. Gil, C. Morhain, X.D. Tang, Superlatt. Microstruct. **41**(5–6), 352–359 (2007)
37. B. Gil, Superlatt. Microstruct. **43**(5–6), 408–416 (2008)
38. T. Makino, N.T. Tuan, H.D. Sun, C.H. Chia, Y. Segawa, M. Kawasaki, A. Ohtomo, K. Tamura, T. Suemoto, H. Akiyama, M. Baba, S. Saito, T. Tomita, H. Koinuma, Appl. Phys. Lett. **78**(14), 1979–1981 (2001)
39. T. Makino, Y. Segawa, M. Kawasaki, H. Koinuma, Semicond. Sci. Technol. **20**(4), S78–S91 (2005)
40. X. Wen, J. Davis, L. Van Dao, P. Hannaford, V. Coleman, H. Tan, C. Jagadish, K. Koike, S. Sasa, M. Inoue, M. Yano, Appl. Phys. Lett. **90**(22), 221914 (2007)
41. X. Wen, J. Davis, D. McDonald, L. Dao, P. Hannaford, V. Coleman, H. Tan, C. Jagadish, K. Koike, S. Sasa, M. Inoue, M. Yano, Nanotechnology **18**(31), 315403 (2007)
42. A.Y. Polyakov, N.B. Smirnov, A.V. Govorkov, E.A. Kozhukhova, A.I. Belogorokhov, D.P. Norton, H.S. Kim, S.J. Pearton, J. Electron. Mater. **39**(5), 601–607 (2010)

43. G. Coli, K.K. Bajaj. Appl.Phys. Lett. **78**(19), 2861–2863 (2001)
44. C. Klingshirn, R. Hauschild, J. Fallert, H. Kalt, Phys. Rev. B **75**(11), 115203 (2007)
45. T. Makino, K. Tamura, C.H. Chia, Y. Segawa, M. Kawasaki, A. Ohtomo, H. Koinuma, Phys. Rev. B **66**(23), 233305 (2002)
46. T. Makino, Y. Segawa, M. Kawasaki, J. Appl. Phys. **97**(10), (2005)
47. J.W. Sun, B.P. Zhang, Nanotechnology **19**(48), 485401 (2008)
48. C. Klingshirn, Phys. Status Solidi B Basic Solid State Phys. **244**(9), 3027–3073 (2007)
49. J.M. Chauveau, J. Vives, J. Zuniga-Perez, M. Laugt, M. Teisseire, C. Deparis, C. Morhain, B. Vinter, Appl. Phys. Lett. **93**(23), 231911 (2008)
50. J. Zuniga-Perez, V. Munoz-Sanjose, E. Palacios-Lidon, J. Colchero, Phys. Rev. Lett. **95**(22), 226105 (2005)
51. P. Waltereit, O. Brandt, A. Trampert, H.T. Grahn, J. Menniger, M. Ramsteiner, M. Reiche, K.H. Ploog, Nature **406**(6798), 865–868 (2000)
52. S. Lautenschlaeger, S. Eisermann, M.N. Hofmann, U. Roemer, M. Pinnisch, A. Laufer, B.K. Meyer, H. von Wenckstern, A. Lajn, F. Schmidt, M. Grundmann, J. Blaesing, A. Krost, J. Cryst. Growth **312**(14), 2078–2082 (2010)
53. J.M. Chauveau, C. Morhain, M. Teisseire, M. Laugt, C. Deparis, J. Zuniga-Perez, B. Vinter, Microelectron. J. **40**(3), 512–516 (2009)
54. H. Matsui, H. Tabata, Appl. Phys. Lett. **94**(16), 161907 (2009)
55. J.M. Chauveau, B. Vinter, M. Laugt, M. Teisseire, P. Vennegues, C. Deparis, J. Zuniga-Perez, C. Morhain, J. Korean Phys. Soc. **53**(5), 2934–2938 (2008)
56. J.M. Chauveau, M. Laugt, P. Vennegues, M. Teisseire, B. Lo, C. Deparis, C. Morhain, B. Vinter, Semicond. Sci. Technol. **23**(3), 035005 (2008)
57. V.A. Coleman, M. Buda, H.H. Tan, C. Jagadish, M.R. Phillips, K. Koike, S. Sasa, M. Inoue, M. Yano, Semicond. Sci. Technol. **21**(3), L25–L28 (2006)
58. C.R. Hall, L.V. Dao, K. Koike, S. Sasa, H.H. Tan, M. Inoue, M. Yano, C. Jagadish, J.A. Davis, Appl. Phys. Lett. **96**(19), 193117 (2010)
59. L. Wang, R. Li, Z. Yang, D. Li, T. Yu, N. Liu, L. Liu, W. Chen, X. Hu, Appl. Phys. Lett. **95**(21), 211104 (2009)
60. H.P. Zhao, G.Y. Liu, X.H. Li, R.A. Arif, G.S. Huang, J.D. Poplawsky, S. Tafon Penn, V. Dierolf, N. Tansu. IET Optoelectron. **3**(6), 283–295 (2009)
61. M. Yano, K. Ogata, F.P. Yan, K. Koike, S. Sasa, M. Inoue, Mat. Res. Soc. Symp. Proc. **744**, M3.1.1.(2003)
62. M. Brandt, M. Lange, M. Stolzel, A. Muller, G. Benndorf, J. Zippel, J. Lenzner, M. Lorenz, M. Grundmann, Appl. Phys. Lett. **97**(5), 052101–052103 (2010)
63. J. Zippel, S. Heitsch, M. Stolzel, A. Muller, H. von Wenckstern, G. Benndorf, M. Lorenz, H. Hochmuth, M. Grundmann, J. Lumin. **130**(3), 520–526 (2010)
64. W.E. Bowen, W. Wang, E. Cagin, J.D. Phillips, J. Electron. Mater. **37**(5), 749–754 (2008)
65. P. Misra, T.K. Sharma, S. Porwal, L.M. Kukreja. Appl. Phys. Lett. **89**(16), 161912–161913 (2006)
66. A. Malashevich, D. Vanderbilt, Phys. Rev. B. **75**(4), 045106 (2007)
67. J. Zippel, J. Lenzner, G. Benndorf, M. Lange, H. Hochmuth, M. Lorenz, M. Grundmann, J. Vac. Sci. Technol. B **27**(3), 1735–1740 (2009)
68. J. Zippel, M. Stolzel, A. Muller, G. Benndorf, M. Lorenz, H. Hochmuth, M. Grundmann, Phys. Status Solidi B Basic Solid State Phys. **247**(2), 398–404 (2010)
69. S. Su, Y. Lu, G. Xing, T. Wu, Superlatt. Microstruct. **48**(5), 485–490 (2010)
70. S.H. Tsang, S.F. Yu, H.Y. Yang, H.K. Liang, X.F. Li, IEEE Photon. Technol. Lett. **21**(21), 1624–1626 (2009)
71. G. Shukla, J. Phys. D-Appl. Phys. **42**(7), 075105 (2009)
72. S. Chu, M. Olmedo, Z. Yang, J.Y. Kong, J.L. Liu, Appl. Phys. Lett. **93**(18), 181106 (2008)
73. J. Cui, S. Sadofev, S. Blumstengel, J. Puls, F. Henneberger, Appl. Phys. Lett. **89**(5), 051108 (2006)
74. T. Gruber, C. Kirchner, R. Kling, F. Reuss, A. Waag, Appl. Phys. Lett. **84**(26), 5359–5361 (2004)

Chapter 15
N-Type Oxide Semiconductor Thin-Film Transistors

Pedro Barquinha, Rodrigo Martins, and Elvira Fortunato

Abstract This chapter gives an overview about GIZO TFTs, comprising an intro-
ductory section about generic TFT structure and operation, different semiconductor
technologies for TFTs – with special emphasis on AOSs and particularly on
GIZO – and then some experimental results obtained for GIZO TFTs fabricated
in CENIMAT.

Thin-film transistors (TFTs) are important electronic devices which are predominantly used as *On/Off* switches in active matrix backplanes of flat panel
displays (FPDs), namely liquid crystal displays (LCDs) and organic light emitting
device (OLED) displays. Even if a-Si:H is still dominating the TFT market in
terms of semiconductor technology, oxide semiconductors are emerging as one
of the most promising alternatives for the next generation of TFTs, bringing the
possibility of having fully transparent devices, low processing temperature, low
cost, high performance and electrically stable properties [1, 2]. Amorphous oxide
semiconductors (AOS) such as Gallium–Indium–Zinc oxide (GIZO) [3, 4], even
if fabricated at temperatures below 150°C, are currently capable of providing
transistors with field-effect mobility (μ_{FE}) exceeding $20\,cm^2\,V^{-1}\,s^{-1}$, threshold
voltage (V_T) close to 0 V, *On/Off* ratios above 10^8, subthreshold swing (S) around
$0.20\,V\,dec^{-1}$ and fully recoverable V_T shift (ΔV_T) lower than 0.5 V after 24 h stress
with constant drain current of $10\,\mu A$.

15.1 Device Structure and Operation

A TFT is a device comprising three electrodes – gate, source and drain, one
semiconductor placed between the source and drain electrodes and an insulator

P. Barquinha · R. Martins · E. Fortunato (✉)
CENIMAT-I3N, Departamento de Ciência dos Materiais and CEMOP/UNINOVA, Faculdade de
Ciências e Tecnologia, FCT, Universidade Nova de Lisboa, 2829–516 Caparica, Portugal
e-mail: elvira-fortunato@fct.unl.pt

S. Pearton (ed.), *GaN and ZnO-based Materials and Devices*,
Springer Series in Materials Science 156, DOI 10.1007/978-3-642-23521-4_15,
© Springer-Verlag Berlin Heidelberg 2012

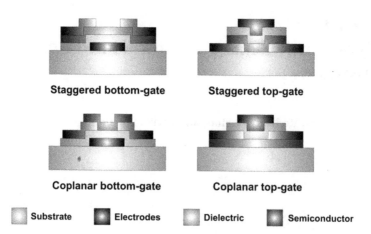

Fig. 15.1 Schematics showing some of the most conventional TFT structures, according to the position of the gate electrode and to the distribution of the electrodes relatively to the semiconductor

(or dielectric) inserted between the gate electrode and the semiconductor. The idea behind it is to have a current flowing between drain and source modulated by varying the potential between the gate and the source electrodes [5]. This modulation, known as field-effect, relies on the capacitive injection of carriers close to the dielectric/semiconductor interface, which is turned possible due to the parallel plate capacitor structure formed by the gate electrode, dielectric and semiconductor.

Figure 15.1 shows some of the most common structures employed to produce TFTs. According to the nomenclature initially defined by Weimer in 1960s, these structures are denominated by staggered or coplanar, depending if the sourcedrain and gate electrodes are on opposite sides or on the same side of the semiconductor. Inside staggered and coplanar structures, two configurations can be distinguished, top-gate (or normal) and bottom-gate (or inverted), depending on whether the gate electrode is on top or bottom of the structure [6, 7].

The ideal operation of an n-type TFT can be described by analyzing the energy band diagram of the capacitor comprised by the gate electrode, dielectric and semiconductor, upon the application of different voltages in the gate electrode (V_G), as shown in Fig. 15.2.

This analysis assumes an ideal case, considering an n-type semiconductor [given that the Fermi level (E_F) is shifted from the midgap toward the conduction-band minimum (CBM)] and that charge accumulation or depletion does not exist close to the dielectric/semiconductor interface or to the semiconductor back-surface in the unbiased state [8]. Under this situation, an upward band-bending results for $V_G < 0$ V, since the negative V_G repels mobile electrons from the dielectric/semiconductor interface, creating a depletion layer near that region that can be extended through the entire semiconductor for $V_G \ll 0$. In (a) and (b) situations, even if a large

Fig. 15.2 Energy band diagrams of an ideal gate electrode/dielectric/n-type semiconductor capacitor for different bias conditions: (**a**) equilibrium ($V_G = 0$ V); (**b**) depletion ($V_G < 0$ V); (**c**) accumulation ($V_G > 0$ V)

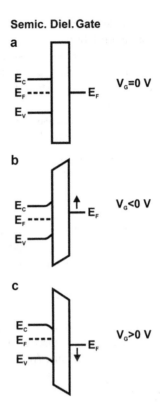

drain-to-source voltage (V_D)[1] is used, a very low current flows between drain and source (I_D), corresponding to the *Off*-state of the transistor. On the other hand, when $V_G > 0$ V (Fig. 15.2c), electrons are accumulated close to the dielectric/semiconductor interface, leading to a downward band-bending in that region, which becomes even more pronounced for $V_G \gg 0$ V. For this condition, a considerable I_D starts flowing upon the application of V_D, corresponding to the *On*-state of the transistor. Based on the description above, a conductive channel is readily formed with a very small increase on V_G. However, in a real case, V_T (corresponding to significant charge accumulation at the dielectric/semiconductor interface, i.e., the V_G necessary to turn on the device) deviates from 0 V, being a function of the gate electrode–semiconductor work function difference, the background carrier concentration of the semiconductor, the charge density residing within the dielectric and the trap density at the interface and within the semiconductor [5, 8]. For an n-type TFT, depending on whether V_T is positive or negative, the devices are

---

[1] V_G, V_D, I_G and I_D are used throughout this chapter, given that the source electrode is always assumed to be grounded, but these notations have exactly the same meaning as V_{GS}, V_{DS}, I_{GS} and I_{DS}.

designated as enhancement or depletion mode, respectively. Both types are useful for circuit fabrication (for instance, NMOS technology requires both enhancement and depletion mode transistors), but for the common function of TFTs as simple electronic switches, enhancement mode is preferable, because no V_G is required to achieve the *Off*-state, turning easier the circuit design and minimizing power dissipation [9].

When the transistor is in the *On*-state, different operation regimes can be distinguished, depending on the value of V_D:

- When $V_D < V_G - V_T$ the TFT is in the pre-pinch-off regime, and I_D is described by: [5]

$$I_D = C_i \mu_{FE} \frac{W}{L} \left[(V_G - V_T) V_D - \frac{1}{2} V_D^2 \right],$$ (15.1)

- where C_i is the gate capacitance per unit area, μ_{FE} is the field-effect mobility, W is the channel width, and L is the channel length. For very low V_D, the quadratic term can be neglected, yielding a linear relation between I_D and V_D. In this situation, the accumulated charges are considered to be uniformly distributed throughout the channel.
- When $V_D > V_G - V_T$ the accumulation layer close to the drain region becomes depleted, leading to the saturation of I_D. This phenomenon is designated by pinch-off and starts to occur when $V_D = V_G - V_T$. At the post-pinch-off or saturation regime, I_D is described by:

$$I_D = C_i \mu_{sat} \frac{W}{2L} (V_G - V_T)^2,$$ (15.2)

where μ_{sat} is the saturation mobility.

The equations describing the operation of field-effect transistors are generally based on the assumption that the rate of variation (i.e., gradient) of the lateral field within the channel is much smaller than the rate of variation of the vertical field, i.e., the channel is seen as one-dimensional [10]. This concept was initially proposed by Shockley and is known as "gradual channel approximation." Even if this approximation is not valid near the drain electrode when the transistor is in saturation regime or for short-channel devices, it describes fairly well most of the operation of TFTs, which generally have L much larger than the dielectric thickness [11]. Still, it is important to mention that (15.1) and (15.2) assume μ_{FE} and μ_{sat} to be constant. For most of the TFTs, with particular relevance for oxide semiconductor-based ones, this is not valid and $\mu_{FE}(V_G)$ and $\mu_{sat}(V_G)$ should be considered instead [12].

Static electrical characterization of these devices comprises output [$I_D(V_D)$ with different steps of V_G] and transfer [$I_D(V_G)$ for a fixed V_D] characteristics. From these, one can extract several quantitative electrical parameters, such as *On/Off* ratio (ratio of the maximum to the minimum I_D, taken from the transfer characteristics), V_T and turn-on voltage (V_{on}, corresponding to the V_G for which a conductive channel is formed close to the dielectric/semiconductor interface)[2], subthreshold swing (S, the necessary V_G to increase I_D by one decade, given by the inverse of the maximum slope of the transfer characteristic), and mobility. μ is a measure of the efficiency of carrier transport in a material, depending on several scattering mechanisms, such as lattice vibrations, ionized impurities, grain boundaries and other structural defects [8, 15]. When μ is measured in a field-effect device, additional scattering arises because the carriers are confined within a narrow region, close to the dielectric/semiconductor interface. Hence, additional mechanisms such as Coulomb scattering from dielectric charges and from interface states or surface roughness scattering contribute to decrease μ [15]. However, it has to be noted that in a TFT μ is modulated by the bias conditions, since the increase on V_G contributes to decrease the barriers associated with grain boundaries in polycrystalline semiconductors and/or allows E_F to be taken close or above CBM, where μ is larger. As will be seen throughout this chapter, this last situation is particularly relevant in oxide TFTs. Different methodologies can be used to extract μ in a TFT:

- Effective mobility (μ_{eff}) – obtained by the conductance (g_d) with low V_D, according to:

$$\mu_{eff} = \frac{g_d}{C_i \frac{W}{L} (V_G - V_T)}. \tag{15.3}$$

This is usually considered the most correct estimation of μ, including the effect of V_G [15]. However, it requires the previous determination of V_T, which can be associated with some error, and is sensitive to contact resistance, since it is extracted at low V_D;

- Field-effect mobility (μ_{FE}) – obtained by the transconductance (g_m) with low V_D, according to:

$$\mu_{FE} = \frac{g_m}{C_i \frac{W}{L} V_D}. \tag{15.4}$$

This is the most commonly used mobility estimation used for TFTs. Although it stills being sensitive to contact resistance effects, it does not require the determination of V_T and it is easily extracted by the derivative of the transfer characteristics;

- Saturation mobility (μ_{sat}) – obtained by the transconductance with high V_D, according to:

$$\mu_{sat} = \frac{\left(\frac{d\sqrt{I_D}}{dV_G}\right)^2}{\frac{1}{2} C_i \frac{W}{L}}, \tag{15.5}$$

---

[2] Given the broad range of methodologies for V_T determination (see, for instance [13]), large ambiguity can arise when comparing different devices using this parameter. As a less ambiguous concept, V_{on} is largely used in literature, simply corresponding to the V_G at which I_D starts to abruptly increase as seen in a $log I_D - -V_G$ plot, or in other words, the V_G necessary to fully turn-off the transistor [14].

μ_{sat} is also very common in TFTs' literature. It does not require V_T and as a high V_D is used is less sensitive to contact resistance. However, it is not physically accurate, as it describes a situation where the channel is pinched-off, i.e., its effective length is smaller than L, which is intrinsically not assumed by the gradual channel approximation used to derive (15.1) and (15.2).

Other methodologies can also be found in literature. From these, particular relevance is assumed by the ones proposed by Hoffman, designated by average and incremental mobility, μ_{avg} and μ_{inc}, respectively [14]. While the former provides an average value of all the carriers induced in the channel, the latter probes the mobility of carriers as they are incrementally added to the channel, thus providing valuable insights into channel carrier transport.

μ is also a very important characteristic since it directly affects their maximum I_D and switching speed, thus defining their range of applications [16]. For instance, it has a direct impact on the maximum operating frequency or cutoff frequency (f_{co}), which can be defined as [8]:

$$f_{co} = \frac{\mu V_D}{2\pi L^2}. \tag{15.6}$$

15.2 Semiconductor Materials for TFTs

15.2.1 The Era of Oxide Semiconductors

Most of the TFTs incorporated in commercially available products rely on a-Si:H as the semiconductor material. a-Si:H is one of the main reasons for the success of TFTs as electronic switches in displays and was initially reported by LeComber, Spear and Ghaith in 1979, as a natural consequence of their work on the analysis of the density of states in disordered structures [17, 18]. In spite of exhibiting considerably lower μ than polycrystalline materials already available by that time, such as CdSe (about 1 against 150–200 $cm^2\,V^{-1}\,s^{-1}$), a-Si:H was and still is perfectly suitable for the application of TFTs as switching elements in most of the currently available LCDs, allowing for low cost, good reproducibility and uniformity in large areas.[3]

But in terms of TFT technologies, the new millennium is marked by the usage of a revolutionary class of transparent semiconductor materials as channel layers: the oxide semiconductors. Although the "big-boom" of this technology is normally associated with reports on n-type[4] ZnO TFTs presented in 2002–2003, there were

---

[3]More details about different semiconductor material technologies for TFTs are given in section 2.2.

[4]Unless otherwise stated, all the oxide semiconductors reported in this chapter are n-type. p-type oxides only recently started to be explored for TFT applications, exhibiting considerably lower performance than n-type ones [19–22].

some tentative applications of oxide semiconductors as channel layers in TFTs 40 years before this, almost coincident with the initial CdS TFTs reported by Weimer. In fact, back in 1964, Klasens and Koelmans proposed a TFT comprising an evaporated SnO_2 semiconductor on glass, with aluminum sourcedrain and gate electrodes and an anodized Al_2O_3 gate dielectric [23]. Few details regarding electrical performance are provided and the transparent semiconductor is essentially used to demonstrate a new self-aligned lift-off process, where the SnO_2 layer allows to expose the photoresist to UV light penetrating from the bottom of the structure in all the areas expect the one shielded by the aluminum gate electrode, defining this way the pattern of the source–drain electrodes (Fig. 15.3a). In 1968, Boesen and Jacobs reported a TFT with a lithium-doped ZnO single crystal semiconductor, with evaporated SiO_x dielectric and aluminum electrodes, but a very small I_D modulation by V_G and no I_D saturation were observed on these devices [24]. Similar poor performance was obtained in SnO_2 TFTs by Aoki and Sasakura in 1970 [25]. In 1996, oxide semiconductors reappeared as channel layers, with two reports on ferroelectric field-effect devices employing SnO_2 : Sb and In_2O_3, by Prins et al. and Seager, respectively [26, 27]. Given that the main intent of the authors was to demonstrate hysteresis associated with the ferroelectric behavior, little information is provided on these papers about device performance, but Prins and co-workers, for instance, report a low *On/Off* ratio of 60. Still, it is noteworthy to observe that full transparency is for the first time persuaded in a TFT and it is only severely affected by the $SrRuO_3$ gate electrode (Fig. 15.3b), although the authors mention that they also fabricated a fully transparent TFT by using a heavily doped SnO_2 gate electrode.

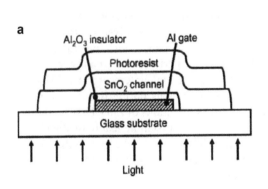

Fig. 15.3 Initial TFTs employing oxide semiconductors as channel layers: (**a**) Schematic of the SnO_2 TFT reported by Klasens and Koelmans in 1964, to show a new self-aligned lift-off process [28]; (**b**) top-view of the ferroelectric Sb : SnO_2 TFT reported by Prins et al. in 1996 [26]

Fig. 15.4 Fully transparent
oxide semiconductor TFTs
produced on glass at
CENIMAT

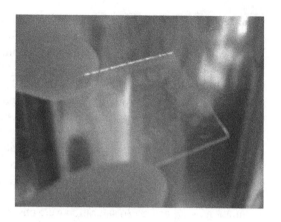

But good performing devices showing that oxide TFTs could be a viable
technology only started to appear in 2002–2003, with the reports on ZnO TFTs by
Masuda et al., Hoffman et al. and Carcia et al. [9, 29, 30]. The first two authors even
report fully transparent devices (comprising transparent conductive oxide (TCO)-
based electrodes), already allowing to obtain respectable performance, comparable
and even surpassing in some aspects the one typically exhibited by a-Si:H and
organic TFTs[5], mostly in terms of μ_{FE}, which could be as high as $2.5\,\mathrm{cm^2\,V^{-1}\,s^{-1}}$.
However, the processing or postprocessing temperatures of the semiconductor
necessary to obtain good devices were still quite high, between 450 and 600°C.
But the work from Carcia et al. showed that using r.f. magnetron sputtering to
deposit ZnO, similar electrical properties could be achieved with room-temperature
processing of the semiconductor layer, even if on this case fully transparent
structures were not demonstrated. During 2003–2004 several reports continued to
appear on oxide TFTs, bringing different innovations to this emerging area. Our
research group at CENIMAT was actively involved in these advances, for instance
by showing the first fully transparent ZnO TFTs with room temperature ZnO
processing (Fig. 15.4), exhibiting improved device performance (mainly regarding
μ_{FE}) even if compared with devices processed at higher temperatures [31–33].
Besides this, some of the most important achievements during this period were:
nonvacuum deposition of ZnO layers [34]; first simulations of ZnO TFTs assuming
that their properties are largely dictated by the polycrystalline nature of ZnO [35];
new methods for extraction of mobility in ZnO TFTs [14]; application of ZnO TFTs
as UV photodetectors [36, 37]; exploration of SnO_2 TFTs [38]; use of In_2O_3 or ZnO
nanowires as channel layers [39–42], among many others.

While most of the research work was being devoted to binary compounds
such as ZnO, In_2O_3 or SnO_2, Nomura et al. suggested in 2003 to use a complex
$InGaO_3(ZnO)_5$ (or GIZO) single-crystalline semiconductor layer in a TFT [43].

---

[5]Most of the organic TFTs are p-type, i.e., the conduction mechanism is due to holes rather than
electrons.

This layer was epitaxially grown on an yttria-stabilized zirconia substrate and allowed to obtain an impressive $\mu_{eff} \approx 80 \, cm^2 \, V^{-1} \, s^{-1}$, $V_{on} \approx -0.5 \, V$ and *On/Off* ratio $\approx 10^6$. Even if a very high temperature of $1,400°C$ was necessary to attain this level of performance, it showed that oxide TFTs possessed a large room for improvement. And in fact, in the next year Nomura et al. presented a work that definitely proved the enormous potential of oxide semiconductors (and multicomponent oxides in particular) for TFT applications, by demonstrating a transparent TFT on a flexible substrate using near-room temperature processing [3]. For this end, they used a PLD-deposited amorphous GIZO layer as the semiconductor. Even if the performance was far away from the single-crystalline TFTs presented by the same authors, $\mu_{sat} \approx 9 \, cm^2 \, V^{-1} \, s^{-1}$, $V_T \approx 1$–$2 \, V$ and *On/Off* ratio $\approx 10^3$ could still be achieved, mostly because the low sensitivity of these multicomponent oxides to structural disorder, attributed to the large and overlapping *s*-orbitals of the cations that constitute their conduction bands.

Nomura's work opened the door for an impressively growing number of publications in the next years regarding the application of amorphous multicomponent oxides as channel layers in TFTs. Several combinations of cations with $(n-1)d^{10}ns^0$ ($n \geq 4$) electronic configuration started to be used for this end, such as ZTO [12,44–46], IZO [47–50], IGO [51, 52], GIZO [4, 53–57], and GTZO [58]. [6] With the continuous improvements verified on these devices, it is now common to obtain remarkable electrical properties, considerably superior to a-Si:H or organic TFTs, such as μ_{FE} above $10 \, cm^2 \, V^{-1} \, s^{-1}$, close to 0 V_{on}, *On/Off* ratio exceeding 10^7 and $S \approx 0.20$–$0.25 \, V \, dec^{-1}$, with the indium-based semiconductors having the added advantage of allowing for very low or even room temperature processing. In fact, nowadays the processing temperature of these TFTs is dictated not by the semiconductor layer, but rather by the dielectric.

15.2.2 Comparison of n-Type Oxide TFTs with Existing TFT Semiconductor Material Technologies

As pointed out before, a-Si:H is the most widely used semiconductor material in TFTs. Together with polycrystalline silicon (poly-Si) and organic materials, they constitute the main semiconductor technologies available nowadays for TFTs. Hence, this section tries to briefly compare them with the amorphous oxide semiconductors (AOS), based on costs, processing and performance of the devices. Table 15.1 gives an overview of these characteristics.

The maturity of the a-Si:H TFTs technology is unquestionably greater than any of the other technologies. In fact, these devices are studied for almost 40 years, and their strengths and limitations are well known. Moreover, there is a huge industrial

---

[6]Note that the given references are just a few examples of the large number of reported works, with many more being available, specially for the period comprised between 2007 and 2009.

Table 15.1 Comparison between most relevant semiconductor material technologies for TFTs

	Inorganic semiconductors			Organic semiconductors	
	a-Si:H	poly-Si	Amorphous oxides		
Maturity/ infrastructures	★★★	★★	★→→★★★	★/★★	★ - Not suitable
Large area deposition	★★★	★	★★★	★★★	★★ - Moderate
Processing temperature	★★	★/★★	★★/★★★	★★★	★★★ - Suitable
Cost	★★★	★	★★/★★★	★★/★★★	
Transparency	★	★	★★★	★★	
Electrical performance	★/★★	★★★	★★★	★/★★	
Electrical stability	★	★★	★★/★★★	★	
Environmental stability	★	★★	★★★	★/★★	

implementation of processing tools for a-Si:H TFTs, mostly caused by its successful use as switching elements in LCDs. But even if oxide TFTs are only now emerging, it can be expected that their industrial implementation will be facilitated, given that they can use the processing and lithographic tools already existent in LCD industry, for instance, the sputtering systems used to fabricate TCO films used as top LCD electrodes. If physical techniques as sputtering are considered, another advantage arises to oxide semiconductors in comparison with silicon-based technologies, since only argon and oxygen are required, rather than explosive and toxic gases such as silane, phosphine or diborane.

Poly-Si presents a great disadvantage regarding large area deposition: due to the polycrystalline structure of this semiconductor, the electrical properties are highly dependent on aspects such as grain size and orientation. Thus, the lack of uniformity and reproducibility in large areas are important drawbacks of this technology. On the contrary, a-Si:H, organic and oxide semiconductors can all exhibit amorphous structures, which are ideal for large area deposition. To obtain larger grain sizes (hence higher performance) in poly-Si, processing temperatures higher than for the other semiconductor technologies are required, but even a-Si:H generally requires temperatures exceeding 250–300°C to reach its best performance. Oxides and mainly organic semiconductors require the lowest processing temperature, which makes them compatible with inexpensive glass or even plastic substrates, turning possible the concept of flexible electronics. This is also traduced in lower costs for these technologies. Also, the advent of oxide semiconductor processing techniques such as ink-jet or spin-coating, until now mostly devoted to organic materials, permits to foreseen further cost reductions. Still, benefiting from its well established industry, a-Si:H TFTs also present nowadays low production costs. Regarding poly-Si, the fact that it requires high temperature or complex/expensive crystallization methods, such as excimer laser annealing (ELA), results in higher-cost TFTs, even if the base material, silicon, is the second most abundant element in the earth's crust [59].

Transparency is naturally a large advantage of oxide semiconductors, although some organic semiconductors also present reasonable transparency in the visible range. This characteristic is important not only for the fabrication of fully transparent electronic circuits or displays but also to increase the aperture ratio in displays, which directly results in improved brightness levels [7]. Furthermore, this means that transparent TFTs can be made with large W/L ratios, allowing them to supply larger currents for a given supply voltage, which is extremely important for applications where the TFT is used as a current driver, as in OLED displays [60].

Regarding electrical performance, poly-Si TFTs have the advantage of exhibiting the highest μ_{FE}, which even for low-temperature poly-Si (LTPS) TFTs can reach $200 \, cm^2 \, V^{-1} \, s^{-1}$ [61]. This feature allows to use them also for the drivers circuitry in LCDs. a-Si:H and organic semiconductor TFTs have very small μ_{FE}, typically less than $1 \, cm^2 \, V^{-1} \, s^{-1}$, while oxide TFTs exhibit intermediate μ_{FE} values between poly-Si and a:Si:H TFTs, but more than one order of magnitude larger than a-Si:H. Still, even if poly-Si TFTs have a clear advantage regarding μ_{FE}, it should be noted that these devices generally exhibit large leakage currents, which are associated with electron–hole generation stimulated by electric fields, via the trap-states on the grain boundaries [62]. This is traduced in reduced *On/Off* ratios when compared with the other TFT technologies, limiting their application as switching elements, specially for large area and high resolution displays. Even if little is known yet about long-term stability of oxide TFTs, the initial results show that V_T shift under constant bias or current stress is the predominant instability effect, but the magnitude of variation can be quite smaller than that exhibited by a-Si:H TFTs, as will be shown in Sect. 15.3.7. Furthermore, given the large bandgap (E_G) of oxide semiconductors, their properties are not as sensitive to visible light as a-Si:H (and also to poly-Si, although to a considerably less extent), where degradation can arise due to the creation of dangling bonds according to the Staebler–Wronski effect, with the initial properties only being reestablished after an annealing treatment [63]. This imposes the usage of light shields in silicon-based TFTs, increasing the process complexity, cost, and contributing to decrease even more the aperture ratio when these TFTs are used in displays. Even if UV filters have to be used for oxide TFTs to block wavelengths lower than 400–450 nm, these filters do not compromise the overall appearance of the devices [64]. Organic materials also present a well known sensitivity to environmental species, such as water and oxygen, which can even permanently affect their properties. Hence, a meticulous passivation step is required for these devices. Even if this passivation step is a requisite for subsequent integration of devices, regardless of their semiconductor technology, it will be shown in Sect. 15.3.4 that the interaction with oxygen can even be explored with advantage for oxide TFTs.

Based on all this, oxide semiconductors provide a solid and viable alternative for the present and future of TFTs, covering important drawbacks of the existing technologies, ending up being a very attractive technology that allows for transparent, low cost, low temperature and high performance devices.

15.3   Multicomponent Oxide TFTs @ CENIMAT

In this section, results regarding GIZO TFTs produced in CENIMAT are presented. Staggered bottom-gate TFTs were fabricated using heavily doped silicon wafers acting as substrate and gate electrode, with a 100 nm thick SiO_2 dielectric deposited by a well established plasma-enhanced chemical vapor deposition (PECVD) process at 400°C. Most emphasis was given to the oxide semiconductor, which was deposited by r.f. magnetron sputtering without intentional substrate heating in a home-made system. The effect of composition inside the Ga–In–Zn oxide system was widely explored by using GIZO ceramic targets with different compositions. Different processing parameters were also tested, namely the percentage of oxygen content in the Ar + O_2 mixture ($\%O_2$, between 0 and 10.0%), the deposition pressure (p_{dep}, between 0.4 and 1.0 Pa), the rf power density (P_{rf}, between 1.1 and 2.7 W cm^{-2}) and the deposited oxide semiconductor thickness (d_s). The TFTs were produced using source–drain electrodes of e-beam evaporated Ti/Au, around 5 and 70 nm thick, respectively. Since a passivation layer would be required for subsequent integration of TFTs, for instance in active matrix backplanes, different materials produced using various deposition techniques were tested for this purpose. All the TFT layers were patterned using lift-off and for most of the devices presented here W/L ratio was 1. Annealing treatments, performed before and after passivation layer processing, were carried out using a tubular furnace, in air atmosphere, a 10°C min^{-1} heating ramp and temperatures (T_A) ranging from 150 to 300°C, for 1 h.

15.3.1   Role of Oxygen During GIZO Sputtering

Besides doping, oxygen vacancies derived from stoichiometry deviations on the deposited films are the main contributors for electrical conduction in oxides. These oxygen vacancies form shallow electron donor levels close to the conduction band and are readily ionized near room temperature, with each doubly charged oxygen vacancy being able to contribute with two free electrons, preserving charge neutrality [65–68]. Hence, the variation of $\%O_2$ is one of the most effective ways to control the electrical properties of oxide semiconductors. Figure 15.5a shows the effect of $\%O_2$ used to fabricate the GIZO layer on the transfer characteristics of oxide TFTs [4]. The semiconductor layer is based on GIZO 2:4:2, produced with $p_{dep} = 0.7$ Pa and $P_{rf} = 1.1$ W cm^{-2}, with $d_s = 40$ nm. A low $T_A = 150$°C is used here in order to have large but reproducible variations on the electrical properties arising from changing $\%O_2$ (higher T_A would lead to less variations, see also Sect. 15.3.5).

Figure 15.5a shows that as $\%O_2$ increases V_{on} is shifted toward more positive values. This effect is easily understandable having in mind the field-effect theory and the way the E_F is shifted with $\%O_2$ on oxide semiconductors. As V_G increases,

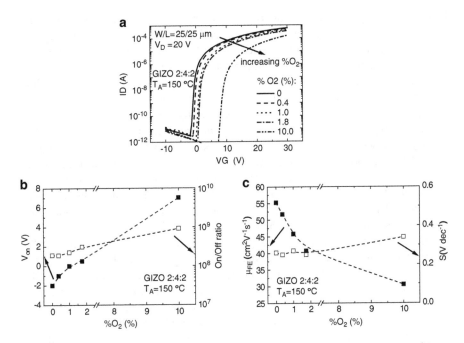

Fig. 15.5 Effect of %O_2 on the electrical properties of GIZO 2:4:2 TFTs annealed at 150°C on: (**a**) transfer characteristics; (**b**) V_{on} and On/Off ratio; (**c**) μ_{FE} and S

the first induced charges need to compensate unfilled traps, including the ones from the bulk semiconductor and the ones arising from its interface with the dielectric. A higher %O_2 translates in lower background carrier concentration (N), since oxygen vacancies are the main source of free electrons in oxide semiconductors. A lower N means that, with zero-gate bias, E_F is deeper inside the bandgap, away from the CBM. The E_F location in energy levels deeper in the bandgap is plausible for all the data presented in Fig. 15.5a, since for all these thin films $N \ll 10^{16}\,\text{cm}^{-3}$. Hence, for lower N a larger number of traps have to be filled by the induced charges supplied by V_G before E_F reaches tail and extended states, where the induced carriers start to be free to increase I_D. This results in larger V_{on} for devices having GIZO films with lower N (higher %O_2). Superimposed to the "natural" free carrier suppression by increased %O_2 mentioned above, the more severe bombardment phenomena by energetic ions for higher %O_2 can also yield a less compact film and generate more defects, both in GIZO's bulk and at its interface with SiO_2, contributing also to the increase of V_{on}. The fact that GIZO films sputtered with high %O_2 have lower indium concentration than GIZO films sputtered with low %O_2 can also affect negatively the performance of the devices, as will be shown in Sect. 15.3.7.

The electrical parameters most affected by %O_2 are V_{on} and μ_{FE}, whose trends seem to be related, with devices with lower V_{on} exhibiting larger μ_{FE} (Fig. 15.5b, c).

This is not strange considering the background given above: for films with higher N (lower V_{on}), as V_G is increased alarger fraction of induced charges is available to drift between source and drain. Since carrier transport on oxide semiconductors is strongly enhanced by a larger concentration of available free charges, μ_{FE} is increased. Similar relations were found by other authors [47, 69, 70]. Regarding *On/Off* ratios, all the devices present values between 10^8 and 10^9, which are comparable or even superior to the best performing a-Si:H TFTs. This is of great importance regarding their application as switching elements [7]. The S value is slightly increased for $\%O_2 = 10.0\%$, which is consistent with a larger defect density on GIZO bulk and/or on its interface with SiO_2 [71]. The obtained results agree quite well with a model where V_{on} control is attributed to deep gap states while subthreshold, threshold and *On*-state current behaviors are attributed to tail states, as proposed by Hsieh et al. In this model, the effect of trap states at the gate–insulator interface is implicitly included in bulk subgap states [72].

The properties' degradation for higher $\%O_2$ is also verified by measuring the V_{on} shift in consecutive transfer characteristics (ΔV_{on}). This simple procedure is useful to analyze early-stage aging of devices and also to infer about the instability mechanisms that might be present. For instance, it is reported that V_T is shifted using this procedure on In_2O_3-based TFTs, which is attributed to the existence of interface defects [73]. The results obtained using this procedure for the GIZO TFTs analyzed in Fig. 15.5 are presented in Table 15.2. For devices where GIZO is produced with low $\%O_2$ (0 and 0.4%) $\Delta V_{on} \approx 0\,V$, which might indicate that a low density of traps exists or that the background N obtained in these films is enough to permit that a low density of unfilled traps exists at zero-gate bias. However, as $\%O_2$ is increased ΔV_{on} starts to deviate from 0 V toward positive values, reaching 6 V for $\%O_2 = 10.0\%$ (Fig. 15.6a). In this case, a higher trap density could be expected and because the background N is lower a large number of unfilled traps would be present in the unbiased device. Note that this positive ΔV_{on} shift is consistent with electron trapping at or near the $GIZO/SiO_2$ interface rather than with more severe instabilities, such as positive or negative ion migration in the insulator, which would lead to a negative ΔV_{on} [28]. Hysteresis analyses are also performed, and although the verified hysteresis is always clockwise, consistent with trap filling by accumulated electrons (rather than counterclockwise, which might indicate ionic drift) [74], this hysteresis is almost negligible for $\%O_2 = 0$ and 0.4% (Fig. 15.6b), but increased to around 6–7 V for $\%O_2 = 10.0\%$. However, even for the high − $\%O_2$ devices, hysteresis almost disappears (or at least is considerably reduced) after the four successive transfer characteristic measurements described above, because after that the accumulated electrons supplied during the previous V_G sweeps are filling most of the existent traps.[7] Further evidence that electron trapping at or near the $GIZO/SiO_2$ interface should be critical for GIZO produced with high $\%O_2$ is

---

[7]Given this effect, which arises as a consequence of the TFT measurement protocol used in this work, hysteresis is not used systematically throughout this chapter to evaluate charge trapping or other instability phenomena on the TFTs, being ΔV_{on} used instead.

Table 15.2 Effect of $\%O_2$ on ΔV_{on} after four consecutive transfer characteristic measurements. Results for GIZO 2:4:2 TFTs annealed at 150°C

$\%O_2(\%)$	ΔV_{on} after four consecutive measurements (V)
0	0
0.4	0
1.0	0.5
1.8	1.0
10.0	6.0

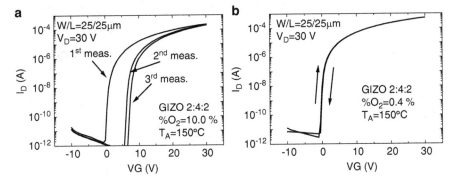

Fig. 15.6 Transfer characteristics showing (**a**) successive measurements and (**b**) hysteresis for $\%O_2 = 10.0$ and 0.4%, respectively. Results for GIZO 2:4:2 TFTs annealed at 150°C

the recovery of the initial V_{on} and large hysteresis after a long period of rest time, typically more than 1 day. This should be related to the slow release of electrons by deep traps and not with some sort of ionic drift because a large amount of energy (only attainable, for instance, by a thermal annealing treatment) would need to be supplied to the device in order for ions to diffuse back to their original positions [28].

15.3.2 Role of Deposition Pressure and rf Power Density During GIZO Sputtering

The effects of p_{dep} and P_{rf} on the electrical properties of oxide semiconductor thin films are somehow related, since both affect the energy of sputtered and plasma species arriving to the substrate. Figure 15.7a shows the transfer characteristics for TFTs employing a semiconductor layer based on GIZO 2:4:2, deposited with $\%O_2 = 0.4\%$, $P_{rf} = 1.1\,W\,cm^{-2}$, $d_s = 40\,nm$ and different p_{dep}. Final devices were annealed at 150°C. Extracted electrical parameters are presented in Table 15.3.

The data show that for $p_{dep} = 0.4\,Pa$, although a non-negligible field-effect and a high $\mu_{FE} = 56.6\,cm^2\,V^{-1}\,s^{-1}$ are obtained, the channel conductivity modulation is rather poor, being the transistor always in the *On*-state, even for negative V_G. This

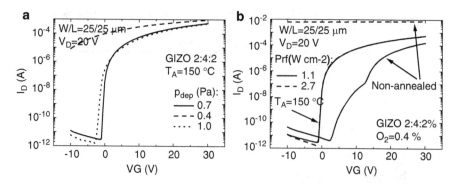

Fig. 15.7 Effect of (**a**) p_{dep} and (**b**) P_{rf} on the transfer characteristics exhibited by GIZO 2:4:2 TFTs (%O_2 = 0.4%)

Table 15.3 Electrical properties obtained for the devices annealed at T_A = 150°C depicted in Fig. 15.7

$p_{dep}(Pa),P_{rf}(Wcm^{-2})$	$\mu_{FE}(cm^2V^{-1}s^{-1})$	On/Off ratio	$V_{on}(V)$	$V_T(V)$	$S(Vdec^{-1})$
0.4, 1.1	56.6	–	–	–	–
0.7, 1.1	51.7	1.9×10^8	$^-1.0$	1.0	0.25
1.0, 1.1	50.2	3.3×10^8	$^-3.0$	0.1	0.37
0.7, 2.7	52.4	3.3×10^8	$^-1.5$	0.3	0.26

behavior is a direct consequence of a very conductive GIZO film ($\rho \approx 10^{-3}\Omega$ cm), hence of a large N that cannot be entirely depleted within the V_G range used here. For devices where GIZO is produced at higher p_{dep}, between 0.7 and 1.0 Pa, typical transistor behavior is obtained, but properties are slightly improved for the 0.7 Pa deposited GIZO, being possible to achieve higher μ_{FE}, closer to 0 V_{on} and smaller S. This should be the consequence of a more compact film structure with a lower number of defects, related to the enhanced sputtered energy and higher diffusion length of species when lower p_{dep} is used. In fact, spectroscopic ellipsometry analysis confirms that short-range order and density are improved when decreasing p_{dep} from 1.0 to 0.7 and to 0.4 Pa. The evolution of electrical properties verified from 0.7 to 0.4 Pa is somehow contradictory compared to what is reported for instance by Jeong et al. for GIZO TFTs, where a large improvement on S was verified for films deposited at lower p_{dep} without comprising the device's Off-current and V_{on} [71]. To understand the differences between the two works we must take into account that different processing conditions were used, namely, %O_2, which is much higher in [71], and target composition, which is 2:2:1 in [71] and 2:4:2 in the present work. These two parameters have a strong effect on the electrical properties of oxide semiconductor thin films, being expected that the ones produced in this work have considerably lower ρ (higher N) than the ones reported by Jeong et al. Thus, even if an improvement on S is plausible for lower p_{dep}, given the higher compactness and lower defect density of the GIZO thin films obtained under these conditions, the effect is overshadowed here by a high N, which does not allow for ideal channel conductivity modulation, with the devices always remaining in the *On*-state. This clearly shows that to tailor the oxide semiconductor properties toward a specific application, one needs to have in mind that these properties are affected simultaneously by a large number of (post-) deposition parameters.

The effect of P_{rf} on the electrical properties of GIZO-based TFTs is shown in Fig. 15.7b and Table 15.3, for GIZO 2:4:2 films deposited with $\%O_2 = 0.4\%$, $p_{dep} = 0.7$ Pa and $d_s = 40$ nm. A low $\%O_2$ is chosen to prevent the effects mentioned in the previous paragraph. For nonannealed devices, large differences are verified, with $P_{rf} = 2.7$ W cm^{-2} resulting in always-on characteristics due to a low ρ and large N ($\approx 10^{-2}\Omega$ cm and 10^{19} cm^{-3}, respectively). On the other hand, GIZO films produced with $P_{rf} = 1.1$ W cm^{-2} have $\rho \approx 10^5\Omega$ cm, yielding devices with well-defined *On*- and *Off*-states but poor overall performance. The properties of both transistors are impressively improved after a low temperature annealing (150°C) and interestingly they converge to the same values, despite the ρ values measured in 200 nm thick films still exhibit a significant difference ($\approx 10^1$ and $10^3\Omega$ cm, for $P_{rf} = 2.7$ and 1.1 W cm^{-2}, respectively). The different effects of the annealing treatment depending on the as-grown properties of the films will be addressed in Sect. 15.3.5.

15.3.3 Role of GIZO Target Composition

In this section, the performance of TFTs is discussed regarding the composition of their active layers within the Ga–In–Zn oxide system. All the films are produced using $\%O_2 = 0.4\%$, $p_{dep} = 0.7$ Pa, $P_{rf} = 1.1$ W cm^{-2} and $d_s = 40$ nm, being the final devices annealed at $T_A = 150$°C. Transfer characteristics are presented in Fig. 15.8 for TFTs with GIZO semiconductor layers sputtered from ceramic targets with 2:8:2, 2:4:1, 2:4:2, 2:2:2 and 2:2:1 compositions (Ga:In:Zn atomic ratios), being the μ_{FE} and V_{on} values presented in the ternary diagram of Fig. 15.9.

The amorphous structure of these multicomponent materials turns grain boundaries unimportant and the carrier transport is mostly limited by potential barriers located around the CBM, associated with the structural randomness, which can easily be surpassed in properly processed films (i.e., with adequate $\%O_2$, p_{dep} and P_{rf}) by increasing V_G. Furthermore, the background N can be adjusted within a broad range simply by changing the relative proportions of the cations, which is of great relevance since this parameter is perhaps the most important one to control in order to obtain good transistor performance.

The main effect readily visible in Figs. 15.8 and 15.9 is that the addition of more gallium (i.e., reduction of In/(In + Ga)) considerably shifts V_{on} toward more positive values and decreases μ_{FE}. This is related to the strong bonds that gallium forms with oxygen [75], helping to suppress the generation of free carriers and raising ρ of the semiconductor material. In fact, Ga incorporation allows to turn devices that cannot be switched off within the V_G range used here (IZO 2:1) into transistors with clear *On*- and *Off*-states and good switching performance (GIZO 2:4:2 and

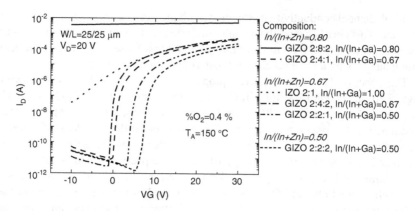

Fig. 15.8 Effect of GIZO target composition on the transfer characteristics of TFTs annealed at 150°C, with %O$_2$ = 0.4%. Results for IZO 2:1 also shown for comparison

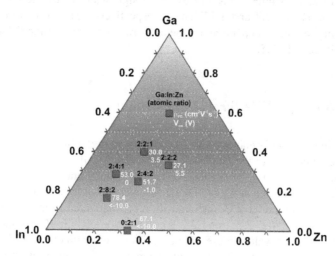

Fig. 15.9 μ_{FE} and V_{on} obtained for GIZO TFTs produced using ceramic targets with different compositions. Results for IZO 2:1 also shown for comparison. Devices annealed at 150°C, with %O$_2$ = 0.4%

2:2:1, which have the same In/(In + Zn) as IZO 2:1). Since for higher In/(In + Zn) N tends to be larger[8], for these cases the Ga incorporation is highly advantageous. However, for smaller In/(In+Zn), a careful choice of In/(In+Ga) has to be made: for In/(In + Ga) = 0.50, corresponding to GIZO 2:2:1 and GIZO 2:2:2 compositions, it is verified that the devices not only exhibit considerably lower μ_{FE} and higher V_{on}, but also larger ΔV_{on}, around 4 and 5 V, respectively, against $\Delta V_{on} \approx 0$ V for

---

[8]During our research work we also verified that for the same sputtering conditions, N is always higher for In$_2$O$_3$ than for ZnO films.

the remaining GIZO compositions. These trends are justified by two factors: first N is lower, making E_F to be shifted toward the midgap, thus a larger fraction of the charges induced by V_G have to fill empty traps before E_F reaches CBM or above it, where μ_{FE} is maximized; second, since the ratio of In/(In + Zn + Ga) starts to decrease, the higher structural disorder close to CBM increases the potential barriers that constrain the movement of free carriers, turning harder the E_F shifting above CBM mentioned in the previous point. Naturally, even if this describes the results obtained here, all this depends on the remaining processing parameters and should not be taken as an entirely strict rule: in fact, a large number of reports regarding stable TFTs with close to 0 V V_{on} exist in the literature, for compositions around 2:2:2 and 2:2:1 [71, 76, 77]. The effect of zinc, besides being a stabilizer of the amorphous structure, is not totally clear regarding TFT performance. Still, it was consistently observed that for In/(In + Zn) $= 0.67$, provided that some gallium is included in the structure to inhibit excessive free carrier generation, improved subthreshold properties were obtained over TFTs with GIZO films having In/(In + Zn) $= 0.80$. Although a clear explanation for this is not conceivable at this time, it seems that zinc can somehow contribute to the modulation of the shallow states below CBM and/or to the reduction of interfacial states between GIZO and the dielectric layer, as initially proposed by Iwasaki et al. [57].

Even if in different compositional ranges, recent papers also show similar results regarding the trends on the properties of TFTs depending on oxide semiconductor composition, both for IZO [78] and GIZO-based devices [55, 57, 79–81]. Note that the Ga–In–Zn oxide region of the ternary diagram explored during our research aims to obtain large μ_{FE} but controllable V_{on} and S, hence the work was more focused on the indium-rich region, with smaller amounts of gallium and zinc.

15.3.4 Role of GIZO Thickness

As seen in the previous sections, some oxide semiconductor layers yield "always-on" transistors that cannot be switched off even using a large negative V_G. This situation arises as a natural consequence of the high N of those layers, rendering the channel conductivity modulation achieved with V_G ineffective to deplete all those free carriers to attain the *Off*-state. For these cases, besides changing the composition and/or deposition conditions of the oxide semiconductor, another possibility that can render good performing devices, with well-defined *On*- and *Off*-states within a reasonable V_G range, is to use very thin oxide semiconductor layers, maintaining all the remaining device structure unchanged. In 2006, we showed that this effect could be highly effective for IZO TFTs [48]. To clearly understand the principle behind this, the surface effects of oxide semiconductors have to be briefly reviewed. The TFTs analyzed here have the staggered bottom-gate structure, so a surface of the semiconductor film (from now on designated by "back surface") is always exposed to the environment. For very thin films, as the ones used here, the control of this back surface has a dramatic effect on the

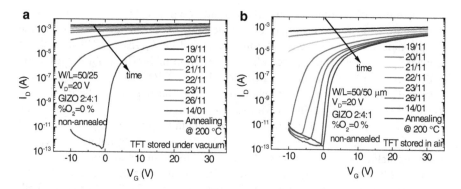

Fig. 15.10 Evolution of transfer characteristics of nonannealed GIZO 2:4:1 TFTs, with $\%O_2 = 0\%$. Devices stored (**a**) in vacuum; (**b**) exposed to air

properties exhibited by the oxide TFTs, because these oxides strongly interact with environmental species such as O_2 and H_2O [82, 83]. Considering that the back surface has a large density of defects such as oxygen vacancies, plausible given that surfaces or interfaces represent discontinuities on the material and oxygen vacancies are the most abundant defect on oxide semiconductors, such defects can work as adsorption sites [5, 42]. The oxygen adsorption at these sites is reported to happen as a two-step process: initially, oxygen is physisorbed, leading to weakly bounded molecules; then, the physisorbed molecules capture electrons from the semiconductor and are converted into chemisorbed and strongly bonded species [84]. This results in the creation of a depletion layer close to the back surface of air-exposed oxide semiconductors. This effect is experimentally shown in Fig. 15.10 for two nonannealed GIZO TFTs, intentionally produced with a more conductive GIZO film (2:4:1 composition and $\%O_2 = 0\%$) than the ideal. The transfer characteristics of the devices were measured during 2 months, but during this period one of them was left in vacuum (only brought to atmospheric pressure to perform the measurements), while the other was left exposed to air, both under dark conditions.

The results show that a highly conductive path is always present in the TFTs left in vacuum, because oxygen adsorption was only possible during the time the devices were removed from vacuum (Fig. 15.10a). Clear *On*- and *Off*-states are only achieved after annealing the device in air at 200°C. On the other hand, the back surface of GIZO in air-exposed devices is free to adsorb oxygen. This way, after only 3 days of air exposure the devices already can be switched off applying $V_G = -10\,V$ and their properties continue to shift until an ideal close to $0V_{on}$ is achieved (Fig. 15.10b). Due to this evolution, there is not a great difference on the characteristics after the annealing treatment. Although the exact rate of oxygen adsorption is not determined in the present experience, it is evident that it is a rather slow process, in line with the results obtained by Lagowski et al. and Kang et al., for ZnO and GIZO films, respectively [82, 84].

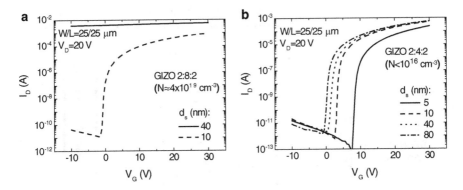

Fig. 15.11 Effect of d_s on the transfer characteristics exhibited by (**a**) GIZO 2:8:2 TFTs; (**b**) GIZO 2:4:2 TFTs. Devices annealed at 150°C, with %$O_2 = 0.4$%

Table 15.4 Electrical properties obtained for the devices depicted in Fig. 15.11

$Target, d_s (nm)$	$\mu_{FE}(cm^2 V^{-1} s^{-1})$	On/Off ratio	$V_{on}(V)$	$V_T(V)$	$S(V dec^{-1})$	$\Delta V_{on}(V)$
1:4:2, 10	73.9	7.3×10^7	−1.5	0.2	0.29	0
1:4:2, 40	78.4	–	–	–	–	–
1:2:2, 5	43.4	1.1×10^9	7.5	11.1	0.21	6.5
1:2:2, 10	51.1	8.0×10^8	2.5	4.4	0.23	2.9
1:2:2, 40	54.9	2.8×10^8	0.5	3.6	0.31	0.8
1:2:2, 80	52.0	4.2×10^8	−1.0	1.8	0.28	0.7

The creation of these *depletion regions close to the back surface and their interaction with the bulk and front surface (dielectric/semiconductor interface) of the oxide semiconductor layer are the effects dictating the extent of the variations on the electrical properties exhibited by TFTs having different d_s. Figure 15.11 shows two extreme cases of d_s variation, one for a high N (> 10^{19} cm^{-3}) semiconductor (Fig. 15.11a), other for a more ideal case, for a semiconductor with N < 10^{16} cm^{-3}* (Fig. 15.11b). To achieve this, the only process variation is the target composition, which is GIZO 2:8:2 in Fig. 15.11a and GIZO 2:4:2 in Fig. 15.11b. The other GIZO process parameters are %$O_2 = 0.4$%, $p_{dep} = 0.7$ Pa and $P_{rf} = 1.1$ W cm^{-2}, being the final devices annealed at 150°C. Table 15.4 summarizes relevant electrical parameters extracted from the transfer characteristics.

It is readily visible that the transfer characteristics of the $N > 10^{19}$ cm^{-3} device are completely changed when d_s is decreased from 40 to 10 nm: properly working devices are obtained using the thinner GIZO layers, with μ_{FE} not being largely affected. This can be explained by the analysis of band and carrier density diagrams for different values of d_s, considering a highly doped semiconductor (Fig. 15.12). On this analysis, some modifications are introduced when comparing with the ideal band diagrams presented in Fig. 15.2, where work function differences between the gate electrode and the semiconductor ($\Phi_G - \Phi_S$) and interface states were neglected. The situation presented here, without accounting for the effect of the back surface,

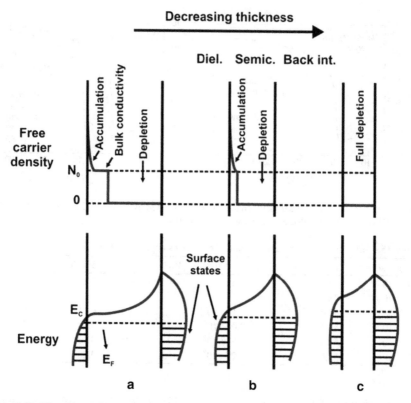

Fig. 15.12 N and band diagrams for different d_s: (**a**) high; (**b**) medium; (**c**) small. Adapted from [5]

represents the worst-case scenario for the depletion of the semiconductor bulk, by considering the existence of an accumulation layer at the dielectric/semiconductor interface even at zero-gate bias, created by negative $\Phi_G - \Phi_S$, donor-like interface states and positive charges contained in the dielectric [5]. The semiconductor region close to the back surface is depleted of free carriers due to the interaction of that surface with oxygen, as explained before. The potential difference across the depletion region (V_{dep}) is given by [5]:

$$V_{dep} = \frac{qNy_d^2}{2\varepsilon_0\varepsilon_s},\tag{15.7}$$

where q is the electronic charge, y_d the width of the depletion layer, ε_0 the permittivity of free space, and ε_s the dielectric constant of the semiconductor.

Considering a high d_s (Fig. 15.12a), a region of the semiconductor's bulk always presents a large N, equal to the background N, turning the *Off*-state of the transistor impossible to achieve. To deplete this region, it would be necessary that the dielectric field could penetrate the accumulation layer at the semiconductor/dielectric interface, which involves removing first all the free and trapped electrons from that interface.[9] This could be achieved using a large negative V_G. However, such a large V_G is unlikely to be possible, given that it would lead to the breakdown of the dielectric layer. The strategy of increasing the dielectric thickness to increase the breakdown voltage is not viable, given that it would also reduce the capacitance, hence the charge density possible to induce/withdraw. Also, using a dielectric with a higher κ would not be a solution to withdraw larger charge densities from the semiconductor, since dielectrics with very large κ fail to present large breakdown fields [85].

By reducing d_s, the thickness of the semiconductor's bulk unperturbed by the accumulation and depletion layers at the interface and back surface, respectively, also decreases, and in a limit situation the accumulation and depletion regions almost touch (Fig. 15.12b). Eventually, by decreasing d_s even more, to a value smaller than y_d, the semiconductor region with the background N is suppressed. By making $d_s \ll y_d$, V_{dep} will be significantly reduced, as it scales with y_d^2. This allows for unfilled traps and remaining physisorbed oxygen on the back surface to capture electrons from the channel, leading to the full depletion of the entire semiconductor thickness (Fig. 15.12c), hence to the *Off*-state of the transistor. Naturally, for a high N, the magnitude of these effects depends on d_s itself, on the back surface state density and on the dielectric/semiconductor interface condition. These parameters have a direct effect on the obtained V_{on} values.

Note that in the discussion above, it is assumed that the high background N allows for filling most of the trap levels present at the semiconductor's bulk at zero-gate bias, since E_F is already around the CBM in the unbiased state. However, if a lower N semiconductor is used, E_F is moved well below CBM and the band diagram of the resulting structure is inevitably changed. Figure 15.11b shows the evolution of the transfer characteristics with d_s when using a lower N GIZO layer. In this case, a region with large conductance inside the semiconductor's bulk does not exist, due to the low background N. Regarding the dielectric/semiconductor interface, it is unlikely that an accumulation layer is present (or at most it should be considerably less evident than for higher N semiconductors) even if $\Phi_G - \Phi_S < 0$ and positive charges exist in the dielectric, because a large number of unfilled traps would exist at this interface, counterbalancing the downward band bending that can exist for the case of a high N semiconductor. Traps would be unfilled at zero-gate-bias because the E_F at the semiconductor would be too low in energy to permit charges to move to those trap states. Concerning the back surface and the associated depletion region, its trend with d_s would be the same as above, i.e., getting closer and eventually reaching the dielectric/semiconductor interface as d_s

---

[9]Note that most of the traps at the dielectric/semiconductor interface should already be filled even if the assumptions of $\Phi_G - \Phi_S < 0$ and positive charges contained in the dielectric are not valid, because electrons from the high N semiconductor can be captured by the lower energy trap states at that interface.

is decreased. Still, this movement would be enhanced since larger y_d is expected here, given the lower N ($y_d \sim N^{-1/2}$). Hence, the depletion layer could even reach the dielectric/semiconductor interface in the unbiased state for a relatively large d_s, depending on N and on the total surface state density.

Given this background, considering a high d_s, the unbiased state results in a low I_D current. Whether this would correspond or not to the *Off*-state of the TFT (i.e., if V_{on} is 0 or slightly negative) depends essentially on the balance between unfilled trap states at the interface and at semiconductor's bulk, $\Phi_G - \Phi_S$ and charges lying in the dielectric layer. For this high d_s, positive V_{on} can also be obtained if the semiconductor and/or its interface with the dielectric layer possess a large defect density that require a positive V_G to be compensated. This is not the case for the largest d_s presented in Fig. 15.11b (80 nm), where a slightly negative V_{on} is obtained.

As d_s is decreased, the depletion layer associated with the back surface reaches the semiconductor/dielectric interface. Hence, before any free charges can be induced by V_G, the depletion layer has to be removed at the interface, resulting in a shift of V_{on} toward positive values. This requires an increasingly higher V_G (higher V_{on}) as d_s is further reduced, because the dielectric/semiconductor interface is now even more depleted.

When d_s is reduced to very small values (5 nm in Fig. 15.11b), the unbiased device is essentially dominated by the properties of both surfaces of the semiconductor film (back surface and interface). Given that V_{dep} isdecreased (y_d decreases), a considerable density of the charges induced by V_G can be captured by the back surface traps. Note that a large amount of these traps should be empty in the unbiased state, because there are not enough electrons available at the semiconductor to fill them. Moreover, there is no accumulation channel at the unbiased state to shield the effect of the back surface traps, because the entire semiconductor is depleted, since $d_s \ll y_d$. All this results in a large increase of V_{on}. Moreover, significant instability also arises for very small d_s, as evidenced by the large ΔV_{on} obtained for the $d_s = 5$ nm devices. This can be attributed to the fact that the interface and back surface traps are largely being filled by the charges induced by V_G and not by the semiconductor itself at the unbiased state. Hence, given that these defect states can behave both as slow and fast traps [5], various rates of trapping/detrapping processes might occur while V_G is being swept. Note that a large ΔV_{on} is not observed for TFTs employing thin GIZO layers with high N (Table 15.4), since in that case most of the traps would already be filled before applying any V_G.

A considerable decrease of μ_{FE} is also observed for the TFTs based on GIZO 2:4:2 when $d_s = 5$ nm. This should be ascribed to the fact that d_s starts to be comparable to the mean free path of electrons within the channel and their movement starts to be confined essentially to a very narrow region close to the dielectric/semiconductor interface, turning the electrons more sensitive to those interface defects [5]. For such a small d_s, another problem can arise, which is the fact that the films may start to become noncontinuous or inhomogeneous, degrading the transport properties [5].

The increased *Off*-current obtained for higher d_s obscures the effect of the V_G-induced current, resulting in an increase of S with d_s [86, 87]. However, the differences in $I_D - Off$ observed in TFTs produced on Si/SiO$_2$ substrates are to some extent artificial, arising due to the 100 ms delay times used for the electrical measurements. If lower delay times are used the devices depicted in Fig. 15.11b present essentially the same *Off*-current values, resulting in less discrepancy among the S values. The residual differences can be attributed to the movement of the charge centroid of the induced accumulation layer away from the dielectric/semiconductor interface, increase in the sheet trap density and an increase in the series resistance for higher d_s, as proposed by Chiang [86].

The results discussed above show that within a large d_s range, thickness variation can be a very effective way to control the V_{on} of an oxide semiconductor TFT without considerably affecting the remaining electrical properties, being the extent of V_{on} variation dependent on the background N. In fact, this practice was successfully used by Park et al. to control V_T on GIZO TFTs between -15 and 0 V [88]. Note that although %O$_2$ also has a significant effect on V_{on} (or V_T), the variation of this processing parameter also changes considerably other electrical properties, as seen in Fig. 15.5.

15.3.5 Role of TFT's Annealing Temperature

Until now, only TFTs annealed at 150°C were analyzed. This T_A is ideal for flexible and low-cost electronics, where the maximum processing temperature has to be kept as low as possible. Even without accounting for pure device performance, the low T_A treatment at the end of device production is useful to remove any residual contaminations due to the involved processing steps, namely the lithographic processes, being crucial to enhance the reproducibility and reliability of the transistors. Still, depending on the properties of the as-deposited semiconductor materials, imposed by the target composition, %O$_2$, p_{dep} and P_{rf} used to produce them, device performance can be greatly affected by T_A. Figures 15.13 and 15.14 show this for TFTs employing different oxide semiconductors, all deposited with $p_{dep} = 0.7$ Pa, $P_{rf} = 1.1$ W cm^{-2} and $d_s = 40$ nm, being the final devices annealed up to 300°C.

For TFTs employing GIZO 2:2:1 channel layers deposited with %O$_2$ = 0.4% (Figs. 15.13a and 15.14a), a significant improvement from nonannealed to $T_A = 150$°C is verified. The most interesting feature between the two transfer characteristics is that the nonideality observed for the nonannealed device, in the form of a kink starting at $V_G \approx 9$ V, disappears after $T_A = 150$°C. This suggests the existence of trap levels on the nonannealed TFT that are holding back the smooth progress of E_F toward (and above) CBM, although the nature of these traps is not clearly understood for now. As the energy range corresponding to these trap levels is surpassed by increasing V_G, all the traps are filled and the newly induced charges become available to raise I_D. This is consistent with the discrete acceptor-trap model proposed by Hong et al. and Wager et al. [28, 89], where kinks are more evident as the trap density increases. The association of these kinks with dielectric/semiconductor trap levels that need to be filled by electrons seems to be

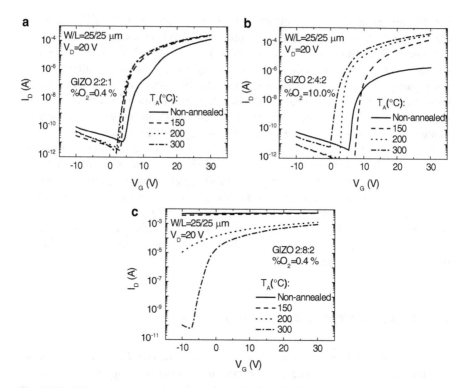

Fig. 15.13 Effect of T_A on the transfer characteristics exhibited by TFTs with different semiconductor layers: (**a**) GIZO 2:2:1, $\%O_2 = 0.4\%$; (**b**) GIZO 2:4:2, $\%O_2 = 10.0\%$; (**c**) GIZO 2:8:2, $\%O_2 = 0.4\%$

plausible because devices produced in parallel but using a high-κ dielectric such as aluminum-titanium-oxide instead of the low-κ SiO$_2$ generally do not present any kinks and have a more abrupt increase of I_D in the subthreshold region (i.e., lower S). This should be a consequence of different dielectric/semiconductor interface properties and of a larger capacitance, thus of higher density of charges induced by the high-κ dielectric, which allows for easier trap filling. Nonidealities in the form of large kinks in the transfer characteristics were also obtained by Hoffman in zinc–tin oxide TFTs [12] and were also found using some sputtered dielectrics during our research work. Returning to the discussion regarding the effect of T_A on the GIZO TFTs presented in Fig. 15.13a, for $T_A > 150°C$ the properties are further improved, suggesting that the density of defect states on the semiconductor and/or at the GIZO/SiO$_2$ interface continues to be reduced and that short-range order is improved. Indeed, this local atomic rearrangement was confirmed by spectroscopic ellipsometry, by the decrease of the broadening parameter [55, 90] for higher T_A. Nevertheless, with $T_A = 300°C$ the ρ of the GIZO 2:2:1 film continues to be high, around $10^6 \Omega$ cm due to its high gallium concentration, resulting in $V_{on} \approx 3$ V.

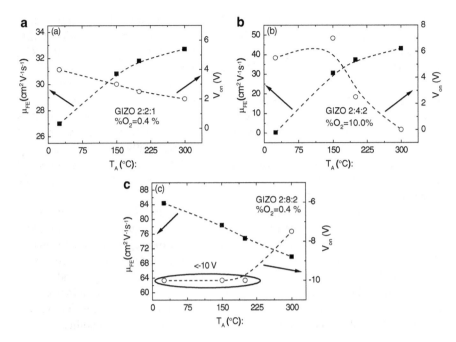

Fig. 15.14 Electrical properties obtained for the devices depicted in Fig. 15.13, having different oxide semiconductor processes: (**a**) GIZO 2:2:1, $\%O_2 = 0.4\%$; (**b**) GIZO 2:4:2, $\%O_2 = 10.0\%$; (**c**) GIZO 2:8:2, $\%O_2 = 0.4\%$

Even if enhanced properties with higher T_A are observed for the previously discussed GIZO TFTs, the magnitude of the improvements is not very large, because the GIZO films are produced with the $\%O_2$, p_{dep} and P_{rf} conditions yielding the most stable films. A different situation is verified for devices employing GIZO films produced under nonideal conditions, such as $\%O_2 = 10.0\%$. Figures 15.13b and 15.14b illustrate this for a TFT based on GIZO 2:4:2. The trends with T_A verified in Figs. 15.13a and 15.14a are also observed here, but the effects are now much more pronounced, because the as-grown semiconductor is highly resistive, having a large defect density and poor compactness, due to the already mentioned oxygen ion bombardment that occurs when such a high $\%O_2$ is used: note, for instance, the large increase of μ_{FE} between the nonannealed and the $T_A = 150°C$ condition, from 0.25 to 30.5 cm^2 V^{-1} s^{-1}. The predominant causes for this improvement are ascribed not only to weakly bonded oxygen desorption but also to enhanced local atomic rearrangement and overall reduction of the trap state densities, both at semiconductor's bulk and at its interface with SiO$_2$. In response to these effects, at $T_A = 300°C$, GIZO's ρ decreases around five orders of magnitude relatively to the as-grown value, being obtained $\rho \approx 10^3 \Omega$ cm. Recently, a reduction of trap density near CBM was experimentally verified using $C-V$ measurements in GIZO films after annealing at 300°C, by Kimura et al. [91].

A distinct behavior with T_A is verified for TFTs having initially highly conductive semiconductor layers. This is evident by using an indium-rich GIZO composition, such as 2:8:2 (Figs. 15.13c and 15.14c). Here, only for $T_A = 200°C$ it is possible to start observing reasonable channel conductivity modulation, although the devices still remain in the *On*-state for the range of V_G used here. After annealing at 300°C, the devices start operating properly, although in depletion mode, with $V_{on} = -7.5$ V. For this composition, because the as-deposited films have a very high oxygen deficiency, resulting in $N \approx 10^{20}$ cm^{-3}, the evolution of properties with increasing T_A is being controlled essentially by the N variation, that only decreases to 10^{17} cm^{-3} range at $T_A = 300°C$. Hence, the improvement of semiconductor/dielectric interface and decrease of the trap states density as T_A increases are less relevant for this case.

Besides overall properties optimization, a T_A below 300°C is also useful to improve the reproducibility of devices (i.e., for semiconductors deposited under intentionally equal conditions) and also to decrease the discrepancies among devices where the semiconductor is intentionally produced under different processing conditions. [92] These two uniformization processes constitute a great advantage regarding the industrialization of AOS technology, since process variations can be significantly attenuated by using a final low temperature treatment, which has the advantage of simultaneously improving device performance and reliability.

15.3.6   Role of Passivation Layer

Passivation is a crucial step to protect the devices from subsequent processes, such as the integration with liquid crystal cells or OLEDs [93]. Passivation also has an important role in isolating the devices from the environment, a concern specially relevant not only for organic semiconductors but also for oxide semiconductors: for instance, the electrical properties of GIZO TFTs are reported to significantly change when the devices are exposed to a high humidity environment [83]. Also, passivation layers are useful to protect the surface of oxide semiconductors from ionic damage due to source–drain reactive ion etching (RIE) processes in staggered bottom gate structures [61, 77]. Additionally, it was also reported by Lee et al. that variation of TFT performance across large area substrates is greatly decreased after passivating devices with SiO$_x$ [94].

Even if passivation brings several advantages, it is imperative to study how it affects the electrical properties of oxide TFTs. Recently, we reported large differences in electrical properties when GIZO TFTs are passivated with SiO$_2$ or SU-8 [95]. In this section we show the effects on TFTs employing a GIZO layer with a 2:4:2 composition, $\%O_2 = 0.4\%$, $p_{dep} = 0.7$ Pa, $P_{rf} = 1.1$ W cm^{-2} and $d_s = 40$ nm, passivated with different materials, deposited with various techniques: SU-8 by spin-coating, MgF$_2$ and SiO$_2$ by e-beam evaporation and SiO$_2$ by sputtering. All the passivation layers were ≈ 200 nm thick, except for the SU-8, which was considerably thicker, ≈ 10–20 μm, due to the available SU-8 formulation at our

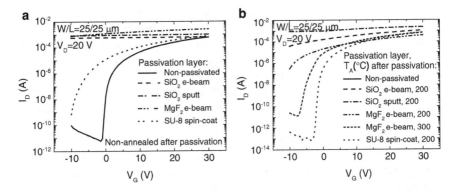

Fig. 15.15 Transfer characteristics of GIZO 2:4:2 TFTs (%O_2 = 0.4%) annealed at 200°C prior depositing different passivation layers: (**a**) nonannealed after passivation layer deposition; (**b**) annealed at 200°C after passivation layer deposition. In (**a**), transfer characteristics referring to SiO_2 e-beam and SiO_2 sputtering are obtained with V_D = 1 V

Table 15.5 Electrical properties obtained for the devices depicted in Fig. 15.15b

Passivation layer	$\mu_{FE}(cm^2 V^{-1} s^{-1})$	$V_{on}(V)$	On/Off ratio	$S(Vdec^{-1})$	$\Delta V_{on}(V)$
Nonpassivated	59.6	−1.0	9.2×10^7	0.30	0.1
SiO_2 (e-beam)	60.9	≪ −10.0	–	–	–
SiO_2 (sputtering)	83.5	≪ −10.0	–	–	–
MgF_2 (e-beam)	61.8	≪ −10.0	–	–	–
MgF_2 (e-beam) @ 300°C	57.8	−7.5	4.9×10^7	0.68	−0.5
SU-8 (spin-coating)	61.2	−3.0	9.4×10^9	0.28	0.2

laboratory. Annealing treatments at 200°C were performed before and after the passivation layer deposition.

Figure 15.15 shows the transfer characteristics obtained before (Fig. 15.15a) and after (Fig. 15.15b) the second annealing step, being relevant electrical parameters presented in Table 15.5.

The results can be understood by investigating the effects occurring at the air-exposed surface of GIZO. As mentioned in Sect. 15.3.4, in a nonpassivated oxide semiconductor oxygen is physisorbed at the back surface, creating acceptor-like surface states that attract electrons from the semiconductor. This gives rise to a depletion region in the semiconductor close to that surface and allows obtaining good performing devices, with well-defined *Off*- and *On*-states, even for large N. Consider now the deposition of a passivation layer on top of the back surface: when this happens, oxygen can no longer be adsorbed and the depletion layer is not formed anymore. Indeed, an accumulation layer can even appear at this surface if oxygen vacancies are created during the passivation layer deposition (for instance, by breaking of metal cation–oxygen bonds [96, 97]) and/or if the passivation layer has a large density of positive charges. Moreover, this effect is even magnified if the semiconductor has a relatively large N, because in that case the depletion layer in the nonpassivated material is thinner ($y_d \sim N^{-1/2}$) and E_F in the unperturbed region of the semiconductor is very close to CBM, so a minor accumulation layer of electrons can lead to a dramatic increase of N and consequently to nondesirable I_D flowing in the zero-gate bias state. This naturally results in the degradation of the electrical properties of the TFTs, leading in extreme cases to devices that cannot be switched off even with a large negative V_G.

Given this background, the different results depicted in Fig. 15.15 and Table 15.5 can be understood based on the passivation material and on its deposition technique. For vacuum deposition techniques (e-beam and sputtering), some of the oxygen lying at GIZO's back surface can be removed during pump down time. This way, the semiconductor region close to the back surface is no longer depleted. Furthermore, during sputtering deposition, the intense substrate bombardment can break weak metal cation–oxygen bonds, generating oxygen vacancies that increase N. This was in fact observed by different authors when exposing a GIZO surface to RIE or to N_2 and NH_3 plasma treatments[10] [61, 77, 96]. If superimposed to this, the deposited passivation layer has a large density of positive charges at its bulk and/or close to its interface with the semiconductor, a significant electrons accumulation layer is created close to its back surface, turning the channel conductivity modulation hard to achieve, even for a large negative V_G. Considering this, the most dramatic effect is observed for sputtered SiO_2, then for e-beam SiO_2 and MgF_2. For these materials, after $T_A = 200°C$, some improvements are observed, mainly for the e-beamed materials, possible due to the decrease of the density of charges on (or close to) their interface with GIZO. Since the annealing treatments were performed in air and due to the relatively large density of pinholes of these passivation layers, it is also plausible to assume that during annealing some oxygen could easily diffuse trough these pinholes and reach the GIZO/passivation layer interface. A second annealing step was also used in other works in literature to turn the properties of passivated devices similar to the ones obtained before passivation [100, 101]. Still, in the present work, all these passivation layers continue to result in always-on devices. The effect of $T_A = 300°C$ is also shown for MgF_2, being visible that the *Off*-state can be achieved, although with a highly negative V_G ($V_{on} = -7.5$ V). Moreover, these transistors exhibit a negative ΔV_{on} and counterclockwise hysteresis, which can mean that ionic drift mechanisms can be present at the device [28, 102, 103]. Mobile ions are generally responsible for large instabilities and failures in transistors, and their presence should be minimized for most of the applications [102].

The results obtained for SU-8 show that this material is considerably more effective regarding device passivation. SU-8 is an epoxy-based negative chemically amplified resist that is commonly used for the fabrication of high aspect ratio features [104]. It is sensitive to near UV-region (365 nm), but it is also compatible with

---

[10]Note that in the case of plasma species containing H^+, the incorporation of this element in oxide semiconductors can also form shallow donor states that contribute to the increase of N, as confirmed by first-principle calculations for GIZO [98]. On the other hand, for oxygen containing plasmas, the incorporation of this element can lead to acceptor-like surface states that reestablish the depletion layer close to the oxide semiconductor's back surface [96, 97, 99].

X-ray and electron beam lithography [105, 106]. It finds applications in numerous fields, such as sensors, micro-fluidic components and inkjet print head nozzles. SU-8 was already used in fully transparent electronic devices: for instance, Gorrn et al. have shown the integration of an OLED with an oxide TFT, where SU-8 is used to define the OLED pixel area [45]. Regarding its application as a passivation material in GIZO TFTs, it can be seen that even without using a second-step annealing a V_{on} close to -10 V can be achieved (Fig. 15.15a and Table 15.5). When compared with the other transfer characteristics in Fig. 15.15a, the improved results with SU-8 can be justified to a great extent by a totally different fabrication process than the one used for the other passivation materials. SU-8 film production involves only a nonvacuum technique (spin-coating), multiple baking steps to evaporate solvents, UV exposure to initiate chemical amplification, postexposure baking to promote cross-linking reaction and development. A subsequent baking process can promote further cross-linking of the polymeric network. Given that a nonvaccum deposition technique is used and due to the absence of back surface bombardment during the fabrication of the passivation layer, oxygen can remain at the GIZO back surface. However, the large depletion layer close to the back surface of the nonpassivated material is not maintained. In fact, some of the weakly bonded oxygen might lost its bonding with GIZO, for instance during the baking steps, UV exposure or due to the rearrangements that take place in the SU-8 film during cross-linking. All this can be superimposed to the existence of positive charges lying in the SU-8 layer, for instance H^+ that arises due to the decomposition of the photoacid generator present in SU-8, which catalyzes the cross-linking reaction. Thus, an accumulation layer can even replace the previous depletion layer close to the back surface of GIZO. However, this should be considerably less significant than for the other analyzed passivation layers, since as shown in Fig. 15.15a the GIZO layer in TFTs passivated with SU-8 can be almost entirely depleted by using a V_G close to -10 V. After a second annealing step at 200°C, electrical properties close to the ones of nonpassivated devices are obtained. At this stage, cross-linking of the SU-8 layer is fully achieved (confirmed by FTIR analysis) and remaining H^+ ions available at the passivation layer, close to its interface with GIZO, can probably capture some electrons from GIZO, re-establishing a depletion layer close to the back surface of GIZO.

To infer about the effectiveness of the SU-8 passivation layer, the nonpassivated and SU-8 devices presented in Fig. 15.15b were measured in high vacuum. As evident from Fig. 15.16a, the nonpassivated TFT presents a very large negative shift of V_{on} when measured under vacuum, consistent with the strong interaction of the back surface with oxygen explained before.[11] For thin films or nanowires, this surface interaction extends to a point where it dominates the overall material properties [39, 41, 42, 82, 101]. Note that the most significant differences on the transfer characteristics are found only after the devices remain in vacuum for

---

[11]Note that in fig. 16 the *Off*-current of both nonpassivated and passivated devices is similar, which seems to contradict the results shown in fig. 15b. However, this is an artifact caused by the higher noise level of the measurement system used to obtain the results depicted in fig. 16.

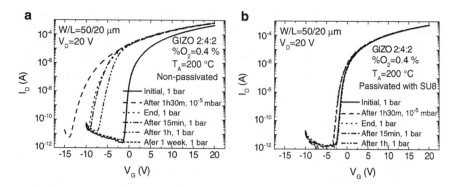

Fig. 15.16 Transfer characteristics of (**a**) nonpassivated and (**b**) SU-8-passivated GIZO 2:4:2 TFTs, with $\%O_2 = 0.4\%$ and $T_A = 200°C$. Measurements performed in air, under high-vacuum and after recovering in air

30–60 min. This confirms the fact that oxygen adsorption/desorption in oxide semiconductor surfaces can be rather slow processes and both slow and fast surface states can play an important role in these phenomena [5,82,84]. In fact, even if some positive shift of V_{on} is verified right after bringing the device again to atmospheric pressure, the initial properties are only recovered after 1 week of air exposure. These results are consistent with the ones obtained by Kang et al. in nonpassivated GIZO TFTs, where V_{on} shifts as large as -50 V were observed under vacuum. The change in GIZO's N due to oxygen adsorption/desorption was measured to be of the order of $10^{20}cm^{-3}$ [82]. On the other hand, SU-8-passivated TFTs do not present significant changes on their electrical properties when measured under vacuum and the small changes are almost instantaneously recovered after bringing the devices to atmospheric pressure (Fig. 15.16b).

A large variety of organic layers present poor diffusion barriers to oxygen and water vapor [107]. This can have severe implications in the properties of oxide semiconductor devices: as shown above, oxygen considerably shifts the transfer characteristics of GIZO TFTs; also, it was shown by Jeong et al. and Park et al. that H_2O molecules can act as electron trap centers as well as electron donors on the GIZO surface, resulting in considerable shifts of the electrical properties and severe instability [83,108]. Although an exhaustive study was not carried out in the present devices, moisture does not seem to affect the properties of SU-8-passivated devices, at least in a relative humidity range between 30 and 70% (Table 15.6).

15.3.7 Constant Drain Current and Constant Gate Bias Stress Measurements

Stress measurements constitute an important starting point to study the instability mechanisms on transistors, as well as to evaluate how they will perform when inte-

Table 15.6 Summary of the most important findings arising from this research work, including optimal values for each studied variable and layer

	Variable	Key comments (regarding material and device performance)
Active layer	Composition	Most important parameter to define the final properties of films and devices.
		Indium-richer – higher μ but also N, thus highly negative V_{on}.
		Gallium inhibits excessive free carrier generation.
		Zinc stabilizes the amorphous structure.
		Best devices obtained with GIZO 2:4:1 and 2:4:2.
	%O_2	Direct control of oxygen vacancies concentration, thus N.
		Low %O_2 results in conductive films (highly negative V_{on}).
		High %O_2 leads to resistive films and increases substrate bombardment effects (lower μ_{FE}, higher V_{on} and ΔV_{on}).
		Best GIZO layers for TFTs obtained with %O_2 = 0.4–1.0%.
	p_{dep}, P_{rf}	Associated with the energy of sputtered and plasma species.
		Higher P_{rf} and lower p_{dep} are favorable to obtain compact and very low-ρ multicomponent oxide films, but can also lead to severe substrate bombardment effects.
		Best compromise for GIZO TFT application obtained with p_{dep} = 0.7 Pa and P_{rf} = 1.1 W cm^{-2}.
	d_s	Controls the predominance of surface effects and surface depletion regions relatively to bulk properties.
		For high-N semiconductors reduced d_s permits to turn always-on TFTs into usable devices.
		For low-to-moderate-N semiconductors, d_s variation allows for V_{on} adjustment without compromising the remaining properties.
		For optimized TFTs, d_s = 20–30 nm provides the best performance.
	T_A	Complex effects, depending on the as-deposited properties.
		Generally, higher T_A leads to improved performance and stability, as well as to less dissimilarity among different films.
		ρ increases/decreases with T_A for initially low-ρ/high-ρ films.
		Even if for optimized TFTs T_A = 150°C already provides good electrical properties, T_A = 200°C always improves stability.
Passivation layer material and deposition process		Vacuum/physical deposited materials – poor TFT performance.
		Spin-coated SU-8 allows for similar properties to nonpassivated devices and improved stability under stress and vacuum measurements.

grated in circuits. Results obtained on GIZO TFTs regarding stress measurements under constant I_D or V_G are briefly discussed in this section.

It is well known, for more than 30 years, that the electrical properties of a-Si:H TFTs are degraded when they are subjected to a constant V_G for long periods of time [6]. A large number of reports regarding gate bias stress measurements exist in the literature of TFTs, for the different semiconductor technologies [109–111]. The most frequent instability mechanism is found to be charge trapping at or close to the dielectric/semiconductor interface, which results in a shift of V_T (ΔV_T). The same mechanism has been found to be predominant for oxide TFTs stressed under constant gate bias, as shown by recent reports [64, 100, 112–115]. However, even if constant gate bias stress constitutes the most used methodology to infer about the electrical stability of TFTs, from an application point of view it is also relevant to study how the devices behave under a constant I_D. A good stability under constant I_D is critical, for instance, when the devices are intended to be used as the "drive TFTs" in an OLED display, since for this application the transistor has to supply a high and stable current to the OLED when the pixel is turned on, otherwise significant variation in the pixel brightness can arise [7, 115]. Essentially, constant I_D stress measurements accurately simulate the stress conditions that a TFT may be subjected to when integrated into an analog circuit environment [116].

Constant I_D stress measurements were performed forcing $I_D = 10\,\mu A$ during 24 h, keeping the drain and gate terminals shorted in a diode-connected configuration, with the source electrode grounded. This configuration permits automatic V_G (thus, V_{on}) adjustment to maintain the constant I_D. After that, the recovery dynamics of the devices was also analyzed, keeping them in the dark. Gate bias stress measurements were also performed. In this case, a constant $V_G = 20\,V$ was applied during 8 h, keeping the drain and source electrodes grounded.

Figure 15.17 shows two extreme examples of the evolution of transfer characteristics with constant I_D stress and recovery time, for nonpassivated GIZO TFTs with different $\%O_2$, 0.4 and 10.0%, with $T_A = 150°C$. It can be seen that $\%O_2 = 10.0\%$ leads to highly unstable devices, with the transfer characteristics being significantly shifted toward higher V_G values during stress. This result is consistent with a large density of defects on the semiconductor and on its interface with the dielectric, which contribute to trap the induced electrons, resulting in a large ΔV_T [10]. Note that this effect is in agreement with the early-stage aging evaluation obtained by measuring the turn-on voltage shift (ΔV_{on}) in four consecutive transfer characteristics (Table 15.2), since the instability mechanism involved in both kinds of measurements is essentially the same (charge trapping) although in constant I_D stress measurements the testing conditions are more severe and extended throughout a longer time scale. Still, the initial properties are recovered for both devices, typically after 4–5 h after the stress measurement (Fig. 15.17b). Besides ΔV_T, the other properties of the devices remain essentially unchanged during stress, meaning that effects such as defect state creation, which could result in a significant increase of S and in nonrecoverable properties [8], are not relevant for the TFTs analyzed herein. Note that these stress measurements were performed around 18 months after device fabrication. As we previously shown in [92], aging can lead to considerable improvements on initially unstable devices, such as the ones where the oxide semiconductor is produced with high $\%O_2$. Nevertheless, the constant I_D stress measurements show that significant differences still exist on the devices depending on the $\%O_2$ used to fabricate the oxide semiconductor layer, being smaller $\%O_2$ preferable to obtain more stable devices.

Optimized GIZO 2:4:2 devices annealed at different temperatures and with a SU-8 passivation layer were also subjected to constant I_D stress measurements.

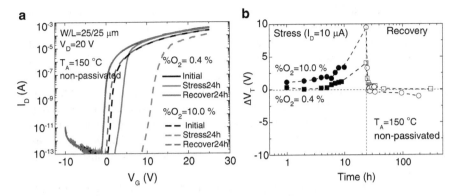

Fig. 15.17 Constant I_D stress measurements of nonpassivated GIZO TFTs with different %O$_2$ used to deposit the GIZO layer, 0.4 and 10.0%, and $T_A = 150°$C: (**a**) transfer characteristics; (**b**) ΔV_T vs. stress/recovery time

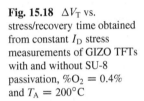

Fig. 15.18 ΔV_T vs. stress/recovery time obtained from constant I_D stress measurements of GIZO TFTs with and without SU-8 passivation, %O$_2$ = 0.4% and $T_A = 200°$C

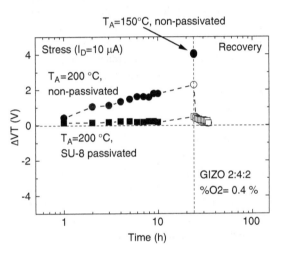

The results show that for $T_A = 200°$C, ΔV_T is reduced (Fig. 15.18). The improved stability can be related with enhanced dielectric/semiconductor interface properties and a reduction in the subgap states of GIZO [76]. A lower ΔV_T on annealed GIZO TFTs exposed to gate bias stress was also observed by Suresh et al. [114].

However, the most impressive optimization arises after device passivation, where ΔV_T as small as 0.46 V could be obtained. This gives new insights about the instability mechanisms that might be occurring in GIZO TFTs. Given that the dielectric/semiconductor interface should have the same properties for both passivated and nonpassivated devices, GIZO's back surface should also have an important contribution to the shifts observed during stress. For ZnO, it is known that oxygen adsorption/desorption processes might be enhanced when the material is subjected to an electric field [117]. Given that the electronegativities of indium, gallium and zinc are quite similar, it is plausible to assume that the same processes can happen at the surface of a GIZO film. As oxygen is adsorbed in the nonpassivated device, the depletion layer close to the back surface assumes more relevance, raising V_T. On the other hand, when GIZO is passivated additional oxygen cannot be provided to the back surface, hence this source of instability is eliminated. Naturally, this assumption is only correct if the passivation material is dense and/or does not have a large concentration of mobile ions that could affect the stability of the transistor. The SU-8 passivation layer used here seems to fulfill these requirements. Jeong et al. observed that GIZO TFTs passivated with organic photoacryl and nondense SiO_x films result in larger ΔV_T under gate bias stress measurements than dense SiO_x films [108]. Improvements on the stability of passivated oxide TFTs are also reported by Levy et al. and Cho et al. [100, 118].

The results obtained here compare quite favorably with the recent reports in literature regarding GIZO TFTs tested under constant I_D stress. In 2007, Jeong et al. reported a $\Delta V_T = 3.4\,V$ after 60h with $I_D = 3\,\mu A$ for a nonpassivated device, while in 2008 the same research group obtained a $\Delta V_T = 2.3\,V$ after 10h with $I_D = 10\,\mu A$ for a SiO_x-passivated device [81, 119]. Concerning other TFT technologies, a-Si:H TFTs were also analyzed by Jahinuzzaman et al., revealing considerably worse stability than GIZO TFTs: for $I_D = 10\,\mu A$, a $\Delta V_T \approx 10\,V$ is obtained after 24h stress, increasing to $\Delta V_T \approx 15$ after more 24h [116].

Typical bias stress measurements were also performed on SU-8-passivated TFTs, by applying a constant $V_G = 20\,V$ during 8h, while keeping the source and drain electrodes grounded. The results are presented in Fig. 15.19, reinforcing the good stability of the passivated devices: in fact, a small and recoverable $\Delta V_T \approx 0.10\,V$ was the only shift observed.

A final remark should also be made regarding the stability of the devices over time, when they are not continuously stressed but rather stored in an unbiased state, in typical ambient conditions, i.e., conventional indoor light, temperature of 20–25°C and relative humidity of 40–45%. Even without any passivation layer and with a T_A as low as 150°C, TFTs employing GIZO layers processed under optimized conditions do not present significant shifts in their electrical properties 2

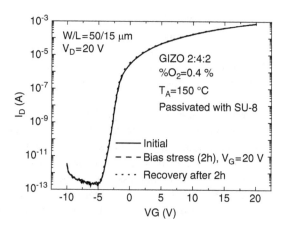

Fig. 15.19 Transfer characteristics evolution obtained from constant bias stress measurements on GIZO TFTs with $\%O_2 = 0.4\%$ and annealed at 150°C, passivated with SU-8

years after their fabrication, being the largest variation observed on V_{on}, that shifts around 1 V. For nonideal GIZO deposition conditions (e.g., high %O_2), the variation is considerably higher, being the main evolution the shift of V_{on} toward 0 V and attenuation of nonidealities in the transfer characteristics [92].

15.4   Conclusions and Outlook

Far from being an exhaustive review envisaging the achievement of optimal (post-) processing conditions of oxide semiconductors for application in oxide TFTs, this chapter tried to provide a generic overview about what are the main effects on the electrical properties of devices when their main processing conditions are varied. Some of the most important findings are summarized in the following table.

Even if not mentioned here, the effect of replacing Ti/Au source–drain electrodes by a sputtered TCO (IZO) was also studied on optimized GIZO TFTs. The results showed that both materials result in similar performance up to $T_A = 200°C$. Above this T_A a degradation of the electrical properties occurs for both electrodes, but is more evident for IZO since its ρ increases around one order of magnitude due to significant oxidation.

The deposition of a low-temperature dielectric is another key issue for low-temperature electronics in general and oxide TFTs in particular, being a requirement for the fabrication of fully transparent (and flexible) devices. Additionally, the dielectric layer dictates, to a large extent, the final performance and stability of the devices. At CENIMAT, we have been working on multicomponent and multilayer dielectrics based on mixtures of Ta_2O_5 or HfO_2 (high-κ) with SiO_2 or Al_2O_3 (high E_G) produced by sputtering without intentional substrate heating. Even if it is still an ongoing work, very good device performance can already be obtained after $T_A = 150°C$, as shown in Fig. 15.20 for a GIZO TFT employing a mixture of Ta_2O_5 and SiO_2 as the dielectric layer. For such a device, properties like $On/Off ratio = 3 \times 10^8$, $\mu_{SAT} = 15\,cm^2\,V^{-1}\,s^{-1}$, $V_{on} =^- 0.4\,V$ and $S = 0.20\,V\,dec^{-1}$ are obtained.

Reports on transparent circuits based on oxide transistors are starting to proliferate, mainly focusing inverters and ring-oscillators [52, 120, 121] or even gate and source drivers for displays [122, 123]. The functionality and speed of these circuits will certainly continue to grow and new opportunities start to arise with the advent of p-type oxide semiconductors, which will allow for transparent CMOS fabrication [19–22]. Although the performance of p-type oxide TFTs is still far from the one obtained with n-type oxides, devices with On/Off ratio $\approx 10^3$ and $\mu_{FE} \approx 1\,cm^2\,V^{-1}\,s^{-1}$ were already fabricated in CENIMAT using sputtered SnO, with $T_A = 200°C$ [19].

The world of oxide TFTs and transparent electronics is growing at an impressive pace. In fact, not only academia but also big players in consumer electronics' market, like LG or Samsung, are investing resources on the development of commercial products using this technology and concepts. The first ones start already to be available, such as the first 22" transparent LCD panels by Samsung (Fig. 15.21),

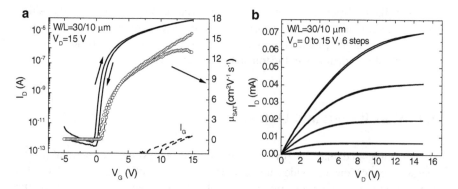

Fig. 15.20 (**a**) Transfer and (**b**) output characteristics of GIZO TFTs with a Ta_2O_5–SiO_2 dielectric layer. Devices annealed at $T_A = 150°C$

Fig. 15.21 Transparent LCD display produced by Samsung [124]

specially relevant for advertising in shop windows and outdoor billboards [124]. And judging by the several prototypes shown by these companies, many more products will certainly arise, such as the 70" LCD display supporting 3D and Ultra-Definition (UD) resolution $(3,840 \times 2,160)$ with a 240 Hz refresh rate, possible due to the high performance of oxide TFTs [125].

Acknowledgements The authors thank the Microelectronic and Optoelectronic Materials Group in CENIMAT. Thanks are also due to the Multiflexioxides Project Consortium (NMP3-CT-2006–032231) and ORAMA Project Consortium (FP7-NMP-2009-LARGE-3). The authors also thank Portuguese Science Foundation (FCT-MCTES) for the funding through the projects PTDC/CTM/73943/2006, PTDC/EEA-ELC/64975/2006 and PTDC/CTM/099124/2008. Thanks are also due to the European Research Council for the ERC 2008 Advanced Grant (INVISIBLE contract number 228144) and IT R&D program of MKE (2006-S079–03, Smart window with transparent electronic devices) from ETRI Korea.

References

1. R. Martins, P. Barquinha, I. Ferreira, L. Pereira, G. Goncalves, E. Fortunato, J. Appl. Phys. **101**, p. 044505 (2007)
2. E. Fortunato, L. Pereira, P. Barquinha, I. Ferreira, R. Prabakaran, G. Goncalves, A. Goncalves, R. Martins, Philos. Mag. **89**, p. 2741 (2009)
3. K. Nomura, H. Ohta, A. Takagi, T. Kamiya, M. Hirano, H. Hosono, Nature **432**, p. 488 (2004)
4. P. Barquinha, L. Pereira, G. Goncalves, R. Martins, E. Fortunato, J. Electrochem. Soc. **156**, p. H161 (2009)
5. A.C. Tickle, *Thin-Film Transistors – A New Approach to Microelectronics*, (Wiley, New York 1969)
6. C.R. Kagan, P. Andry, Thin-Film Transistors, (Marcel Dekker, Inc., New York 2003)
7. J.-H. Lee, D.N. Liu, S.-T. Wu, *Introduction to Flat Panel Displays*, (Wiley, West Sussex, UK 2008)
8. E.S. Yang, *Microelectronic Devices*, (McGraw-Hill, Singapore 1988)
9. R.L. Hoffman, B.J. Norris, J.F. Wager, *Appl. Phys. Lett.* **82**, p. 733 (2003)
10. S.M. Sze, *Physics of Semiconductor Devices*, (Wiley, New York 1981)
11. T. Ytterdal, Y. Cheng, T.A. Fjeldly, *Device Modeling for Analog and RF CMOS Circuit Design*, (Wiley , New York 2003)
12. R.L. Hoffman, Solid-State Electron **50**, p. 784 (2006)
13. A. Ortiz-Conde, F.J.G. Sanchez, J.J. Liou, A. Cerdeira, M. Estrada, Y. Yue, Microelectron. Reliabil. **42**, p. 583 (2002)
14. R.L. Hoffman, J. Appl. Phys. **95** p. 5813 (2004)
15. D.K. Schroder, *Semiconductor Material and Device Characterization*, (Wiley, New Jersey 2006)
16. W. Lim, S.H. Kim, Y.L. Wang, J.W. Lee, D.P. Norton, S.J. Pearton, F. Ren, I.I. Kravchenko, J. Vac. Sci. Technol. B **26**, p. 959. (2008)
17. P.G. Lecomber, W.E. Spear, A. Ghaith, Electron. Lett. **15**, p. 179 (1979)
18. A. Madan, R. Martins, Philos. Mag. **89**, p. 2431 (2009)
19. E. Fortunato, R. Barros, P. Barquinha, V. Figueiredo, S.H.K. Park, C.S. Hwang, R. Martins, Appl. Phys. Lett. **97**, p. 052105 (2010)
20. E. Fortunato, V. Figueiredo, P. Barquinha, E. Elamurugu, R. Barros, G. Goncalves, S.H.K. Park, C.S. Hwang, R. Martins, Appl. Phys. Lett. **96**, p. 239902 (2010)
21. Y. Ogo, H. Hiramatsu, K. Nomura, H. Yanagi, T. Kamiya, M. Hirano, H. Hosono, Appl. Phys. Lett. **93**, p. 032113 (2008)
22. K. Matsuzaki, K. Nomura, H. Yanagi, T. Kamiya, M. Hirano, H. Hosono, Appl. Phys. Lett. **93**, p. 202107 (2008)
23. H.A. Klasens, H. Koelmans, Solid-State Electron **7**, p. 701 (1964)
24. G.F. Boesen, J.E. Jacobs, Proc. Institute Electrical Electronics Eng. **56**, p. 2094 (1968)
25. A. Aoki, H. Sasakura, Jpn. J. Appl. Phys. **9**, p. 582 (1970)
26. M.W.J. Prins, K.O. GrosseHolz, G. Muller, J.F.M. Cillessen, J.B. Giesbers, R.P. Weening, R.M. Wolf, Appl. Phys. Lett. **68**, p. 3650 (1996)
27. C.H. Seager, D.C. McIntyre, W.L. Warren, B.A. Tuttle, Appl. Phys. Lett. **68**, p. 2660 (1996)
28. J.F. Wager, D.A. Keszler, R.E. Presley, *Transparent Electronics*, (Springer, New York 2008)
29. S. Masuda, K. Kitamura, Y. Okumura, S. Miyatake, H. Tabata, T. Kawai, J. Appl. Phys. **93**, p. 1624 (2003)
30. P.F. Carcia, R.S. McLean, M.H. Reilly, G. Nunes, Appl. Phys. Lett. **82**, p. 1117 (2003)
31. J. Nishii, F.M. Hossain, S. Takagi, T. Aita, K. Saikusa, Y. Ohmaki, I. Ohkubo, S. Kishimoto, A. Ohtomo, T. Fukumura, F. Matsukura, Y. Ohno, H. Koinuma, H. Ohno, M. Kawasaki, Jpn. J. Appl. Phys. Part 2 – Lett. **42**, p. L347 (2003)
32. E.M.C. Fortunato, P.M.C. Barquinha, A. Pimentel, A.M.F. Goncalves, A.J.S. Marques, R.F.P. Martins, L.M.N. Pereira, Appl. Phys. Lett. **85**, p. 2541 (2004)

33. E. Fortunato, P. Barquinha, A. Pimentel, A. Goncalves, A. Marques, L. Pereira, R. Martins, *Zinc Oxide Thin-Film Transistors*, In: N.H.T.E. Nickel, Editor, NATO Advanced Research Workshop on Zinc Oxide as a Material for Micro- and Optoelectronic Applications, (Springer, St Petersburg, Russia 2004)
34. B.J. Norris, J. Anderson, J.F. Wager, D.A. Keszler, J. Phys. D-Appl. Phys. **36**, p. L105 (2003)
35. F.M. Hossain, J. Nishii, S. Takagi, A. Ohtomo, T. Fukumura, H. Fujioka, H. Ohno, H. Koinuma, M. Kawasaki, J. Appl. Phys. **94**, p. 7768 (2003)
36. H.S. Bae, M.H. Yoon, J.H. Kim, S. Im, Appl. Phys. Lett. **83**, p. 5313 (2003)
37. H.S. Bae, S. Im, Thin Solid Films **469–470**, p. 75 (2004)
38. R.E. Presley, C.L. Munsee, C.H. Park, D. Hong, J.F. Wager, D.A. Keszler, J. Phys. D-Appl. Phys. **37**, p. 2810 (2004)
39. D. Zhang, C. Li, S. Han, X. Liu, T. Tang, W. Jin, C. Zhou, Appl. Phys. Lett. **82**, p. 112 (2003)
40. Y.W. Heo, L.C. Tien, Y. Kwon, D.P. Norton, S.J. Pearton, B.S. Kang, F. Ren, Appl. Phys. Lett. **85**, p. 2274 (2004)
41. Q.H. Li, Q. Wan, Y.X. Liang, T.H. Wang, Appl. Phys. Lett. **84**, p. 4556 (2004)
42. Z. Fan, D. Wang, P.-C. Chang, W.-Y. Tseng, J.G. Lu, Appl. Phys. Lett. **85**, p. 5923 (2004)
43. K. Nomura, H. Ohta, K. Ueda, T. Kamiya, M. Hirano, H. Hosono, Science **300**, p. 1269 (2003)
44. H.Q. Chiang, J.F. Wager, R.L. Hoffman, J. Jeong, D.A. Keszler, Appl. Phys. Lett. **86**, p. 013503 (2005)
45. P. Gorrn, M. Sander, J. Meyer, M. Kroger, E. Becker, H.H. Johannes, W. Kowalsky, T. Riedl, Adv. Mater. **18**, p. 738 (2006)
46. W.B. Jackson, R.L. Hoffman, G.S. Herman, Appl. Phys. Lett. **87**, p. 193503 (2005)
47. N.L. Dehuff, E.S. Kettenring, D. Hong, H.Q. Chiang, J.F. Wager, R.L. Hoffman, C.H. Park, D.A. Keszler, J. Appl. Phys. **97**, p. 064505 (2005)
48. P. Barquinha, A. Pimentel, A. Marques, L. Pereira, R. Martins, E. Fortunato, J. Non-Cryst. Solids **352**, p. 1749 (2006)
49. B. Yaglioglu, H.Y. Yeom, R. Beresford, D.C. Paine, Appl. Phys. Lett. **89**, p. 062103 (2006)
50. D.C. Paine, B. Yaglioglu, Z. Beiley, S. Lee, Thin Solid Films **516**, p. 5894 (2008)
51. G. Goncalves, P. Barquinha, L. Pereira, N. Franco, E. Alves, R. Martins, E. Fortunato, Electrochem. Solid State Lett. **13**, p. II20 (2010)
52. R.E. Presley, D. Hong, H.Q. Chiang, C.M. Hung, R.L. Hoffman, J.F. Wager, Solid-State Electron **50**, p. 500 (2006)
53. A. Suresh, P. Wellenius, A. Dhawan, J. Muth, Appl. Phys. Lett. **90**, p. 123512 (2007)
54. J.S. Park, J.K. Jeong, Y.G. Mo, H.D. Kim, S.I. Kim, Appl. Phys. Lett. **90**, p. 262106 (2007)
55. D. Kang, I. Song, C. Kim, Y. Park, T.D. Kang, H.S. Lee, J.W. Park, S.H. Baek, S.H. Choi, H. Lee, Appl. Phys. Lett. **91**, p. 091910 (2007)
56. H. Hosono, K. Nomura, Y. Ogo, T. Uruga, T. Kamiya, J. Non-Cryst. Solids **354**, p. 2796 (2008)
57. T. Iwasaki, N. Itagaki, T. Den, H. Kumomi, K. Nomura, T. Kamiya, H. Hosono, Appl. Phys. Lett. **90**, p. 242114 (2007)
58. E.M.C. Fortunato, L.M.N. Pereira, P.M.C. Barquinha, A.M.B. do Rego, G. Goncalves, A. Vila, J.R. Morante, R.F.P. Martins, Appl. Phys. Lett. **92**, p. 222103 (2008)
59. http://www.qmc.ufsc.br/qmcweb/artigos/historia_descoberta.htm
60. D. Redinger, V. Subramanian, IEEE Trans. Electron Devices **54**, p. 1301 (2007)
61. H.D. Kim, J.K. Jeong, H.J. Chung, Y.G. Mo, Invited paper: Technological challenges for large-size AMOLED display, *International Symposium of the Society-for-Information-Display (SID 2008)*, (Soc Information Display, Los Angeles, CA 2008)
62. J.H. Jeon, M.C. Lee, K.C. Park, S.H. Jung, M.K. Han, *International Electron Devices Meeting 2000, Technical Digest*, IEEE, New York (2000)
63. D.L. Staebler, C.R. Wronski, Appl. Phys. Lett. **31**, p. 292 (1977)
64. P. Gorrn, P. Holzer, T. Riedl, W. Kowalsky, J. Wang, T. Weimann, P. Hinze, S. Kipp, Appl. Phys. Lett. **90**, p. 063502 (2007)
65. P.F. Carcia, R.S. McLean, M.H. Reilly, J. Soc. Inf. Disp. **13**, p. 547 (2005)
66. B.G. Lewis, D.C. Paine, MRS Bull. **25**, p. 22 (2000)

67. J.R. Bellingham, W.A. Phillips, C.J. Adkins, J. Phys.-Condens. Matter **2**, p. 6207 (1990)
68. I. Hamberg, C.G. Granqvist, J. Appl. Phys. **60**, p. R123 (1986)
69. H.Q. Chiang, D. Hong, C.M. Hung, R.E. Presley, J.F. Wager, C.H. Park, D.A. Keszler, G.S. Herman, J. Vac. Sci. Technol. B **24**, p. 2702 (2006)
70. H.Q. Chiang, B.R. McFarlane, D. Hong, R.E. Presley, J.F. Wager, J. Non-Cryst. Solids 354, p. 2826 (2008)
71. J.H. Jeong, H.W. Yang, J.S. Park, J.K. Jeong, Y.G. Mo, H.D. Kim, J. Song, C.S. Hwang, Electrochem. Solid State Lett. **11**, p. H157 (2008)
72. H.H. Hsieh, T. Kamiya, K. Nomura, H. Hosono, C.C. Wu, Appl. Phys. Lett. **92**, p. 133503 (2008)
73. G. Lavareda, C.N. de Carvalho, E. Fortunato, A.. Ramos, E. Alves, O. Conde, A. Amaral, J. Non-Cryst. Solids **352**, p. 2311 (2006)
74. M.S. Grover, P.A. Hersh, H.Q. Chiang, E.S. Kettenring, J.F. Wager, D.A. Keszler, J. Phys. D-Appl. Phys. **40**, p. 1335 (2007)
75. H. Hosono, J. Non-Cryst. Solids **352**, p. 851 (2006)
76. K. Nomura, T. Kamiya, H. Ohta, M. Hirano, H. Hosono, Appl. Phys. Lett. **93**, p. 192107 (2008)
77. M. Kim, J.H. Jeong, H.J. Lee, T.K. Ahn, H.S. Shin, J.S. Park, J.K. Jeong, Y.G. Mo, H.D. Kim, Appl. Phys. Lett. **90**, p. 212114 (2007)
78. M.G. McDowell, I.G. Hill, IEEE Trans. Electron Devices **56**, p. 343 (2009)
79. G.H. Kim, B.D. Ahn, H.S. Shin, W.H. Jeong, H.J. Kim, H.J. Kim, Appl. Phys. Lett. **94**, p. 233501 (2009)
80. H. Kumomi, K. Nomura, T. Kamiya, H. Hosono, Thin Solid Films **516**, p. 1516 (2008)
81. J.K. Jeong, J.H. Jeong, H.W. Yang, J.S. Park, Y.G. Mo, H.D. Kim, Appl. Phys. Lett. **91**, p. 113505 (2007)
82. D. Kang, H. Lim, C. Kim, I. Song, J. Park, Y. Park, J. Chung, Appl. Phys. Lett. **90**, p. 192101 (2007)
83. J.S. Park, J.K. Jeong, H.J. Chung, Y.G. Mo, H.D. Kim, Appl. Phys. Lett. **92**, p. 072104 (2008)
84. J. Lagowski, E.S. Sproles, Jr., H.C. Gatos, J. Appl. Phys. **48**, p. 3566 (1977)
85. J. Robertson, MRS Bull. **27**, p. 217 (2002)
86. H.Q. Chiang, *Development of Oxide Semiconductors: Materials, Devices, and Integration, Electrical And Computer Engineering*, (Oregon State University, Oregon 2007)
87. C.S. Chuang, T.C. Fung, B.G. Mullins, K. Nomura, T. Kamiya, H.P.D. Shieh, H. Hosono, J. Kanicki, *2008 Sid International Symposium, Digest of Technical Papers, Vol Xxxix, Books I-III*, (Soc Information Display, Playa Del Rey 2008)
88. J.S. Park, J.K. Jeong, Y.G. Mo, H.D. Kim, C.J. Kim, Appl. Phys. Lett. **93**, p. 033513 (2008)
89. D. Hong, G. Yerubandi, H.Q. Chiang, M.C. Spiegelberg, J.F. Wager, Crit. Rev. Solid State Mater. Sci. **33**, p. 101 (2008)
90. H. Aguas, V. Silva, E. Fortunato, S. Lebib, P.R.I. Cabarrocas, I. Ferreira, L. Guimaraes, R. Martins, Jpn. J. Appl. Phys. Part 1 - Regul. Pap. Short Notes Rev. Pap. **42**, p. 4935 (2003)
91. M. Kimura, T. Nakanishi, K. Nomura, T. Kamiya, H. Hosono, Appl. Phys. Lett. **92**, p. 133512 (2008)
92. E. Fortunato, P. Barquinha, G. Goncalo, L. Pereira, R. Martins In: A. Facchetti and T.J. Marks, Editors, *Transparent Electronics: From Synthesis to Applications*, (Wiley, UK 2010)
93. A. Sato, K. Abe, R. Hayashi, H. Kumomi, K. Nomura, T. Kamiya, M. Hirano, H. Hosono, Appl. Phys. Lett. **94**, p. 133502 (2009)
94. H.N. Lee, J. Kyung, M.C. Sung, D.Y. Kim, S.K. Kang, S.J. Kim, C.N. Kim, H.G. Kim, S.T. Kim, J. Soc. Inf. Disp. **16**, p. 265 (2008)
95. A. Olziersky, P. Barquinha, A. Vila, L. Pereira, G. Goncalves, E. Fortunato, R. Martins, J.R. Morante, J. Appl. Phys. **108**, p. 064505 (2010)
96. K.S. Son, T.S. Kim, J.S. Jung, M.K. Ryu, K.B. Park, B.W. Yoo, J.W. Kim, Y.G. Lee, J.Y. Kwon, S.Y. Lee, J.M. Kim, *2008 Sid International Symposium, Digest of Technical Papers, Vol XXXIX, Books I-III*, Soc Information Display, Playa Del Rey (2008)

97. J. Park, S. Kim, C. Kim, S. Kim, I. Song, H. Yin, K.K. Kim, S. Lee, K. Hong, J. Lee, J. Jung, E. Lee, K.W. Kwon, Y. Park, Appl. Phys. Lett. **93**, p. 053505 (2008)
98. H. Omura, H. Kumomi, K. Nomura, T. Kamiya, M. Hirano, H. Hosono, J. Appl. Phys. **105**, p. 8 (2009)
99. K. Remashan, D.K. Hwang, S.D. Park, J.W. Bae, G.Y. Yeom, S.J. Park, J.H. Jang, Electrochem. Solid State Lett. **11**, p. H55 (2008)
100. I.T. Cho, J.M. Lee, J.H. Lee, H.I. Kwon, Semicond. Sci. Technol. **24**, p. 015013 (2009)
101. D. Hong, J.F. Wager, J. Vac. Sci. Technol. B **23**, p. L25 (2005)
102. M. Egginger, S. Bauer, R. Schwödiauer, H. Neugebauer, N. Sariciftci, *Monatshefte für Chemie/Chemical Monthly***140**, p. 735 (2009)
103. S. Sasa, M. Ozaki, K. Koike, M. Yano, M. Inoue, Appl. Phys. Lett. **89**, p. 053502 (2006)
104. J.M. Shaw, J.D. Gelorme, N.C. LaBianca, W.E. Conley, S.J. Holmes, IBM J. Res. Dev. **41**, p. 81 (1997)
105. A.L. Bogdanov, S.S. Peredkov, Microelectron. Eng. **53**, p. 493 (2000)
106. A.K. Nallani, S.W. Park, J.B. Lee In: J.C. Chiao, V.K. Varadan and C. Cane, Editors, *Smart Sensors, Actuators, and Mems, Pts 1 and 2*, (Spie-Int Soc Optical Engineering, Bellingham (2003)
107. M.S. Weaver, L.A. Michalski, K. Rajan, M.A. Rothman, J.A. Silvernail, J.J. Brown, P.E. Burrows, G.L. Graff, M.E. Gross, P.M. Martin, M. Hall, E. Mast, C. Bonham, W. Bennett, M. Zumhoff, Appl. Phys. Lett. **81**, p. 2929 (2002)
108. J.K. Jeong, H.W. Yang, J.H. Jeong, Y.G. Mo, H.D. Kim, Appl. Phys. Lett. **93**, p. 123508 (2008)
109. M.J. Powell, C. Vanberkel, I.D. French, D.H. Nicholls, Appl. Phys. Lett. **51**, p. 1242 (1987)
110. S.G.J. Mathijssen, M. Colle, H. Gomes, E.C.P. Smits, B. de Boer, I. McCulloch, P.A. Bobbert, D.M. de Leeuw, Adv. Mater. **19**, p. 2785 (2007)
111. N.D. Young, A. Gill, Semiconduct. Sci. Technol. **5**, p. 72 (1990)
112. Y.K. Moon, S. Lee, W.S. Kim, B.W. Kang, C.O. Jeong, D.H. Lee, J.W. Park, Appl. Phys. Lett. **95**, p. 013507 (2009)
113. M.E. Lopes, H.L. Gomes, M.C.R. Medeiros, P. Barquinha, L. Pereira, E. Fortunato, R. Martins, I. Ferreira, Appl. Phys. Lett. **95**, p. 063502 (2009)
114. A. Suresh, J.F. Muth, Appl. Phys. Lett. **92**, p. 033502 (2008)
115. J.M. Lee, I.T. Cho, J.H. Lee, H.I. Kwon, Appl. Phys. Lett. **93**, p. 093504 (2008)
116. S.M. Jahinuzzaman, A. Sultana, K. Sakariya, P. Servati, A. Nathan, Appl. Phys. Lett. **87**, p. 023502 (2005)
117. D.H. Zhang, Mater. Chem. Phys. **45**, p. 248 (1996)
118. D.H. Levy, D. Freeman, S.F. Nelson, P.J. Cowdery-Corvan, L.M. Irving, Appl. Phys. Lett. **92**, p. 192101 (2008)
119. J.K. Jeong, J.H. Jeong, J.H. Choi, J.S. Im, S.H. Kim, H.W. Yang, K.N. Kang, K.S. Kim, T.K. Ahn, H.J. Chung, M. Kim, B.S. Gu, J.S. Park, Y.G. Mo, H.D. Kim, H.K. Chung, Distinguished paper: 12.1-inch WXGA AMOLED display driven by indium-gallium-zinc oxide TFTs array, *International Symposium of the Society-for-Information-Display (SID 2008)*, Soc Information Display, (Los Angeles, CA 2008)
120. M. Ofuji, K. Abe, H. Shimizu, N. Kaji, R. Hayashi, M. Sano, H. Kumomi, K. Nomura, T. Kamiya, H. Hosono, IEEE Electron Device Lett. **28**, p. 273 (2007)
121. H. Frenzel, F. Schein, A. Lajn, H.v. Wenckstern, M. Grundmann, Appl. Phys. Lett. **96**, p. 113502 (2010)
122. P. Görrn, F. Ghaffari, T. Riedl, W. Kowalsky, Solid-State Electron. **53**, p. 329 (2009)
123. T. Osada, K. Akimoto, T. Sato, M. Ikeda, M. Tsubuku, J. Sakata, J. Koyama, T. Serikawa, S. Yamazaki, SID Digest, p. 184 (2009)
124. http://www.gizmag.com/samsungs-transparent-lcd-display/18283/picture/132495/
125. http://www.samsunghub.com/2010/11/08/samsung-develops-70-inch-lcd-panel-using-oxide-semiconductor-tech/

Index

S. Pearton (ed.), *GaN and ZnO-based Materials and Devices*,
Springer Series in Materials Science 156, DOI 10.1007/978-3-642-23521-4,
© Springer-Verlag Berlin Heidelberg 2012

CPSIA information can be obtained
at www.ICGtesting.com
Printed in the USA
LVHW061104170520
655852LV00003B/241